Grzimek's
Animal Life Encyclopedia

Second Edition

••••

Grzimek's
Animal Life Encyclopedia

Second Edition

●●●●

Volume 3
Insects

Arthur V. Evans, Advisory Editor
Rosser W. Garrison, Advisory Editor
Neil Schlager, Editor

Joseph E. Trumpey, Chief Scientific Illustrator

Michael Hutchins, Series Editor
In association with the American Zoo and Aquarium Association

GALE®

THOMSON
TM
GALE

Detroit • New York • San Diego • San Francisco • Cleveland • New Haven, Conn. • Waterville, Maine • London • Munich

Grzimek's Animal Life Encyclopedia, Second Edition
Volume 3: Insects
Produced by Schlager Group Inc.
Neil Schlager, Editor
Vanessa Torrado-Caputo, Associate Editor

Project Editor
Melissa C. McDade

Editorial
Madeline Harris, Christine Jeryan, Kate Kretschmann, Mark Springer

Indexing Services
Synapse, the Knowledge Link Corporation

Permissions
Margaret Chamberlain

Imaging and Multimedia
Mary K. Grimes, Lezlie Light, Christine O'Bryan, Barbara Yarrow, Robyn V. Young

Product Design
Tracey Rowens, Jennifer Wahi

Manufacturing
Wendy Blurton, Dorothy Maki, Evi Seoud, Mary Beth Trimper

For permission to use material from this product, submit your request via Web at http://www.gale-edit.com/permissions, or you may download our Permissions Request form and submit your request by fax or mail to: The Gale Group, Inc., Permissions Department, 27500 Drake Road, Farmington Hills, MI, 48331-3535, Permissions hotline: 248-699-8074 or 800-877-4253, ext. 8006, Fax: 248-699-8074 or 800-762-4058.

Cover photo of American hover fly (*Metasyrphus americanus*) by E. R. Degginger, Bruce Coleman, Inc. Back cover photos of sea anemone by AP/Wide World Photos/University of Wisconsin-Superior; land snail, lionfish, golden frog, and green python by JLM Visuals; red-legged locust © 2001 Susan Sam; hornbill by Margaret F. Kinnaird; and tiger by Jeff Lepore/Photo Researchers. All reproduced by permission.

While every effort has been made to ensure the reliability of the information presented in this publication, The Gale Group, Inc. does not guarantee the accuracy of the data contained herein. The Gale Group, Inc. accepts no payment for listing; and inclusion in the publication of any organization, agency, institution, publication, service, or individual does not imply endorsement of the editors and publisher. Errors brought to the attention of the publisher and verified to the satisfaction of the publisher will be corrected in future editions.

ISBN   0-7876-5362-4 (vols. 1–17 set)
            0-7876-5779-4 (vol. 3)
This title is also available as an e-book.
ISBN   0-7876-7750-7 (17-vol set)

Contact your Gale sales representative for ordering information.

LIBRARY OF CONGRESS CATALOGING-IN-PUBLICATION DATA

Grzimek, Bernhard.
  [Tierleben. English]
  Grzimek's animal life encyclopedia.— 2nd ed.
      v. cm.
  Includes bibliographical references.
  Contents: v. 1. Lower metazoans and lesser deuterosomes / Neil Schlager, editor — v. 2. Protostomes / Neil Schlager, editor — v. 3. Insects / Neil Schlager, editor — v. 4-5. Fishes I-II / Neil Schlager, editor — v. 6. Amphibians / Neil Schlager, editor — v. 7. Reptiles / Neil Schlager, editor — v. 8-11. Birds I-IV / Donna Olendorf, editor — v. 12-16. Mammals I-V / Melissa C. McDade, editor — v. 17. Cumulative index / Melissa C. McDade, editor.
  ISBN 0-7876-5362-4 (set hardcover : alk. paper)
  1. Zoology—Encyclopedias. I. Title: Animal life encyclopedia. II. Schlager, Neil, 1966- III. Olendorf, Donna IV. McDade, Melissa C. V. American Zoo and Aquarium Association. VI. Title.
  QL7 .G7813 2004

                                                        590'.3—dc21
                                                        2002003351

Printed in Canada
10 9 8 7 6 5 4 3 2 1

Recommended citation: *Grzimek's Animal Life Encyclopedia,* 2nd edition. Volume 3, *Insects,* edited by Michael Hutchins, Arthur V. Evans, Rosser W. Garrison, and Neil Schlager. Farmington Hills, MI: Gale Group, 2003.

Contents

• • • • •

Foreword

Earth is teeming with life. No one knows exactly how many distinct organisms inhabit our planet, but more than 5 million different species of animals and plants could exist, ranging from microscopic algae and bacteria to gigantic elephants, redwood trees and blue whales. Yet, throughout this wonderful tapestry of living creatures, there runs a single thread: Deoxyribonucleic acid or DNA. The existence of DNA, an elegant, twisted organic molecule that is the building block of all life, is perhaps the best evidence that all living organisms on this planet share a common ancestry. Our ancient connection to the living world may drive our curiosity, and perhaps also explain our seemingly insatiable desire for information about animals and nature. Noted zoologist, E. O. Wilson, recently coined the term "biophilia" to describe this phenomenon. The term is derived from the Greek *bios* meaning "life" and *philos* meaning "love." Wilson argues that we are human because of our innate affinity to and interest in the other organisms with which we share our planet. They are, as he says, "the matrix in which the human mind originated and is permanently rooted." To put it simply and metaphorically, our love for nature flows in our blood and is deeply engrained in both our psyche and cultural traditions.

Our own personal awakenings to the natural world are as diverse as humanity itself. I spent my early childhood in rural Iowa where nature was an integral part of my life. My father and I spent many hours collecting, identifying and studying local insects, amphibians and reptiles. These experiences had a significant impact on my early intellectual and even spiritual development. One event I can recall most vividly. I had collected a cocoon in a field near my home in early spring. The large, silky capsule was attached to a stick. I brought the cocoon back to my room and placed it in a jar on top of my dresser. I remember waking one morning and, there, perched on the tip of the stick was a large moth, slowly moving its delicate, light green wings in the early morning sunlight. It took my breath away. To my inexperienced eyes, it was one of the most beautiful things I had ever seen. I knew it was a moth, but did not know which species. Upon closer examination, I noticed two moon-like markings on the wings and also noted that the wings had long "tails", much like the ubiquitous tiger swallow-tail butterflies that visited the lilac bush in our backyard. Not wanting to suffer my ignorance any longer, I reached immediately for my *Golden Guide to North American Insects* and searched through the section on moths and butterflies. It was a luna moth! My heart was pounding with the excitement of new knowledge as I ran to share the discovery with my parents.

I consider myself very fortunate to have made a living as a professional biologist and conservationist for the past 20 years. I've traveled to over 30 countries and six continents to study and photograph wildlife or to attend related conferences and meetings. Yet, each time I encounter a new and unusual animal or habitat my heart still races with the same excitement of my youth. If this is biophilia, then I certainly possess it, and it is my hope that others will experience it too. I am therefore extremely proud to have served as the series editor for the Gale Group's rewrite of *Grzimek's Animal Life Encyclopedia*, one of the best known and widely used reference works on the animal world. *Grzimek's* is a celebration of animals, a snapshot of our current knowledge of the Earth's incredible range of biological diversity. Although many other animal encyclopedias exist, *Grzimek's Animal Life Encyclopedia* remains unparalleled in its size and in the breadth of topics and organisms it covers.

The revision of these volumes could not come at a more opportune time. In fact, there is a desperate need for a deeper understanding and appreciation of our natural world. Many species are classified as threatened or endangered, and the situation is expected to get much worse before it gets better. Species extinction has always been part of the evolutionary history of life; some organisms adapt to changing circumstances and some do not. However, the current rate of species loss is now estimated to be 1,000–10,000 times the normal "background" rate of extinction since life began on Earth some 4 billion years ago. The primary factor responsible for this decline in biological diversity is the exponential growth of human populations, combined with peoples' unsustainable appetite for natural resources, such as land, water, minerals, oil, and timber. The world's human population now exceeds 6 billion, and even though the average birth rate has begun to decline, most demographers believe that the global human population will reach 8–10 billion in the next 50 years. Much of this projected growth will occur in developing countries in Central and South America, Asia and Africa-regions that are rich in unique biological diversity.

Finding solutions to conservation challenges will not be easy in today's human-dominated world. A growing number of people live in urban settings and are becoming increasingly isolated from nature. They "hunt" in supermarkets and malls, live in apartments and houses, spend their time watching television and searching the World Wide Web. Children and adults must be taught to value biological diversity and the habitats that support it. Education is of prime importance now while we still have time to respond to the impending crisis. There still exist in many parts of the world large numbers of biological "hotspots"—places that are relatively unaffected by humans and which still contain a rich store of their original animal and plant life. These living repositories, along with selected populations of animals and plants held in professionally managed zoos, aquariums and botanical gardens, could provide the basis for restoring the planet's biological wealth and ecological health. This encyclopedia and the collective knowledge it represents can assist in educating people about animals and their ecological and cultural significance. Perhaps it will also assist others in making deeper connections to nature and spreading biophilia. Information on the conservation status, threats and efforts to preserve various species have been integrated into this revision. We have also included information on the cultural significance of animals, including their roles in art and religion.

It was over 30 years ago that Dr. Bernhard Grzimek, then director of the Frankfurt Zoo in Frankfurt, Germany, edited the first edition of *Grzimek's Animal Life Encyclopedia*. Dr. Grzimek was among the world's best known zoo directors and conservationists. He was a prolific author, publishing nine books. Among his contributions were: *Serengeti Shall Not Die*, *Rhinos Belong to Everybody* and *He and I and the Elephants*. Dr. Grzimek's career was remarkable. He was one of the first modern zoo or aquarium directors to understand the importance of zoo involvement in *in situ* conservation, that is, of their role in preserving wildlife in nature. During his tenure, Frankfurt Zoo became one of the leading western advocates and supporters of wildlife conservation in East Africa. Dr. Grzimek served as a Trustee of the National Parks Board of Uganda and Tanzania and assisted in the development of several protected areas. The film he made with his son Michael, *Serengeti Shall Not Die*, won the 1959 Oscar for best documentary.

Professor Grzimek has recently been criticized by some for his failure to consider the human element in wildlife conservation. He once wrote: "A national park must remain a primordial wilderness to be effective. No men, not even native ones, should live inside its borders." Such ideas, although considered politically incorrect by many, may in retrospect actually prove to be true. Human populations throughout Africa continue to grow exponentially, forcing wildlife into small islands of natural habitat surrounded by a sea of humanity. The illegal commercial bushmeat trade—the hunting of endangered wild animals for large scale human consumption—is pushing many species, including our closest relatives, the gorillas, bonobos and chimpanzees, to the brink of extinction. The trade is driven by widespread poverty and lack of economic alternatives. In order for some species to survive it will be necessary, as Grzimek suggested, to establish and enforce

a system of protected areas where wildlife can roam free from exploitation of any kind.

While it is clear that modern conservation must take the needs of both wildlife and people into consideration, what will the quality of human life be if the collective impact of short-term economic decisions is allowed to drive wildlife populations into irreversible extinction? Many rural populations living in areas of high biodiversity are dependent on wild animals as their major source of protein. In addition, wildlife tourism is the primary source of foreign currency in many developing countries and is critical to their financial and social stability. When this source of protein and income is gone, what will become of the local people? The loss of species is not only a conservation disaster; it also has the potential to be a human tragedy of immense proportions. Protected areas, such as national parks, and regulated hunting in areas outside of parks are the only solutions. What critics do not realize is that the fate of wildlife and people in developing countries is closely intertwined. Forests and savannas emptied of wildlife will result in hungry, desperate people, and will, in the long-term lead to extreme poverty and social instability. Dr. Grzimek's early contributions to conservation should be recognized, not only as benefiting wildlife, but as benefiting local people as well.

Dr. Grzimek's hope in publishing his *Animal Life Encyclopedia* was that it would "...disseminate knowledge of the animals and love for them," so that future generations would "...have an opportunity to live together with the great diversity of these magnificent creatures." As stated above, our goals in producing this updated and revised edition are similar. However, our challenges in producing this encyclopedia were more formidable. The volume of knowledge to be summarized is certainly much greater in the twenty-first century than it was in the 1970's and 80's. Scientists, both professional and amateur, have learned and published a great deal about the animal kingdom in the past three decades, and our understanding of biological and ecological theory has also progressed. Perhaps our greatest hurdle in producing this revision was to include the new information, while at the same time retaining some of the characteristics that have made *Grzimek's Animal Life Encyclopedia* so popular. We have therefore strived to retain the series' narrative style, while giving the information more organizational structure. Unlike the original *Grzimek's*, this updated version organizes information under specific topic areas, such as reproduction, behavior, ecology and so forth. In addition, the basic organizational structure is generally consistent from one volume to the next, regardless of the animal groups covered. This should make it easier for users to locate information more quickly and efficiently. Like the original Grzimek's, we have done our best to avoid any overly technical language that would make the work difficult to understand by non-biologists. When certain technical expressions were necessary, we have included explanations or clarifications.

Considering the vast array of knowledge that such a work represents, it would be impossible for any one zoologist to have completed these volumes. We have therefore sought specialists from various disciplines to write the sections with which they are most familiar. As with the original *Grzimek's*, we have engaged the best scholars available to serve as topic editors, writers, and consultants. There were some complaints about inaccuracies in the original English version that may have been due to mistakes or misinterpretation during the complicated translation process. However, unlike the original *Grzimek's*, which was translated from German, this revision has been completely re-written by English-speaking scientists. This work was truly a cooperative endeavor, and I thank all of those dedicated individuals who have written, edited, consulted, drawn, photographed, or contributed to its production in any way. The names of the topic editors, authors, and illustrators are presented in the list of contributors in each individual volume.

The overall structure of this reference work is based on the classification of animals into naturally related groups, a discipline known as taxonomy or biosystematics. Taxonomy is the science through which various organisms are discovered, identified, described, named, classified and catalogued. It should be noted that in preparing this volume we adopted what might be termed a conservative approach, relying primarily on traditional animal classification schemes. Taxonomy has always been a volatile field, with frequent arguments over the naming of or evolutionary relationships between various organisms. The advent of DNA fingerprinting and other advanced biochemical techniques has revolutionized the field and, not unexpectedly, has produced both advances and confusion. In producing these volumes, we have consulted with specialists to obtain the most up-to-date information possible, but knowing that new findings may result in changes at any time. When scientific controversy over the classification of a particular animal or group of animals existed, we did our best to point this out in the text.

Readers should note that it was impossible to include as much detail on some animal groups as was provided on others. For example, the marine and freshwater fish, with vast numbers of orders, families, and species, did not receive as detailed a treatment as did the birds and mammals. Due to practical and financial considerations, the publishers could provide only so much space for each animal group. In such cases, it was impossible to provide more than a broad overview and to feature a few selected examples for the purposes of illustration. To help compensate, we have provided a few key bibliographic references in each section to aid those interested in learning more. This is a common limitation in all reference works, but *Grzimek's Encyclopedia of Animal Life* is still the most comprehensive work of its kind.

I am indebted to the Gale Group, Inc. and Senior Editor Donna Olendorf for selecting me as Series Editor for this project. It was an honor to follow in the footsteps of Dr. Grzimek and to play a key role in the revision that still bears his name. *Grzimek's Animal Life Encyclopedia* is being published by the Gale Group, Inc. in affiliation with my employer, the American Zoo and Aquarium Association (AZA), and I would like to thank AZA Executive Director, Sydney J. Butler; AZA Past-President Ted Beattie (John G. Shedd Aquarium, Chicago, IL); and current AZA President, John Lewis (John Ball Zoological Garden, Grand Rapids, MI), for approving my participation. I would also like to thank AZA Conservation and Science Department Program Assistant, Michael Souza, for his assistance during the project. The AZA is a professional membership association, representing 205 accredited zoological parks and aquariums in North America. As Director/William Conway Chair, AZA Department of Conservation and Science, I feel that I am a philosophical descendant of Dr. Grzimek, whose many works I have collected and read. The zoo and aquarium profession has come a long way since the 1970s, due, in part, to innovative thinkers such as Dr. Grzimek. I hope this latest revision of his work will continue his extraordinary legacy.

Silver Spring, Maryland, 2001
Michael Hutchins
Series Editor

· · · · ·

How to use this book

Grzimek's Animal Life Encyclopedia is an internationally prominent scientific reference compilation, first published in German in the late 1960s, under the editorship of zoologist Bernhard Grzimek (1909–1987). In a cooperative effort between Gale and the American Zoo and Aquarium Association, the series has been completely revised and updated for the first time in over 30 years. Gale expanded the series from 13 to 17 volumes, commissioned new color paintings, and updated the information so as to make the set easier to use. The order of revisions is:

Volumes 8–11: Birds I–IV
Volume 6: Amphibians
Volume 7: Reptiles
Volumes 4–5: Fishes I–II
Volumes 12–16: Mammals I–V
Volume 3: Insects
Volume 2: Protostomes
Volume 1: Lower Metazoans and Lesser Deuterostomes
Volume 17: Cumulative Index

Organized by taxonomy

The overall structure of this reference work is based on the classification of animals into naturally related groups, a discipline known as taxonomy—the science in which various organisms are discovered, identified, described, named, classified, and cataloged. Starting with the simplest life forms, the lower metazoans and lesser deuterostomes, in Volume 1, the series progresses through the more complex classes, concluding with the mammals in Volumes 12–16. Volume 17 is a stand-alone cumulative index.

Organization of chapters within each volume reinforces the taxonomic hierarchy. In the case of the volume on Insects, introductory chapters describe general characteristics of all insects, followed by taxonomic chapters dedicated to order. Species accounts appear at the end of order chapters.

Introductory chapters have a loose structure, reminiscent of the first edition. Chapters on orders, by contrast, are highly structured, following a prescribed format of standard rubrics that make information easy to find. These chapters typically include:

Thumbnail introduction
   Scientific name
   Common name
   Class
   Order
   Number of families
Main chapter
   Evolution and systematics
   Physical characteristics
   Distribution
   Habitat
   Behavior
   Feeding ecology and diet
   Reproductive biology
   Conservation status
   Significance to humans
Species accounts
   Common name
   Scientific name
   Family
   Taxonomy
   Other common names
   Physical characteristics
   Distribution
   Habitat
   Behavior
   Feeding ecology and diet
   Reproductive biology
   Conservation status
   Significance to humans
Resources
   Books
   Periodicals
   Organizations
   Other

Color graphics enhance understanding

Grzimek's features approximately 3,500 color photos, including nearly 130 in the Insects volume; 3,500 total color maps, including approximately 100 in the Insects volume; and approximately 5,500 total color illustrations, including approximately 300 in the Insects volume. Each featured species of animal is accompanied by both a distribution map and an illustration.

All maps in *Grzimek's* were created specifically for the project by XNR Productions. Distribution information was provided by expert contributors and, if necessary, further researched at the University of Michigan Zoological Museum library. Maps are intended to show broad distribution, not definitive ranges.

All the color illustrations in *Grzimek's* were created specifically for the project by Michigan Science Art. Expert contributors recommended the species to be illustrated and provided feedback to the artists, who supplemented this information with authoritative references and animal specimens from the University of Michigan Zoological Museum library. In addition to illustrations of species, *Grzimek's* features drawings that illustrate characteristic traits and behaviors.

About the contributors

All of the chapters were written by entomologists who are specialists on specific subjects and/or taxonomic groups. Topic editors Arthur V. Evans and Rosser W. Garrison reviewed the completed chapters to insure consistency and accuracy.

Standards employed

In preparing the volume on Insects, the editors relied primarily on the taxonomic structure outlined in *The Insects of Australia: A Textbook for Students and Research Workers*, 2nd edition, edited by the Division of Entomology, Commonwealth Scientific and Industrial Research Organisation (1991). Systematics is a dynamic discipline in that new species are being discovered continuously, and new techniques (e.g., DNA sequencing) frequently result in changes in the hypothesized evolutionary relationships among various organisms. Consequently, controversy often exists regarding classification of a particular animal or group of animals; such differences are mentioned in the text.

Grzimek's has been designed with ready reference in mind, and the editors have standardized information wherever feasible. For **Conservation status,** *Grzimek's* follows the IUCN Red List system, developed by its Species Survival Commission. The Red List provides the world's most comprehensive inventory of the global conservation status of plants and animals. Using a set of criteria to evaluate extinction risk, the IUCN recognizes the following categories: Extinct, Extinct in the Wild, Critically Endangered, Endangered, Vulnerable, Conservation Dependent, Near Threatened, Least Concern, and Data Deficient. For a complete explanation of each category, visit the IUCN Web page at <http://www.iucn.org/themes/ssc/redlists/categor.htm>.

In addition to IUCN ratings, chapters may contain other conservation information, such as a species' inclusion on one of three Convention on International Trade in Endangered Species (CITES) appendices. Adopted in 1975, CITES is a global treaty whose focus is the protection of plant and animal species from unregulated international trade.

In the species accounts throughout the volume, the editors have attempted to provide common names not only in English but also in French, German, Spanish, and local dialects.

Grzimek's provides the following standard information on lineage in the ***Taxonomy*** rubric of each species account: [First described as] *Raphidia flavipes* [by] Stein, [in] 1863, [based on a specimen from] Greece. The person's name and date refer to earliest identification of a species, although the species name may have changed since first identification. However, the entity of insect is the same.

Readers should note that within chapters, species accounts are organized alphabetically by family name and then alphabetically by scientific name.

Anatomical illustrations

While the encyclopedia attempts to minimize scientific jargon, readers will encounter numerous technical terms related to anatomy and physiology throughout the volume. To assist readers in placing physiological terms in their proper context, we have created a number of detailed anatomical drawings. These can be found on pages 18 to 33 in the "Structure and function" chapter. Readers are urged to make heavy use of these drawings. In addition, many anatomical terms are defined in the ***Glossary*** at the back of the book.

Appendices and index

In addition to the main text and the aforementioned ***Glossary,*** the volume contains numerous other elements. ***For further reading*** directs readers to additional sources of information about insects. Valuable contact information for ***Organizations*** is also included in an appendix. An exhaustive ***Insects family list*** records all families of insects as recognized by the editors and contributors of the volume. And a full-color ***Geologic time scale*** helps readers understand prehistoric time periods. Additionally, the volume contains a ***Subject index.***

Acknowledgements

Gale would like to thank several individuals for their important contributions to the volume. Dr. Arthur V. Evans and Dr. Rosser W. Garrison, topic editors for the Insects volume, oversaw all phases of the volume, including creation of the topic list, chapter review, and compilation of the appendices. Neil Schlager, project manager for the Insects volume, and Vanessa Torrado-Caputo, associate editor at Schlager Group, coordinated the writing and editing of the text. Dr. Michael Hutchins, chief consulting editor for the series, and Michael Souza, program assistant, Department of Conservation and Science at the American Zoo and Aquarium Association, provided valuable input and research support.

• • • • •

Advisory boards

Series advisor

Michael Hutchins, PhD
Director of Conservation and Science/William Conway Chair
American Zoo and Aquarium Association
Silver Spring, Maryland

Subject advisors

Volume 1: Lower Metazoans and Lesser Deuterostomes
Dennis A. Thoney, PhD
Director, Marine Laboratory & Facilities
Humboldt State University
Arcata, California

Volume 2: Protostomes
Sean F. Craig, PhD
Assistant Professor, Department of Biological Sciences
Humboldt State University
Arcata, California

Dennis A. Thoney, PhD
Director, Marine Laboratory & Facilities
Humboldt State University
Arcata, California

Volume 3: Insects
Arthur V. Evans, DSc
Research Associate, Department of Entomology
Smithsonian Institution
Washington, DC

Rosser W. Garrison, PhD
Research Associate, Department of Entomology
Natural History Museum
Los Angeles, California

Volumes 4–5: Fishes I– II
Paul V. Loiselle, PhD
Curator, Freshwater Fishes
New York Aquarium
Brooklyn, New York

Dennis A. Thoney, PhD
Director, Marine Laboratory & Facilities
Humboldt State University
Arcata, California

Volume 6: Amphibians
William E. Duellman, PhD
Curator of Herpetology Emeritus
Natural History Museum and Biodiversity Research Center
University of Kansas
Lawrence, Kansas

Volume 7: Reptiles
James B. Murphy, DSc
Smithsonian Research Associate
Department of Herpetology
National Zoological Park
Washington, DC

Volumes 8–11: Birds I–IV
Walter J. Bock, PhD
Permanent secretary, International Ornithological Congress
Professor of Evolutionary Biology
Department of Biological Sciences,
Columbia University
New York, New York

Jerome A. Jackson, PhD
Program Director, Whitaker Center for Science, Mathematics, and Technology Education
Florida Gulf Coast University
Ft. Myers, Florida

Volumes 12–16: Mammals I–V
Valerius Geist, PhD
Professor Emeritus of Environmental Science
University of Calgary
Calgary, Alberta
Canada

Devra G. Kleiman, PhD
Smithsonian Research Associate

National Zoological Park
Washington, DC

Library advisors

James Bobick
Head, Science & Technology Department
Carnegie Library of Pittsburgh
Pittsburgh, Pennsylvania

Linda L. Coates
Associate Director of Libraries
Zoological Society of San Diego Library
San Diego, California

Lloyd Davidson, PhD
Life Sciences bibliographer and head, Access Services
Seeley G. Mudd Library for Science and Engineering
Evanston, Illinois

Thane Johnson
Librarian

Oklahoma City Zoo
Oklahoma City, Oklahoma

Charles Jones
Library Media Specialist
Plymouth Salem High School
Plymouth, Michigan

Ken Kister
Reviewer/General Reference teacher
Tampa, Florida

Richard Nagler
Reference Librarian
Oakland Community College
Southfield Campus
Southfield, Michigan

Roland Person
Librarian, Science Division
Morris Library
Southern Illinois University
Carbondale, Illinois

Contributing writers

Insects

Elisa Angrisano, PhD
Universidad Nacional de Buenos Aires
Buenos Aires, Argentina

Horst Aspöck, PhD
Department of Medical Parasitology,
Clinical Institute of Hygiene and
Medical Microbiology
University of Vienna
Vienna, Austria

Ulrike Aspöck, PhD
Natural History Museum of Vienna
and University of Vienna
Vienna, Austria

Axel O. Bachmann, Doctor en Cien-
cias Biológicas
Universidad de Buenos Aires
Conicet
Buenos Aires, Argentina

Günter Bechly, PhD
Staatliches Museum für Naturkunde
Stuttgart, Germany

Andrew F. G. Bourke, PhD
Zoological Society of London
London, United Kingdom

Paul D. Brock
Slough, United Kingdom

Reginald Chapman, DSc
University of Arizona
Tucson, Arizona

Jeffrey A. Cole, BS
Natural History Museum
Los Angeles, California

Eduardo Domínguez, PhD
Universidad Nacional de Tucumán
Tucumán, Argentina

Arthur V. Evans, DSc
Smithsonian Institution
Washington, DC

Rosser W. Garrison, PhD
Natural History Museum
Los Angeles, California

Michael Hastriter, MS
Monte L. Bean Life Science Museum
Brigham Young University
Provo, Utah

Klaus-Dieter Klass, PhD
Museum für Tierkunde
Dresden, Germany

Marta Loiácono, DSc
Facultad de Ciencias Naturales y
Museo
La Plata, Buenos Aires, Argentina

Cecilia Margaría, Lic
Facultad de Ciencias Naturales y
Museo
La Plata, Buenos Aires, Argentina

Cynthia L. Mazer, MS
Cleveland Botanical Garden
Cleveland, Ohio

Silvia A. Mazzucconi, Doctora en
Ciencias Biológicas
Universidad de Buenos Aires
Buenos Aires, Argentina

Juan J. Morrone, PhD
Museo de Zoología, Facultad de
Ciencias
UNAM
Mexico City, Mexico

Laurence A. Mound, DSc
The Natural History Museum
London, United Kingdom

Timothy George Myles, PhD
University of Toronto
Toronto, Ontario, Canada

Piotr Naskrecki, PhD
Museum of Comparative Zoology
Harvard University
Cambridge, Massachusetts

Timothy R. New
La Trobe University
Melbourne, Australia

Hubert Rausch
Scheibbs, Austria

Martha Victoria Rosett Lutz, PhD
University of Kentucky
Lexington, Kentucky

Louis M. Roth, PhD
Museum of Comparative Zoology
Harvard University
Cambridge, Massachusetts

Michael J. Samways, PhD
University of Stellenbosch
Maiteland, South Africa

Vincent S. Smith, PhD
University of Glasgow
Glasgow, United Kingdom

Kenneth Stewart, PhD
University of North Texas
Denton, Texas

S. Y. Storozhenko
Institute of Biology and Soil Science,
Far East Branch of Russian Academy
of Sciences
Vladivostock, Russia

Natalia von Ellenrieder, PhD
Natural History Museum
Los Angeles, California

Shaun L. Winterton, PhD
North Carolina State University
Raleigh, North Carolina

Kazunori Yoshizawa, PhD
Hokkaido University
Sapporo, Japan

Contributing illustrators

Drawings by Michigan Science Art

Joseph E. Trumpey, Director, AB, MFA
Science Illustration, School of Art and Design, University of Michigan

Wendy Baker, ADN, BFA

Ryan Burkhalter, BFA, MFA

Brian Cressman, BFA, MFA

Emily S. Damstra, BFA, MFA

Maggie Dongvillo, BFA

Barbara Duperron, BFA, MFA

Jarrod Erdody, BA, MFA

Dan Erickson, BA, MS

Patricia Ferrer, AB, BFA, MFA

George Starr Hammond, BA, MS, PhD

Gillian Harris, BA

Jonathan Higgins, BFA, MFA

Amanda Humphrey, BFA

Emilia Kwiatkowski, BS, BFA

Jacqueline Mahannah, BFA, MFA

John Megahan, BA, BS, MS

Michelle L. Meneghini, BFA, MFA

Katie Nealis, BFA

Laura E. Pabst, BFA

Christina St. Clair, BFA

Bruce D. Worden, BFA

Kristen Workman, BFA, MFA

Thanks are due to the University of Michigan, Museum of Zoology, which provided specimens that served as models for the images.

Maps by XNR Productions

Paul Exner, Chief cartographer
XNR Productions, Madison, WI

Tanya Buckingham

Jon Daugherity

Laura Exner

Andy Grosvold

Cory Johnson

Paula Robbins

. . . . .

Topic overviews

What is an insect?

Evolution and systematics

Structure and function

Life history and reproduction

Ecology

Distribution and biogeography

Behavior

Social insects

Insects and humans

Conservation

What is an insect?

Overview

We live in the "age of insects." Humans have walked on Earth for only a mere fraction of the 350 million years that insects have crawled, burrowed, jumped, bored, or flown on the planet. Insects are the largest group of animals on Earth, with over 1.5 million species known to science up to now, and represent nearly one-half of all plants and animals. Although scientists do not know how many insect species there are and probably will never know, some researchers believe the number of species may reach 10 to 30 million. Even a "typical"

backyard may contain several thousand species of insects, and these populations may number into the millions. It is estimated that there are 200 million insects for every human alive today. Just the total biomass of ants on Earth, representing some 9,000 species, would outweigh that of humans twelve times over. Insect habitats are disappearing faster than we can catalog and classify the insects, and there are not enough

A nut weevil (*Curculio nucum*) larva emerging from a hole in a hazel nut. (Photo by Kim Taylor. Bruce Coleman, Inc. Reproduced by permission.)

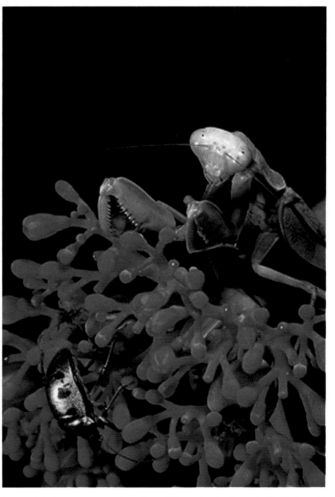

A mantid about to eat a jewelbug. (Photo by A. Captain/R. Kulkarni/ S. Thakur. Reproduced by permission.)

A leaf-footed bug (family Coreidae) caring for young, in Indonesia. (Photo by Jan Taylor. Bruce Coleman, Inc. Reproduced by permission.)

trained specialists to identify all the insect specimens housed in the world's museums.

The reproductive prowess of insects is well known. Developing quickly under ideal laboratory conditions, the fruit fly (*Drosophila melanogaster*) can complete its entire life cycle in about two weeks, producing 25 generations annually. Just two flies would produce 100 flies in the next generation—50 males and 50 females. If these all survived to reproduce, the resulting progeny would number 5,000 flies! Carried out to the 25th generation, there would be 1.192×10^{41} flies, or a ball of flies (1,000 per cubic inch) with a diameter of 96,372,988 mi (155, 097, 290 km), the distance from Earth to the Sun. Fortunately this population explosion is held in check by many factors. Most insects fail to reproduce, suffering the ravages of hungry predators, succumbing to disease and parasites, or starving from lack of suitable food.

Physical characteristics

Insects are at once entirely familiar, yet completely alien. Their jaws work from side to side, not up and down. Insect eyes, if present, are each unblinking and composed of dozens, hundreds, or even thousands of individual lenses. Insects feel, taste, and smell the world through incredibly sensitive receptors borne on long and elaborate antennae, earlike structures

on their legs, or on incredibly responsive feet. Although they lack nostrils or lungs, insects still breathe, thanks to small holes located on the sides of their bodies behind their heads, connected to an internal network of finely branched tubes.

Like other members of the phylum Arthropoda (which includes arachnids, horseshoe crabs, millipedes, centipedes, and crustaceans), insects have ventral nerve cords and tough skeletons on the outside of their bodies. This external skeleton is quite pliable and consists of a series of body divisions and plates joined with flexible hinges that allow for considerable movement.

As our knowledge of insects has increased, their classification has inevitably become more complex. They are now classified in the subphylum Hexapoda, and are characterized by having three body regions (head, thorax, and abdomen) and a three-segmented thorax bearing six legs. The orders Protura, Collembola, and Diplura, formerly considered insects, now make up the class Entognatha. Entognaths have mouthparts recessed into the head capsule, reduced Malpighian tubules (excretory tubes), and reduced or absent compound eyes.

The remaining orders treated in this volume are in the class Insecta. Insects have external mouthparts that are exposed from the head capsule, lack muscles in the antennae beyond the first segment, have tarsi that are subdivided into tarsomeres, and females are equipped with ovipositors. The word "insect" is derived from the Latin word *insectum*, meaning notched, and refers to their body segmentation. The second and third segments of the adult thorax often bear wings, which may obscure its subdivisions.

Insects are one of only four classes of animals (with pterosaurs, birds, and bats) to have achieved true flight, and were the first to take to the air. The evolution of insect wings was altogether different from that of the wings of other flying creatures, which developed from modified forelimbs. Instead, insect wings evolved from structures present in addition to their legs, not unlike Pegasus, the winged horse of Greek mythology. Long extinct dragonflies winged their way through Carboniferous forests some 220 million years ago and had wings measuring 27.6 in (700 mm) or more across. Today the record for wing width for an insect belongs to a noctuid moth from Brazil whose wings stretch 11 in (280 mm) from tip to tip. Insects are limited in size by their external skeletons and their mode of breathing. While most species range in length from 0.04 to 0.4 in (1 to 10 mm), a few are smaller than the largest Protozoa. The parasitic wasps that attack the eggs of other insects are less than 0.008 in (0.2 mm) long, smaller than the period at the end of this sentence. Some giant tropical insects, measuring 6.7 in (17 cm), are considerably larger than the smallest mammals.

Behavior

The small size of insects has allowed them to colonize and exploit innumerable habitats not available to larger animals. Most species live among the canopies of lush tropical forests. Some species are permanent residents of towering peaks some 19,685 ft (6,000 m) above sea level. Others live in eternal darkness within the deep recesses of subterranean caves. Some occupy extreme habitats such as the fringes of boiling hot springs, briny salt lakes, sun-baked deserts, and even thick pools of petroleum. The polar regions support a few insects that manage to cling to life on surrounding islands or as parasites on Arctic and Antarctic vertebrates. Fewer still have conquered the oceans, skating along the swelling surface. No insects have managed to penetrate and conquer the depths of freshwater lakes and oceans.

The feeding ecologies of insects are extremely varied, and insects often dominate food webs in terms of both population size and species richness. Equipped with chewing, piercing/sucking mouthparts, or combinations thereof, insects cut, tear, or imbibe a wide range of foodstuffs, including most plant and animal tissues and their fluids. Plant-feeding insects attack all vegetative and reproductive structures, while scavengers plumb the soil and leaf litter for organic matter. Some species collect plant and animal materials—not to eat, but to feed to their young or use as mulch to grow fungus as food. Many ants

Zebra butterfly (*Heliconius charitonia*) feeding on flower nectar. (Photo by Jianming Li. Reproduced by permission.)

A lanternfly in Koyna, Japan. (Photo by A. Captain/R. Kulkarni/S. Thakur. Reproduced by permission.)

"keep" caterpillars or aphids as if they were dairy cattle, milking them for fluids rich in carbohydrates. Predatory species generally kill their prey outright; parasites and parasitoids feed internally or externally on their hosts over a period of time or make brief visits to acquire their blood meals.

Resources

Books
Borror, D. J., C. A. Triplehorn, and N. F. Johnson. *An Introduction to the Study of Insects*. Philadelphia: Saunders College Publishing, 1989.

CSIRO, ed. *The Insects of Australia: A Textbook for Students and Research Workers*, 2nd ed. Carlton, Australia: Melbourne University Press, 1990.

Periodicals
Hogue, C. L. "Cultural Entomology." *Annual Review of Entomology* 32 (1987): 181–199.

Arthur V. Evans, DSc

Evolution and systematics

Fossil insects and their significance

Given the tiny and delicate bodies of most insects, it is perhaps surprising that remains of these organisms can be preserved for millions of years. After all, most fossils represent only hard parts of other organisms such as bones of vertebrates or shells of mollusks. Fossil remains of soft-bodied animals such as worms or jellyfish are extremely rare and can only be preserved under very special circumstances. In contrast to the large number of living insect species, fossil insects are rare compared to other groups. One obstacle for the fossilization of insects is that most insect species do not live in water. Because they can usually only be preserved as fossils in subaquatic sediments (amber is an exception to this rule), they thus have to be accidentally displaced into the water of an ocean or a lake.

Since most insects are terrestrial animals, the fossil record for these species is poor. Freshwater groups such as water-bugs and water-beetles, as well as the larvae of mayflies, dragonflies, stoneflies, alderflies, and the vast majority of caddisflies, live in rivers or lakes, and their fossil record is much better. Comparatively few insect species live in brackish water and in the tidal area of seashores, and only a single small group of water-bugs has evolved to conquer marine habitats: it is the extant (i.e., living) sea skater, or water strider, genus *Halobates* of the family Halobatidae, which only recently in Earth's history evolved to live on the surface of the ocean.

The first and most important prerequisite for fossilization is the embedding of the insect body in a subaquatic environment with stagnant water that allows undisturbed formation of layered sediments on the ground. Terrestrial insects can be washed into lakes by floods, and flying insects can be blown onto the surface of lakes or the sea during heavy storms. Dwellers of rivers and brooks must also be washed into lakes, lagoons, or the sea to become fossilized, because there are no suitable sedimentation conditions in running water. Aquatic insects that live in lakes and ponds can be preserved in sediments on the ground of their habitat, a type of preservation known as "autochthonous preservation."

Further conditions must be fulfilled for an insect to be fossilized. First, the insect must penetrate the water surface and sink to the bottom. This is achieved most easily if the insect is displaced alive into the water and drowns, so that its inter-nal cavities become filled with water. Insects that have been entangled in floating mats of algae can easily sink down more rapidly with it. However, if dead or even desiccated insects are blown onto the water surface, they may float for a very long time and will start to rot or be eaten by fish, enhancing disarticulation of their bodies (especially wings), which will have a chance to sink down and be preserved as isolated fossil remains. Dead terrestrial insects washed into water bodies by rivers or floods can become completely fossilized depending on the length of time and distance of specimen transport and drift. Consequently, the state of preservation and the completeness of fossil insects are good indicators for the conditions of embedding. A further important factor is the chemical makeup of the water where the insect is embedded. When the water body includes an oxygen-rich zone with abundant fish life, sinking insect bodies may be eaten by fish and never reach the ground. In contrast, hostile conditions such as hypersalinity, digested sludge with poisonous hydrogen sulfide, and low oxygen content prohibit the presence of ground-dwelling scavengers (e.g., worms, mollusks, and crustaceans) and make the preservation of insect fossils more likely. Such conditions near the bottom of the water body usually are present only in deep, calm water without any significant water exchange.

Finally, a dead insect or other carcass must be rapidly covered with new sediments, so that the body can be preserved as a fossil when these sediments are later consolidated into rock. Very often such sedimentation events occur in regular intervals. The resulting rocks are then fissionable in plates (e.g., lithographic limestone) along the former interfaces between two sedimentations. When the fossils are situated directly on the surface of these plates, they are immediately visible after the rock has been split and need only minor preparation. However, when the fossil insects are concealed within the plates, they can only be recognized by an inconspicuous bulge and/or discoloration of the plate surface, and must be prepared with great care and suitable tools (e.g., pneumatic graver and needles) in order to remove the covering rock without damaging the fossil.

Types of fossils

Impressions

The particular way the fossil insect is preserved strongly depends on the types of sediments and the chemical compo-

Fossil of a dragonfly in limestone matrix, from Sohnhofen, Germany, Jurassic era. The wingspan is approximately 8 in (20.3 cm). (Photo by Jianming Li. Reproduced by permission.)

sition of the water as well as the circumstances of the transformation of the sediments into rock. Most often the insect bodies completely decay in the course of time and only an impression of the animal remains as fossil. This is the case with fossil insects from Carboniferous coal layers, the Lower Jurassic oil shales of Middle Europe, and most limestones throughout the world. Even though these fossils are impressions, some body parts may be accentuated and traced with a secondary coloration if diluted metal oxides (e.g., iron oxide or manganese oxide) penetrate the body cavities in dendritic fashion. Dendrites can be reddish to brown (iron oxide) or black (manganese oxide). This phenomenon is exemplified in wing venation of fossil dragonflies from the Solnhofen lithographic limestones from the Upper Jurassic of Germany.

The finer the sediments, the greater the number of details that may be preserved in the fossil insects, so that even delicate bristles or facets of complex eyes are still visible. Sediments exposed to strong pressure and compaction during the transformation into rock often result in completely flattened impressions. However, if layers harden relatively fast, impressions can retain a three-dimensional profile of parts of the former insect body, for example the corrugation and pleating of the wings.

Under particular chemical circumstances, the organic matter of the insect body can be impregnated or replaced by mineral substances and therefore preserved in the original shape with all of its three-dimension properties (e.g., the pleating of the wing membrane). This occurs in all fossil insects from the Lower Cretaceous Crato limestones from northeastern Brazil, where insect bodies were preserved as iron-oxide-hydroxide (limonite). These fossil insects are tinted reddish brown and are often very distinct from the bright yellowish limestone. This special mode of fossilization has even permitted the preservation of soft parts such as muscles or internal organs. In some cases, the color pattern of the wings of cockroaches, bugs, beetles, and lacewings is still visible. This rare phenomenon provides information on the appearance of extinct animals that is usually not available in fossils.

Incrustation

A second mode of preservation involves decay processes in which the insect body induces a chemical reaction that leads to the precipitation of minerals around the dead insect. This process can produce bulbs of rock (geodes or concretions) in which the fossil insect is preserved three-dimensionally. Fossil insects from Mazon Creek, a famous locality from the Carboniferous era of North America, are preserved this way. Incrustation with sinter, the chalk generated by hot wells, can preserve dead insects as three-dimensional impressions that have fallen into this mineral water. There, hollow molds can be filled with composition rubber to obtain perfect copies of the original insect bodies.

Embedding

The third and rarest method of fossilization involves the embedding of insects within crystals, for example dragonfly larvae in gypsum crystals from the Miocene of Italy. These crystals developed in a desiccating coastal water body in the Tertiary age, when the Mediterranean Sea was separated from the Atlantic Ocean by a barrier at the Strait of Gibraltar. However, this hypersalinity of the water was not the habitat of the enclosed dragonfly larvae, because they are close relatives of extant dragonflies that never live in such environments. The enclosed dragonfly larvae are not the animals themselves, but only dried skins from the final molting of the larvae into adult dragonflies. Such skins (exuviae) are very robust and are transported during storms to habitats such as that mentioned above.

Insect inclusions in amber represent the most important exception from the rule that insects can only be fossilized in subaquatic environments. These animals are preserved in fossil resins with their natural shape with all details in a quality that is unmatched by any other kind of fossilization. The oldest known fossil insect inclusions in amber were discovered in Lebanon and are of Lower Cretaceous age (about 120 million years old [mya]). The insects of the famous Baltic amber and the Dominican amber from the Caribbean are much younger (45–15 mya) and have been dated to the early to mid-Tertiary. Insects enclosed in the latter two fossil resins are already much more modern than those of the Lower Cretaceous amber, which were contemporaries of dinosaurs and pterosaurs. The novel and subsequent movie *Jurassic Park*, in which scientists revive dinosaurs by using the DNA of dinosaur blood imbibed by mosquitoes fossilized in amber is highly unlikely, since no suitable DNA has ever been discovered in insects fossilized in amber.

Even preservation of more imperishable exoskeleton (chitin) comprises relatively recent insect fossils, and then only under very favorable circumstances. More frequently, chitin is preserved in subfossil insects from relatively recent layers, for example from the Pleistocene asphalt lakes of Rancho La Brea near Los Angeles, which are only 8,000–40,000 years old. The oldest known fossil insects with preservation of chitin are of Tertiary age. However, the preservation of metallic colors in some small damselflies from the Lower Cretaceous Crato limestones of Brazil could indicate that the original exoskeleton was preserved in these cases, but confirmation of this would require chemical analysis.

A trained eye is necessary to discover and recognize many insect fossils. They are often not situated on easily cleavable places but instead are concreted in stone matrix. Once fossil insects are found, their features such as wing venation may become more visible when submerged in alcohol.

Paleoentomology can be cumbersome and hard work, but discoveries of fossil insects can provide us with knowledge of the history of life on Earth. Study of insect fossils increases our knowledge about past biodiversity, past climate and habitats, extinction events, changes in the geographical distribution of groups, sequence of anatomical changes in the course of evolution, minimum age of origin of extant groups or the lifespan of extinct groups, and types of organisms and adaptations that do not exist anymore. For example, extant snakeflies (Rhaphidioptera) are restricted to the Northern Hemisphere and only live in temperate (cooler) areas, but fossil snakeflies from the Lower Cretaceous Crato limestones of Brazil correspond to a warm and arid area with savannah-like vegetation. The extinction of all tropical snakeflies at the end of the Cretaceous could be related to climatic consequences of the meteorite impact that also led to the extinction of dinosaurs. Only those snakeflies that were adapted to cooler climates survived.

Subtropical and tropical areas not only differ in climate from temperate or cooler regions, but also in the composition of their flora and fauna. This is observable in insect fauna: praying mantids, termites, cicadas, walkingsticks, and many other insect groups are adapted and restricted to warm climate zones. Earth's appearance and its climate have changed dramatically over time. The position and shape of continents have changed, oceans have emerged and vanished, cold or warm streams have changed their course, and the polar caps have disappeared and reappeared and expanded dramatically during ice ages. Freezing, barren regions like the Antarctic formerly had a warm climate with a rich vegetation and fauna. Areas of North America and Middle Europe also supported tropical or arid climates as well as cold periods with extensive glaciation.

Fossils often provide clues to reconstructing climatic changes during Earth's history. When extant relatives of a fossil organism are strictly confined to tropical or desert areas, it is tempting to assume that this was also the case with their fossil relatives. This assumption will be correct in most cases, but in other instances extant groups such as snakeflies have adapted to a cooler climate within their evolutionary history. Thus, their fossils may be poor indicators for a certain type of climate. It is therefore important to compare the complete fossil record of a certain locality with the modern relatives and their habitats. Many freshwater deposits yield a variety of fossil plants, vertebrates, and arthropods. If several of these species belong to faunal and floral assemblages that are clearly indicators for a certain climate, it is possible to reconstruct the past climate with confidence (as long as other species present are generalists or had unknown preferences). In Messel near Darmstadt in Germany, for example, lacustrine sediments of the Eocene have yielded several fossil insects such as walking sticks that suggest a previously warm climate. This is in accord with evidence from vertebrate fossils such as prosimians and crocodiles.

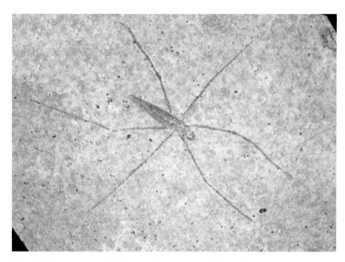

Fossil of a water strider in mudstone matrix from Sohnhofen, Germany, Jurassic era. The span between the legs is about 4 in (10 cm). (Photo by Jianming Li. Reproduced by permission.)

Baltic amber has also yielded numerous insects (e.g., web-spinners, walkingsticks, praying mantids, termites, and palm bugs) that indicate a warm and humid climate. Palm bugs indirectly demonstrate the presence of palm trees in the amber forest. The presence of the preserved insects is in accord with the fauna from the Messel fossils that lived in about the same period. Thus, the climate in Middle Europe was much warmer in the early Tertiary (45 mya) than today.

Fossil insects not only contribute to the reconstruction of past climates, they also provide evidence of the prevailing vegetation types and landscape. For example, the insect fauna of the Crato limestones from the Lower Cretaceous of Brazil includes not only certain species (e.g., cicadas, ant lions, nemopterids, termites) that suggest a warm climate, but also numerous insect groups (cockroaches, locusts, bugs, robber flies) that presently live in very different habitats and climatic conditions. However, their relative frequency in the fossil record from this site is in perfect agreement with insect communities in modern savanna areas and is further supported by fossils of other arthropods (e.g., sun spiders) and plants (order Gnetales). Nevertheless, this Cretaceous savanna must have been dissected by rivers and brooks, because of the presence of numerous fossils of aquatic insect larvae of mayflies (Euthyplociidae) and dragonflies (Gomphidae) that belong to modern families that are strictly riverine. Geological evidence (e.g., dolomite and salt pseudomorphs) and other evidence (e.g., fossils of marine fishes) clearly show that the Crato limestones originated as sediments in a brackish lagoon, in which the terrestrial and aquatic insects were transported by flowing water or wind. Taken together, this evidence allows for a nearly complete reconstruction of the habitat, landscape, climate, flora, and fauna of this locality in South America 120 mya.

The ancestry of insects

Insects belong to the large group of arthropods that also includes arachnids, crustaceans, and myriapods. For many decades, insects were generally considered close relatives of myriapods, and the ancestor of insects was consequently believed to have been a myriapod-like terrestrial arthropod. However, this hypothetical assumption was not supported by any fossil evidence. It was first challenged by the finding that respiratory organs (tracheae) of various myriapod groups and insects were superficially similar but quite different in their construction, so that they more likely evolved by convergent evolution from a common aquatic ancestor that did not possess tracheae at all. The close relationship between insects and myriapods was strongly challenged by new results from molecular, ontogenetic, and morphological studies that revealed congruent evidence towards a closer relationship between insects and higher crustaceans (Malacostraca), which would also suggest a marine ancestor of insects, but of much different appearance than previously believed. The hypothetical reconstruction of the most recent common ancestor of all insects thus strongly depends on the correct determination of the position of insects in the tree of life and whether their closest relatives were terrestrial or aquatic organisms.

The discovery of genuine fossils from the stem group of insects would allow a much more profound reconstruction and also would represent an independent test for the hypothetical reconstructions and their underlying phylogenetic hypotheses. The oldest fossils that can be identified as true hexapods were discovered in the Middle Devonian Rhynie chert of Scotland (400 mya). This chert originated when a swamp of primitive plants was flooded with hot volcanic water in which many minerals were dissolved. These fossil hexapods are morphologically more or less identical with some extant species of springtails and can therefore easily be placed in the extant order Collembola. Since two most closely related groups of organisms, so-called sister groups, originated by the splitting of one common stem species, they must be of the same age. Together with the small wingless orders Protura and maybe Diplura, springtails belong to the subclass Entognatha. Consequently, the second subclass of hexapods, Insecta, which includes all modern insects with ectognathous (exposed) mouthparts, must also be of Devonian age at least. The most primitive and probably oldest members of ectognathous insects are the two wingless orders Microcoryphia (bristletails) and Thysanura (silverfish), often still known as thysanurans. No Devonian fossils of these insects have yet been discovered, except for some fragments of compound eyes and mouthparts that have been found by dissolving Devonian cherts from North America with acid.

Except for those few Devonian fossils mentioned above, the oldest fossil insects occur in layers from the lower Upper Carboniferous (320 mya). These rocks show a surprising diversity of various insect groups: not only wingless insects such as bristletails and silverfish, but also the oldest known insects with wings, such as ancestors of mayflies and dragonflies, as well as primitive relatives of cockroaches and orthopterans. Within 80 million years between the Middle Devonian and the Upper Carboniferous, the evolution of insects resulted in a great diversity of different insect groups and also allowed for the conquest of the airspace by generating a remarkable new structure: two pairs of large membranous wings with complex articulation and musculature.

Before the Devonian period, there must have been a long period of slow evolution for the ancestral line of insects, because well-preserved fossils of other arthropod groups such as chelicerates and crustaceans are known from Cambrian sediments, which are about 200 million years older than the oldest insect fossils. If insects (or insects together with myriapods) are most closely related to crustaceans, their early marine ancestors must have existed in the Cambrian as well. However, no fossils of these early ancestors have been discovered yet. These ancestors simply may have been overlooked or even misidentified because they do not look like insects but rather have a more crustacean-like general appearance. Therefore, it is important to evaluate which combination of characters would characterize an ancestor, based on the current knowledge of the relationship of insects and the morphology of the most primitive extant representatives of insects and their suggested sister group.

One of the most conspicuous characters in many modern insects, such as dragonflies, bugs, beetles, bees, and butterflies, is the presence of wings. However, the most primitive and basal hexapod orders such as springtails, diplurans, bristletails, and silverfish, as well as their fossil relatives, all lack wings. Since the closest relatives of insects, myriapods and/or crustaceans, also lack wings, it is obvious that the absence of wings in those primitive orders is not due to reduction but rather due to their branching from the insect phylogenetic tree before the evolution of wings. Consequently, ancestors of all insects must also have been wingless.

Besides numerous other anatomical details that are often not preserved or visible in fossils, all insects are characterized by a division of the body into three distinct parts: head, thorax with three segments—each with a pair of legs—and abdomen with a maximum of 11 segments that contains internal organs and genital organs but includes no walking legs. The division into three body parts is a clear distinction of insects from other arthropod groups: myriapods also have a head, but their trunk is not divided into thorax and abdomen, and all of their segments bear one or two pairs of legs of about the same size. Due to the presumed close relationship of insects to myriapods and crustaceans, it is likely that ancestors of insects still had legs (maybe already of reduced size) on the abdominal segments. Like myriapods, all insects only have one pair of antennae, while extant crustaceans have two pairs and extant arachnids have none. Unlike other arthropods, insects have a single pair of appendages on the terminal body segment.

These considerations allow for the prediction that the ancestor of all insects most probably had the following combination of characters besides the usual character set of all arthropods (compound eyes, exoskeleton, articulated legs, thorax, etc.): a distinctly delimited head with only one pair of antennae; a three-segmented wingless thorax with three pairs of large walking legs; and a longer abdomen with at least 11 segments, a pair of terminal appendages, and perhaps a pair of smaller leglets on most abdominal segments. Furthermore, it is likely that this ancestor was an aquatic marine animal.

A fossil organism (*Devonohexapodus bocksbergensis*) with exactly this combination of characters was discovered in the Lower Devonian slates of Bundenbach (Hunsrück) in Germany in 2003. Its head bears only one pair of long antennae, the thorax has three pairs of long walking legs, and the abdomen has about 30 segments, each with a pair of small leglets, while the terminal segment bears a pair of curious appendages that are unlike walking legs and directed backwards. It seems to be closely related to (or more probably even identical with) another fossil organism, *Wingertshellicus backesi*, that was previously described as an enigmatic arthropod but has a very similar general appearance and combination of characters. The presence of legs on the abdominal segments is compatible with both possible sister groups of insects, because crustaceans and myriapods both possess legs or leg derivatives on the trunk segments. In myriapods these legs are more or less identical in their anatomy and size on all segments, while in crustaceans there is a difference between the anterior walking legs and posterior trunk appendages that are shorter and often of different shape. Therefore, the aforementioned Devonian fossils suggest a closer relationship of insects with crustaceans. In extant insects the abdominal leglets are either reduced or transformed into other structures (e.g., genital styli, jumping fork of springtails). However, in bristletails and some primitive silverfish, there are still so-called styli present on the abdominal segments that are quite similar to the short abdominal leglets of *Devonohexapodus*.

As is often the case in evolutionary biology, there exists conflicting evidence that poses some as yet unsolved problems for scientists. The Upper Carboniferous fossil locality Mazon Creek in North America has yielded several fossil wingless insects, similar to extant thysanurans, that possessed true legs with segments and paired claws on eight abdominal segments just like the three pairs of walking legs on the thorax. The fossils are also smaller in size than *Devonohexapodus* and seem to have been terrestrial organisms, thus rather pointing to a myriapod relationship and origin of insects. Since they are much younger than the oldest true insects, they may already have been living fossils in their time, just like *Devonohexapodus*, which was contemporaneous with the first true terrestrial insects.

Devonohexapodus was found in a purely marine deposit, but it could have been a terrestrial animal that was washed into the sea by rivers or floods. However, if that were the case, one would expect to find other terrestrial animals and plants as well. The Hunsrück slates yielded a large diversity of marine organisms but no terrestrial plants or animals at all. Consequently, *Devonohexapodus* was probably a marine animal; the crustacean-like appearance and structure also suggest an aquatic lifestyle. *Devonohexapodus* thus seems to be the first record of a marine ancestor of insects, or considering its age, an offshoot from the ancestral line of insects that survived into the Devonian, when more advanced and terrestrial insects had already evolved from their common ancestors. This fossil, as well as evidence from phylogenetic and comparative morphological research, supports the hypothesis that insects evolved directly from marine arthropods (either related to crustaceans or myriapods) and not from a common terrestrial ancestor of myriapods and insects. Ancestors of arachnids (e.g., trilobites) and the most primitive extant relatives of arachnids (horseshoe crabs) also are marine animals, just like most crustaceans (all crustaceans in freshwater and terrestrial

environments are thought to be derived from marine relatives). The anatomical differences within the respiratory (tracheal) system in various myriapod groups suggest that these myriapods did not have a common terrestrial ancestor but that different groups of myriapods conquered land several times independently. Their ancestors may have been amphibious, which facilitated their final transition to a completely terrestrial lifestyle. Some crustaceans, such as woodlice (Isopoda), managed this transition via amphibious ancestors; the most primitive woodlouse still has an amphibious lifestyle on seashores. Since certain organs like tracheae for breathing air have clearly evolved independently in some terrestrial arachnids (and even velvet worms), apparent similarities between terrestrial myriapods and insects could simply be due to convergent evolution. Different unrelated arthropod groups obviously developed similar structures when they left the ocean and became terrestrial animals, so that all structures related to a terrestrial lifestyle may be poor indicators for a close relationship despite overall similarity.

The conquest of the land

About 400 mya during the Upper Silurian and Lower Devonian, one of the most significant events happened in the evolution of life on Earth: an increased oxygen level in the atmosphere coupled with the correlated generation of an ozone layer offered protection against harmful ultraviolet radiation, and the first primitive green plants colonized the continents. The first terrestrial arthropods appeared soon after, followed by tetrapod vertebrates. Before that time, a highly diverse ecology existed in the world's oceans, especially along continental shelves and coastal regions with shallow water, but the continents themselves were stony deserts that resembled the surface of Mars. The ancestors of insects still inhabited the oceans at this time, as evidenced by discovery of their fossils.

The first pioneers of terrestrial habitats were various algae and primitive vascular plants such as rhyniophytes (*Rhynia*) and psilophytes (*Psilophyton*), which were naked stalks lacking any leaves or roots. These primitive herbaceous plants were confined to the edges of shallow coastal waters and swamps and were not yet "true" terrestrial plants. The oldest fossil insects as well as ancient amphibians strongly adapted to aquatic habitats have been found together with fossils of these early plants. As explained above, various terrestrial groups of arthropods (e.g., velvet worms, arachnids, centipedes, millipedes, insects, and some crustaceans) conquered the dry land several times independently and are not derived from a common terrestrial ancestor, even though they show similar adaptations for a terrestrial mode of life. The emergence of plants on land was a necessary prerequisite for the first arthropods to make the transition to terrestrial life. Early land plants provided nutrition for the first terrestrial arthropods. In the Rhynie cherts of the Lower Devonian from Scotland, fossils have been discovered that provide direct evidence for the feeding on plants by myriapods and unknown arthropods with sucking mouthparts.

The earliest terrestrial insects were wingless and tiny ground-dwellers such as springtails, diplurans, bristletails, and silverfish. Just like their modern relatives, they probably fed on detritus—organic substances on the ground composed of decaying plant material mingled with fungal meshworks and bacterial colonies. Other early terrestrial arthropods such as centipedes and arachnids were predators that fed on those small insects or on each other. As soon as the environmental conditions became suitable due to changes in the atmosphere and the evolution of land plants, the multiple conquest of the land by previously aquatic arthropods was facilitated by the evolution of certain features of the arthropod structural design. This design, which had evolved 600 mya during the Cambrian era in the ancestor of all aquatic arthropods, included the exoskeleton that later provided protection against dehydration by evaporation of body fluids, and the mechanical support for a body that was no longer supported by the water. Another important pre-adaptation was the presence of walking legs that also allowed for an active and swift locomotion on dry land.

Ancestors of most terrestrial arthropod groups during the time of the transition from aquatic to terrestrial life may have been very small amphibious creatures. They could have breathed under water and in air through simple diffusion of oxygen through their skin, which is not a very effective way of respiration. With increased demands for the efficiency of the respiratory system in completely terrestrial animals, various groups independently developed complex systems of ramified tubular invaginations (tracheae) to increase the oxygen supply for internal organs and muscles.

The origin of wings and flight

The colonization of totally new habitats represented an important step in the history of evolution. This is the case not only for the colonization of the dry land by plants and animals in the Devonian period, but also for the later conquest of the air by the four groups of animals that developed the ability for active flight: insects, pterosaurs, bats, and birds. Of these groups, insects were the first to acquire organs of flight.

Although researchers are not sure at which point in Earth's history insects developed wings and the ability to fly, a number of fossil winged insects (dragonflies, mayflies, cockroaches, and several extinct groups) are known from the lowermost Upper Carboniferous period (c. 320 mya). The oldest-known winged insect, *Delitzschala bitterfeldensis,* was described from a drilling core from Delitzsch in the vicinity of Bitterfeld in eastern Germany. This fossil is dated from the uppermost Lower Carboniferous and is about 325 million years old. It belongs to the extinct group Paleodictyoptera, which also included other primitive winged insects. The evolution of insect wings with complex wing venation and sophisticated articulation therefore must have taken place by the Lower Carboniferous if not in the Upper Devonian.

Unfortunately, there are only a few fossil insects known from the Devonian, and they all represent primarily wingless insects (e.g., springtails and bristletails). The fossil *Eopterum devonicum* from the Middle Devonian of Russia was long believed to be the most ancient winged insect, but the apparent wings have been shown to represent not an organ for flight but rather only the isolated tail fan of a crustacean.

Scientists have relied on hypothetical reconstructions of this important step in evolution, based on indirect evidence and plausible speculations. This has resulted in numerous different, and often conflicting, hypotheses about the evolution of insect wings and flight. Two alternative theories of wing development dominate the discussion among scientists: the exite theory and the paranotal theory.

The exite theory

Proponents of the exite theory believe that wings evolved as derivatives of lateral appendages (exites) of the bases of the walking legs that are present in one extant group of wingless insects, the bristletails. This theory is largely dependent on disputed fossil evidence and on the fact that the wings of all insects are supplied with oxygen by a branch of the leg trachea. Furthermore, there are functional arguments, because these exites are flexible structures and therefore better pre-adapted to be transformed into mobile appendages such as wings. The first protowings could not yet have served as flight organs but must have had a different function that later changed in the course of evolution. These mobile appendages may have served primarily as gill plates in aquatic larvae just as in extant mayflies. Wing venation systems later evolved as structures supporting the transport of oxygen. Such gill plates are present as paired dorsolateral appendages on the abdomen of fossil and extant mayfly larvae and bear a striking similarity to developing wing buds on the thorax of these insects. Some fossil insect larvae from the Carboniferous and Permian in North America have abdominal gills that are indistinguishable from thoracic wing buds. Wing buds are known to have been mobile in those Paleozoic insect larvae, while they are fused with the thorax in all extant larvae and only become mobile after the final molt to adult.

The presence of a third pair of smaller but mobile winglets on the first thoracic segment has been discovered in early fossil winged insects (paleodictyopterans, dragonflies, and protorthopterans) from the Carboniferous. (All extant winged insects possess only two pairs of wings on the two posterior thoracic segments.) This third pair of winglets is characteristic of all winged insects and has been reduced in modern insects. Their presence could also support the hypothesis that wings were derived from paired mobile appendages that were originally present on more segments than today, and that the thoracic wings represent the equivalents of the abdominal gills of mayfly larvae.

One strong argument against the exite theory exists: if wings and abdominal gills of mayfly larvae are corresponding structures of the same origin, as is strongly suggested by the fossil evidence, then the thoracic exites and abdominal styles that would have been their predecessors must be of the same origin and cannot be derivatives of walking legs because they occur together with legs on the thorax. However, there is morphological and paleontological evidence that the abdominal and thoracic styles of bristletails are different: thoracic exites of bristletails lack muscles, contrary to their abdominal styles; fossil wingless insects still have short segmented legs with paired claws on the abdomen, which strongly indicates that the abdominal styles are reduced legs and therefore of completely different origin from thoracic exites. Since only bristletails possess thoracic exites, these structures do not seem to belong to the common structure of insects. Conversely, they may represent a derived feature of bristletails alone, because they occur nowhere else among insects and myriapods. The alleged presence of thoracic exites in other fossil insect groups is contentious, because it cannot be confirmed by independent studies. Consequently, it is unlikely that the thoracic exites of bristletails represent vestiges of the biramous (forked) leg of crustaceans and trilobites, as was previously believed by many scientists. Altogether, the exite theory is poorly supported and in conflict with much of the other evidence.

The paranotal theory

The paranotal theory is endorsed in most popular books about insects and textbooks of entomology. This theory states that wings originated from lateral stiff and flat expansions (paranota) of the sclerite plate (notum) on the upper side of the thoracic segments. This view is strongly supported by the ontogenetic development of wing buds in modern insect larvae, which are immobile and fused with the thorax up to the final molt. Another argument is the presence of paranotal lobes in silverfish, which are the closest relatives of winged insects among the primarily wingless insect groups. In silverfish these paranotal lobes are supplied with oxygen by a branch of the leg trachea just as for wings of winged insects. A further argument could be that the wing articulation of primitive winged insects (e.g., mayflies and dragonflies) is less sophisticated and does not allow these animals to flex and/or fold their wings flat over the abdomen. In contrast, all remaining winged insects (Neoptera) possess this ability. Most proponents of the paranotal theory believe that the lateral expansions originated as airfoils that improved the ability for long jumps followed by gliding, and that the mobility of these airfoils was a later achievement in evolution. However, the exite hypothesis—that the protowings did not evolve as organs of flight but as larval gill plates—would also be compatible with a paranotal origin of these structures. Therefore, the paranotal theory would not conflict with the interpretation of wings and abdominal gills of mayfly larvae as corresponding structures of the same origin.

No one knows why only insects, alone of all invertebrates, developed the powers of flight. It may be that other invertebrate groups did not have the chance to evolve structures such as wings. Acquisition of flight offered exploitation of an unfilled niche. The ability to fly allowed for the colonization of a new habitat (i.e., air) and movement to new habitats when local environmental conditions became less favorable; acquisition of food; ability to escape predation; and more readily enhanced gene flow between previously remote populations. There could have been a coevolution between spiders and insects, in which the predatorial threat of spiders could have exerted pressure reinforcing the development and refinement of active flight in insects, while the latter forced spiders to evolve more and more sophisticated strategies to catch them (e.g., web building).

The age and end of the giants

About 300 mya, during the Carboniferous period, many parts of the world consisted of vast swamp forests with giant horsetails and primitive lycopod trees (e.g., *Sigillaria* and *Lepidodendron* that reached heights of up to 131 ft [40 m]). Since all of these plants had long stems with no leaves or only small crowns on top, these Carboniferous swamp forests allowed for understory insolation. Fossil remains of these forests show that the swamps were inhabited by primitive amphibians and various arthropods, such as arachnids, myriapods, and many insects such as the extinct paleodictyopterans as well as ancestors of mayflies, dragonflies, cockroaches, and orthopterans. Many of the winged insects attained giant size. Even though the average wingspan of Carboniferous species of paleodictyopterans, mayflies, and dragonflies was only 3.9–7.9 in (10–20 cm), the biggest paleodictyopterans and mayflies (e.g., *Bojophlebia prokopi*) reached maximum wingspans of 15.7–19.7 in (40–50 cm). The biggest Carboniferous dragonflies of the extinct family Meganeuridae reached a maximum wingspan of 25.6 in (65 cm). By the onset of the Permian period, a few giant species of the North American dragonfly genus *Meganeuropsis* had a wingspan of more than 29.5 in (75 cm) and thus represented the biggest insect ever known.

The largest extant insects include the longhorn beetle, *Titanus gigantea*, from the Amazon rainforest with a body length of up to 6.5 in (16.5 cm); the African goliath beetle, *Goliatus goliatus*, which is the heaviest extant insect with a weight of up to 2.5 oz (70 g) and a wingspan of up to 9.8 in (25 cm); the South American owlet moth, *Thysania agrippina*, with a wingspan of more than 11.8 in (30 cm); or the stick insect *Phobaeticus kirbyi* from Southeast Asia, which is the longest extant insect with a maximum length of 13.0 in (33 cm). The biggest dragonflies living today have a wingspan of only 6.7–7.9 in (17–20 cm) and thus are significantly smaller than their giant fossil relatives of the Carboniferous and Permian.

The loss of gigantism in insects has been attributed to changes in the composition of the atmosphere (e.g., increased oxygen levels) or climate, but none of these hypotheses are really convincing. Another more plausible hypothesis is that lack of aerial vertebrate predators allowed these insects to evolve to maximum sizes during the Carboniferous and Permian periods. These insects could therefore reach the maximum size that was physically allowed by their general body plan. Respiration with tracheae, by diffusion and weakly effective active ventilation, and constructional constraints of the exoskeleton and the muscle apparatus were the major factors that posed an upper limit of growth, so that insects could not evolve to have a wingspan of more than 3.3 ft (1 m). There may have been a competitive evolutionary race for the increase in body size between plant-feeding paleodictyopterans with sucking mouthparts and their predators, dragonflies. No comparatively large ground-dwelling insects are known from fossils, perhaps because predators such as large amphibians, early reptiles, and large arachnids prohibited such a dramatic size increase.

Early pterosaurs such as *Eudimorphodon* from the Upper Triassic of Italy are the oldest known flying vertebrates that have a typical insect-feeding dentition. Because these early pterosaurs had the same perfectly developed wing apparatus as successive pterosaurs, the group probably evolved significantly earlier in Earth's history, possibly in the early Triassic. It is tempting to assume that the extinction and permanent disappearance of giant flying insects right after the Permian is directly correlated with the predatorial threat by the first pterosaurs in the early Triassic. The high air drag of the large wings and the limited power of the flight muscles compared to the size of the wings did not allow these insects a fast and swift flight, as some modern insects are capable of. These clumsy giants could not escape the new aerial predators that were faster, swifter, stronger, and more intelligent and were thus doomed to extinction. Even before the extinction of pterosaurs, birds started their successful history to become the pterosaurs' successors as rulers of the air, and in the Tertiary the evolution of bats made even the night a dangerous time for flying insects, so that after the Triassic there was no chance for insects to evolve giant flying forms ever again.

The coevolution of insects and flowers

The relationship between flowering plants and pollinating insects was first described only 200 years ago by the German teacher and theologian Christian Konrad Sprengel. Sprengel presented his discoveries in his 1793 book *Das Entdeckte Geheimnis der Natur im Bau und in der Befruchtung der Blumen* (The unraveled secret of nature about the construction and pollination of flowers). A long history of evolution was necessary to create and advance such wonderful symbioses between the myriad types of flowers and their pollinators. The most primitive plants such as mosses, clubmosses, horsetails, and ferns still possess flagellate male germ cells that need rainwater for them to reach the female gametes for pollination. The famous maidenhair tree *Gingko biloba*, which is considered a living fossil, has retained this type of water-bound pollination. Within the gymnosperms, which include conifers, pollination by wind evolved. In the Gnetales, the closest relatives of flowering plants, pollination is achieved by the wind as well but is also accomplished with the help of various insects such as beetles and flies. Angiosperms, the genuine flowering plants, are predominantly pollinated by insects. However some tropical flowering plants are specialized for pollination by birds (e.g., humming birds), bats, or other mammals (e.g., monkeys, marsupials), but this must be a relatively recent and secondary phenomenon because these vertebrate pollinators appeared much later in evolution than flowering plants. Only angiosperms have developed sophisticated adaptations of their inflorescences, such as particular attractive color patterns and scents, nectar glands, and highly complex types of blossoms that are often only accessible for a single species of insect that is specialized and dependent on them.

The first pollinators may have been beetles that fed on pollen and secondarily acted as pollinators when they visited succeeding conspecific flowers while having some pollen attached to their body. Pollination by beetles is still common among primitive flowering plants such as water lilies (and cycads, one of the few nonflowering plants that are still pollinated by insects). Pollination in these plants is probably costly to the plant because the pollen contains numerous nutrients and substances that are energetically expensive to produce. This may be one reason why plants later evolved better strategies to attract and satisfy their pollinators, for example by offering bees and butterflies relatively cheap sources of food such as watery sugar solutions produced by special nectar glands.

The oldest fossil flowering plants are known from deposits from the Lower Cretaceous (130 mya). Alleged fossil angiosperms from the Lower Jurassic of China are also of Lower Cretaceous age. Most modern insect orders and many suborders are also known from the Lower Cretaceous fossil record. For example, the Crato limestones from the Lower Cretaceous of northeastern Brazil have not only yielded various early flowering plants but also early putative pollinators such as bees, certain flies, and moths, but no diurnal butterflies. Butterflies appeared much later in Earth's history in the Moler-Fur formation from Denmark and in Baltic amber, both of Lower Tertiary age (40–50 mya).

The enormous diversity of flowering plants and insects is a result of coevolution between these two groups. The specialization among various groups of pollinators on certain flowers has allowed multiple species in the same habitat. Most modern insect subgroups (e.g., bees, moths, flies, beetles) were present after the coevolution of plants and their pollinators. The diverse insect fauna of various Tertiary amber localities (e.g., Baltic and Dominican amber) is therefore not greatly different from the modern one, except for changes in the distribution of some groups due to climatic changes in the Tertiary.

Resources

Books

Boudreaux, H. B. *Arthropod Phylogeny with Special Reference to Insects.* New York: J. Wiley, 1979.

Carpenter, F. M. "Superclass Hexapoda." In *Treatise on Invertebrate Paleontology (R), Arthropoda* 4, 3–4, edited by R. C. Moore and R. L. Kaesler. Boulder, CO, and Lawrence, KS: Geological Society of America and University of Kansas Press, 1992.

Frickhinger, K. A. *Die Fossilien von Solnhofen.* 2 vols. Korb: Goldschneck, 1994–1999.

Greenslade, P., and P. E. S. Whalley. "The Systematic Position of *Rhyniella praecursor* Hirst & Maulik (Collembola), the Earliest Known Hexapod." In *2nd International Symposium Apterygota,* edited by R. Dallai. Siena, Italy: University of Siena, 1986.

Grimaldi, D. A. *Amber: Window to the Past.* New York: American Museum of Natural History, 1996.

Gupta, A. P. *Arthropod Phylogeny.* New York: Van Nostrand Reinhold, 1979.

Handlirsch, A. *Die Fossilen Insekten und die Phylogenie der Rezenten Formen.* Leipzig: Engelmann, 1906–1908.

Hennig, W. *Insect Phylogeny.* New York: Wiley and Sons, 1981.

———. *Phylogenetic Systematics*. Urbana: University of Illinois Press, 1966.

Kristensen, N. P. "The Ground Plan and Basal Diversification of the Hexapods." In *Arthropod Relationships*, edited by R. A. Fortey and R. H. Thomas. London: Chapman and Hall, 1997.

———. "Insect Phylogeny Based on Morphological Evidence." In *The Hierarchy of Life*, edited by B. Fernholm, K. Bremer, and H. Jörnvall. Amsterdam: Elsevier, 1989.

———. "Phylogeny of Extant Hexapods." In *The Insects of Australia*, 2nd ed., edited by I. D. Naumann. Melbourne: Melbourne University Press, 1991.

Krzeminska, E., and W. Krzeminski. *Les fantomes de l'ambre: Insectes fossiles dans l'ambre de la Baltique*. Neuchâtel, Switzerland: Muséum d'Histoire Naturelle de Neuchâtel, 1992.

Kukalová-Peck, J. "Arthropod Phylogeny and 'Basal' Morphological Structures." In *Arthropod Relationships*, edited by R. A. Fortey and R. H. Thomas. London: Chapman and Hall, 1997.

———. "Fossil History and the Evolution of Hexapod Structures." In *The Insects of Australia*, 2nd ed., edited by I. D. Naumann, Melbourne: Melbourne University Press, 1991.

Poinar, G. O. *The Amber Forest*. Princeton: Princeton University Press, 1999.

———. *Life in Amber*. Stanford: Stanford University Press, 1992.

Rasnitsyn, A. P., and D. L. J. Quicke. *History of Insects*. Dordrecht, The Netherlands: Kluwer, 2002.

Rohdendorf, B. B., ed. *Fundamentals of Paleontology. Vol. 9: Arthropoda, Tracheata, Chelicerata*. New Dehli: Amerind Publ., 1991.

Ross, A. *Amber: The Natural Time Capsule*. London: The Natural History Museum, 1998.

Ross, A. J., and E. A. Jarzembowski. "Arthropoda (Hexapoda; Insecta)." In *The Fossil Record*, vol. 2, edited by M. J. Benton. London: Chapman and Hall, 1993.

Schmitt, M. *Wie sich das Leben entwickelte: Die faszinierende Geschichte der Evolution*. Munich: Mosaik, 1994.

Weitschat, W., and W. Wichard. *Atlas of Plants and Animals in Baltic Amber*. Munich: Pfeil, 2002.

Willmann, R. "Advances and Problems in Insect Phylogeny." In *Arthropod Relationships*, edited by R. A. Fortey and R. H. Thomas. London: Chapman and Hall, 1997.

Periodicals

Ansorge, J. "Heterophlebia buckmani (Brodie 1845) (Odonata: 'Anisozygoptera'): Das erste Insekt aus dem untertoarcischen Posidonienschiefer von Holzmaden (Württemberg, SW Deutschland)." *Stuttgarter Beiträge zur Naturkunde* Serie B 275 (1999): 1–9.

Bechly, G. "Mainstream Cladistics versus Hennigian Phylogenetic Systematics." *Stuttgarter Beiträge zur Naturkunde* Serie A 613 (2000): 1–11.

———. "Santana: Die Schatzkammer fossiler Insekten aus der Unterkreide Brasiliens." *Fossilien* 2 (1998): 95–99, and 3 (1998): 148–156.

Bechly, G., F. Haas, W. Schawaller, H. Schmalfuss, and U. Schmid. "Ur-Geziefer: Die faszinierende Evolution der Insekten." *Stuttgarter Beiträge zur Naturkunde* Serie C 49 (2001): 1–94.

Bergström J., W. Dohle, K.-E. Lauterbach, and P. Weygoldt. "Arthropoden-Phylogenie." *Abhandlungen des Naturwissenschaftlichen Vereins in Hamburg* NF 23 (1980): 1–327.

Brauckmann, C. "Arachniden und Insekten aus dem Namurium von Hagen-Vorhalle (Ober-Karbon; West-Deutschland)." *Veröffentlichungen aus dem Fuhlrott-Museum* 1 (1991): 1–275.

Brauckmann, C., B. Brauckmann, and E. Gröning. "The Stratigraphical Position of the Oldest Known Pterygota (Insecta. Carboniferous, Namurian)." *Annales de la Société Géolique Belgique* 117, no. 1 (1996): 47–56.

Briggs, D. E. G., and C. Bartels. "New Arthropods from the Lower Devonian Hunsrück Slate (Lower Emsian, Rhenish Massif, western Germany)." *Palaeontology* 44 (2001): 275–303.

Carpenter, F. M. "Fossil Insects." *Gamma Alpha Record* 40, no. 3 (1950): 60–68.

Greenslade, P. "Reply to R. A. Crowson's 'Comments on Insecta of the Rhynie chert.'" *Entomologia Generalis* 13 (1988): 115–117.

Grimaldi, D. A., ed. "Insects from the Santana Formation, Lower Cretaceous, of Brazil." *Bulletin of the American Museum of Natural History* 195 (1990): 1–191.

Grimaldi, D. A. "Insect Evolutionary History from Handlirsch to Hennig, and Beyond." *Journal of Paleontology* 75 (2001): 1152–1160.

Haas, F., D. Waloszek, and R. Hartenberger. "*Devonohexapodus bocksbergensis*, a New Marine Hexapod from the Lower Devonian Hunsrück Slates, and the Origin of Atelocerata and Hexapoda." *Organisms Diversity and Evolution* 3 (2003): 39–54.

Handlirsch, A. "Neue Untersuchungen über die fossilen Insekten mit Ergänzungen und Nachträgen sowie Ausblicken auf Phylogenetische, Palaeogeographische und allgemeine biologische Probleme." *Annalen des Naturhistorischen Museums in Wien* 49 (1939): 1–240.

Hilken, G. "Vergleich von Tracheensystemen unter phylogenetischem Aspekt." *Verhandlungen des Naturwissenschaftlichen Vereins in Hamburg* 37 (1998): 5–94.

Klass, K. D., and N. P. Kristensen. "The Ground Plan and Affinities of Hexapods: Recent Progress and Open Problems." *Annales de la Société Entomologique de France* 37 (2001): 265–298.

Kristensen, N. P. "Forty Years' Insect Phylogenetic Systematics. Hennig's 'Kritische Bemerkungen...', and Subsequent Developments." *Zoologische Beiträge* N.F. 36, no. 1 (1995): 83–124.

———. "Phylogeny of Insect Orders." *Annual Review of Entomology* 26 (1981): 133–157.

Kukalová-Peck, J. "Ephemeroid Wing Venation Based upon New Gigantic Carboniferous Mayflies and Basic Morphology, Phylogeny, and Metamorphosis of Pterygote Insects (Insecta, Ephemerida)." *Canadian Journal of Zoology* 63 (1985): 933–955.

———. "New Carboniferous Diplura, Monura, and Thysanura, the Hexapod Ground Plan, and the Role of Thoracic Side Lobes in the Origin of Wings (Insecta)." *Canadian Journal of Zoology* 65 (1987): 2327–2345.

———. "Origin of the Insect Wing and Wing Articulation from the Arthropodan Leg." *Canadian Journal of Zoology* 61 (1983): 1618–1669.

Labandeira, C. C. "A Compendium of Fossil Insect Families." *Milwaukee Public Museum Contributions in Biology and Geology* 88 (1994): 1–71.

Labandeira, C. C., and J. J. Sepkoski, Jr. "Insect Diversity in the Fossil Record." *Science* 261 (1993): 310–315.

Labandeira, C. C., and D. M. Smith. "Forging a Future for Fossil Insects: Thoughts on the First International Congress of Paleoentomology." *Paleobiology* 25, no. 1 (1999): 154–157.

Lutz, H. "Taphozönosen terrestrischer Insekten in aquatischen Sedimenten: Ein Beitrag zur Rekonstruktion des Paläoenvironments." *Neues Jahrbuch für Geologie und Paläontologie* Abhandlungen 203 (1997): 173–210.

Malz, H., and H. Schröder. "Fossile Libellen: Biologisch betrachtet." *Kleine Senckenberg-Reihe* 9 (1979): 1–46.

Martill, D. M., and E. Frey. "Color Patterning Preserved in Lower Cretaceous Birds and Insects: The Crato-Formation of N.E. Brazil." *Neues Jahrbuch für Geologie und Paläontologie* Monatshefte (1995): 118–128.

Richter, S. "The Tetraconata Concept: Hexapod-Crustacean Relationships and the Phylogeny of Crustacea." *Organisms Diversity and Evolution* 2 (2002): 217–237.

Riek, E. F., and J. Kukalová-Peck. "A New Interpretation of Dragonfly Wing Venation Based upon Early Carboniferous Fossils from Argentina (Insecta: Odonatoidea) and Basic Character States in Pterygote Wings." *Canadian Journal of Zoology* 62 (1984): 1150–1166.

Schlüter, T. "Fossil Insect Localities in Gondwana." *Entomologia Generalis* 15, no. 1 (1990): 61–76.

Tischlinger, H. "Bemerkungen zur Insekten-Taphonomie der Solnhofener Plattenkalke." *Archaeopteryx* 19 (2001): 29–44.

Whalley, P., and E. A. Jarzembowski. "A New Assessment of *Rhyniella*, the Earliest Known Insect, from the Devonian of Rhynie, Scotland." *Nature* 291 (1981): 317.

Wheeler, W. C., M. Whiting, Q. D. Wheeler, and J. M.Carpenter. "The Phylogeny of the Extant Hexapod Orders." *Cladistics* 17 (2001): 113–169.

Wichard, W., and W. Weitschat. "Wasserinsekten im Bernstein: Eine paläobiologische Studie." *Entomologische Mitteilungen aus dem Löbbecke Museum und Aquazoo* Beihefte 4 (1996): 1–122.

Wooton, R. J. "Paleozoic Insects." *Annual Review of Entomology* 26 (1981): 319–344.

Organizations

The International Palaeoentomological Society. Web site: <http://www.cwru.edu/affil/fossilinsects/>

The Willi Hennig Society. Web site: <http://www.cladistics.org>

Other

Bechly, Günter. "Glossary of Phylogenetic Systematics with a Criti of Mainstream Cladism." 1998 [May 30, 2003]. <http://www.bechly.de/glossary.htm>.

———. "Synoptic Timetable of Earth History." 1998 [May 30, 2003]. <http://www.bechly.de/paleodat.htm>.

Beckemeyer, Roy J. "Fossil Insects." April 28, 2003 [May 30, 2003]. <http://www.windsofkansas.com/fossil_insects.html>.

"Entomology Index of Internet Resources." Dept. of Entomology, Iowa State University [May 30, 2003]. <http://www.ent.iastate.edu/List/>.

Evenhuis, Neal L. "Catalog of Fossil Flies of the World." Bishop Museum. October 17, 2002 [May 30, 2003]. <http://hbs.bishopmuseum.org/fossilcat/>.

"A Guide to Online Insect Systematic Resources." North Carolina State University AgNIC. December 14, 1998 [May 30, 2003]. <http://www.lib.ncsu.edu/agnic/sys_entomology/index_flash.html>.

"Internet Resource Guide for Zoology: Insecta." BIOSIS [May 30, 2003]. <http://www.biosis.org/zrdocs/zoolinfo/grp_ins .htm>.

"Lagerstatten." Peabody Museum of Natural History, Yale University [May 30, 2003]. <http://www.yale.edu/ypmip/lagrlst.html>.

Maddison, David R., ed. "The Tree of Life Web Project." 2002 [May 30, 2003]. <http://tolweb.org/tree?group=Hexapoda&contgroup=Arthropoda>.

Martínez-Delclòs, Xavier, and Günter Bechly, eds. "Meganeura Palaeoentomological Newsletter." January 3, 2000 [May 30, 2003]. <http://www.ub.es/dpep/meganeura/meganeura.htm>.

Günter Bechly, PhD

· · · · ·

Structure and function

Introduction

Insects are segmented animals with an external skeleton (cuticle) in which the segments are grouped in three sections: a head, formed from the protocephalon and seven post-oral segments; a thorax of three segments; and an abdomen of eleven segments plus the telson. The external signs of segmentation are largely lost in the head, except for the segments bearing the mouthparts. In the abdomen the number of visible segments often is reduced, because segments have fused together. The head is the sensory/neural and feeding center of the insect. The thorax is the locomotor center, with three pairs of legs and, in adults, two pairs of wings. The abdomen holds the structures concerned with food processing and reproduction and, externally, the genitalia.

Cuticle

The cuticle is secreted by the epidermis and covers the whole of the outside of the body as well as lining the foregut and hindgut and the tracheal system, which are formed as invaginations of the epidermis. Most of the cuticle is composed of a mixture of proteins and the polysaccharide chitin. Outside this chitinous cuticle is a chemically complex epicuticle that does not contain chitin. It is only a few microns thick.

Chitinous cuticle

Chitin occurs as long molecules that are bound together to form microfibrils. These microfibrils lie parallel to the plane of the surface and, at any depth below the surface, to each other. In successive layers the orientation changes, usually giving rise to a helicoid (spiral) arrangement through the thickness of the cuticle. This gives strength to the cuticle in all directions. Sometimes layers of helicoidally arranged microfibrils alternate with layers in which the microfibrils have a consistent orientation. These layers differ in their refractive indexes, and the metallic colors of insects typically are the result of differences in the optical properties of the successive layers, so that only specific wavelengths of light are reflected.

The helicoid arrangement of microfibrils provides strength to the cuticle, but it does not impart hardness or rigidity. Hardness in insect cuticle derives from the linking together of proteins. The process of linking the proteins is called sclerotization, and the hardened cuticle that results is said to be sclerotized or

tanned. Hardening is restricted to the outer parts of chitinous cuticle, so that the cuticle becomes differentiated into the outer sclerotized exocuticle and an inner endocuticle that remains unsclerotized. Sclerotization does not take place until the cuticle is expanded fully after a molt and depends on the transport of chemicals from the epidermis. This is achieved via a series of slender processes of the epidermal cells that extend through the chitinous cuticle, creating canals in the cuticle that run at right angles to the surface. These are called pore canals.

Sclerotization affords some rigidity in addition to hardness, but in many areas of the cuticle this rigidity is enhanced by shallow folds in the cuticle. Their effect is comparable to that of a T-girder. The folds are seen as grooves, called sulci (singular: "sulcus"), on the outside of the cuticle. Sulci are most common on the head and thorax, where they define areas of cuticle that are given specific names. Additional rigidity is achieved where fingerlike inpushings of the cuticle, called apodemes, meet internally, forming an endophragmal skeleton. This occurs in the head of all insects, where two pairs of apodemes, originating anteriorly and posteriorly on the head, join beneath the brain to form the tentorium, which provides the head with great rigidity in the horizontal plane. In winged insects lateral and ventral apodemes in the thorax may join or be held together by muscles forming a strut that holds the sides (pleura) of the thorax rigid with respect to the ventral surface (sternum). This is essential for the movement of the wings in flight. The tubular form of the legs and other appendages makes them rigid.

Flexibility in the cuticle, which allows different parts of the body to move with respect to each other, depends on regions of movable cuticle between the hardened plates (sclerites). Sclerotization does not occur in this flexible cuticle, which is referred to as "membranous." It is most extensive in the region of the neck, between the abdominal segments, and between segments of the appendages. Membranous cuticle also is found where the wings join the thorax and at the bases of the antennae, mouthparts, and other appendages, giving them freedom to move. Precision of movement is achieved by points of articulation at which there is only a very small region of membrane between adjacent sclerites.

A rubberlike protein, called resilin, also is known to be present in some insects and may occur more widely. When it is

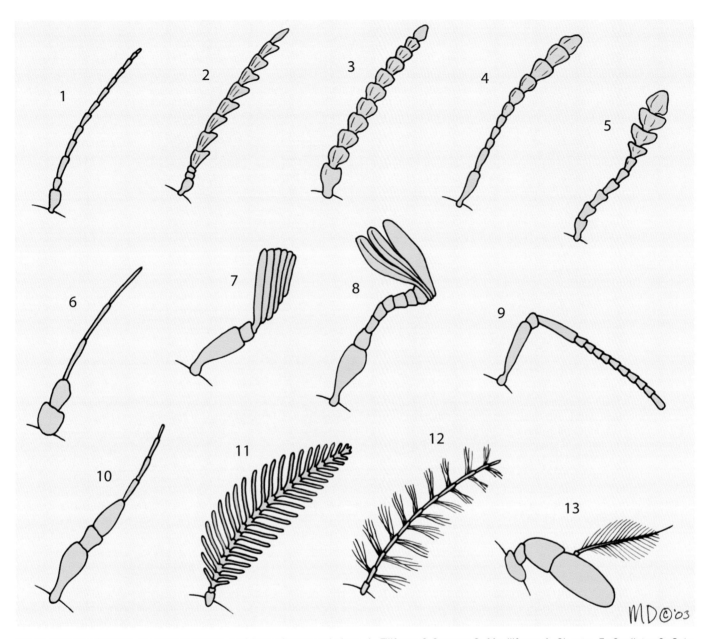

Insect antennae function as sensory organs, and have shapes and sizes. 1. Filiform; 2 Serrate; 3. Moniliform; 4. Clavate; 5. Capilate; 6. Setaceous; 7. Flabellate; 8. Lamellate; 9. Geniculate; 10. Stylate; 11. Pectinate; 12. Plumose; 13. Aristate. (Illustration by Marguette Dongvillo)

distorted, it retains the energy imparted to it and, like a rubber ball, returns to its original shape when the tension is released. There is a pad of resilin in the hind wing hinge of the locust and also in the side of the thorax of the flea, where the release of stored energy gives rise to the jump. Small amounts also are present in the hinge of the labrum in the locust and in the abdomen of some beetles.

The strength, rigidity and articulations of the cuticle provide the insect with support, protection, and precision of movement. In larval forms, such as caterpillars and fly larvae, most of the cuticle remains unsclerotized. In these cases, the hemolymph (insects' blood) functions as a hydrostatic (held by water pressure) skeleton, and movements are much less precise.

Epicuticle

Three or, in some species, four chemically distinct layers are present in the epicuticle. The innermost layer (inner epicuticle) contains lipoproteins but is chemically complex. Its functions are unknown. The next layer, the outer epicuticle, is made of polymerized lipid, though it probably also contains some protein. It is believed to be inextensible, such that it can unfold but not stretch. It defines the details of patterns on the surface of the cuticle. Outside the outer epicuticle is a layer of wax. This comprises a mixture of chemical compounds whose composition varies considerably between insect taxa. The wax limits water loss through the cuticle and so is a major feature contributing to the success of insects as terrestrial organisms, for whom water is at a premium. Because this layer becomes abraded (worn away) during normal activities, it has

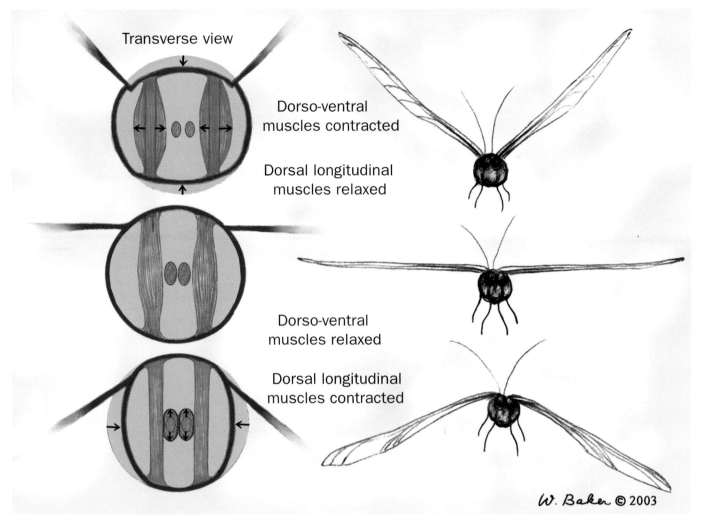

Muscles involved in insect flight. (Illustration by Wendy Baker)

to be renewed continually. New compounds are synthesized in the epidermis and are thought to be transported to the surface via wax canal filaments that run through the pore canals and the inner and outer epicuticles. A fourth layer sometimes occurs outside the wax, but its functions are unknown.

Epidermis

The epidermis is a single layer of cells. In addition to producing the cuticle, it contains many glands that secrete chemicals to the outside of the insect. These chemicals include many pheromones, involved in communication with other members of the same species, and defensive compounds that often are repellent to potential enemies. In the latter case, the glands frequently include a reservoir in which the noxious substances are accumulated until they are needed.

Feeding and digestion

Mouthparts

The appendages of four segments of the head form the insect's mouthparts, the structures involved in manipulating food and passing it back to the alimentary canal. Although the mouthparts functionally resemble the jaws of vertebrates, they differ fundamentally in being outside the mouth. They retain their greatest resemblance to the leglike structures from which they are derived in the more basal groups of insects, the Microcoryphia, Thysanura, Blattodea, Mantodea, and Orthoptera, although they also occur throughout the Coleoptera, in many Hymenoptera, and in larval Lepidoptera. These insects are said to possess "biting and chewing" mouthparts.

Suspended immediately in front of the mouth is the labrum. It is unpaired, and, unlike the remaining mouthparts, its origin from appendages is not obvious. It forms a lip that prevents food from falling out from the mandibles as it is moved toward the mouth. Upwardly pointing cuticular spines on its inner face help keep the food moving in the right direction. On the inside of the labrum, just outside the mouth, are taste receptors that presumably make the final decision concerning the acceptability of food before it is ingested.

The mandibles are the most anterior of the post-oral appendages. They consist of a single, unsegmented unit, which, in all but the Microcoryphia, has two points of articulation

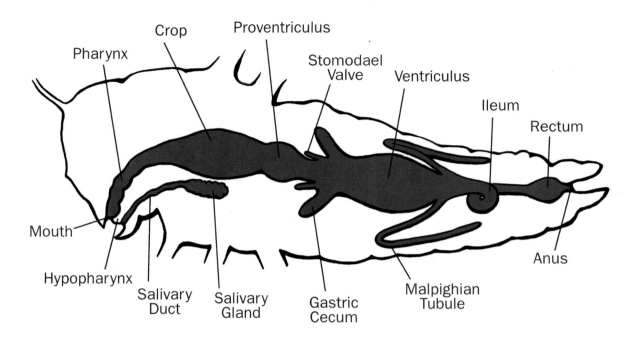

Basic structure of the alimentary canal. (Illustration by Jarrod Erdody)

with the head capsule. This restricts their movement to the transverse plane and, because of the rigidity of the head capsule, gives them the ability to cut through hard objects. Their power is provided by large adductor muscles that occupy much of the space within the head. The cutting surface of the mandibles bears a series of cusps, whose form and arrangement vary according to the nature of the food. The cusps are hardened with heavy metals, commonly zinc and manganese, in addition to being heavily sclerotized.

Behind the mandibles are the maxillae and labium. They often retain a jointed appendage in the form of a palp that has large numbers of contact chemoreceptors at its tip. Each maxilla has a single articulation with the head capsule, giving it great mobility. The primary function of the maxillae is to manipulate food toward the mouth, although their sensory structures also are involved in food selection. The third pair of appendages forms the labium. The labium resembles the maxillae, but the structures on either side are fused together so that it forms a lower lip behind the mouth. The duct from the salivary glands opens immediately in front of the base of the labium. Consequently, saliva reaches the food before the food enters the mouth, and in some species pre-oral digestion by the salivary enzymes is more important than digestion within the alimentary canal.

Many insects are fluid feeders, and in these insects the mouthparts form tubular structures through which liquid is drawn into the alimentary canal. The Lepidoptera, bees, and many flies feed on nectar that is freely available in floral nectaries. Other fluid-feeding insects, such as the Hemiptera, fleas, and blood-sucking flies, feed on fluids that are contained within their food plants or animals. Consequently, in these groups some components of the mouthparts are modified for

piercing the host tissues, whereas others form the tubes through which food is taken in and saliva is injected into the wound. The tubular and piercing components of the mouthparts of different taxa are derived from different components of the basic biting and chewing mouthparts.

Alimentary canal

Developmentally, the alimentary canal is formed as three units: foregut, midgut, and hindgut. The foregut and hindgut develop as invaginations (in-foldings) of the epidermis and so are lined with cuticle; the midgut has a separate origin and has no cuticular lining. The most anterior part of the alimentary canal (pharynx) has extrinsic muscles that draw food into the mouth and pass it backward. These muscles form a powerful sucking pump in fluid-feeding insects. From the pharynx, the food passes along the esophagus, which often is expanded posteriorly to form a temporary storage chamber, the crop. The cuticle lining the crop is impermeable, so food can be stored without affecting hemolymph composition.

The midgut is involved with enzyme synthesis and secretion and with digestion and absorption of nutrients. The principal cells of which it is formed are large and metabolically active, requiring replacement at relatively frequent intervals. New principal cells are produced from groups of undifferentiated cells at the base of the epithelium. There are also endocrine cells in the midgut wall. They probably regulate enzyme synthesis. The surface area of the midgut often is increased by a number of diverticula (sacs), called "midgut caeca." Where this occurs, the central tubular part of the midgut is called the ventriculus. Posteriorly the ventriculus connects with the hindgut, and at this point the Malpighian tubules of the excretory system also connect with the hindgut.

The hindgut is differentiated into a tubular ileum and a barrel-shaped rectum. A major function of the latter is the removal of water from the urine and feces so that water loss from the body is kept to a minimum. The rectum is lined by a very delicate, freely permeable cuticle.

Excretion

Malpighian tubules are the main excretory organs of most insects. They are long, slender, blindly ending tubes that arise from the hindgut close to its junction with the midgut. The number of tubules varies in different species, ranging from two in scale insects to more than 200 in some grasshoppers. They extend through the hemocoel (body cavity) and are in continual writhing motion.

Ammonia is the primary end product of nitrogen metabolism. It is highly toxic and must be removed from the body, but its safe elimination requires large amounts of water. Because terrestrial insects must conserve water, they eliminate much of their waste nitrogen as uric acid, which has low toxicity. This compound is synthesized in the fat body and transported to the Malpighian tubules, where it is pumped into the primary urine, which also contains inorganic ions that are essential for urine production. Urine flows down the tubules and into the hindgut, joining undigested food as it passes from the midgut. In the rectum, salts and water are removed from the fluid, because it is important for the insect to conserve them, and the uric acid passes out in the feces. Fluid urine, without fecal material, is produced only when insects have too much water.

This bumblebee is equipped with a long tongue for collecting nectar. (Photo by Dwight Kuhn. Bruce Coleman, Inc. Reproduced by permission.)

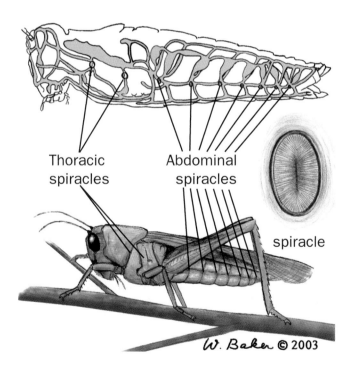

Insect respiratory system. Oxygen and carbon dioxide move through a system of tubes (trachea) that branch to all parts of the body. Air enters via the spiracles on the insects' bodies. (Illustration by Wendy Baker)

Gas exchange

Gas exchange in insects takes place via a system of tubes, the tracheae, that carry air directly to the tissues; there is no respiratory pigment in the blood, as there is in most other animals. The tubes arise as invaginations of the epidermis, one on either side of each body segment. The invaginations from adjacent segments join to form longitudinal trunks running the length of the body; from these trunks, and from transverse connections, finer branches extend to all the tissues. At their innermost ends, the tracheae continue as fine intracellular tubes—tracheoles—less than a micron in diameter; it is here that gas exchange with the tissues occurs. In flight muscles, which have huge demands for oxygen when the insect flies, the tracheoles indent the muscle membrane so that they become functionally intracellular, ending adjacent to the muscle mitochondria, where oxidation occurs. In this way, the tissue diffusion path is reduced to only a few microns. This is important, because the rate of diffusion of oxygen is 100,000 times greater in air than in the tissues.

The segmental openings of the tracheal system are called spiracles. Dragonflies, cockroaches, grasshoppers, and the lar-

A leaf-footed bug (*Diactor bilineatus,* family Coreidae) from Brazil showing the three pairs of legs, one pair of antennae, and three body parts typical of insects. (Photo by Rosser W. Garrison. Reproduced by permission.)

vae of some Diptera and Hymenoptera have 10 pairs of spiracles, two thoracic and eight abdominal. Most other terrestrial insects have eight or nine pairs. In the immature stages of aquatic insects the number of spiracles is greatly reduced, and they may be absent altogether in insects that obtain oxygen directly from the water, such as dragonfly and mayfly nymphs. These insects are said to be "apneustic," but even in them the tracheal system is retained. This allows for much more rapid diffusion of oxygen around the body than if oxygen were dissolved in the hemolymph.

The spiracles of most terrestrial insects have valves that close. Closure minimizes water loss from the tracheal system, and insects keep the spiracles closed as long as is consistent with efficient respiration. With the spiracles closed, the removal of oxygen from the tracheae causes a reduction in pressure. This is not offset by the production of carbon dioxide, because this gas is much more soluble and much goes into solution in the hemolymph. The tracheae do not collapse as the pressure decreases. Because they are formed from epidermis, they are lined with cuticle, and this is made into thickened spiral ridges, called taenidia, running along all the tracheae. This spiral thickening resists collapse, just as the spiral construction of the wall of a vacuum hose does. Consequently, when the spiracles are opened, air flows into the tracheae.

Diffusion alone is sufficient to account for the oxygen requirements of the tissues of small insects at rest. Larger insects and active insects, however, require some form of forced ventilation of the tracheal system. This is made possible by sections of the tracheae that are expanded into balloon-like air sacs. Unlike the tracheae themselves, the air sacs are sub-

ject to expansion and collapse. During expansion, air is drawn into the tracheal system through the spiracles; when the air sacs collapse, the air is forced out again. The changes in volume of the air sacs result from changes in the pressure of the hemolymph in which they lie. In active insects the pressure changes result from changes in body volume, often involving changes in the length of the abdomen. Ventilation in some resting insects also may take place without changes in body volume, by movement of hemolymph between the thorax and the abdomen so that the air sacs in the thorax expand while those in the abdomen collapse and vice versa.

Wings and flight

Most adult insects have two pairs of wings, one pair on each of the second and third thoracic segments, or the mesothoracic and metathoracic segments. A wing consists of a double layer of cuticle that is continuous with the cuticle of the thorax. In most insects the cuticle of the wing is unsclerotized, although in Orthoptera, Blattodea, and Mantodea the forewings are weakly sclerotized, and in Coleoptera they are heavily sclerotized. These harder forewings provide protection for the more extensive hindwings, which furnish most of the power for flight in these groups. The flexibility of the membranous wings allows them to be folded at rest and also permits changes in shape during flight, which are important aerodynamically. The production of power, however, requires the wings to be rigid to some extent, and rigidity is conferred by the wing veins. These veins are tubular, and their cuticle is sclerotized, so that they provide girders to support the wing membrane. Differences in cross-sectional shape and the degree of sclerotization, as well as small breaks in the veins, allow the wing to bend in certain directions during parts of the wing stroke. These details are critical in generating the forces that keep the insect airborne. There are

The two sets of wings on this brown-spotted yellow wing dragonfly (*Celithemis eponina*) are clearly visible. (Photo by Larry West. Bruce Coleman, Inc. Reproduced by permission.)

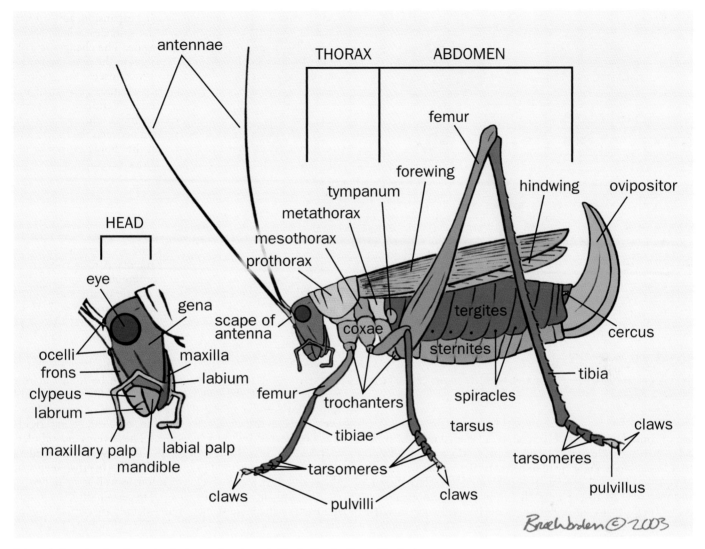

A lateral view showing the major features of an insect. (Illustration by Bruce Worden)

broad similarities in the arrangement of the wing veins in the different orders of insects, but there are also many differences that reflect the ways in which the wings are used. At the base of the wing the veins articulate with the cuticle of the thorax via several axillary sclerites. These give a degree of mobility somewhat analogous to the carpal bones in the human wrist, so that the wing can be folded and unfolded and its camber changed during flight.

The movements of wings that produce aerodynamic forces result, in most insects, from changes in the shape of the thorax and not, primarily, from the action of muscles attached directly to the wings. Downward movement of the wings (depression) is produced when the upper surface of the thoracic segment (notum) is raised relative to the sides (pleura). Upward movement (levation) occurs when the notum is lowered. These changes in shape are produced by the indirect flight muscles in the mesothoracic and metathoracic segments. Dorsal longitudinal muscles extend from the front of one segment to the front of the next. When they contract, they raise the notum and cause wing depression. Dorsoven-

tral longitudinal muscles, running from the notum to the sternum in the wing-bearing segment, pull the notum down and cause wing levation. Because the power needed for flight is so great, these muscles are very large and occupy the greater part of the thorax. Direct flight muscles, which are attached to the underside of the wing at its base, produce changes in the shape of the wing during the downstroke. In Odonata and Blattodea, however, these muscles are the main wing depressors, and the indirect dorsoventral muscles are only weakly developed.

The wings move up and down at a high frequency in flight, to provide sufficient power to support the insect in the air. In general, smaller insects have a higher wing-beat frequency than larger ones. Odonata, Orthoptera, and most Lepidoptera have relatively low wing-beat frequencies, usually less than 40 cycles per second. Many Diptera and Hymenoptera, and some Hemiptera, on the other hand, have wing-beat frequencies greater than 200 cycles per second. These very high frequencies require anatomical and physiological specializations of the indirect flight muscles. Because the muscles use so

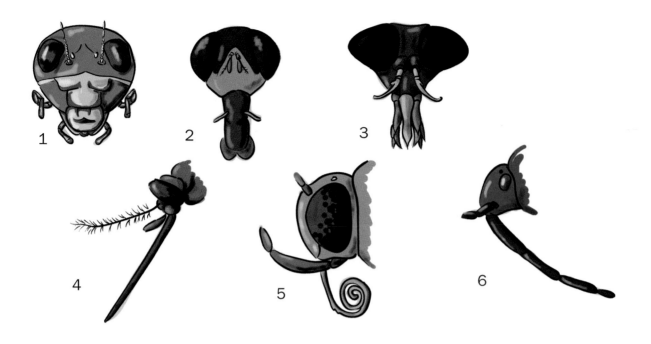

Insects have different mouth parts for feeding. 1. Cricket (chewing); 2. House fly (mopping); 3. Horse fly (piercing and sucking); 4. Mosquito (piercing and sucking); 5. Moth (sucking); 6. Froghopper (piercing and sucking). (Illustration by Ryan Burkhalter)

much energy, they also require a great deal of oxygen, and the tracheal system of the thorax is modified to ensure good airflow to the muscles while not depriving other parts of the body, and especially the nervous system, of oxygen.

Circulatory system

The insect's body cavity is called a hemocoel because of its mode of formation, and the liquid it contains is called hemolymph rather than blood. The hemolymph bathes the organs of the body directly and is not, in general, enclosed in vessels.

Circulation

Circulation of the hemolymph within the body cavity results from the action of a vessel lying dorsally in the hemocoel and extending the length of the body. The posterior part of this vessel pulsates and is known as the heart. As it expands, in diastole, hemolymph is drawn in through a series of valved openings, ostia. When it contracts, in systole, the valves close, and hemolymph is driven forward through the noncontractile anterior section of the vessel, the aorta, that opens into the hemocoel in the head. This creates a sluggish circulation, forcing hemolymph back from the head to the abdomen, where it moves up and into the heart to be pumped forward again. In adult Coleoptera, Diptera, and Lepidoptera, however, the hemocoel is partially occluded between the thorax and the abdomen, and hemolymph is shunted back and forth between the two as the heart undergoes reversals in its direction of pumping.

Accessory pulsatile organs maintain the circulation in the wings and antennae and sometimes also in the legs. One or two pulsatile organs in the thorax pull hemolymph from the trailing edges of the wings, causing new hemolymph to be drawn into the veins anteriorly. Other structures at the bases of the antennae pump hemolymph through a vessel that opens close to the tip of the antenna. Hemolymph returns to the body cavity in the antennal lumen outside the vessel.

Hemolymph

Hormones and nutrients are circulated in the hemolymph, but in most insects the hemolymph is not involved in gas exchange. The water in the hemolymph provides a reservoir for the tissues, and in desiccated insects its volume is greatly reduced. In insects with a soft cuticle, such as caterpillars, and in newly molted insects before the cuticle has hardened, the hemolymph serves as a hydrostatic skeleton.

Amino acids are stored in the hemolymph as proteins. These storage proteins are synthesized in the fat body as the insect accumulates nitrogenous nutrients. The proteins are broken down at molts and, especially, at metamorphosis, when the amino acids are re-synthesized into tissue proteins. The hemolymph also contains cells known as hemocytes. There are several different types of hemocytes, all of which are nucleated. A principal function of the hemocytes is defense against potential pathogens. Bacteria are engulfed by single cells; larger organisms are encapsulated by large numbers of hemocytes.

Fat body

The fat body is an ill-defined structure in the hemolymph between the alimentary canal and the body wall. It is made

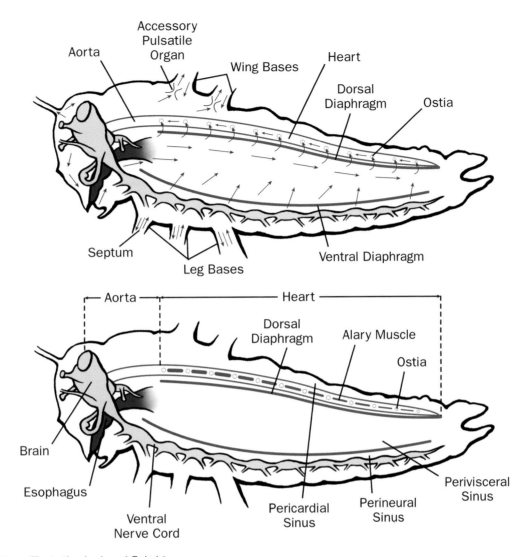

Insect circulatory system. (Illustration by Jarrod Erdody)

up of sheets of cells, and its development varies. In a newly molted insect, the fat body may be barely visible, but as the insect feeds and accumulates nutrients, the fat body comes to occupy a large part of the hemocoel. The cells become filled with lipids and carbohydrates in the form of glycogen, a poly-saccharide comparable to starch in plants. The fat body is also a major site for the synthesis of storage and yolk proteins. In addition, it is involved in detoxification of toxic compounds that may be derived from food or, like insecticides, may penetrate the cuticle.

Nervous system and sense organs

Central nervous system

The insect central nervous stem consists of a brain, lying dorsally within the head above the esophagus, and a chain of ventral ganglia joined together by a pair of connectives. Fundamentally, each body segment has a ganglion, but some degree of fusion always occurs, so that the number of visible ganglia invariably is less than the number of body segments. The ganglia of the mandibular, maxillary, and labial segments are fused into a single sub-esophageal ganglion, and the last four abdominal ganglia also fuse into one. In the most extreme situation, all the thoracic and abdominal ganglia combine into a single ganglion. This occurs in the adult house fly, for example. Even where fusion takes place, the segmental components usually remain discrete within the compound ganglion.

The basic unit of the insect nervous system is the nerve cell, or neuron. The neuron is produced into a (usually) short process, the dendrite, that receives information from other neurons or from the environment, and a much longer process, the axon, that conveys information to other nerve cells and effector organs. There are three types of neurons: sensory neurons that receive information from the outside environment; motor neurons that convey instructions to effector organs, muscles, or glands; and interneurons that connect sensory input with motor output and provide for flexibility of response. The cell bodies of sensory neurons are

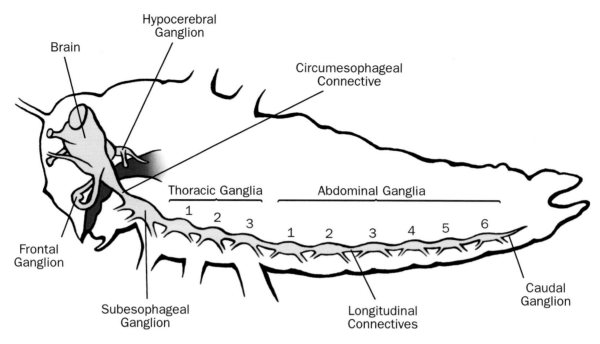

Insect central nervous system, including the brain and segmental ganglia. (Illustration by Jarrod Erdody)

within or immediately beneath the epidermis. The cuticle above them is modified to allow environmental stimuli—light, touch, or a chemical—to reach the dendrite, and the information is conveyed to the central nervous system along the axon. The cell bodies of interneurons and motor neurons are grouped together in the ganglia, where all the synapses between neurons occur. From each ganglion, peripheral nerves radiate outward to the various structures of the appropriate body segment. The nerves typically include the axons of sensory cells carrying information to the central nervous system and the axons of motor neurons carrying information to the effector organs.

Insect neurons transmit information in the same way as neurons in other animals, by the passage of changes in the electrical potential across the cell membrane. Because the movement of ions into and out of the neuron produces these changes, it is critical that the ionic environment of the neurons should remain constant. The ionic composition of the insect's blood is subject to change, however. To ensure that such changes do not affect the functioning of the neurons, the central nervous system is separated by a barrier of cells called the perineurium, known more generally as the "blood-brain barrier." These cells regulate the ionic composition of the fluid immediately surrounding the neurons.

Mechanoreception

Insects have many different kinds of mechanoreceptors concerned either with the perception of mechanical signals from the environment (exteroception) or with providing positional information about the parts of the body (proprioception). Most of the larger hairs on the cuticle are stimulated by contact, giving the insect its sense of touch. They are

present on all parts of the body. Sometimes the hairs are very long and slender, and their articulation with the cuticle is so sensitive that they move with airborne vibrations. Such hairs on the prothorax of some caterpillars are stimulated by the sound of the wing beat of a wasp that might be a predator. Similar hairs are present on the cerci of crickets. Essentially similar but very small hairs occur in groups called hair plates at some joints in the cuticle, where they function as proprioceptors. The hairs are bent over and stimulated when one segment of cuticle moves toward another. Hair plates are present on the basal segments of the antenna and on the small sclerites in the neck membrane. They give information on the position of the antenna relative to the head and the head relative to the thorax, respectively.

Many other types of mechanoreceptors are present in the cuticle, detecting stress on the cuticle and within the body. They have a variety of functions. Some may be involved in the detection of airborne or substrate-borne vibrations; others are stretch receptors that give the insect information about the degree of distension of its foregut, for instance.

Taste

Taste in insects is more appropriately called contact chemoreception or gustation, because similar receptors are involved not only in detecting food but also in the perception of oviposition substrates and sometimes pheromones. Contact chemoreceptors are short, hairlike or conical projections of the cuticle with a terminal pore through which chemicals in the environment reach the dendrites of the sensory neurons. These receptors occur in groups on the mouthparts and also commonly on the tarsi. Small numbers are present on the ovipositors of some insects.

There often are four sensory neurons in a single contact chemoreceptor. These are functionally of two kinds: they may initiate behavior, such as feeding or oviposition, or they may inhibit it. Each cell responds to a different range of compounds with varying degrees of sensitivity. Nearly all insects examined have neurons that respond to some sugars and amino acids, but not all the essential nutrients are tasted. Similarly, cells that inhibit behavior respond differentially to compounds that inhibit feeding or oviposition. Whether an insect feeds or oviposits on a particular substrate depends on the balance between the sensory input from neurons signaling acceptability and those signaling unacceptability.

Olfaction

The antennae are the principal olfactory organs of insects, and their length or extent of branching is an indication of the numbers of olfactory receptors and the importance of olfaction in the insects' lives. The number of receptors typically is large. The antenna of a large grasshopper may have about 6,000 receptors; that of a cockroach may have as many as 250,000. These very large numbers increase the likelihood that odor molecules, often in low concentration, will reach a receptor and so be detected by the insect. Often there is sexual dimorphism in antennal form, again reflecting the relative importance of olfaction to the two sexes. Some male moths that are attracted to females by a sex pheromone have plumose antennae, whereas those of the female are filiform.

Olfactory receptors have large numbers of small pores in the cuticle through which odor molecules can reach the neurons inside. The form of the receptors varies. Often they are hairlike; sometimes they are flat plates in the cuticle, and sometimes they are short pegs in a depression of the cuticle. Each receptor has numerous associated sensory neurons that convey olfactory information directly to antennal lobes, which are connected to the centers for learning and memory. The antennal lobe is similar to the olfactory lobe in vertebrates, and the processing of olfactory information probably is broadly similar. Insects have the ability to respond to a wide range of odors in addition to those that are of particular importance to the species, such as pheromones and specific host-associated odors.

Vision

Adult insects and the larvae of apterygote and endopterygote insects have a pair of compound eyes on the head. A compound eye comprises a number of similar units called ommatidia, each of which has a light-gathering system (lens), made partly from the transparent cuticle covering the eye, and a light receptor consisting of light-sensitive neurons called retinula cells. In dragonflies each eye contains about 10,000 ommatidia, the eye of a worker honey bee about 5,500 and that of a fruit fly about 800. The ommatidia of day-flying insects are separated from each other by cells containing a screening pigment that prevents light from passing from one ommatidium to another. Consequently, the retinula cells of one ommatidium receive light only from the lens system of the same ommatidium. The image formed by the eye as a whole consists of a series of spots of light, one from each ommatidium, varying in intensity and giving a fairly sharp rep-

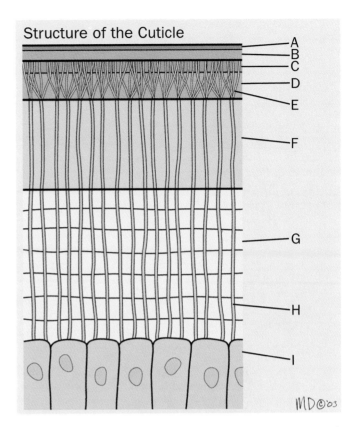

Structure of the Cuticle

The structure of the cuticle forms a rigid casing that supports the soft inner body parts. A. Cement; B. Wax; C. Outer cuticle; D. Inner epicuticle; E. Epicuticular filament; F. Exocuticle; G. Endocuticle; H. Pore canal; I. Epidermis. (Illustration by Marguette Dongvillo)

resentation of the object. The image has a superficial resemblance to an old newsprint photograph. This type of eye is called an apposition eye.

Night-flying insects experience low light intensities, and their compound eyes have a structure that permits maximal use of the available light. The screening pigment between ommatidia is withdrawn to the extreme outer and inner parts of the eye, where it does not interfere with the movement of light between ommatidia. As a consequence, the retinula cells of one ommatidium are illuminated by light from several lens systems. As a result, the image formed by the eye is much less sharp than that produced by an apposition eye. This type of eye is called a superposition eye.

Some insects are known to have color vision. In most species that have been studied, the visual spectrum is slightly different from that of humans. Insects can detect ultraviolet light, whereas humans cannot, but many insects cannot distinguish red. The fact that insects typically do not pollinate red flowers is a reflection of this; birds usually pollinate red flowers. Some insects, notably bees and ants, have the ability to distinguish the plane of polarization of light. Light from the sky is partially polarized, and the dominant plane of polarization is determined by the position of the sun. Bees and ants can learn the pattern of polarization of light from the sky and can use this information to navigate, usually back to the nest, even if the sun is not visible.

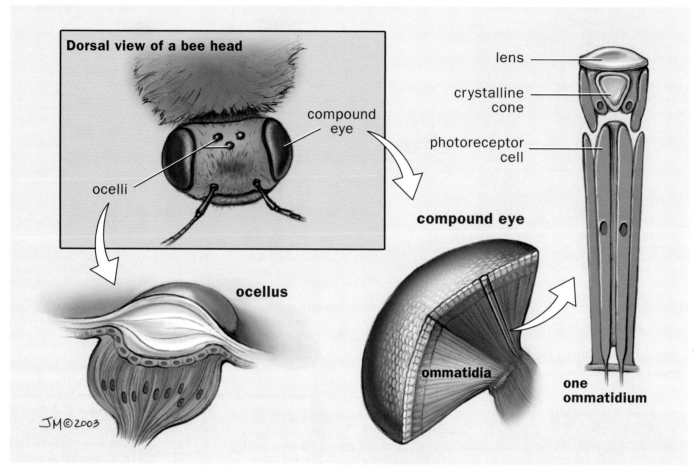

Compound insect eye. (Illustration by Jacqueline Mahannah.)

In addition to compound eyes, adult insects and exopterygote larvae typically have three ocelli arranged in a triangle on the front of the head between the compound eyes. Ocelli have single lenses and are sensitive to rapid changes in light intensity. Larval endopterygote insects have no compound eyes but often have stemmata on the side of the head. Sawfly larvae have only one stemma on each side, but caterpillars and tiger beetle larvae have six. Like ocelli, each stemma has only a single lens, and although the arrangement of sensory cells varies, the connections within the insect's brain are quite different from those of ocelli. Even the single stemma of sawfly larvae is capable of image formation, but the six stemmata of caterpillars seem capable of only crude form perception.

Endocrine system and hormones

Normal metabolism and major events, such as molting and metamorphosis, are controlled by hormones. Insects have three endocrine glands: the prothoracic gland, a diffuse structure in the prothorax; the corpora allata, small round bodies, one on either side of the foregut just behind the brain; and the corpora cardiaca adjacent to the heart and also just behind the brain. In addition, there are numerous isolated endocrine cells in the midgut epithelium. Many hormones also are produced by neurosecretory cells in the central nervous system. These cells release their products (neurohormones) into the hemolymph at numerous neurohemal organs, where the cell axons are exposed to the hemolymph.

The prothoracic gland produces the molting hormone, ecdysone. Because insects do not molt after becoming adult, these glands generally are absent from adult insects. The corpora allata produce juvenile hormone, which plays a major role in metamorphosis and also regulates yolk production in many insect species. The corpora cardiaca produce adipokinetic hormone, which controls the mobilization of lipids from the fat body. This is especially important in insects, such as locusts, that use lipids as the major fuel in flight. The corpora cardiaca also act as neurohemal organs for many neurohormones produced by cells in the brain.

Reproductive system

The paired gonads of both sexes lie in the dorsal part of the abdomen above the alimentary canal. From the gonads, a duct on either side passes around the gut, and the two ducts join beneath the gut to form a single median duct from which the gametes are discharged. In addition, both sexes have storage organs and accessory glands associated with the reproductive system.

Close-up of horsefly head showing its compound eyes. (Photo by Animals Animals © J. A. L. Cooke, OSF. Reproduced by permission.)

diploid, having a set of chromosome from each parent, but their gametes have only one set (haploid). The reduction in number (meiosis) occurs at two divisions of the spermatocytes, so that each spermatocyte gives rise to four spermatozoa. These events occur in sequence along the length of each sperm tube. From the testes, the sperm move to seminal vesicles, where they are stored until they are transferred to a female. At the time of transfer, in many insects, the sperm are packaged in a structure called a spermatophore, which protects them from desiccation during transfer. The spermatophore is produced by material from the male accessory glands, which usually open separately into the median ejaculatory duct. Other glands, at least in some species, produce chemicals that affect the female's physiology or behavior. These chemicals may cause her to reject subsequent mating attempts by males (fruit flies), they may provide protein that is added to the protein put into eggs by the female (many insects), or they may confer protection against predators to the female and her eggs (some moths).

Male

Each testis is made up of tubes varying in number from only one in some beetles to more than 100 in grasshoppers. Germ cells at the tip of each testis tube divide, giving rise after several divisions to spermatocytes. Most insects are

Female

Each ovary is made up of ovarioles. The number of ovarioles varies in different taxa. Lepidoptera have four in each ovary, fruit flies have 10–30; and large grasshoppers have more than 100. Stem cells at the tip of each ovariole divide

A mayfly larva with its cast skin (top). (Photo by P. A. Hinchliffe. Bruce Coleman, Inc. Reproduced by permission.)

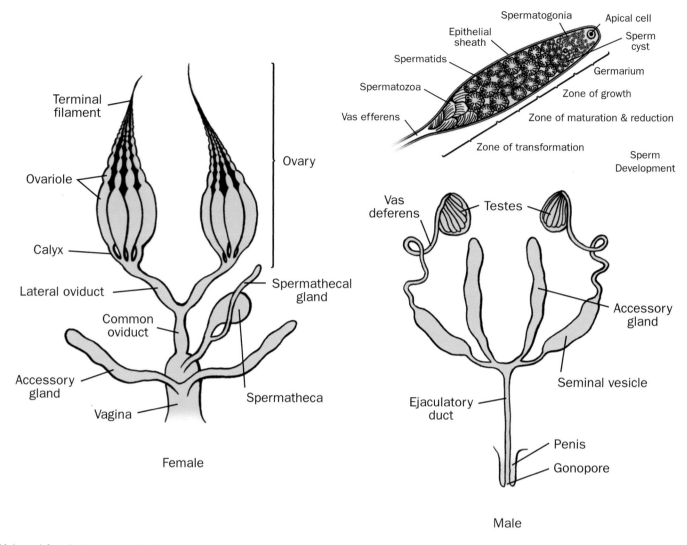

Male and female insect reproductive systems and sperm development. (Illustration Jarrod Erdody)

to produce oocytes (eggs). This parallels the process in males, but in females the cell divisions at which chromosome number is reduced do not occur until the oocyte is laid and penetrated by a sperm. In the ovariole each oocyte is provisioned with yolk, the proteins of which are synthesized in the fat body and then carried to the ovaries in the hemolymph. Finally, a layer of cells surrounding the oocyte produces the chorion (eggshell). As in the testis tubes, the developmental stages occur successively along the ovarioles, with the most advanced stages at the greatest distance from the tip.

The unfertilized eggs move from the ovarioles and are stored in the oviducts. The female accessory glands make the glue that is used by many insects to stick eggs to the substratum as they are laid. In a few cases, such as grasshoppers, cockroaches, mantids, and tortoise beetles, they provide material for the production of an ootheca, in which the eggs are concealed and protected after laying. The female reproductive organs include a spermatheca, in which sperm are stored after the female is inseminated. Fertilization of the eggs does not take place until they are laid, which may be months after insemination, especially in social insects.

Resources

Books

Chapman, R. F. *The Insects: Structure and Function.* 4th edition. Cambridge: Cambridge University Press, 1998.

Reginald F. Chapman, DSc

Life history and reproduction

Growth

Development of an insect from egg to adult proceeds through a series of larval stages separated by molts. Growth in weight is a more or less continuous process, as it is in other animals, but the parts of the body that are hardened (sclerotized), such as the head capsule and leg segments, do not increase in size except when the insect molts. At the molt these hard parts are replaced by new, larger ones. As a result, they increase in size quite suddenly and then remain unchanged until the next molt. Growth thus takes place in a series of steps. There is a tendency for the relative increase in linear dimensions of a sclerite to be similar in successive molts. Often, it is in the range 1.2 to 1.4. This constancy of relative increase is expressed as Dyar's law. Although this may be true for some parts of the body and for some molts, other parts may grow at different rates, and the relative increase may vary with the stage of development.

The stages between molts are referred to as larval instars. The number of instars varies between species, and there has been a tendency for the number of instars to decrease in the course of insect evolution. The most primitive, wingless insects (Apterygota) usually have 10 or more larval instars. In cockroaches and mantids and in orthopteroid insects (grasshoppers, crickets, bush crickets), the number usually is in the range five to 10, although the two aquatic groups, mayflies and stoneflies, often have more than 20 larval instars. Hemipteroids (sucking bugs and lice) and most holometabolous insects (see later discussion) have smaller numbers, often less than five, and the number of instars tends to be constant within a species.

Molting

Cuticle is a secretion of the epidermal cells, and the first step in producing a new cuticle is for the epidermal cells to divide. Insect epidermal cells do not divide at any other time. This cell division coincides with, and may be the cause of, the epidermis's becoming separated from the existing (old) cuticle, a process known as apolysis. From this time on, the insect does not feed again until the old cuticle is shed and a new one is fully expanded and hardened. This period of nonfeeding may last for hours or even two to three days, depending on the species.

Following cell division, the epidermal cells grow so that the surface area of the epidermis increases. Because the old

cuticle outside the epidermis prevents expansion, the epidermis becomes folded. As this occurs, the epidermal cells secrete enzymes that digest the proteins and chitin of the old cuticle from the inside, but at first the enzymes are not active. Before they become active, the epidermal cells start to secrete the new cuticle, beginning with the outer epicuticle and followed by the inner epicuticle. The former defines the ultimate size that the next stage of development can reach, and, because it follows the folds of the epidermis, its surface area is greater than that of the old cuticle. Once these layers are formed, the enzymes beneath the old cuticle are activated and begin to digest the old cuticle.

This digestion is important for two reasons. First, it weakens the cuticle so that the insect is able to break out of it; second, the cuticle contains valuable resources, especially proteins, and by digesting them and the chitin the insect is able to recover and reuse the nutrients. Because the cuticle on the outside of the body is continuous with that lining the foregut and hindgut, the space between the epidermis and the cuticle is continuous with the lumen of the alimentary canal. Consequently, the fluid containing the digested remains of the old cuticle, known as molting fluid, can be drawn directly into the alimentary canal and passed to the midgut. Here, the amino acids from the proteins and the sugar residues from the chitin are absorbed and become available for reuse.

Only the old endocuticle is digested. The exocuticle is impervious to attack by the enzymes because of the linking between protein chains (sclerotization) that occurs in its production. Simultaneously with digestion of the old endocuticle, the secretion of proteins and chitin of the new cuticle begins. These are not digested, despite the fact that sclerotization has not yet taken place, because they are protected by the new epicuticle layers. This new chitinous cuticle, still undifferentiated into exocuticle and endocuticle, is called procuticle.

Now the insect is ready to shed its old cuticle, a process known as ecdysis. (The term "molt" is commonly used to refer to this part of the process alone.) Exerting pressure on the weakened old cuticle from within brings about shedding. To accomplish this, the insect increases its body volume. While it is still feeding, before the molt begins, the insect

Emergence

Molting is a process by which the outermost layer of skin is cast off, allowing for the growth of the insect. This takes place in stages.

As an insect grows, its exoskeleton is replaced through molting, as the exoskeleton does not grow with the insect. (Illustration by Christina St. Clair)

retains more water from its food, so that its hemolymph (blood) volume is increased. Then, at the time of ecdysis, it swallows air so that the alimentary canal becomes expanded to a large air-filled sausage-shaped structure. Waves of muscle contraction starting posteriorly and moving forward compress the abdomen and force blood into the thorax and head, so the new cuticle presses hard against the inside of the old cuticle. Repetition of this process causes the old cuticle to split along predetermined lines of weakness (ecdysial lines) where there is no exocuticle. These lines often run along the dorsal side of the thorax and over the top of the head. Once the old cuticle is split, the insect completes its escape by further pumping.

As soon as the legs are free, the insect uses them to crawl out of the old cuticle. This is possible only because, before beginning ecdysis, the insect has attached its old cuticle to the substrate. If it did not do this, it would have great difficulty breaking free of the abdomen and would continue to drag the old cuticle with it. Caterpillars spin a small pad of silk and attach themselves to it by the hooks (crochets) on the posterior prolegs. It also is common for insects to suspend themselves vertically, head down, so that once the old cuticle is split, their escape is aided by gravity. The whole cuticle is shed, including the linings of the foregut and hindgut, tracheae, and apodemes. All these are replaced at each molt. Just before ecdysis, a wax layer is formed on the outside of the new cuticle. This timing is critical, because it ensures that the insect

will have its new waterproofing layer when the old cuticle is shed. The wax is thought to reach the surface via the wax canal filaments that extend from the epidermal cells through the outer epicuticle. Wax production continues throughout the period between molts.

When the insect breaks free of the old cuticle, the new cuticle is still soft and heavily wrinkled. It is typically very pale in color, almost white. Expansion of the new cuticle by flattening out of the folds results from further body compressions, which force hemolymph into the regions to be expanded. This process is most obvious as the wings expand in adult insects. When expansion of the cuticle is complete, hardening occurs. If it took place earlier, the insect would be unable to expand completely. Permanently crumpled wings, occasionally seen in adult insects, are a result of hardening taking place too early. Hardening results from the binding together of proteins in the procuticle, a process analogous to the tanning of leather and known as sclerotization. Procuticle contains not only structural proteins and chitin but also enzymes called phenol oxidases. The enzymes are inactive, because no substrate (phenols) is present. When cuticle expansion is complete, phenolic compounds produced in the epidermal cells pass into the cytoplasmic processes in pore canals and so reach the outer layers of the procuticle. As the phenols are secreted into the procuticle, they become oxidized by the enzyme already present, and the oxidized products bind with proteins, linking them together so that the cuticle

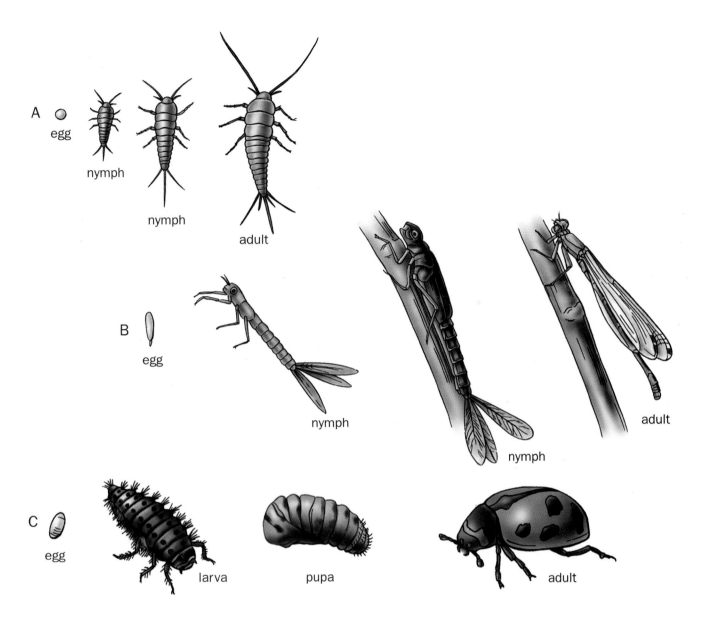

Insect metamorphosis: A. Ametabolous development; B. Incomplete metamorphosis; C. Complete metamorphosis. (Illustration by Patricia Ferrer)

becomes hard. This process occurs only in those parts of the cuticle that become hard plates (sclerites); membranous areas remain unsclerotized. Sclerotization is accompanied by darkening of the cuticle, and the sclerotized procuticle subsequently is known as exocuticle. Cuticle production continues after the molt, in some species right through to the time of the next molt. This cuticle remains unsclerotized and forms the endocuticle.

Development

Primitively wingless insects (Apterygota) hatch from the egg in a form that essentially resembles the adult, except for being smaller and lacking external genitalia. At each molt the insect grows, and the genitalia become progressively larger, until finally the insect is able to reproduce. There is no marked

change of form (metamorphosis) between the last larval stage and the adult. These insects are said to be *ametabolous* (without metamorphosis). They continue to molt as adults. This does not occur in winged insects, except for the single additional molt of the sub-imago of mayflies.

Winged insects exhibit much more marked differences between larva and adult. In cockroaches and mantids, orthopteroids, and hemipteroids, the first instar larva resembles the adult but lacks wings as well as genitalia. These features develop progressively at successive molts, the wings appearing as outgrowths, known as wing pads or wing buds, of the dorsolateral region of the mesothoracic and metathoracic segments. The wing pads come to lie over the insect's back. At the molt to adult, however, expansion of the wings to their fully developed form is accompanied by relatively extensive changes in other parts of the body, especially the

An aphid gives birth to live young. (Photo by Dwight Kuhn. Bruce Coleman, Inc. Reproduced by permission.)

thorax. These insects are said to undergo a partial metamorphosis; they are hemimetabolous. The feeding habits of these insects generally are similar as larvae and adults, and there are no major changes in head or mouthparts. Dragonflies, mayflies, and stoneflies that have aquatic larvae have a basically similar development, although metamorphosis is slightly more marked because of the loss of larval characters specifically associated with the aquatic environment, notably the gills. Because of the similarity of larval and adult forms, larvae of hemimetabolous insects often are called nymphs to distinguish them from the larval forms of holometabolous insects.

In many insects, however, the larval form is markedly different from the adult, and often the feeding habits of larva and adult also differ. This is obvious in butterflies and moths—the caterpillar is soft-bodied and uses biting and chewing mouthparts to feed on solid plant material, whereas the adult is well sclerotized and typically has sucking mouthparts and feeds on nectar. There is no outward sign of adult structures in the larva. The final instar larva is simply a larger version of the first. It does not molt directly to the adult stage; instead, it molts to a stage known as a pupa, which is immobile and does not feed. It possesses adult structures, such as wings; compound eyes (lacking in the larva); and, in a butterfly, long antennae and a long proboscis. Sometimes

these structures are stuck down to the rest of the body, but they are clearly visible. The adult finally emerges (ecloses) from the pupa. Insects in which larvae and adults differ markedly and that have a pupal stage are said to undergo complete metamorphosis; they are holometabolous. Apart from Lepidoptera, beetles, flies, fleas, caddis flies, bees and their relatives (Hymenoptera), scorpion flies, and lacewings are holometabolous.

Although the larvae of holometabolous insects exhibit no outward sign of adult structures, the adult wings, eyes, antennae, legs, and mouthparts are developing within them in pockets formed from the larval epidermis. The imaginal discs, groups of epidermal cells destined to form adult (that is, imago) structures, may be evident in the first instar, but their development becomes marked in the last instar larva. Because the space in the pockets is restricted, the adult structures become extensively folded. When the insect pupates, the epidermal pockets are turned inside out, so that the adult structures come to lie on the outside of the insect instead of being hidden within the pockets. At the same time, the folds flatten out before the pupal cuticle is secreted. The wings do not reach their final size at this molt but increase in size at the pupa-to-adult molt. During the pupal period, extensive reconstruction of internal systems, such as the nervous system and alimentary canal, takes place. Some muscles retain

A Puerto Rican walkingstick (*Lamponius portoricensis*) molting to adult. The species transforms at night to lessen its chances of being eaten. (Photo by Rosser W. Garrison. Reproduced by permission.)

their larval functions, but others are lost, and new ones develop in the pupa for use in the adult. In particular, the flight muscles, which are present as small dorsoventral muscles in the larva, become greatly enlarged.

The evolution of the pupal stage represents a major advance in insect evolution, because it frees the larva to adapt to new environments completely independently of the adult. Nutrient accumulation, accompanied by growth, occurs primarily during the larval stages. Feeding by adults of many holometabolous species is concerned primarily with obtaining energy for flight and nutrients for egg production, although even these nutrients are derived from larval feeding in many species. The reduction or complete loss of external appendages and the soft-bodied form of many holometabolous larvae facilitates invasion of substrates so that they can feed from within instead of being confined to the outside like hemimetabolous insects. Many of these insects bore into the tissues of plants and other animals, and many flies and Hymenoptera are parasitic on other insects. The four orders Coleoptera (beetles), Diptera (flies), Hymenoptera (bees and their allies), and Lepidoptera (butterflies and moths) contain more than 75% of the species of all living animals, and this is probably to a large extent a consequence of holometabolous development.

In most animals the term "adult" refers to sexual maturity, but this is not the case for insects. In this case, "adult" refers to the final stage of development, which, in most species, is winged. Some insects are sexually mature when they first become adult, and they can start to reproduce immediately, but in many other species sexual development is delayed. An extreme example is the red locust of central Africa. Adults emerge March to April, when the habitat is drying up. Sufficient grass remains for the adults to feed and maintain themselves, but they do not become sexually mature and lay eggs until late October to November, at the beginning of the rainy season that produces a new crop of grass on which the larvae feed after hatching.

Social insects

The development of social insects is not fundamentally different from that of other insects, but the stages we normally encounter are not sexually reproducing adults. Termites have hemimetabolous development, but within a colony, for most of the year, only one female, the queen, is an adult and lays eggs. Unlike most hemimetabolous insects, the fate of the larvae that hatch from these eggs varies considerably. Any larva in its early stages has the potential to become a worker, a soldier, or a reproducing adult, depending on the season and the needs of the colony. Both workers and soldiers are larval forms

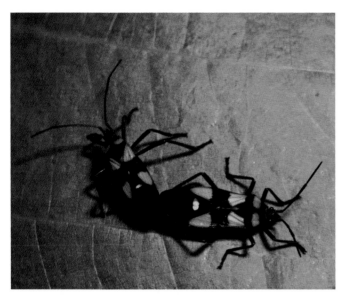

Heteropteran insects mating in a Nigerian rainforest. (Photo by P. Ward. Bruce Coleman, Inc. Reproduced by permission.)

and, in some species, can undergo regressive molts that enable them to contribute to the reproductive pool. Winged adults are produced only at certain times of year, often coinciding with the onset of a rainy season. As soon as conditions are suitable, they leave the colony to find mates and found new colonies. The original colony retains its original king and queen with workers and soldiers.

Bees, wasps, and ants, in contrast, are holometabolous insects. They differ from termites in two important respects: all the stages normally encountered outside the nest are sterile adult females, and all the workers and soldiers, if they are produced, also are females. When a female egg is laid, it has the potential to develop into a queen or a worker (or soldier), but the manner in which it develops depends on the quality and quantity of food it receives. For most of the year, nearly all the larvae become workers. Males are produced only seasonally and leave the nest without contributing to it.

It is noteworthy that among the social insects, all the eggs of one sex are genetically similar, yet they have the capacity to develop into different castes. The production of divergent forms from genetically similar individuals is called phenotypic variation. This is a widespread phenomenon outside the social insects as well as within them. For example, butterflies often produce seasonally different forms, even though they are genetically very similar.

Life histories

Insect life histories are determined largely by two factors: temperature and the availability of food resources. Although insects raise body temperature while flying, larval insects are cold-blooded (poikilothermal). Only among social insects do adults regulate the body temperature of their offspring. Despite the fact that the larval insects are cold-blooded, the metabolic processes that underlie insect development are tem-

perature dependent and operate most efficiently at temperatures close to 95°F (35°C). As a result, they function efficiently only when the insect's body temperature is above 86°F (30°C), and this is true only when the environmental temperature is high or when the insect is able to warm itself by solar radiation even though the air temperature is low. Thus, in the tropics development rarely is temperature limited, but it may be affected by resource availability. In more temperate regions insect activity is seasonal, becoming markedly so closer to the polar regions.

In the humid tropics, where the availability of moisture permits plant growth throughout the year, insects are able to breed continuously and may have several generations in the course of a year, a condition known as multivoltine (= "flight"). The body surfaces of warm-blooded vertebrates also provide a stable and suitable temperature, and this environment has been exploited by the lice that never leave their hosts except to transfer from one individual to another.

With increasing distance from the equator, insects become progressively less able to remain active during the cooler months. Sometimes they can produce several generations during the warm period, but as summers grow shorter and winters longer with increasing latitude, the number of annual generations is reduced to one (univoltine); in a few instances a species may require several summer seasons to complete development. In a most extreme example, the caterpillars of *Gynaephora groenlandica* that live within the Arctic Circle take 13 years to develop. This species is able to raise its body temperature sufficiently high to feed by basking in the sun, but the time available for feeding each day is very short, as is the period each year when air temperature is sufficiently high for the insect to reach an operating temperature even by basking.

If a nonmigrating species extends over a broad latitudinal range, its life history may change with latitude. A good example of this is the dagger moth in Russia. In the south (40° north latitude) it is bivoltine. Larvae of the first generation hatch in April and feed until the end of May, when they pupate. Adults soon emerge and give rise to a second generation in August. At St. Petersburg (60°N latitude), however, the species is univoltine. The larvae hatch in July and develop to pupae that remain in diapause until June of the following year.

Where periods suitable and unsuitable for insect development follow each other in a regular manner, as with summer and winter and wet and dry seasons, insects have evolved the capacity to survive the inclement periods and optimize their use of suitable times. They survive unsuitable periods either by a period of delayed development, known as diapause, or by migrating to more suitable environments. In either case, it is important that the insect has previous warning of the coming adversity. Both diapause and migration involve preparation in the form of physiological changes in the insects; if they were to wait until the temperature drops, for example, it might be too late to make the necessary preparations. Consequently, the response does not depend on the prevailing conditions but rather is an evolved adaptation. The most reliable indicator of seasonal events is the change in day length, and this provides many insects and other animals with a signal. Shortening days presage the coming of winter. Even in the tropics, where changes in day length are small, insects can perceive and respond to the changes as their particular environment requires.

Before diapause, insects accumulate nutrient reserves to sustain them during the period of diapause and in the period immediately following it. During diapause, metabolic activity is greatly reduced, and the reserves are depleted only very slowly. Normal development stops and is not resumed simply as a result of the return of suitable conditions. If this were the case, warm days in winter could lead to a premature resumption of activity. Diapause, then, is not simply a decrease in activity that is related quantitatively to environmental conditions, but instead is a qualitative change in the insect's physiology. For many species, a period of exposure to low temperatures is necessary for the termination of diapause. Following this period, a return to suitable conditions leads to a resumption of normal development. Diapause may occur in the egg, larva, pupa, or adult, but in any one species it typically happens in one of these stages that is specific for the species. Adults in diapause may continue to be active and even feed if food is available, but they remain reproductively immature.

Temperatures substantially below freezing kill most insects, because ice forming within the cells damages cell membranes. Many insects at high latitudes, however, commonly experience such conditions for substantial periods of time. These insects must have the capacity to survive under these extreme conditions, and two different strategies have evolved. Some insects avoid freezing by using the equivalent of antifreeze; others have evolved methods of tolerating freezing by slowing the formation of ice crystals within the cells. Freeze-susceptible insects, as the first are called, often accumulate glycerol in the hemolymph before diapause and the cold conditions occur. The glycerol permits supercooling of the hemolymph so that the insect can be cooled to $-4°F$ ($-20°C$) or even lower without freezing. Examples of insects using this strategy are found in all the major orders, such as the pupa of the cabbage white butterfly, the larvae of bark beetles, the pupae of the flesh fly, and the eggs of some sawflies. Freeze-tolerant insects have ice-nucleating agents in the hemolymph or cells that bring about freezing in a controlled manner that limits tissue damage. These insects often can tolerate periods below $-58°F$ ($-50°C$). The larva of *Gynaephora groenlandica* can withstand temperatures as low as $-94°F$ ($-70°C$). This species remains frozen for about nine months in each year of its 13-year development.

In general, insects have not evolved comparable means of enduring the effects of excessively high temperatures. Instead they avoid such temperatures behaviorally, by moving into shade and exposing themselves to air movement or, in the long term, by seeking appropriately cool microhabitats in which to undergo diapause. There is, however, one remarkable exception of a midge, *Polypedilum*, in Nigeria. Its larva lives in exposed freshwater pools on rocks. The pools dry up in the dry season, and the exposed surfaces of the rocks reach temperatures above those at which life normally occurs. Slow dehydration of the larvae (experimentally) produces individuals that can withstand one minute at the boiling point of wa-

Shield bug (*Eysacoris fabricii*) eggs hatching. (Photo by Kim Taylor. Bruce Coleman, Inc. Reproduced by permission.)

ter (212°F [100°C]), and totally dehydrated individuals have been kept in the laboratory for three years and recover when rehydrated. The phenomenon of an organism exhibiting no visible signs of life yet remaining able to recover is called cryptobiosis (hidden life).

Examples of insects that avoid adverse conditions by migrating also are found in all the major groups. One of the best known is the American monarch butterfly, which moves south from areas in southern Canada and most of the United States to a site in Mexico, where it survives the winter. The complete journey is made by individual insects that do not breed on the way south or at the overwintering site. The return movement that starts in the spring is spread over several generations that breed en route. Some tropical African butterflies also use migration as a means of avoiding seasonally occurring dry periods. Certain aphids feeding on plant annuals produce winged sexual forms in response to declining day length; on the other hand, during the summer, reproduction is asexual, and most of the insects are wingless. Being winged enables these insects to leave the annuals before they die and move to perennial plants, where they produce eggs that overwinter.

Within the limits set by temperature, the life histories of most insects are determined by the availability of food resources. This is most obvious in tropical and subtropical areas with seasonal rainfall. Breeding can occur continuously during the rainy season but often not at all during the dry season, primarily because plants will not grow without water and plant-feeding insects are at the bottom of the food chain for the vast majority of insects. For many species, the timing of development is determined by a narrow window of time in which a suitable food resource is available. In Britain larvae of the winter moth feed on oak. The first-stage larvae can eat oak leaves only around the time of bud break. Consequently, the larvae must hatch within a two-week period in early April. If they did not, they would virtually all die. Feeding is complete by the second half of May, and the larvae drop to the ground and pupate. In theory, there is time for a second generation, but the lack of appropriate food for the early larvae prohibits this, and so the species is univoltine, and the adults are active in early winter. The life history of a tachninid fly, which parasitizes larvae of the winter moth, is, in turn, determined by the time of occurrence of the larvae. It is certainly true for a very large number of insects that their life histories are governed by the availability of some essential resource.

Certain insects have adapted their life histories to cope with irregularly occurring changes in their local environment. Under such circumstances, the insects use cues other than day length. Aphids, for example, produce winged forms in response to crowding. This enables them to leave a deteriorating habitat and start colonies on new plants. Unlike those produced in the fall in response to changing day length, these insects are all females and give rise only to females.

Nutritionally poor resources also affect the life history, since it takes the insect longer to acquire the nutrients necessary for development than is true on a high-quality substrate. The 17-year cicada is an example. Its larvae feed on the xylem of plant roots. Xylem contains only very low concentrations of amino acids, and the larva must process very large volumes of xylem to obtain an adequate quantity. Some species overcome problems of nutritional inadequacy by associating with microorganisms, such as bacteria, that supply the limiting nutrients. This is true of all aphids, enabling them to produce many generations within a season.

Reproduction

Most insects, like the majority of other organisms, are diploid, that is, they have two sets of chromosomes, one derived from each parent. This requires sexual reproduction, which is important because it maintains a high level of variability within the population, giving it the power to adapt to new environmental circumstances.

Sex determination

Sex among insects, as in most other animals, is determined by sex chromosomes. Females typically have two X chromosomes (XX), whereas males have one X and one Y chromosome (XY). The Lepidoptera (butterflies and moths) are exceptional; females are XY and males XX. The Hymenoptera (bees and their allies) are unusual in having haploid males with only a single set of chromosomes (XO). Although development of the gametes is regulated by hormones in insects, these hormones do not control the development of secondary sexual characters, as they do in birds and mammals.

Sperm transfer and fertilization

Insects were, evolutionarily, one of the first major groups of animals to have become terrestrial, and the terrestrial environment necessitates special methods of sperm transfer in addition to the other problems that are associated with living on land. Aquatic animals frequently simply discharge their gametes into the water, but sperm transfer between male and female on land requires a more sophisticated approach. In some primitively wingless insects (Apterygota), males deposit sperm-containing droplets on the ground and spin silk threads to restrict movement by the female; she is guided over the droplet and inserts it into her genitalia.

Among the winged insects, many produce a sperm-containing structure, called a spermatophore, in which to transfer sperm to the female. Direct insemination, not involving a spermatophore, occurs in some sucking bugs (Heteroptera), caddis flies, beetles, flies, and Hymenoptera. Where there is a spermatophore, it may be produced by the male before copulation, as in crickets, but more commonly it is formed while the male and female are coupled. The spermatophore, whose primary function is to protect the sperm during transfer to the female, is formed from secretions of the male accessory glands. In some instances, it has assumed additional functions. It may provide the female with nutrients or with chemicals that impart protection against predators for her and her eggs, or it may contain chemicals that modify the female's physiology and behavior.

In stick insects, crickets, and mantids, only the neck of the spermatophore enters the female duct, the rest remaining outside. Females eat this portion of the spermatophore, even though it initially contains sperm; sperm transfer to the female's spermatheca is complete before the spermatophore is eaten. This provides the female with protein that contributes to yolk formation and allows her to produce bigger eggs and sometimes more eggs than would otherwise be possible. In some bush crickets the weight of the spermatophore may exceed 20% of the male's weight, indicating the importance of the resources he is contributing to the female and his offspring. Protein transfer also occurs in some species where the spermatophore is formed entirely within the female ducts, although the quantities are much more limited than in exposed spermatophores.

Males of many butterfly species are known to drink from rain puddles (a process known as "puddling"). In a species of skipper this is known to enable the male to accumulate sodium, which, because of the food the species eats, is in relatively short supply. The sodium is transferred to the female in the spermatophore, which may weigh as much as 10% of the body weight, and this increases her fertility. The same thing probably occurs in other puddling species. In addition to nutrients, the male may transfer chemicals that protect the female and her eggs from potential predators, as is known in some tiger moths. Chemicals that affect female fertility by acting via the nervous system also are transferred in a few cases. In others, the chemicals reduce the readiness of the female to mate again. This is clearly an advantage for the male that transfers the sperm, since it increases the likelihood that his genes, and not those of another male, will be conveyed to the female's offspring. Although few examples of these effects are so far known, it is probable that they are common.

Fertilization of the eggs does not take place at the time of sperm transfer (insemination of the female). Between insemination and fertilization of the eggs, the sperm are stored in the spermatheca, part of the female reproductive system. Sometimes the period of storage is relatively brief, because the eggs are laid soon after insemination; in other cases, and notably in some social insects, live sperm are retained within the female for months or even years. Hence, it is important to distinguish between insemination of the female and fertilization of the egg. In most other organisms, these events occur within a short time of each other.

As each egg passes down the oviduct, sperm are released from the spermatheca and effect fertilization. In Hymenoptera, females are produced from fertilized eggs, becoming diploid in the process, but unfertilized eggs develop into males, which consequently are haploid. Fertilization of the egg is not a haphazard occurrence; it is regulated by the female. This is why bees and wasps produce males only at certain times of the year. Some parasitic wasps determine whether an egg should be fertilized from the size of the potential host.

Parthenogenesis

Occasionally, eggs develop without fertilization, a phenomenon called parthenogenesis. This happens erratically in many insect species, but in some groups it is the rule. Aphids are the best-known example. They may reproduce parthenogenetically for many generations during the summer, giving rise only to females that are diploid. Combined with viviparity, this enables them to produce large numbers of young in a short time. Lack of sexuality, however, can lead to loss of variability. Aphids meet this problem by periodically (usually at the onset of winter) producing sexual forms, so that mating restores a degree of genetic diversity.

Bacterial infections affecting reproduction

A bacterium, *Wolbachia*, has a profound effect on reproduction in many insects. The most common effect of infection is cytoplasmic incompatibility (incompatibility between the male sperm and female oocyte resulting from abnormalities in the cells outside the nucleus) between males and females such that viable offspring are not produced. Sometimes the incompatibility is asymmetrical: infected females mated with uninfected males give rise to viable offspring, but uninfected females mated with infected males do not. Cytoplasmic incompatibility due to the bacterium occurs in many insect orders. In parasitic Hymenoptera, the bacterium sometimes prevents male formation so that the insect reproduces parthenogenetically. If the wasps are cured experimentally with antibiotics, they produce normal males and females. It is believed that *Wolbachia* has had major effects on insect evolution.

Oviposition and the egg

Oviposition habits vary greatly. Some species lay eggs in clusters, and others lay them singly, and these differences may occur even in closely related species. The large white butterfly (*Pieris brassicae*), for example, lays its eggs in clusters of 100 or more, while the cabbage white (*Pieris rapae*) lays its eggs singly. Cockroaches, mantids, and grasshoppers always lay their eggs in groups, creating a protective envelope around the eggs. The whole structure is called an ootheca, and the material of which it is made is produced by the female accessory glands and hardens on contact with air.

Many insects deposit their eggs on a surface, commonly the outside of a food resource, sticking the eggs in position with an adhesive secretion. Others lay eggs inside structures; in these insects the ovipositor is modified for piercing. Bush crickets provide a good example, with their long, swordlike ovipositor. The long, slender ovipositors of ichneumonid wasps allow them to lay eggs within the insects they parasitize, even if this insect is boring inside wood. The ovipositor of the grasshopper is used to dig a hole in the ground, in which the eggs are laid. Because the eggs of many insects are laid in exposed positions, control of water loss is critical, just as it is at later stages of development. Waterproofing often depends on a layer of wax on the inside of the shell and on the production of a special cuticle, the serosal cuticle, by the developing embryo. The degree of desiccation that an egg can withstand varies greatly with the environment to which the insect is adapted and the stage of embryonic development.

Viviparity

Although the vast majority of insects lay eggs, some are viviparous, giving birth to live young. In the simplest form of viviparity, fertilized eggs are retained within the female instead of being laid. The yolk in the egg provides all the nourishment available to the developing embryo, and the larva is deposited soon after it hatches. This process, known as ovoviviparity, occurs among some species in several orders of insects but is especially common in parasitic flies in the family Tachinidae.

Viviparity in which the female provides nourishment to the embryo in addition to or instead of yolk occurs in relatively few insects. Aphids are the most widely known and abundant of such insects. Their embryos receive nutrients from a modified ovariole sheath as well as directly from the female's hemolymph, and the offspring are deposited when they reach the first larval stage.

The most extreme development of viviparity in insects is seen in a few closely related families of flies, which, as adults, feed on the blood of vertebrates. The best-known example is the tsetse fly. Each ovary of a tsetse fly has only two ovarioles, and only one oocyte (developing egg) is produced at a time. The fertilized egg is retained within the female, and the embryo uses yolk for development. When the larva hatches, it stays in the female and is fed on "milk," rich in amino acids and lipids, secreted by the female accessory glands. The contents of the "milk" are derived directly from the blood ingested by the adult female. The larva grows and molts within the uterus and stays there until development is complete. Then it is deposited by the female and immediately seeks a place in which to pupate. It does not feed while it is free living.

Resources

Books

Chapman, R. F. *The Insects: Structure and Function.* 4th edition. Cambridge: Cambridge University Press, 1998.

Eisner, T., C. Rossini, A. González, V. K. Iyengar, M. V. S. Siegler, and S. R. Smedley. "Paternal Investment in Egg Defense." In *Chemoecology of Insect Eggs and Egg Deposition,* edited by M. Hilker and T. Meiners. Berlin: Blackwell Verlag, 2002.

Periodicals

Stouthamer, R., J. A. J. Breeuwer, and G. D. D. Hurst. "*Wolbachia pipientis*: Microbial Manipulator of Arthropod Reproduction." *Annual Review of Microbiology* 53 (1999): 71–102.

Reginald F. Chapman, DSc

• • • • •

Ecology

Introduction to insect ecology

Imagine a world without insects. At first thought, this might seem wonderful. No biting flies. No termites to destroy homes. No malaria, yellow fever, sleeping sickness, or any of the multitudinous diseases vectored by insects. No need to worry about insects eating crops and no need to take special precautions to keep insect pests out of stored food or other supplies.

Certainly, there would be tradeoffs. There would be no elegant butterflies, including the migratory monarch *Danaus plexippus* that has captured the imagination of generations of humans. We would not hear crickets chirping at night. We would have no bees and so no honey. These are trivial losses compared with the benefits we would experience, however, and they might seem like a reasonable price to pay for a world without insect pests. Unfortunately, there would be other tradeoffs. Earth is home to approximately 950,000 described species of insects, more than half of the 1.82 million known species of plants and animals combined. Estimates of the total number of insect species on Earth, including undiscovered species, range from two million to 80 million, with five million commonly accepted as a reasonable estimate. Insects play indispensable roles in maintaining many familiar aspects of life on our planet. The story of these diverse and numerous roles is the story of insect ecology.

The niche theory of ecology

Ecology is the study of interactions between living organisms and both the living and nonliving components of their environment. Any introduction to ecology would be incomplete without an explanation of the concept of a niche, which is the organizing theme for the study of ecology. Although the term niche often is used in its historic sense to refer to a physical space in a habitat, that interpretation is restrictive and incomplete. There is a better definition, particularly for understanding the niches of insects. A niche is the suite of ecological strategies by which a given species maintains a balanced energy budget while converting available energy to the maximum number of offspring. These offspring carry genes into the next generation, perpetuating successful ecological strategies. Thus, an ecological niche includes not only the choice of a physical place to live and forage but also behavioral strategies for avoiding enemies, finding food, locating

and selecting mates, and producing offspring. The niche of a given species is defined by the specific ecology of that species, embracing all the interactions between that species and its biotic (living) and abiotic (nonliving) environment.

This definition of an ecological niche has both predictive and explanatory powers. For example, niche theory predicts that under adverse environmental conditions an attempt at reproduction will be abandoned if the energy cost of producing offspring could deplete the energy budget so severely that both offspring and parent are likely to die. This explains why, in adverse conditions, plants abort seeds and animals abort or abandon their young. If parent and offspring both die as the result of environmental stress, neither of them will pass genes to the next generation; thus, under stressful conditions, some parents sacrifice their offspring, conserve energy, and make another attempt at reproduction later, when conditions are more favorable.

Insects exhibit several variations on this theme of sacrificing individuals to balance the energy budget. One variation occurs when immature insects eat one another, including their own siblings. The larvae of monarch butterflies do this. Another grisly example of balancing the energy budget via cannibalism is that of the female praying mantis consuming her mate, sometimes during the mating act. The male's body provides nutrition for development and provisioning of the eggs he fertilized. This behavior makes perfect sense in the context of niche theory. The ultimate goal of life is to pass on genes; therefore the strategies that make up an ecological niche are centered on producing the maximum number of offspring without going into energy debt. Maintaining a balanced energy budget is a crucial facet of ecology, even if it means forfeiting some offspring—or even oneself.

The importance of insect eating habits

The key to understanding insect ecology lies in analyzing the energy budget of each species of interest, and this starts with their dietary habits. All the energy available to insects is derived from their food, so the first entry in the energy budget of a species depends on the food resources exploited by that species. Insects feed on nectar, pollen, leaves, fruits, and many other plant parts, including wood. They also eat dung, fungi, and both vertebrate and invertebrate animals.

A farmer in a tractor spraying fruit trees with pesticide. Pesticides are a requirement of modern intensive agriculture because single crops provide a large uniform environment in which pests can rapidly spread. (Photo by ©David Nunuk/Science Photo Library/Photo Researchers, Inc. Reproduced by permission.)

Insects are extraordinarily numerous, both in terms of numbers of species and numbers of individuals, and so their eating habits have a significant influence on the global ecosystem.

Some insects, such as bees, butterflies, and some flies, feed on nectar. In the process they often pollinate the plants supplying nectar. Imagine a world without insects, and you will have to imagine a world without many of the foods we take for granted. Approximately 60% of all crop plants require animal pollination agents, most of which are insects. There is no doubt that the human diet would be impoverished dramatically without insect-pollinated plants.

The species of insects that feed on nectar and transport pollen are vastly outnumbered by the species that feed directly on the leaves, stems, roots, flowers, and fruits of plants. The feeding habits of insects have had profound effects on most of the major ecosystems on Earth. Plants evolve adaptations in response to insect damage, and it is believed that insect feeding behavior has affected the diversity of plant species dramatically. Insects have had a huge influence on the biodiversity of this planet.

Other kinds of ecological interactions, also related to insect feeding habits, would be lost if there were no insects.

For example, insects remove carrion and dung. When an animal dies, the body often is still warm when the first insects arrive and begin dismantling and recycling the body, returning its materials and stored nutrients to Earth's ecosystem. Flies lay eggs on the body, and the resulting maggots are an important factor in the rapid conversion of dead flesh to new life. Some beetles specialize in feeding on carrion. Some dead animals, such as mice, are so small that sexton beetles (family Silphidae) dig a hole under the body, bury it, and prepare it as food for themselves and their offspring. Decomposition of dead animals would take much longer if it were dependent solely on the action of bacteria and fungi. Insect undertakers speed the process and increase the efficiency of nutrient recycling.

Some kinds of insects eat dung. This takes place in most climates, but it is particularly spectacular in the tropics, where dung is so coveted by insects that often it is still fresh and warm from the animal when the first scarab beetles converge and claim portions as their own. The world would be piled high with dung if insects did not remove and recycle animal wastes.

The importance of insects on Earth

There are innumerable specific examples of the integral roles of insects in life as we know it on this planet. One distressing consequence of a complete loss of all insects from Earth would be the loss of many vertebrate animals that eat insects. Birds, reptiles, and many mammals, including some primates, rely on insects for part or all of their diets. Some human societies consume insects, which are an excellent source of protein and fat. Although biting flies are a bane of human existence, both because of the nuisance and, more important, because of their capacity to carry disease, the immature stages of these flies often are water dwelling and are vital sources of food for fish. Our streams, rivers, lakes, and ponds would be depleted of fish and other aquatic life if all insects vanished from the Earth.

Ecology is the study of interactions between living organisms and their environments, and comprehending the critical and integrated role of insects on Earth is the philosophical underpinning for understanding insect ecology. Insight into both the classic and the unique roles of insects in Earth's ecological systems enhances an understanding and appreciation of the intricacy of life on this planet.

Unique characteristics of insects

The ecological interactions of insects are dictated by their unique characteristics, many of which are completely different from the human experience. Comprehending these differences, which are based on size, physiological characteristics, and morphological features, lays the groundwork for understanding insect ecological interactions. Essential characteristics of insects that dictate their interactions with the environment are small size, exoskeleton, metamorphosis, high reproductive capacity, genetic adaptability, and, for most kinds of insects, their ability to fly. Only insects have this unique suite of characteristics. These features act in concert to define the niches of insects. Some fundamental concepts related to these characteristics are worth reviewing in the context of a discussion of insect ecology.

Small size

Insects are tiny; this fact is enormously significant in their ecology. Among the smallest known insects are beetles in the family Ptiliidae. Adults are less than 0.04 in (1 mm) in length and could crawl comfortably through the eye of a sewing needle. Adult wasps in the family Mymaridae are less than 0.02 in (half a millimeter) in length. At the other end of the insect size scale are the tropical walkingsticks, such as the Australian phasmid *Palophus* (order Phasmatodea), which can grow to 10 in (25 cm) in length. The largest of the giant silk moths (family Saturniidae) are the atlas moths, with wingspreads up to 12 in (30 cm). In terms of mass, the Goliath beetle, an African scarab (family Scarabaeidae), is one of the largest, weighing about 3.5 oz (100 g). The largest insects are diminutive, however, when compared with most mammals and other vertebrates.

Central to understanding insect ecology is recognizing that their absolute energy requirement is small, commensurate with their size. An organism that requires only a modest energy budget to be successful has an advantage over an organism that demands a large budget; there are more opportunities for insects with small energy budgets than for

An American hover fly (*Metasyrphus americanus*) pollinates a flower. (Photo by E. R. Degginger. Bruce Coleman, Inc. Reproduced by permission.)

those with large energy budgets and thus more niches for insects than for vertebrates. This consequence of their small size helps explain why approximately 80% of all known animal species are insects.

Another important consequence of small size is that insects have a relatively large ratio of surface area to volume. This means that they lose water and heat much more rapidly than would a larger animal. Many of their ecological strategies have evolved because of these specific vulnerabilities. To a large extent, insect ecology is structured around the need to obtain and conserve water, and this challenge is enhanced by their small size. In contrast to sophisticated morphological and behavioral controls on water loss, insects have adapted to heat loss by not having control of body temperature. The energy cost of maintaining a constant warm body temperature for such a small animal would strain the energy budget, so in contrast to mammals which maintain a constant body temperature, insects are poikilothermic. This means that a critical feature of their ecology is that their bodies can function at approximately the ambient environmental temperature.

The dung beetle (*Circellium bacchus*) rolls a ball of elephant dung in South Africa. (Photo by Ann & Steve Toon Wildlife Photography. Reproduced by permission.)

Another ramification of this aspect of insect ecology is that developmental rate is determined by temperature. Each insect species has a lower temperature threshold below which development will not proceed and an upper temperature threshold above which the insect dies. Between these two limits, the developmental rate increases with rising temperature. Immature insects grow faster in warm conditions. This variable development time has implications for allocation of energy and thus is an essential concept in insect ecology.

Physical forces

Because of their small size, insects live in a world dominated by forces we rarely contemplate, while at the same time they are comparatively free of some of the forces and influences that beset humans and other large vertebrates. For example, the force of gravity rules many aspects of human life, and gravity is a significant factor in determining the energetic cost of locomotion for large animals. In contrast, the tiny force exerted by the surface tension of water is relatively insignificant for humans and other large vertebrates, particularly in the context of locomotion.

The world experienced by insects is exactly the reverse: gravity is comparatively less influential than are the cohesion forces between molecules. It is partly because of this reversal that insects are able to have exoskeletons. Most insects do not have a large enough physical mass to be injured in a fall, but tiny insects can be trapped and rendered helpless by a drop of water because the cohesion forces between water molecules are stronger than their legs and wings. Therefore, cohesion forces, which barely affect human ecology, play a dominant

role in insect ecology. Molecular cohesion allows insects to walk upside down on ceilings, and if an insect happens to fall from the ceiling, the relatively smaller effect of gravity means that the insect is unlikely to get hurt. Adaptations related to these physical forces play key roles in the energy budget.

Metamorphosis

Insects are notorious for changing form as they develop from immature stages to reproductive adults. Termed "metamorphosis," this change in morphological features during development also may entail modifying habitat, food preferences, and behavior. An understanding of insect metamorphosis is fundamental to understanding their ecology, because a dramatic change of form allows insects to switch their ecological interactions from one life stage to the next. It is an evolutionary advantage to be able to make a radical shift in ecological strategies to best suit the needs and goals of each life stage.

Reproductive capacity

Insects have phenomenal powers of reproduction. Female insects routinely produce hundreds of eggs. Although abandoning one's offspring is considered the height of irresponsibility for humans, it is widespread among insect species. Insects typically produce a batch of eggs and immediately desert them, though sometimes with minimal provisions for their nourishment and safety. Nutritional provisioning may take the form of leaving the egg on an appropriate host plant selected by the mother. Or, in some odd cases, provisioning takes the form of arranging for another organism to provide care. There are walking stick insects that provide their eggs with a lipid-rich appendage called a "capitulum." This appendage attracts ants, which eat the capitulum and disperse the eggs to relatively safe locations, where they can incubate and hatch.

However, parental care among insects is uncommon. The few exceptions appear in diverse taxonomic groups, including the orders Hemiptera (family Belostomatidae), Orthoptera (family Gryllidae, subfamily Gryllotalpinae), Coleoptera, Isoptera, and Hymenoptera. The last two orders listed are of particular interest, as they include social species. Social insects represent a group placed in an extraordinary situation in which the energy budget of the ecological niche includes a large allotment for care of the young. For most species of insects, the energy budget for reproduction does not include caring for the young. Producing huge numbers of offspring, sometimes many hundreds per female, is one of the special ecological features of insects. Because of their enormous reproductive potential, insects in favorable environments can generate huge populations with stultifying rapidity. Anyone who has seen a plant colonized by aphids or a container of grain invaded by weevils or grain moths is familiar with this phenomenon.

Genetic adaptability

The genetic adaptability of insects is tied closely to their innate capacity to reproduce in great numbers and also is a function of their typically short life spans. This concept requires an understanding of Darwin's principle of "survival of the fittest," one of the most profoundly misunderstood ideas in biology. It frequently is interpreted as a struggle between individuals, with those that are stronger, faster, and more aggressive surviving. In the Darwinian sense, however, "fitness" means number of offspring. This concept relates the definition of fitness closely to the definition of an ecological niche.

An individual that produces many offspring has a high fitness rating, whereas one with no offspring has a fitness of zero. Strength, speed, and aggression are relevant only if they contribute to an individual's ability to reproduce. Nature essentially provides a selective breeding program in which the natural hardships of the environment and the successful strategies of the ecological niche act instead of a sentient breeder. This process of differential survival and reproduction is called "natural selection."

From one generation of insects to the next, changes in the genetic make-up of the population are wrought by the processes of natural selection and take place with relative speed. Think of it this way: in less time than it takes for a newborn human to grow old enough to learn to walk and talk, a mosquito species that transmits malaria can go through more than 40 generations. Under favorable conditions, each female mosquito is capable of producing 100 to 150 eggs, and each individual could complete its life cycle in about two weeks.

With the potential to produce more than 100 eggs per female and more than a dozen generations per year, mosquito populations can build up with staggering speed. Each generation is subjected to natural weeding out of the unfit individuals, such as those susceptible to the poisons that humans use to control their populations. Many mosquitoes succumb to insecticides and may die without ever reproducing; their genes simply vanish from the population. Those individuals able to detoxify insecticides, however, will survive and ultimately generate a population of insects resistant to the insecticides that have been used against them. These genetic changes take place over many generations of mosquitoes, but dozens of mosquito generations go by in considerably less time than the span of one human lifetime. Therefore, from the human perspective, insecticide resistance evolves rapidly.

Flight

Remarkably, few groups of animals have evolved wings and the capacity for flight, despite several obvious advantages of flying. Like fitness, flight is connected closely to the energy budget and therefore to the ecological niche concept. Flight is useful for escape, foraging, and dispersal. Although flight has special advantages, it also comes with disadvantages. Larger animals have to contend with the relationship between weight and stalling speed; they have to accelerate more quickly and fly faster to take off and remain aloft. A mosquito can stay aloft at a speed of 0.2 mph (0.3 kph). The largest insects need achieve only about 15 mph (24 kph), whereas many small birds have to reach a minimum speed of about 20 mph (32 kph) to stay in the air. This increases the demands on the energy budget, making unassisted flight too expensive for most larger animals.

Insects have a special advantage: their particular combination of small size and relatively low energy investment per individual offspring means that they can easily afford both flight

and reproduction without straining their energy budget. For example, some beetles complete their entire larval development while feeding on one seed. In contrast, a single seed probably would not supply the energy expense of takeoff for a large bird.

There is another fuel-related consideration: because insects are poikilothermic, they do not expend large amounts of energy trying to keep warm in a cold place. This has a double advantage in terms of the power of flight. First, a poikilothermic lifestyle allots less energy to staying warm, freeing up part of the energy budget for other investments, such as flight. Second, the general trend toward cooler temperatures higher in the atmosphere is less of a concern for insects. Instead of needing additional energy to keep warm while flying high, insects can simply let their body temperature drop and thereby conserve energy.

Insects often exploit their small size and simply allow themselves to be caught in wind drafts, reducing the cost of flight almost to zero by permitting themselves to become the aerial equivalent of plankton. It is common to find tiny insects being swept along high in the air. This efficient form of dispersal spreads insects from one habitat to another with relatively little energy expenditure.

Classic themes in ecology

Most ecological strategies can be reduced to two basic goals: the acquisition of energy from the environment and the balanced allocation of energy for all life processes. Within each of these two goal areas, there are broad patterns common to all living organisms. These patterns are classic themes in ecology, often studied using insects as paradigms. Insects also have special ecological features not common among other living organisms; those aspects are reviewed under "Special features of insect ecology." Classic themes in ecology include competition, feeding ecology, reproductive strategies, succession, biodiversity, predator-prey interactions, and parasite-host interactions, which are discussed in the following sections. Mutualism, another classic ecological pattern, is considered under "Ant-plant interactions."

Competition

The concept of a species-specific energy budget is an organizing principle in ecology. Sometimes the strategies involved in acquiring and allocating energy are modified as the result of overlap of resource exploitation by one or more species; this is specifically true in the case of a limited resource. When one or more individuals suffer negative effects on fitness from niche overlap and subsequent niche modification, the situation is referred to as "competition." There cannot be competition if there is no niche overlap or if the shared resource is unlimited. Under the defining conditions, however, competition has the potential to have a negative impact on the energy budget of the competing organisms. Competition takes two broad forms: intraspecific competition, or competition among members of the same species, and interspecific competition, or competition among individuals belonging to different species. The energy budget of any living organism begins with food; therefore, an obvious arena for competition is communal exploitation of a limited source of food energy. Other potentially limiting resources include water, mates, and habitat space.

Intraspecific competition can include overlap in the use of both energy resources and mates. Competition for mates often is an evolutionary force driving behavioral characteristics, such as territoriality among dragonflies. Generally, intraspecific competition affects members of one life stage (often adults), but it can embrace competition between insects at different life stages. It is worth noting that metamorphosis has the potential to minimize or eliminate competition between life stages for those insect species that utilize different niches at distinct developmental stages.

Interspecific competition can be an evolutionary force driving niche specialization or speciation. As competition increases, niche specialization also may expand, resulting in fine-tuned partitioning of resources. A classic example is the variety of different feeding guilds among the many insect species found on a single tree: gall formers, leaf miners, leaf eaters, and root feeders all can share the same plant, each specializing in a particular mode of obtaining energy.

Feeding ecology

Insects can be divided into different groups based on their feeding habits, and this categorization can be accomplished via two distinct but overlapping systems. One system classifies insects on the basis of the kinds of food they eat: carnivores, herbivores, or omnivores. A second system categorizes insects as either generalists or specialists, depending on the degree of specialization of their feeding habits. Each of these categorization systems provides insight into ecological interactions, but no system presently in use aligns perfectly with current taxonomy, so there is no easy way to integrate these categories with phylogenetic groupings.

Herbivores, for example, the larvae of moths and butterflies (order Lepidoptera), feed almost exclusively on plants. A special consideration related to feeding on plants is the tendency of plants to produce defensive chemical compounds intended to deter feeding or even to poison aspiring herbivores. Insects can evolve their own chemicals to surmount these defenses, usually in the form of detoxifying enzymes, although they may store toxins in their bodies without detoxification and even recruit these chemicals for their own defense. Insects also can evolve types of behavior to avoid exposure to plant toxins. An example of this is trenching, in which insects chew through plant veins, allow sap to drain out, and then consume the plant material that thereby has been deprived of toxin-containing sap.

Another special consideration of feeding on plants is that, with the exception of seeds, most plant material is relatively rich in carbohydrates and low in proteins. To be successful, plant-feeding insects need to evolve mechanisms for ensuring adequate nitrogen intake. These mechanisms must fit into the ecological niche of the species without resulting in energy-budget imbalances.

Carnivorous insects face a reverse problem: their animal prey tends to be replete with protein, so nitrogen is not a limiting resource, but an imbalance between nitrogen and other nutrients can cause problems with internal osmotic balance and also with the energy requirements of processing amino acids. Some amino acids, such as arginine and histidine, are rich in nitrogen atoms. These amino acids require such a large energetic investment during digestion that it is more efficient to discard the excess than to attempt to break these amino acids down chemically. For example, blood-feeding tsetse flies (family Glossinidae) discard the amino acids arginine and histidine from their food because the cost of metabolizing these molecules is greater than the energy yield from digestion.

Omnivorous insects have the advantage of the potential to balance their nutritional requirements, particularly nitrogen and sugars, but have a special disadvantage: they may need to maintain a larger array of digestive and detoxifying enzymes, and this makes large demands on the energy budget. This issue leads to the second major system for categorizing feeding habits: generalists versus specialists. In the context of the energy budget, specialists have restrictive feeding habits and thus can be more efficient in their production and use of digestive and detoxifying enzymes. The tradeoff is that specialist feeders are limited in their repertoire of possible food sources and may expend too much energy searching for appropriate food. Generalists, in contrast, have the advantage of a much larger repertoire of food resources, but there is an associated expense: these insects may have to invest more energy in digestive and detoxifying enzymes. Some insects employ inducible enzymes. They have the genetic capabilities to synthesize appropriate enzymes but do so only when induced by exposure to relevant chemicals in their food. This is a compromise that allows exploitation of a wide variety of food resources while retaining an energy-efficient strategy for digestion and detoxification.

Reproductive strategies

A key part of the definition of a niche is the need to produce the maximum number of offspring while maintaining a balanced energy budget. Only those offspring who survive to produce offspring of their own, however, can count as successful reproductive efforts. This caveat has implications for reproductive strategies. It is necessary but not sufficient to produce offspring. They have to survive and successfully reproduce.

Reproductive strategies fall into two broad patterns. The reproductive strategy of producing many offspring with a relatively tiny investment in each is referred to as an r strategy, with the r representing the intrinsic rate of increase of a population. This recognizes that population increase has the potential to follow an exponential curve when population growth is unchecked.

In contrast, the reproductive strategy of producing only a few offspring but putting a large investment into each is referred to as a K strategy. The K represents the carrying capacity of the environment for that species and implies a limiting factor suppressing population growth. This recognizes the limits of the energy budget that defines the niche of a particular species. Very specifically, in the case of a species that produces few offspring, the designation K reflects the heavy emphasis on reproductive costs in the total energy budget. These organisms consign large amounts of energy into producing few offspring and enhancing their survival. Thus, K

species tend to have relatively low mortality rates early in life. In contrast, *r* species allot only small amounts of energy to producing and nurturing individual offspring, and this is reflected in relatively high mortality rates early in life.

Insects characteristically use an *r* strategy. For most insects, sheer numbers take precedence over parental care: if enough offspring are produced, some are almost certain to survive. At one extreme, insects such as giant silk moths (family Saturniidae) may produce as many as 300 eggs, each with minimal provisioning. Most of these offspring die as eggs or larvae. Very few survive to become reproductively mature. At the other extreme, insects such as tsetse flies and bat flies (family Streblidae) produce only a few young but invest energy in parental care, thereby ensuring that a high percentage of the offspring survive to maturity.

There are rare and special cases in which certain insects can reproduce as larvae. This phenomenon is called paedogenesis and has an ecological correlation. Species exhibiting paedogenesis usually live in transient and somewhat isolated habitats with limited resources as well as limited time to exploit those resources. The ability to reproduce as larvae allows these insects to produce a maximum number of offspring rapidly before the source of their energy budget is used up. This unusual phenomenon is known to occur in only a few kinds of insects and some mites (class Arachnida).

Succession

Ecologists have long observed that the kinds of organisms found in a particular ecosystem changes over time. This process of change tends to be predictable and directional and is called succession. Although sequences of succession vary among different habitats, the broad features include initial rapid colonization by species that use predominantly *r* reproductive strategies, followed by the slower establishment of species that use predominantly *K* reproductive strategies. Ultimately, this leads to the establishment of an ecological system known as a "climax community." An entomological example of succession is the directional and predictable series of insects that arrive at, colonize, and exploit a large animal carcass, such as a dead deer in the woods. Another example is the change of component species of the insect community on a plant as the plant ages.

Early stages of succession are characterized by large numbers of a few species, so the emphasis is on colonization with low biological species diversity. Some of the initial colonists depart during the process of succession and are not found at later stages. The climax community, however, often includes a mix of *r* and *K* species, with the *r* species exploiting resources that tend to be transient or patchy, while the *K* species invest in permanent structures that allow them to exploit more constant resources. The hallmark of a climax community is its species richness, known as biodiversity.

Biodiversity

The measure of richness of different species of living organisms per unit of geographic area is summarized as biodiversity. Usually, biodiversity is correlated directly with the complexity of the habitat and the geologic topography as well as climax community structure. Complex habitats with a high index of biodiversity generally are more ecologically stable

and resistant to disturbances. Less complex habitats tend to have a low index of biodiversity and typically are less ecologically stable. Monoculture croplands are prime examples of the latter type of ecosystem. Human manipulation often creates ecosystems with a low index of biodiversity, and this has serious implications both for loss of species and for the spread of pest species. Encouragement of biodiversity and preservation of endangered species revolves around preservation of their required habitats with enough geographic area to allow for the retention of ecosystem complexity.

Predator-prey interactions

The consumption of one animal by another is a classic theme in ecology and derives much of its importance from the concept of the energy budget. When an animal is killed and consumed, it is referred to as predation. Predators contribute significantly to the flow of energy through an ecosystem: when they eat, they annex and mobilize the entire energy content of a living animal, and they do this many times during their life span as they consume many prey individuals. Insects can be predators, prey, or both.

Insects also fill a special subcategory of predators known as "parasitoids." Parasitoids differ from predators because they generally live inside a single prey individual and consume it from within, ultimately killing it. Unlike predators, parasitoids typically are highly specialized with regard to prey species. Both predators and parasitoids can operate in groups. A pack of predatory lions—or even ants—can eat a single prey individual, or a group of parasitoid fly larvae can eat a single caterpillar. Predators and parasitoids are critical in the regulation of population growth and therefore are important in evolutionary selection. Examples of predatory insects include various families of beetles and true bugs as well as the praying mantids familiar to organic gardeners.

Parasite-host interactions

Like predators, parasites consume living animals. A crucial difference is that parasites do not usually kill the "prey" outright, although they may debilitate a host seriously enough to contribute to its death. Parasites live either on or inside their hosts and cause generalized weakening by draining the host of nutrients. In general, a parasite is a more specialized feeder and has a smaller energy budget than a predator, and this is significant in terms of numbers of species. There are relatively more ecological niches available for parasites than for predators. Many insect species have adopted a parasitic lifestyle. The combination of small size, modest energy requirement, and propensity for rapid speciation predisposes insects to this ecological strategy. Examples of parasitic insects include flies, fleas, and lice.

Special features of insect ecology

The unique suite of behavioral, morphological, and physiological features possessed by insects results in highly specialized ecological phenomena. Special features of insect ecology include marine insects, plant-insect interactions, sociality, and diapause. In addition, the special application of insect ecology to the human issue of pest control is discussed in the context of integrated pest management (IPM).

A dead leaf mantid (*Deroplatys* sp.) showing effective camouflage as a defense against predators. (Photo by Bob Jensen. Bruce Coleman, Inc. Reproduced by permission.)

Marine insects

One of the few ecosystems in which insects are notably scarce is the marine ecosystem, although there are many freshwater aquatic species of insects. A brief review of the unique characteristics of insects, analyzed in light of the need for a balanced energy budget, provides a ready explanation for this dearth of marine insect species.

Marine habitats have salty water. For insects, with their small size and high surface-to-volume ratio, this translates into an enormous risk of severe dehydration: the salt in this environment pulls fluid out of their bodies. Moreover, salts can invade the insects' bodies and upset their internal osmotic balance. Some insects use energetically expensive physiological mechanisms, actively excreting excess salt from their bodies to reduce dehydration in marine habitats. These adaptations, however, drain the energy budget of insects that live in saltwater. Although a few species have found a way to maintain a balanced budget despite this expensive lifestyle, marine insects are quite rare. The small size that gives insects an advantage in most other habitats is a bane in the sea.

Plant-insect interactions

Interactions between insects and plants can be positive or negative. Negative interactions occur when insects eat plants; this is called herbivory. More than half of all known insect species feed on plants, with most of these species coming from eight major orders. For example, approximately 99% of all known species of moths and butterflies (order Lepidoptera) are herbivorous. Plants often evolve toxic chemicals, which are believed to serve primarily as either feeding deterrents or outright poisons intended to reduce herbivory. Insects, in turn, evolve ways to detoxify or avoid ingesting the toxins. When this happens, the negative interaction between insects and plants is reciprocal.

A landmark paper by Paul Ehrlich and Peter Raven, published in 1964, speculated that these reciprocal negative interactions between insect herbivory and the evolution of plant toxins constituted a special subcategory of evolution. Using a term first coined by C. J. Mode in 1958, they referred to this special form of evolution as coevolution. Their theory states that herbivory by insects is an evolutionary pressure that selects for plants with toxins, and plant toxins, in turn, provide evolutionary selection for varieties of insects capable of detoxifying the plant substances. This process is believed to result in evolution of new species of plants and insects. The theory of coevolution largely explains the great diversity of flowering plants (angiosperms) and insects found on Earth.

Coevolutionary interactions between insects and plants also can be positive. Pollination is an obvious example of a positive interaction and includes cases of coevolution between plants and their insect pollinators. Plants produce nectar specifically to attract insects, because often when an insect lands on a flower to partake of nectar, it either inadvertently or deliberately collects pollen from the flower. Plant fertility is tied closely to ecological niches of insect pollinators.

Insect pollination is unquestionably the most efficient pollination system other than self-fertilization. Insects benefit by obtaining nectar and pollen as food, and plants benefit when pollen is transferred efficiently between flowers. This positive ecological interaction can become reciprocal and can direct the processes of evolutionary change. Coevolution could produce new species that would not have evolved without these tight associations of reciprocal positive interactions. It seems likely that without insects, there would be fewer species of plants. In particular, there would be only a fraction of the number of species of flowering plants currently found in the wild. Not all pollination systems, however, are examples of coevolution in the true sense of long-term reciprocal evolutionary selection.

The efficiency of insect pollination is related directly to insects' unique suite of characteristics and their energy budget. Pollen is not indefinitely viable; it dies, sometimes within hours, once it leaves the parent plant. This puts a premium on pollination systems that emphasize speed. Utilizing a living pollination agent with the power of directed flight is like sending pollen air mail rather than surface mail: it goes faster but costs more. Because insects can easily afford flight in their energy budget, they require only a small allotment of nectar and are the cheapest and most efficient available living pollen carriers. Plants that use insect pollinators expend less energy producing nectar and therefore have an easier time balancing their energy budgets while producing offspring. Many species of plants rely on the efficiency of insect pollination, and the diversity of plants on Earth would be reduced drastically if insects were eliminated.

Mutualism and ant-plant interactions

Ecologists define a variety of interspecific interactions, including the broad categories of competition, predation, parasitism, and mutualism. Mutualism is a special case of symbiosis in which individuals from different species participate in an association that is beneficial to both. Insects provide some of the prime examples of ecological mutualism. Prime among these are ant-plant associations, which are examples of tight coevolution as well as mutually beneficial symbiotic relationships.

One classic example of ant-plant interactions comes from the associations of plants in the genus *Acacia* and ants in the genus *Pseudomyrmex*. An estimated 10% of Central American *Acacia* species recruit ants as their bodyguards. These plants have enlarged, modified thorns in which the ants make their homes. The *Acacia* plants provide the ants with nectar and protein, the latter produced in special structures that grow at the tips of new leaves. In exchange for room and board, ants aggressively protect *Acacia* plants. They fend off herbivores

that might otherwise eat the *Acacia* plants, and they also chew off the growing tips of nearby plants, thus suppressing the growth of potential competitors of their hosts. For both *Acacia* plants and ants, the investment of energy in maintaining their association yields a sufficiently high return to justify this expense in the energy budget. A similarly favorable return on the energetic investment is the evolutionary force driving other types of mutualism, and other insect examples can be found among fig wasps, ants that cultivate aphids for honeydew, termites and their intestinal microorganisms, among many other associations of species.

Social insects

Division of labor is a strategy that increases efficiency of energy use. Insects that live solitary lives do not have the luxury of specializing in one or just a few tasks, but some insects have evolved social systems that feature division of labor. Other key characteristics of insect sociality are overlap of generations, cooperation in care of young, and intraspecific communication. Overlap of generations is found in many insect species, and both sexual signaling and aggregation chemical signals are common between males and females of the same species. Only social insects, however, have a combination of generation overlap, communication of information, and communal brood care. Social insects are in the orders Hymenoptera and Isoptera.

The success of sociality as an ecological strategy can be measured by the number of individuals produced by the system: approximately 75% of insect biomass on Earth comes from social insects—ants and termites. This success is largely due to the extreme efficiency of cooperative care of young. The investment of care increases the proportion of young that survive, and this investment is made possible by the increased efficiency that comes with division of labor. Social insects have individual colony members specialized for reproduction, defense, foraging, brood care, and even housekeeping. Sometimes this job specialization depends on age, so that younger adults have one set of chores and older adults have a different set of chores. For example, the common honey bee (*Apis mellifera*) worker spends time cleaning the hive and maintaining the comb or caring for brood or foraging at different stages of adult life.

Other species of insects, notably ants and termites, have morphologically distinct castes. Each caste has its own job and a set of morphological characteristics to match; soldiers have large jaws and are specialized for protecting the colony; some kinds of ants are specialized for storing liquid food in their bodies. These ants, called honey-pot ants, have enormously swollen abdomens and, when given the correct communication, will regurgitate an edible droplet of sweet liquid for their nest mates. Some ants domesticate other insects, such as aphids or scale insects, and are rewarded with sweet liquid in exchange for protection. This herding behavior is unique to ants.

Diapause

Insects, like other living organisms, must evolve strategies for adapting to climatic challenges. Some insects adapt to specific and predictable climatic conditions, such as cold seasons or dry seasons, with a strategy called diapause. Insect diapause is a genetically hardwired set of physiological changes and adaptive behaviors that serve to synchronize the life cycle of an insect species with predictable seasonal changes in its environment. Diapause is characterized by reduced metabolic activity and suppressed reproductive behavior, but it varies from one insect species to another and is not simply an insect version of hibernation. The unifying feature of this genetic program occurs during a species-specific stage of the insect's life cycle. Once it has begun, it proceeds according to a programmed plan and continues until it has run its course; it cannot be interrupted or shut off.

Diapause is initiated by triggers called token stimuli. These stimuli usually are environmental cues that herald a seasonal change that presents an ecological challenge. For example, winter is invariably preceded by a period of about three months during which each day is slightly shorter than the day before. In temperate and Arctic zones, winter is a period of harsh ecological challenge. Temperatures drop sharply, food becomes scarce, and both the biotic and abiotic environments undergo dramatic alterations. Insects that experience winter diapause detect shortening day length and use it as a cue to begin initiating physiological and behavioral changes in preparation for diapause. The physiological control and the visible manifestations of diapause, however, are as complex and diverse as the many species, life cycles, and environments of insects. The ability to undergo diapause confers on an individual insect a higher probability of surviving predictable adverse environmental challenges and thereby mediates evolutionary adaptation of that species.

The role of ecological information in pest control

There is no question that insects rank high on the list of animal pests, but the definition of what constitutes a pest species is entwined inextricably with ecology. A butterfly might be a desired visitor in a flower garden, but if its larva feeds on a crop plant is it actually a pest? A beetle that helps recycle dead trees in a forest, speeding degradation of dead wood and thereby returning nutrients to the soil, probably would not be considered a pest. The same species of beetle degrading the dead wood of a fence or a building definitely would be thought of as a pest. A bee pollinating a fruit crop is not a pest; the same species of bee nesting in a school playground would be unwelcome. Thus the definition of a pest is based solely on context and human values. Pest control is, in

essence, an ecological problem. It is built on guiding principles that are grounded in an understanding of the ecology of the pest: its natural history and its interactions with both the biotic and abiotic features of the environment.

All the governing principles of pest control can be reduced to a single concept: pests are controlled by making the habitat so hostile that it is incompatible with the maintenance of a balanced energy budget. While this concise statement is an excellent distillation, it leaves much to be desired in terms of providing guidance on how control is to be accomplished. If the guiding principle is to alter the environment to make it totally unsuitable for the insect pest, then the corollary is to manage this without also rendering the environment unsuitable for those living organisms that are considered desirable. Thus, the basic issue in pest control is how to manipulate many environmental interactions in such a way as to create a hostile habitat for the pest while still allowing desirable species to flourish. This technique is called integrated pest management (IPM). It uses a variety of control methods (chemical, mechanical, cultural, and biological) that target a pest species' ecological interactions. The success of any IPM program is dependent upon a thorough knowledge of the pests' life history and ecology.

For example, a farmer practicing IPM could use information about the lower threshold temperature required for development of an insect pest species to predict when that pest will become active in the spring. A plant variety that germinates early in the season can mature and be past its vulnerable stage before pests are active. Introduction of pest-specific predators and parasites into the field can further deplete the pest population. Removing all plant debris at the end of the growing season can eliminate overwintering pests that have entered diapause. In all cases, effective pest control begins with acquiring basic life history information and studying the ecology of the species for which control is desired.

Summary

Insect ecology could be called the study of the energy budget of insects. The unique characteristics of insects provide a framework for understanding their interactions with both the biotic and abiotic components of their environments. Insects can maintain a balanced energy budget in most of the habitats available on Earth, but they are understandably most common in ecosystems where they can obtain large supplies of energy with relatively minimal expenditures. Their unique characteristics and special ecological adaptations play an indispensable role in almost every ecosystem on the planet.

Resources

Books

Agosta, William. *Thieves, Deceivers, and Killers: Tales of Chemistry in Nature.* Princeton, NJ: Princeton University Press, 2001.

Dent, David. *Insect Pest Management.* 2nd edition. New York: CABI Publishing, 2000.

Gillott, Cedric. *Entomology.* 2nd edition. New York: Plenum Press, 1995.

Goddard, Jerome. *Physician's Guide to Arthropods of Medical Importance.* 4th edition. Boca Raton, FL: CRC Press, 2002.

Gordth, Gordon, and David Headrick, compilers. *A Dictionary of Entomology.* New York: CABI Publishing, 2001.

Resources

Price, Peter W. *Insect Ecology.* 3rd edition. New York: John Wiley and Sons, Inc., 1997.

Romoser, William S., and John G. Stoffolano, Jr. *The Science of Entomology.* 4th edition. Boston: WCB/McGraw-Hill, 1998.

Speight, Martin R., M. D. Hunter, and A. D. Watt. *Ecology of Insects: Concepts and Applications.* Oxford: Blackwell Science, 1999.

Periodicals

Ehrlich, Paul R., and Peter H. Raven. "Butterflies and Plants: A Study in Coevolution." *Evolution* 18 (1964): 586–608.

Mode, C. J. "A Mathematical Model for the Coevolution of Obligate Parasites and Their Hosts." *Evolution* 12 (1958): 158–165.

Martha Victoria Rosett Lutz, PhD

Distribution and biogeography

Insect distribution

Insects, the most successful creatures on Earth, are found in virtually all habitats with the exception of the ocean depths. Even in the oceans, an area dominated by crustaceans among the arthropods, they are well represented in coastal and saline habitats (such as mangroves and salt marshes), and the open sea surfaces have been colonized by oceanic sea skaters (*Halobates* spp.). Hexapods are known from all continents from the arctic to the antarctic, and various regions house different numbers and varieties of insects. The warm, humid tropic zones have long been known to support the greatest biodiversity of hexapods. The mid-nineteenth and first half of the early twentieth centuries saw the golden age of exploration by naturalists to far-flung foreign lands. These trips resulted in a tremendous number of new discoveries and probably first invoked questions about observed patterns of distribution of insects. This study of the historical and ecological components resulting in the present distribution patterns of insects is called biogeography.

Biogeography

Biogeography is the biological discipline in which scientists study the geographical distribution of animals, plants, algae, fungi, and microorganisms. It describes distributional patterns of specific groups and attempts to explain how they have come about, by hypothesizing historical and/or ecological causes. In spite of the existence of two distinct subdisciplines, known as historical and ecological biogeography, it is evident that biogeographic explanations should include both historical and ecological components. Biogeographers study such issues as why a certain species is confined to its present range; what enables a species to live where it does; the roles that climate, landscape, and interactions with other organisms play in limiting the distribution of a species; why animals and plants from large, isolated regions such as Australia, New Caledonia, and Madagascar are so distinctive; why some groups of closely related species are confined to the same region, whereas others are found on opposite sides of the world; and why there are so many more species in the tropics than at temperate or arctic latitudes. Although insect biogeography is fundamental to any understanding of global distributional patterns of the biosphere, the inadequate knowledge of the distribution and phylogeny of insects and their excep-tionally high diversity have greatly impeded the progress of insect biogeography.

Historical background

Of the historical processes that have shaped global biogeographic patterns, the most important is continental drift. This theory states that Earth's crust is not composed of fixed ocean basins and continents, as supposed in the nineteenth century, but instead is a changing landscape in which continents and continental portions overlay a liquid core and thus drift across the surface of the planet. The first hexapods are known from the late Devonian period. Over the following 400 million years, several major geotectonic events occurred resulting in important biotic patterns. Following is a list of major geologic periods following the Devonian and the events that characterized each:

- Carboniferous-Permian (345 million years ago [mya]): The supercontinent of Pangea (Asia plus North America, Europe, and Gondwana) is assembled and the Panthalassa Ocean appears. There is a homogeneous land biota. The glaciers wane in Gondwana. A major biotic extinction (approximately 70% of all terrestrial species of animals) at the end of the Permian marks the end of the Paleozoic era. Several fossil orders of insects became extinct by the end of the Permian, including several orders of paleopteran insects (of which the only extant representatives are dragonflies and mayflies) that constituted half of the insect diversity at that time, such as Paleodictyoptera, Permotemisthida, Megasecoptera and Diaphanopterodea.

- Triassic (225 mya): Pangea begins to break up after 160 million years of stability. A relatively modern insect fauna is found from the Triassic onwards.

- Jurassic (190 mya): Pangea breaks up into a northern continent, Laurasia, and a southern continent, Gondwana. Later, Gondwana begins to break up into smaller continents (Africa, South America, Australia, and India-Madagascar), which drift. There are global transgressions of shallow seas.

- Cretaceous (136 mya): India and Madagascar as well as Australia drift from Antarctica, thus completing the fragmentation of Gondwana. Laurasia is still joined but is divided by epicontinental seas. The Tethys circum-equatorial seaway opens. This period marks the beginnings of the boreotropical flora in Laurasia. A major extinction event resulting in the disappearance of 50–70% of all biodiversity (including dinosaurs and ammonites) occurs, marking the end of the Mesozoic era.

- Paleocene (65 mya): Epicontinental seas recede. Laurasia is reunited as a circumpolar landmass. The southern continents are widely separated.

- Eocene-Oligocene (54 mya): Eurasia and North America separate through the Atlantic Ocean but continue to be united by the Bering land bridge. Africa closes with Australasia and India collides with Asia, beginning the rise of the Himalayas.

- Miocene (25 mya): Africa closes with Europe, forming the Alps. Australia moves closer to southeastern Asia, and South America to North America. The shallow sea separating Europe and Asia dries up. The first extensive grasslands provide habitats for the evolution of several plant-eating animals. There are biotic exchanges between Europe, North America, Asia, and Africa.

- Pliocene (10 mya): Formation of the Central American land bridge uniting North and South America. Australia approaches its present position. There is a major biotic exchange between North and South America. A major immigration of South American insects towards North America takes place, reaching 49° N in western North America and lower latitudes eastward, and progressively retracts afterwards, with relictual forms being found today in southern parts of the western and eastern United States. This explains, for example, the distribution of several Neotropical insects related to South American elements as far north as Alberta, New York, and Michigan (e.g., *Triatoma* [Heteroptera: Reduviidae], several species of Scarabaeidae and Cerambycidae [Coleoptera]).

- Pleistocene (2 mya): Continental glaciers develop in the Arctic regions worldwide, which advance and recede in cycles. The Bering land bridge remains open much of the time, allowing dispersion of boreal groups between Eurasia and North America. There is a severe extinction of many large mammals in North America and Europe.

Biogeography in the nineteenth and twentieth centuries

As data on the distribution of insect and other groups accumulated over the nineteenth and twentieth centuries, it became evident that global generalizations about them could be made. In 1876 in his work *The Geographical Distribution of Animals*, Alfred Russel Wallace developed many basic concepts that still influence biogeographers today. For example, Wallace noted that distance by itself does not determine the degree of biogeographic affinity between two regions, because widely separated areas may share many similar organisms at the generic or familial level, whereas close areas may show biotas with marked taxonomic differences. He also confirmed that climate has a strong effect on the taxonomic similarity between two regions, but the relationship is not always linear. For Wallace, prerequisites for determining biogeographic patterns were a detailed knowledge of distributions of organisms throughout the world, a natural classification of organisms, acceptance of the theory of evolution, and detailed knowledge of the ocean floor and stratigraphy to reconstruct past geological connections between land masses. He stated that competition, predation, and other biotic factors play determining roles in the distribution, dispersal, and extinction of animals and plants; that discontinuous ranges may come about through extinction in intermediate areas or through the patchiness of habitats; and that speciation may occur through geographic isolation of populations that subsequently become adapted to local climate and habitat.

Wallace divided the world into six biogeographic realms or regions: Nearctic (North America), Neotropical (South America, Central America, and southern Mexico), Ethiopian (Africa south of the Sahara), Oriental (southern Asia), Palearctic (Eurasia and northern Africa), and Australian (Australia and New Zealand). Wallace used dispersal of species that arise on small centers of origin as the basic process leading to current distributional patterns. This dispersal explanation assumes that the major features of Earth were stable during the evolution of recent life. During what has been named the Wallacean period, lasting until the 1960s, many authors worked under Wallace's framework.

The modern period of biogeography, which began about 1960, is characterized by a rejection of dispersal as an *a priori* explanation. This was prompted by Leon Croizat's ideas of biogeography, as well as the development of phylogenetic systematics, also known as cladistics (constructions of phylogenetic trees or hypotheses of genealogy based on shared derived character states), and modern continental drift theory. Based on his metaphors "space, time, form: the biological synthesis" and "life and earth evolve together"—which imply the coevolution of geographic barriers and biotas, known as vicariance—Croizat developed a new biogeographic methodology, which he named panbiogeography. It consists basically of plotting distributions of organisms on maps and connecting disjunct distribution areas or collection localities together with lines called tracks. Croizat found that individual tracks for unrelated groups of organisms were highly repetitive, and he considered the resulting summary lines as generalized or standard tracks, which indicated the preexistence of ancestral biotas, subsequently fragmented by tectonic and/or climatic changes. Some authors have considered Croizat as one of the most original thinkers of modern comparative biology and believe that his contributions advanced the foundations of a new synthesis between earth and life sciences. Furthermore, following its synthesis with phylogenetic systematics, Croizat's panbiogeography has emerged as being central to vicariance or cladistic biogeography.

Cladistic biogeography assumes that the agreement among phylogenetic trees for different groups of organisms that inhabit the same areas may indicate the sequence of vicariant events that fragmented the studied areas. A cladistic biogeographic analysis comprises three basic steps: (1) construction of several area trees by replacing the terminal groups of organisms in phylogenetic trees by the area(s) where they occur; (2) examination of these area trees, looking for disagreements due to widespread groups, redundant distributions, and missing areas; and (3) construction of general area tree(s) representing the most parsimonious (i.e., shortest) solution based on all the groups analyzed. For example, four phylogenetic lineages of related mayflies from the families Siphlonuridae and Oligoneuriidae have genera restricted to New Zealand, Australia, and South America. When the genera in the phylogenetic tree for these four lineages are replaced by the areas where they occur, four area trees are obtained. These four trees are in agreement in this particular case and indicate that the South American and Australian land masses share a more recent history than either do with New Zealand (the mayflies that inhabit Australia and South America are more closely related with each other than with the mayflies inhabiting New Zealand). If the area tree is compared with theories of Earth history, a possible explanation of the processes that shaped the current distribution of these mayflies can be drawn. New Zealand, Australia, and South America were joined in the supercontinent Gondwana, from which New Zealand broke off first, South America and Australia remaining in contact through Antarctica for a longer period. Thus the original widespread ancestral group of these mayflies was first separated into a New Zealand and a South American/Australian group, and the latter was divided when South America and Australia drifted apart.

Both panbiogeography and cladistic biogeography have challenged traditional biogeographic systems, showing that some of the units recognized in them by Wallace and subsequent authors, such as the Neotropical or Ethiopian regions, do not represent natural units, because parts of them show relationships with different areas. A new biogeographic system was proposed in 2002. It consists of three kingdoms (Holarctic, Holotropical, and Austral) and 12 regions.

Holarctic kingdom

The Holarctic kingdom comprises the northern temperate areas: Europe, Asia north of the Himalayan mountains, Africa north of the Sahara, North America (excluding southern Florida and central Mexico), and Greenland. From a paleogeographic viewpoint, it corresponds basically to the paleocontinent of Laurasia. The Holarctic kingdom comprises two regions. The Nearctic region corresponds to the New World—Canada, most of the United States, and northern Mexico. The Palearctic region corresponds to the Old World—most of Eurasia and Africa north of the Sahara.

Several insect groups characterize the Holarctic kingdom, among them the family Siphlonuridae (order Ephemeroptera); family Cordulegastridae (order Odonata); families Capniidae, Leuctridae, Nemouridae, Taeniopterygidae, Scopuridae, and Chloroperlidae (order Plecoptera); order Grylloblatodea; families Adelgidae, Phylloxoridae, and Aepophilidae, several

This male forest-dwelling damselfly (*Drepanosticta anscephala*) from Thailand is representative of a family confined to the Holotropical region. (Photo by Rosser W. Garrison. Reproduced by permission.)

genera of Saldidae, and families Gerridae, Miridae, and Aphidae (order Hemiptera); order Raphidioptera; genus *Sialis* (Megaloptera); families Amphizoidae, Spheritidae, Eulichadidae, Leiodidae [subfamilies Anistominae, Catopinae, Coloninae, Cholevinae, Leptininae and Leptoderinae], Carabidae [tribes Nebriini, Opisthiini, Pelophilini, Elaphrini and Zabrini] (order Coleoptera); family Ptychopteriidae, tribe Blepharicerini, most of the family Piophilidae, and family Opomyzidae (order Diptera); most genera of the family Limnephilidae (order Trichoptera); family Eriocraniidae, genera *Speyeria*, *Euphydryas*, and *Colias* (order Lepidoptera); and superfamily Cephoidea, superfamily Pamphiloidea, and families Roproniidae and Renyxidae (order Hymenoptera).

Holotropical kingdom

The Holotropical kingdom comprises basically the tropical areas of the world, between 30° south latitude and 30° north latitude. The Holotropical kingdom corresponds to the eastern portion of the Gondwana paleocontinent. Some experts also include the northwestern portion of Australia in the Holotropical region. The Holotropical kingdom comprises four regions. The Neotropical region corresponds to the tropics of the New World: most of South America, Central America, southern and central Mexico, the West Indies, and southern Florida. The Ethiopian or Afrotropical region comprises central Africa, the Arabian Peninsula, Madagascar, and the West Indian Ocean islands. The Oriental region comprises India, Burma, Malaysia, Indonesia, the Philippines, and the Pacific islands. The Australotropical region corresponds to northwestern Australia.

In contrast with the Holarctic and Austral kingdoms, there are not many endemic insect groups characterizing the Holotropical kingdom. Among the endemic groups are the family Platystictidae (order Odonata); family Termitidae (Isoptera); family Aphrophoridae (Hemiptera); tribe Morionini and subfamily Perigoninae (order Coleoptera: family Carabidae); and family Uzelothripidae (Thysanoptera).

Austral kingdom

The Austral kingdom comprises the southern temperate areas in South America, South Africa, Australia, New Zealand, New Guinea, New Caledonia, and Antarctica, and it corresponds to the western portion of the paleocontinent of Gondwana. The Austral kingdom includes six regions. The Andean region comprises southern South America below 30° south latitude, extending through the Andean highlands north of this latitude, to the Puna and North Andean Paramo. The Antarctic region includes Antarctica. The Cape or Afrotemperate region corresponds to South Africa. The Neoguinean region includes New Guinea and New Caledonia. The Australotemperate region includes southeastern Australia. The Neozelandic region includes New Zealand.

There are several insect groups distributed in the Austral kingdom, such as the families Oniscigastridae, Nesameletidae, Rallidontidae, and Ameletopsidae (order Ephemeroptera); family Austropetaliidae (order Odonata); family Austroperliidae (order Plecoptera); family Cylindrachetidae (order Orthoptera); family Australembiidae (order Embioptera); families Peloridiidae, Myerslopiidae, and Tettigarctidae (order Hemiptera); family Nymphidae (order Neuroptera); family Belidae (order Coleoptera), tribe Phrynixini (order Coleoptera: family Curculionidae), tribes Migadopini and Zolini (order Coleoptera: family Carabidae); genus *Austrosimulium* (order Diptera: family Simuliidae), genus *Austroclaudius* (order Diptera: family Chironomidae); family Nannochoristidae (order Mecoptera); families Agathiphagidae and Heterobathmiidae (order Lepidoptera); and families Austrocynipidae, Austroniidae, Maamingidae, Monomachiidae, and Peradeniidae (order Hymenoptera).

Several studies on insect distribution support these new divisions. Sanmartín and others recently undertook a cladistic biogeographic analysis of the Holarctic kingdom. They analyzed 57 animal groups, of which 41 were insect genera or species groups. They found a basic separation between the Nearctic and Palearctic regions due to vicariance. In one analysis of the plant bug family Miridae (order Heteroptera), researchers created a general area tree based on the phylogenetic trees of the genera *Auricillocoris, Dioclerus, Myocapsus, Mertila, Harpedona, Prodromus, Thaumastocoris,* as well as other

species groups. It shows two major biotic components, which correspond to the Laurasia and Gondwana supercontinents. Within the latter, there is a close relationship among the Indo-Pacific (tropical Africa plus the Oriental region), tropical America, and the southern temperate areas (South Africa, temperate South America, and Australia). These results highlight the composite nature of the Neotropical region as well as the closer relationship between the Oriental and Ethiopian regions than previously thought. In another analysis, this time of the Dipteran families Olbiogasteridae, Anisopodidae, and Mycetobiidae, researchers presented a general area tree that supported the basic separation between Laurasia and Gondwana, with a distinction within the latter of a circumtropical and a circumantarctic component—the Holotropical and Austral kingdoms, respectively.

Brundin was the first entomologist to document clearly the relationships among the Austral continents, in his phylogenetic analysis of some groups of the family Chironomidae (order Diptera) from New Zealand, Australia, Patagonia, and South Africa. Edmunds corroborated these connections, based on phylogenetic evidence from mayflies (order Ephemeroptera). More recent biogeographic studies searched for congruence between distributional patterns of insects and other animal and plant groups and concluded that South America is a composite area, because southern South America is closely related to the southern temperate areas (Australia, Tasmania, New Zealand, New Guinea, and New Caledonia) that correspond to the Austral kingdom; and tropical South America is closely related to Africa and North America, for example the genus *Iridictyon* from the Guyanan shield sister group of *Phaon* from tropical Africa (order Odonata); genera of the subfamilies Rutelinae and Melolontinae (order Coleoptera). Other cladistic and panbiogeographic studies also support the hypothesis that South America is a composite area, with its southern portion closely related to the southern temperate areas and the northern portion closely related to the Old World tropics.

Dispersal

Although vicariance offers the best explanation for the distribution patterns of major biotas, dispersion is known to occur and is especially important in accounting for the distribution of organisms that inhabit islands. Isolated islands that are formed by volcanic or coral activity and are originally devoid of life are eventually colonized by plants and animals, including numerous insects that arrive by active flight, riding on birds or floating debris, or passively carried by wind currents. A well-known example is found in the Hawaiian Islands, of volcanic and recent geological origin, where it is estimated that an original pool of 350–400 insect colonizations accounts for the current insect fauna of the islands, comprising about 10,000 described species. The large number of new species that arose from the few colonists to fill in the "empty" ecological niches of the islands is reflected in the 98% of endemic Hawaiian species. The most striking example of explosive adaptive radiation in that archipelago is seen in the more than 600 species of vinegar flies (family Drosophilidae), which diversified here more successfully than in any other part of the world.

A female snakefly (*Aguila* spp.) from California, USA, belongs to a group of insects in the Holarctic kingdom. (Photo by Rosser W. Garrison. Reproduced by permission.)

The number of species on an island depends on its size and its distance from the continent. This idea was developed by MacArthur and Wilson in the 1960s, and is known as the "equilibrium theory of island biogeography." Although it was developed for islands, this theory can be applied to the study of fragmented terrestrial habitats, such as wetlands and forest patches, where ecological and evolutionary processes are also determined largely by isolation, time, and the dispersal capabilities of the organisms that inhabit them. It is especially useful in the design of protected areas for conservation of species that become extinct in parts of their ranges. For example, Schaus's swallowtail butterfly (*Papilio aristodemus ponceanus*) is an endangered species found in Florida, threatened by hurricanes and human destruction of its habitat. Once distributed from Miami down to the Florida Keys, it is restricted today to the Upper Florida Keys. There is currently a program to recolonize the mainland, which includes protection of appropriate patches of its habitat (hardwood hammocks).

Ecological background

Ecological interactions among species sometimes make it difficult to understand a specific geographic distribution in isolation from other species. Distributional patterns of phytophagous (plant-eating) insects are greatly influenced by the geographical distribution of their host plants. Plants may be limited by their disperser or pollinator insects. One way to look at the organization of plant/animal assemblages is to regard them, together with the nonliving components of the environment, as integral parts of ecosystems. There are characteristic assemblages of plants and animals that develop under certain climatic conditions, known as biomes, that are not necessarily coincident with biogeographic units delimited under the viewpoint of historical biogeography: different biomes exist within the same region, and the same biome may occur in different regions. Ten different biomes have been characterized as follows:

Tundra

The tundra biome exists around the Arctic Circle north of the tree line, and also in smaller areas on some subantarctic islands of the Southern Hemisphere. Winter temperatures are as low as −70°F (−57°C). There is a very short growing season for the vegetation, and only cold-tolerant plants can survive. Typical plants are mosses, lichens, sedges, and dwarf trees. Insects are scarce, although black flies, mosquitoes, and midges can be abundant in the temporary ponds formed during summer.

Northern coniferous forest

The northern coniferous forest biome, or taiga, forms a belt across the whole of North America and Eurasia south of the tundra; its southern limit is less definite and grades into deciduous woodland. Winters are long and very cold, and summers are short and often very warm. Trees are mostly evergreen conifers, with needle-shaped waxy leaves. These boreal forests host a great variety of insects, especially wood-boring and bark-feeding species (e.g., bark beetles).

Temperate forest

The temperate forest biome includes four types of temperate forest: (1) mixed forest of conifers and broad-leaf deciduous trees, which was the original vegetation of much of central Europe, eastern Asia, and northeastern North America; (2) mixed forest of conifers and broad-leaf evergreens, which once covered much of the Mediterranean lands and several southern areas, such as Chile, New Zealand, Tasmania, and South Africa; (3) broad-leaf forest consisting almost entirely of deciduous trees, which formerly covered much of Eurasia and eastern North America; and (4) broad-leaf forest consisting almost entirely of evergreens. Temperate forests have warm summers and cold winters. This biome is very rich in insect species.

Tropical rainforest

The tropical rainforest biome occurs between the Tropics of Cancer and Capricorn, in areas where temperatures and light intensity are always high, and rainfall is greater than 78.5 in (200 cm) a year. It has a great diversity of trees, which form an extremely dense canopy. The crowns of the trees are covered by epiphytes (plants that use trees only as support; they are not parasites) and lianas (vines rooted in the ground but with leaves and flowers in the canopy). The tropical rainforest contains the greatest diversity of insects worldwide, with possibly 30 million species.

Temperate grassland

The temperate grassland biome is also known locally as prairie in North America, steppe in Asia, pampas in South America, and veld in South Africa. It occurs in regions where rainfall is intermediate between that of a desert and of a temperate forest, and is characterized by a fairly long dry season. The dominant plants are grasses, the most successful group of land plants, which are hosts for several insect species (e.g., stem-boring flies, root-feeding wireworms and scarabaeid grubs, leatherjackets, armyworms, grass worms, grass bugs, planthoppers, and grasshoppers).

Tropical grassland

The tropical grassland biome, also known as savanna, corresponds to a range of tropical vegetation from pure grassland to woodland with many grass species. It covers a wide belt on either side of the equator between the tropics of Cancer and Capricorn. The climate is warm, and there is a long dry season. Grass is much longer than on the temperate grassland biome, growing to 10–13 ft (3–4 m). Dominant insects are those that graze (e.g., grasshoppers, ants, and chrysomelid beetles).

Chaparral

The chaparral biome occurs where there are mild wet winters and pronounced summer droughts, known as Mediterranean climate, and in areas with less rain than in the grasslands. Vegetation consists of sclerophyllous (hardleaf) scrub of low-growing woody plants. This biome occurs in countries fringing the Mediterranean basin, northwestern Mexico, California, Australia, and central Chile. Insects are often active in winter to correspond with rainy seasons. Early spring species coincide with late winter–early spring plant growth.

Desert

The desert biome is formed in areas experiencing extreme droughts, with rainfall less than 9.8 in (25 cm) per year. There are hot deserts (such as the Sahara) with very high daytime temperatures, often over 122° F (50° C); and cold deserts (such as the Gobi desert in Mongolia) with severe winters and long periods of extreme cold. Desert insects hide under stones or in burrows and are largely adapted to these extreme habitats.

Freshwater

Freshwater biomes are less self-contained than those of the surrounding land or sea, receiving a continuous supply of nutrients from the land, and generally are less productive than either sea or land environments. They include a wide range of environments, such as small ponds and streams, vast lakes, and wide rivers. Vegetation includes floating and rooting plants, which are eaten by a great variety of insects.

Marine

Marine biomes include three principal types: (1) oceanic biome of open water, away from the influence of the shore; (2) rocky shore biome, dominated by large brown seaweeds; and (3) muddy or sandy shore biome, which constantly receives a supply of mud or sand that provide an unstable substrate for attachment. Insects are uncommon in the marine biomes, although several groups have been able to colonize the muddy or shore biome (e.g., springtails and bristletails, shore bugs, water striders and water boatmen, midges, kelp flies, tiger beetles, and rove beetles).

Human activity as a factor affecting distribution patterns

Vicariance and dispersal are not the only factors shaping the distribution of insects in the world today; human activity in recent times constitutes an important factor to be added to the equation. Man-induced changes in the environment, such as deforestation, pollution, agriculture, and construction of towns and cities, alter or destroy considerable extensions of habitat worldwide each year, leading to the local or complete (especially in tropical forests) extinction or displacement of insect species. Some known examples are the reduction of the distribution range of the brown and ringlet butterflies in Europe due to coniferous monocultures, the extinction of Tobias's caddisfly in the Rhine River due to pollution and of the Antioch dunes shield-back katydid on the coast of California due to destruction of the sand dunes where it used to live. Some insect species have adapted to urban environments and are found worldwide accompanying humans, such as German and Oriental cockroaches, cat fleas, flour and granary weevils, house flies, Formosan subterranean termites, European earwigs, and Indian house crickets, among others.

Several species from certain biogeographic regions have spread and become established in other areas of the world through commerce or where their host plants are grown for agriculture or landscape purposes (innumerable "pests" such as lerp psyllids, aphids, scale insects, grain and pantry beetles, leaf beetles, and fruit flies). In many cases, humans have then purposefully introduced other insects, predatory or parasitoids of these target pests, in order to control their populations, which also became established outside of their native range of distribution. Some insects that cause injuries or are vectors of human or cattle diseases or that feed on crops have been eradicated from certain areas of their natural range. For example, the New World primary screwworm, which produces myiasis in horses and occasionally humans, was eradicated from Central America using the sterile male technique (inundating the area with irradiated males that are sterile and that then mate with wild females, which thus produce unfertilized eggs). Likewise, the Rocky Mountain grasshopper, an important agricultural pest during the eighteenth century in the western United States, became extinct at the beginning of the nineteenth century through a combination of factors involving zealous eradication programs and habitat alteration.

Resources

Books

Brown, J. H., and M. V. Lomolino. *Biogeography*, 2nd ed. Sunderland, MA: Sinauer Associates, Inc., 1998.

Cox, C. B., and P. D. Moore. *Biogeography: An Ecological and Evolutionary Approach*. Oxford: Blackwell Science, 1998.

Craw, R. C., J. R. Grehan, and M. J. Heads. *Panbiogeography: Tracking the History of Life*. Oxford Biogeography Series 11. New York: Oxford University Press, 1998.

Craw, R. C., and R. D. M. Page. "Panbiogeography: Method and Metaphor in the New Biogeography." In *Evolutionary Processes and Metaphors*, edited by M.-W. Ho and S. W. Fox. Chichester, U.K.: John Wiley and Sons Ltd., 1988.

Croizat, L. *Panbiogeography*. Vols. 1, 2a, and 2b. Caracas, Venezuela: [n.p.], 1958.

———. *Space, Time, Form: The Biological Synthesis*. Caracas, Venezuela: [n.p.], 1964.

Resources

Hennig, W. *Insect Phylogeny*. New York: John Wiley and Sons, 1981.

Humphries, C. J. "Biogeographical Methods and the Southern Beeches (Fagaceae: *Nothofagus*)." In *Advances in Cladistics*, edited by V. A. Funk and D. R. Brooks. Proceedings of the First Meeting of the Willi Hennig Society. New York: New York Botanical Garden, 1981.

Humphries, C. J., and L. R. Parenti. *Cladistic Biogeography: Interpreting Patterns of Plant and Animal Distributions*. Oxford Biogeography Series 12. Oxford: Oxford University Press, 1999.

Kuschel, G. "Biogeography and Ecology of South America Coleoptera." In *Biogeography and Ecology in South America*, vol. 2, edited by E. J. Fittkau, J. Illies, H. Klinge, G. H. Schwabe, and H. Sioli. The Hague, Netherlands: Junk, 1969.

————. "Problems Concerning an Austral Region." In *Pacific Basin Biogeography: A Symposium*, edited by J. L. Gressitt, C. H. Lindroth, F. R. Fosberg, C. A. Fleming, and E. G. Turbott. Honolulu: Bishop Museum Press, 1964.

Morrone, J. J. *Biogeografía de América Latina y el Caribe*, vol. 3. Zaragoza, Spain: M. & T-Manuales and Tesis SEA, Sociedad Entomológica Aragonesa, 2001.

Nelson, G., and N. I. Platnick. *Systematics and Biogeography: Cladistics and Vicariance*. New York: Columbia University Press, 1981.

Patterson, C. "Methods of Paleobiogeography." In *Vicariance Biogeography: A Critique*, edited by G. Nelson and D. E. Rosen. New York: Columbia University Press, 1981.

Rapoport, E. H. "Algunos problemas biogeográficos del Nuevo Mundo con especial referencia a la región Neotropical." In *Biologie de l'Amérique Australe*, vol. 4, edited by C. Delamare Deboutteville and E. H. Rapoport. Paris: CNRS and CNICT, 1968.

Wallace, A. R. *The Geographical Distribution of Animals*. London: MacMillan and Co., 1876.

Periodicals

Almirón, A., M. Azpelicueta, J. Casciotta, and A. López Cazorla. "Ichthyogeographic Boundary Between the Brazilian and Austral Subregions in South America, Argentina." *Biogeographica* 73 (1997): 23–30.

Amorim, D. S., and S. H. S. Tozoni. "Phylogenetic and Biogeographic Analysis of the Anisopodoidea (Diptera, Bibionomorpha), with an Area Cladogram for Intercontinental Relationships." *Revista Brasileira de Entomologia* 38 (1994): 517–543.

Archbold, N. W., C. J. Pigram, N. Ratman, and S. Hakim. "Indonesian Permian Brachiopod Fauna and Gondwana-South East Asia Relationships." *Nature* 296 (1982): 556–558.

Artigas, J. N., and N. Papavero. "Studies of Mydidae (Diptera). V. Phylogenetic and Biogeographic Notes, Key to the American Genera and Illustrations of Spermathecae." *Gayana Zoología* 54 (1990): 87–116.

Audley-Charles, M. G. "Reconstruction of Eastern Gondwana." *Nature* 306 (1984): 48–50.

Brundin, L. "Transantarctic Relationships and Their Significance." *Kungliga Svenska Vetenskapsakademiens Handligar, series 4*, 11, no. 1 (1966): 1–472.

Cox, C. B. "The Biogeographic Regions Reconsidered." *Journal of Biogeography* 28 (2001): 511–523.

Crisci, J. V., M. M. Cigliano, J. J. Morrone, and S. Roig-Juñent. "Historical Biogeography of Southern South America." *Systematic Zoology* 40 (1991): 152–171.

Crisci, J. V., M. S. de la Fuente, A. A. Lanteri, J. J. Morrone, E. Ortiz Jaureguizar, R. Pascual, and J. L. Prado. "Patagonia, Gondwana Occidental (GW) y Oriental (GE), un modelo de biogeografía histórica." *Ameghiniana* 30 (1993): 104.

Croizat, L., G. Nelson, and D. E. Rosen. "Centres of Origin and Related Concepts." *Systematic Zoology* 23 (1974): 265–287.

Edmunds, G. F., Jr. "Biogeography and Evolution of Ephemeroptera." *Annual Review of Entomology* 17 (1972): 21–42.

Erwin, T. L. "Tropical Forests: Their Richness in Coleoptera and Other Arthropod Species." *Coleopterists Bulletin* 36 (1982): 74–75.

Gressitt, J. L. "Insect Biogeography." *Annual Review of Entomology* 19 (1974): 293–321.

José de Paggi, S. "Ecological and Biogeographical Remarks on the Rotifer Fauna of Argentina." *Revue d'Hydrobiologie Tropicale* 23 (1990): 297–311.

Katinas, L., J. J. Morrone, and J. V. Crisci. "Track Analysis Reveals the Composite Nature of the Andean Biota." *Australian Systematic Botany* 47 (1999): 111–130.

Lopretto, E. C., and J. J. Morrone. "Anaspidacea, Bathynellacea (Syncarida), Generalised Tracks, and the Biogeographical Relationships of South America." *Zoologica Scripta* 27 (1998): 311–318.

Morrone, J. J. "Austral Biogeography and Relict Weevil Taxa (Coleoptera: Nemonychidae, Belidae, Brentidae, and Caridae)." *Journal of Comparative Biology* 1 (1996): 123–127.

————. "Beyond Binary Oppositions." *Cladistics* 9, no. 4 (1993): 437–438.

————. "Biogeographic Regions Under Track and Cladistic Scrutiny." *Journal of Biogeography* 29 (2002): 149–152.

Morrone, J. J., and J. V. Crisci. "Historical Biogeography: Introduction to Methods." *Annual Review of Ecology and Systematics* 26 (1995): 373–401.

Sanmartín, I., H. Enghoff, and F. Ronquist. "Patterns of Animal Dispersal, Vicariance and Diversification in the Holarctic." *Biological Journal of the Linnean Society* 73 (2001): 345–390.

Schuh, R. T., and G. M. Stonedahl. "Historical Biogeography in the Indo-Pacific: A Cladistic Approach." *Cladistics* 2 (1986): 337–355.

Juan J. Morrone, PhD
Natalia von Ellenrieder, PhD

Behavior

Insect behavior can be understood as the result of the complex integrated actions of insects in response to changes to their external (e.g., light, temperature, humidity, other insects) and internal (e.g., the level of a particular hormone) environments. In broad terms, insects exhibit two basic kinds of behavior: innate and learned. Innate behavior, commonly referred to as instinct, is based on inherited properties of the nervous system, whereas learned behavior is acquired through interaction with the environment involving adaptive changes due to experience. Whether a particular behavior pattern is inherited or learned is not always easy to determine, because some inherited behaviors may be modified by experience.

Innate behavior is responsible for simple reflexes, such as extending the proboscis to feed or turning over when falling on the back; for orientation mechanisms, such as leaving an unsuitable environment (for example, water bugs flying away from a drying pond); and for appetitive behavior, such as going toward a potential prey (e.g., female mosquitoes approaching a source of carbon dioxide to get a blood meal). Learned behavior involves the acquisition and storage (memory) of environmental information and the effects that this stored information has on the behavior of the insect. An example of learned behavior is the building of cognitive maps in several wasps, bees, and ants, which use landmarks to establish specific foraging routes or to locate their nests.

Sexual behavior

Insects show diverse and remarkable modes of reproduction. Parthenogenesis, the development from an unfertilized egg, is common among many species of aphids and orthopteroid insects, such as walking sticks. In these species, the males are unknown or rare, and females maintain the population by cloning themselves. Some moths and one beetle practice gynogenesis, in which females and males copulate but the sperm is used only to activate the development of the eggs, not to fertilize them. However, most insects reproduce sexually, males producing sperm that unite with eggs developed within females. There are some unusual types of sexual reproduction found in insects. A few scale insects (coccids) are hermaphrodites, producing both sperm and eggs and fertilizing themselves. All female ants, wasps, and bees display sexual parthenogenesis, in which unfertilized gametes generate only males with half the normal number of chromosomes. Numerous insects alternate between sexual and asexual reproduction at different stages of their life cycle.

Sexual behavior involves the location of a potential mate, usually followed by courtship, mating, and oviposition (egg laying). Males exhibit a drive to secure mates, which leads to competition for access to females. They typically are more colorful than females and can have showy structures to attract the females' attention, for example, large horns on the head of the atlas beetle, *Chalcosoma caucasus*; elongated mandibles in the Chilean stag beetle, *Chaisognathus granti*; and a pair of capitate setae on the head of the Mediterranean fruit fly, *Ceratitis capitata*. Females generally have duller colors and are larger in size. This phenomenon is called sexual dimorphism. An extreme case of sexual dimorphism is found in scale insects—females lack wings and are sessile (remain immobile and attached to the substrate), with reduced legs, whereas males look like normal winged insects. Females can choose among many potential partners, and their preferences are expected to raise their genetic success and, in turn, exert pressure on males favoring traits desirable by females. This is known as sexual selection.

Insect mating systems can be classified into three basic types: polygyny, polyandry, and monogamy. Polygyny results when some males copulate with more than one female in a breeding season, polyandry is when one female mates with more than one male, and monogamy refers to the male and female's having a single partner per breeding season. The most common form of mating in insects (and in other animals too) is polygyny. This is probably due to the vast supply of sperm the male possesses for fertilizing females, whereas each female has a relatively small number of eggs. Females mate simply to acquire enough gametes to fertilize their eggs, and one mating is usually sufficient.

Location of a potential mate

Insects use various communication strategies to locate mates, including mechanical, visual, and chemical tactics. Several families of insects use acoustic signals to locate mates. Calling sounds can be generated simply by the wings as an indirect result of flight, as in mature mosquito females and other flies, or by rubbing together parts of the body—a process called stridulation. Grasshoppers, crickets, cicadas, some bark

A pair of damselflies (*Telebasis salva*) mating. Male (red) is above; female (brown) is below. (Photo by Rosser W. Garrison. Reproduced by permission.)

beetles, and water boatmen stridulate to call for mates. In *Magicicada*, the males sing in chorus, and this aggregation song is responsible for assembling both females and males. Grasshoppers rub the hind legs against a ridge in the forewing, causing the wing to vibrate. In cicadas, males have an area of thin cuticle in the abdomen, called a tymbal, underlined by several air sacs that amplify the click produced when a muscle pulls the tymbal in; the calls consist of a rapid succession of clicks. Some beetles produce mechanical signals by banging the head or abdomen against the ground to attract females, and water striders and some water beetles generate waves in the water to communicate with their potential mates.

Visual signals can be passive, as when a variety of colors are transmitted in a single distinctive message using body surfaces as signal generators, or active, as when body parts are moved in a variety of positions, thus creating a rapid sequence of signals. With the exception of bioluminescence in fireflies (beetles belonging to the families Lampyridae, Phengodidae, and Elateridae), visual signals are restricted to diurnal use. In the case of several swarming insects, such as various flies, mayflies, and caddisflies, individual males in a swarm attract females' attention by fluttering up several meters and then dropping down, reflecting the light with their wings. Firefly males usually fly around the habitat emitting flashing lights that vary from species to species with respect to the color, rate, length,

and intensity of flash pulse. Females, perched on plants or rocks, return the message. In some species, females glow continuously or respond only to continuously glowing males, whereas in others only the male produces the light signals.

Olfactory signals to locate mates are widespread among insects; several moths, some flies, bumblebees, harvester ants, boll weevils, scorpion flies, and some bark beetles, among others, use them. Males, females, or both produce attractant molecules, called sex pheromones, from specialized glands. Pheromones are released into the air and dispersed by the wind, or, as in some types of territorial species, they can be used to scent mark a plant in the territory. Antenna receptors are able to detect just a few molecules of the pheromone in the air, allowing an insect to follow the trail leading to the opposite sex even when the insect is located at considerable distance from the source.

Some plants mimic the shape and color of certain insects to attract them to their flowers and to use them as pollinators. A well-known example is that of bees and certain orchids that resemble bee females and even produce bee pheromones. Pheromones also are produced synthetically and used as lures to trap or control the reproduction of some pest insects, such as fruit flies (*Bactrocera dorsalis* and *Ceratitis capitata*), yellow jackets (*Vespula* species), and gypsy moths (*Lymantria dispar*).

Ants (*Crematogaster scutellaris*) "milking" aphids. (Photo by Bartomeu Borrell. Bruce Coleman, Inc. Reproduced by permission.)

Mating

Many of the cues that act in mate location serve as releasers of mating or courtship behavior as well, leading eventually to copulation. In all insects, fertilization of eggs takes place inside the female reproductive ducts. In primitive insects, such as springtails, there is no mating; the male deposits sperm in packets called spermatophores and scatters them on the ground. Competition among males takes the form of males eating the spermatophores of other males. The female must locate the spermatophore to fertilize her eggs. In silverfishes, the male guides the female to his spermatophore by building a net of silken threads converging on it.

During courtship, escape and attack responses are momentarily inhibited. Insects display an amazing variety of mating patterns. Some are simple and consist in the coming together and copulation of the male and female; others involve elaborate courtship patterns. For example, in some *Panorpa* scorpion flies, a male offers a nuptial gift in the form of prey food to a female and copulates with her while she eats.

In most dragonflies and damselflies, females and males mate several times. A male can contribute to a substantial percentage of the progeny if he is the first one to grab and inseminate the female. In some species a female does not mate more than once in a particular oviposition episode, so that the male that is able to grasp her and mate with her before she oviposits is the most likely to fertilize the eggs that are laid. In numerous dragonfly and damselfly species, males are territorial, guarding a suitable oviposition site from other males. It has been found that in some species the male penis is used not only for insemination but also to remove sperm deposited by previous males from the female sperm storage organ, ensuring the fertilization of the eggs with his own sperm. This is known as sperm precedence or sperm competition. There are several mechanisms that help prevent the female from copulating with other males. For example, dung flies, dragonflies, and damselflies guard or protect the female after copulation, love bugs copulate for a prolonged period of time, male honeybees detach the genitalia and leave it inserted in the female genital opening after mating, and vinegar flies of the genus *Drosophila* transfer a chemical substance that makes the female unreceptive to other mates.

Egg laying

Insects are unique in the possession of an ovipositor, a specialized organ on the female abdominal tip used to lay her eggs, which usually consists of three pairs of plates. This device allows the female to deposit eggs in safer places, in crevices or inside plant tissues or other substrates. Oviposi-

Belostoma male with eggs hatching in water. (Photo by Alan Blank. Bruce Coleman, Inc. Reproduced by permission.)

tors contain mechanosensilla and chemosensilla (sensory hairs), which provide sensory input with respect to hardness and quality of the substrate. Eggs are laid with a view to the future needs of the young. Thus, species with aquatic larvae lay their eggs in or near the water, and species in which larvae have a specialized diet, for example, a particular plant or a particular host in the case of parasitic insects, deposit the eggs in the appropriate environment (e.g., the parasitoid wasp *Entedononecremnus krauteri* in larvae of the giant whitefly *Aleurodicus dugesi*).

Some insects simply drop their eggs at the oviposition site, as is the case with some dragonflies, which can be seen touching the water surface with the tip of the abdomen at the ponds where they breed. Others paste them to a certain substrate. In certain mayflies, the gravid female drops into the water, where her abdomen breaks, setting her eggs free. Other aquatic insects, such as aquatic beetles and damselflies, insert their eggs into the mud or place them in the leaves or stems of plants growing on the margins of ponds or streams; still others, like some mosquitoes, make egg rafts.

Parental care

Once the eggs have been laid, most parent insects simply leave. Among cockroaches and praying mantids, the female secretes a cover around the eggs that provides protection against predators and dehydration; some cockroach females carry this egg case until the larvae emerge. Female mealybugs, scale insects, and some beetles protect their eggs by shielding them with their bodies, other beetles may carry them beneath their bodies, and some carabid beetles construct depressions in the soil for the eggs and clean them regularly until they hatch, to prevent fungi from growing on them. In the case of the chrysomelid beetle, *Acromis sparsa*, the female stays with her offspring until they reach adulthood; the larvae remain aggregated, feeding on the same leaf, so that the mother can shield them with her body in case of danger. Earwig females oviposit in burrows in the ground and guard the eggs until they hatch.

Insect males are mainly polygynous, and for this reason it is not advantageous for them to invest effort in parental care. There are some exceptions. The male of the scarab beetle, *Lethrus apterus*, helps the female construct a burrow and gather leaves to provision the brood cells. In the carrion beetle *Necrophorus* the male assists the female in regurgitating liquefied carrion to feed their offspring, which reside in a nest of carrion. Males of certain bark beetle and sphecid wasp species stay close to the nests of their mates and repel parasites and rival females that try to enter the burrows. The best-known examples of paternal care in insects are found among the water bugs of the family Belostomatidae. In the genera *Abedus* and *Belostoma* the female lays her eggs on the back of the male after mating. The male takes exclusive care of the progeny; he ventilates the eggs, prevents fungus from growing on them, and assists in the emergence of the larvae. In *Lethocerus*, another genus of the same family, the female lays eggs on a stick or plant stem at the level of the water surface, and the male is stationed close by, to guard them against predators.

A digger wasp female builds one or several burrows in the ground and provides the eggs with a certain type of prey, usually spiders or other insects; she then inserts an egg into each prey item and closes the burrow. The prey is alive but paralyzed with a substance injected together with the egg. When the wasp maggot emerges, it eats the prey from the inside out. In some species, the female keeps bringing prey to the larva until it pupates.

Feeding behavior

Some insects are surrounded by their food from the time of hatching, as a result of the oviposition habits of the parent. In the case of many insects that feed on plants, the mother places the eggs on the host plants or, in scavenger or parasitic insects, on suitable detritus or hosts. Among social insects, the larvae are incapable of searching for their own food, and the workers are in charge of feeding them. Most insects, however, must search for their food.

Insects feed on an almost endless variety of food and in many different ways. About half of insects are herbivores, feeding on plants. The most common plant hosts are flowering plants, but some insects feed on ferns, fungi, and algae too. Certain insects are polyphagous and show no preference for any particular host plant. An example is the desert locust, *Schistocerca gregaria*, which, when migrating, feeds on all the plants it finds along its way. Other insects are oligophagous, specializing in certain groups of plants, for example, *Pieris* butterflies, which feed on Cruciferae and other plants with mustard oils. Still others are monophagous, feeding on a single plant species, for example, the weevil *Scyphophorus yuccae*, which feeds only on *Yucca*.

Variation also exists as to the part of the plant that is consumed. Soil mealy bugs, wireworms, cicada larvae, and white grubs, among others, specialize on roots; bark beetles, carpenter ants, and termites on woody parts; psyllids, aphids, leafhoppers, leaf-mining larvae of moths, flies, beetles, and sawflies on leaves; and bees, wasps, beetles, butterflies, and moths on nectar and pollen of flowers. Some herbivorous insects inject a chemical into the plant that induces it to grow abnormally and form a gall. The feeding of the insect usually stimulates the formation of a gall, though in some cases it is initiated by oviposition. A plant gall may have (e.g., galls of psyllids, aphids, and scales) or lack (e.g., galls of moths, beetles, flies, wasps) an opening to the outside.

A few insects are specialized in the production of "fungus gardens." This peculiar habit is found in some ants, termites, and ambrosia beetles. These fungus-growing insects nest in the ground, where they excavate a complex system of galleries and chambers. They cut up leaves and take them to special chambers in the nest, where they are chewed up and seeded with a fungus, which they tend and eat.

An insect can be attracted to the food source from a distance by visual or olfactory clues, but the final selection occurs when the insect is in direct contact with the host. Physical characteristics of the plant, olfaction, and contact chemoreception play a part in this process. For example, in a leafhopper the first attraction to a plant is through color; thus, they land on host and non-host plants of similar color. The leafhopper finally determines the identity of the host plant by touching the surface with its proboscis or by inserting the proboscis into the plant for a short distance. Chemical substances characteristic of the plant then are detected, and the leafhopper stops feeding activities if the plant is not an appropriate host and keeps feeding if it is one.

Carnivorous insects feed on other animals, which are mainly other insects. There are two general sorts of carnivores: predators and parasites. Predator insects are active and seek their prey, usually consisting of one or more smaller insects per meal, whereas parasites live in or on the body of their hosts during at least part of their life cycle, take successive meals from the much larger host, and typically do not kill the host or do so only gradually.

Damselflies and dragonflies, both during larval and adult life; tiger, ground, and ladybird beetles; lacewings and some true bugs; robber flies and larvae of syrphid flies are examples of predators. Most predators actively forage for their prey, but some are ambush hunters. Antlion larvae, for example, dig a pit with sloping sides in the sand and bury them-

Monarch butterflies (*Danaus plexippus*) wintering in Mexico. (Photo by Laura Riley. Bruce Coleman, Inc. Reproduced by permission.)

selves at the bottom with only the head exposed. When an ant or other small insect walks on the margin of the pit, the antlion provokes a landslide, throwing sand with its head, which causes the insect to roll down to its open mandibles. Larvae of tiger beetles live in a vertical burrow in the ground; when an insect comes within range, the larva extrudes about half of the body and grabs the prey with its long, sickle-shaped mandibles. Larvae of caddis flies weave a net of silken threads with which they capture small organisms in the water. Praying mantids and water bugs sit and wait for their prey to come their way; when the prey is close enough, it is taken and held by the fast extension of the raptorial forelegs. In a similar way, some dragonfly larvae wait motionless and strike at passing prey, extending the labial mask with incredible speed and accuracy.

Most hunters have large eyes, since only visual stimuli are fast enough to allow them a rapid reaction to a moving prey. After the capture, recognition is mainly tactile. In predaceous forms with subterranean habits and poorly developed eyes, the localization of prey is largely olfactory. Once they capture a prey, many predators restrain the prey's move-

Queen and worker honey bees (*Apis mellifera*). The worker bees feed and care for the queen and her larvae. (Photo by Kim Taylor. Bruce Coleman, Inc. Reproduced by permission.)

ments by mechanical strength and tear it to pieces with the mandibles. Predators with sucking mouthparts, such as true bugs, robber flies, and some lacewing and beetle larvae, inject salivary secretions that paralyze and kill the prey and then suck the liquefied organs, discarding the empty cuticular shell.

There are parasitic insects on vertebrates and on other insects and arthropods. Some are ectoparasites, living on the outside of their hosts, and others are endoparasites, living inside the host's body. Some parasites feed on only one host species and others on a group of related host species; still others have a wide range of hosts. Lice are an example of ectoparasites of vertebrates; both larva and adult are completely dependent on the host, on whom they spend their entire life cycle. Fleas, which also are ectoparasites on vertebrates, are less dependent on the host; they frequently change hosts and spend some time away from the host as adults. The larval stage is not spent on the host's body. Mosquitoes and some other blood-sucking insects, such as "no-seeums," bed bugs, and assassin bugs, feed from the host only for brief periods; often it is only the females, which require a blood meal to produce eggs.

Ectoparasites of other insects, such as certain "no-seeums," suck blood from the wing veins of lacewings and dragonflies. In many insects the first larval stage is active, whereas the older larvae are parasitic or fixed predators of a specific host. For example, larvae of meloid beetles, called "triangulins," wait in flowers that the parent bee may visit. The active triangulins climb onto the bee, attaching themselves to the bee's hairs with their claws, and are carried in this way to the bee's nest, where they become internal parasites of the bee's eggs or larvae.

Several larvae of flies are endoparasites of vertebrates, invading open sores, alimentary ducts, and nasal cavities; this phenomenon is called myasis. *Gasterophilus* bot flies lay their eggs on the fur coats of horses. When the horse licks its hair, the larvae hatch and affix themselves to the horse's tongue. They then pass into the digestive canal and attach themselves to the stomach or intestinal wall, causing ulcers. The fly *Oestrus ovis* places its larvae on the nostrils of sheep, where they crawl and enter the frontal sinus, causing vertigo. *Cordylobia anthropophaga* maggots produce ulcerated ridges in the skin of humans, dogs, and mice.

Most endoparasitic insects of other insects differ from endoparasites of vertebrates in that they reach a size equal to that of the host and eventually kill it: they are called parasitoids. Most parasitoids live and feed as larvae inside the host and become free-living adults after killing the host. Tachinid and sarcophagid fly larvae attack grasshoppers, caterpillars, true bugs, and wasp larvae, but most parasitoids are found among the ichneumonoid, chalcidoid, and proctrotrupoid wasps. Some parasitoids are very specific to their hosts—restricted to a single or a few species of insects. Parasitoids often regulate numbers of pest insects and are therefore important components of biological control programs.

Some insects are detritivores, feeding on decaying materials, such as carrion, leaf litter, and dung, and are important in the progressive breakdown of organic matter into their basic components to be returned to the soil, where they become available for plants. They also remove unhealthy and obnoxious materials from the landscape. Bow flies, carrion beetles, and skin beetles feed on dead animal tissue, skin, feathers, fur, and hooves and are of great importance in the removal of carrion from the environment. Dung beetles cut and shape vertebrate dung into balls that serve as a food source and a brood chamber for a single larva. Dermestid skin beetles feed on tissue and skin from vertebrate bones. Silverfish feed on dry organic debris and have a taste for paper, especially that containing starch or glue. Cockroaches and other insects are another example of scavengers that feed on dead plants and animals. Wood-boring beetles, termites, carpenter ants, and other wood feeders are important agents in facilitating the conversion of fallen trees and logs to soil.

Defensive behavior

Insects use many means of defense; these can be passive, where an insect relies on its appearance or location, or active, where an insect tries to escape, threatens a predator, or attacks it with chemical weapons. The habitat of numerous insects by itself provides them with a defense mechanism; insects that burrow into plant tissues or in the soil or live under rocks gain protection against predators. Some insects build a protective case or shelter that they carry around. Caddisfly larvae cement sand grains, small twigs and leaves, or other materials together to form the case inside which they live. Some chrysomelid beetles attach their feces to their backs to form a protective shield. Larvae of froghoppers use the excess fluid from the sap they suck to form a mass of bubbles that hides them from their enemies, and larvae of some psyllids construct a protective cover (a "lerp") of a sweet, crystallized substance called honeydew, under which they live.

Several beetles "play dead" when disturbed, dropping to the ground, folding up their legs, and remaining motionless for a while. Some insects blend with their backgrounds by closely resembling leaves, twigs, flowers, thorns, or bark. They are so well camouflaged in their normal surroundings that they become invisible to their predators in what is known as crypsis. In the forests of Papua New Guinea some weevils have modified wing covers that favor the growing of "miniature gardens" of mosses, lichens, algae, and fungi on their backs, which they carry around. The physical similarity with the surroundings is enhanced by the behavior of cryptic insects: inchworms freeze, holding their bodies upright like a twig, grasshoppers orient their bodies as if they were leaves, moths and walking sticks colored like bark remain immobile on their host trees.

On the contrary, the color patterns of several insects are striking in their attempts to intimidate predators or deflect their attention to parts of the body that are least vulnerable. This aposematic, or warning, coloration is found in several butterflies, grasshoppers, lanternflies, praying mantids, walking sticks, true bugs, and homopterans, among others. Some have brightly colored spots on the abdomen or hind wings that are hidden while the insect is at rest and are exposed suddenly when the animal is threatened. These flash colors may cause enemies to become startled, at least for a moment, allowing the insect to find a new hiding place and cover its conspicuous spots, rendering them invisible to the eyes of the predator. Other insects have a pair of eyespots on the upper surface of a pair of wings. If they are disturbed, they fully expose the wings with the eyespots. In some butterflies, this display is accompanied by a hissing or clicking sound produced by rubbing wing veins against each other and against the body. This behavior may elicit an escape response in birds, or else the attacks of birds may be directed at the eyespots and not at other, more vulnerable parts of the body. Other types of color advertisement, usually reds, yellows and blacks, are related to the palatability of the insect; predators associate a particular color with a bad flavor and learn to avoid insects displaying that color. Thus, the black and yellow of bees and wasps are associated with a sting and the red on a black or green background of certain butterflies is connected to a disagreeable taste. Several caterpillars show a striking color combination of yellow, orange, and green, usually associated with the presence of irritant hairs.

Some insects resemble or mimic other insects or even vertebrates. Predators learn to shun distasteful insects with striking colors, and certain insects take advantage of this behavior and avoid being eaten by displaying the warning color pattern of a distasteful or dangerous organism. A few noctuid moths move their legs in the manner of a bristly spider. Many beetles and true bugs mimic wasps not only in color but also in behavior, holding their wings upright, waving their legs and antennae, and bending the tips of their abdomens upward like wasps. The anterior part of the body of some swallowtail caterpillars is enlarged and painted with two eyespots, resembling the head of a snake. If the caterpillar is threatened, it will evert a scent gland that, besides producing an unpleasant odor, looks like the bifid tongue of a snake.

This kind of mimesis, where a harmless insect mimics a dangerous organism, is called Batesian mimetism. For exam-

An owl butterfly (*Caligo memnon*) showing adaptive patterning on its wings to deter predators. (Photo by Jianming Li. Reproduced by permission.)

ple, the South American butterfly *Episcada salvinia rufocincta* feeds on plants of the family Solanaceae and incorporates toxic substances (alkaloids) from its food plants, which make it distasteful to predators. *Paraphlebia zoe* is a damselfly that frequents the same forest clearings where the butterfly flies. Some males of the damselfly imitate the flight of the butterfly and have a white spot in the same position on the wings. Other males in the same population do not have any spots on the wings and appear invisible when they fly. There is also a form of Müllerian mimetism, in which an unpalatable or poisonous insect resembles another distasteful or harmful organism; predators learn to avoid only one color pattern, resulting in advantage to both mimic and model.

Many insects display an escape reaction when threatened, by flying, jumping, running, or diving. Certain noctuid moths detect the ultrasonic sounds produced by foraging bats and perform evasive actions by turning around, flying in zigzag patterns, or dropping to the ground and remaining motionless.

Chemical warfare provides defense for several types of insects. Numerous insects have repugnatorial glands, which secret noxious substances. Some of these secretions also act as alarm pheromones, warning other insects about the proximity of danger. Many cockroaches, stink bugs, beetles, ants, and walking sticks are capable of forcibly spraying an odoriferous secretion, sometimes for several feet. Caterpillars of swallowtail butterflies evert repugnatorial glands located behind the head, liberating a secretion that is effective against ants. The venomous secretions associated with the sting of wasps and bees and the bite of predatory true bugs also can be considered chemical defenses. Coccinelid, chrysomelid, lycid, lampyrid, and meloid beetles discharge blood when threatened, which is called reflex bleeding. In meloid beetles this blood contains substances that produce blisters or sores when it comes in contact with the skin of vertebrates.

Aphids, mealybugs, psyllids, whiteflies, and scale insects have developed a different defense strategy, in which they recruit ants as personal bodyguards. These insects produce a sweet secretion called "honeydew" that is attractive to ants. Thus, ants guard and tend colonies of aphids and aggressively attack any organism trying to feed on them.

Migration

Most species of insects disperse at some time during their life cycles in an attempt to populate new areas, though only a few have migratory mass movements similar to those of birds. In those insects with mass migrations, the movements usually are one way; one generation makes the trip in, and the following generation makes the return trip. Migration is accomplished mainly by flight, typically following the prevailing wind currents.

The best-known example of migrating insects is probably that of the monarch butterfly *Danaus plexippus.* After one or two generations in Canada and the northern United States, a combination of factors in the autumn, probably a shortening photoperiod and a drop in temperature, induces a generation of monarchs not to develop gonads and to migrate south to wintering places in California, Mexico, and Florida, where they congregate in great numbers on certain kinds of trees.

In the spring they begin their journey back north, laying eggs along the way before dying. The subsequent generation completes the flight back, and the flight south of 2,000 mi (3,200 km) or more is repeated the following fall.

Swarms of migratory locusts have been known since biblical times. The migratory locust *Schistocerca gregaria* is known throughout the world for its mass migrations in which hundreds to billions of locusts swarm and advance, eating all vegetation in their sight. This species has two different phases. In its solitary phase the specimens are sedentary, selective in their food choices, and colored pale green or reddish, whereas in the migratory phase the specimens are gregarious, devour any kind of plant, and display contrasting colors. A young larva in either phase can be switch to the other one.

Not all migrating insects fly. Army ants (*Eciton hamatum*) migrate on the ground; army worms of the North American moth *Pseudaletia unipuncta* march onward, devouring every green thing in their path; and maggots of the mourning gnat *Neosciara* crawl in snakelike processions glued to each other in a slimy secretion, searching for an appropriate place for pupation in the forests of Europe.

Factors initiating migrations are not yet understood fully, but an onset of adverse environmental factors, such as crowding, lack of food, and short days, probably plays a part in the generation of endocrine changes leading to migration. For example, a sudden increase in population numbers may induce the production of winged forms in aphids that normally produce wingless forms, and this is correlated with the activation of certain hormone-producing organs.

Resources

Books

Evans, Arthur V., and Charles L. Bellamy. *An Inordinate Fondness for Beetles.* Berkeley: University of California Press, 2000.

Holldobler, B., and E. O. Wilson. *Journey to the Ants: A Story of Scientific Exploration.* Cambridge, MA: Belknap of Harvard University Press, 1995.

McGavin, George C. *Bugs of the World.* London: Blandford Press, 1993.

Preston-Mafham, K. *Grasshoppers and Mantids of the World.* London: Blandford Press, 1998.

Thornhill, Randy, and John Alcock. *The Evolution of Insect Mating Systems.* Cambridge, MA: Harvard University Press, 1983.

Organizations

Animal Behavior Society, Indiana University. 2611 East 10th Street, no. 170, Bloomington, IN 47408-2603 United States. Phone: (812) 856-5541. Fax: (812) 856-5542. E-mail: aboffice@indiana.edu Web site: <http://www.animalbehavior.org/>

Other

Journal of Insect Behavior [cited December 23, 2002]. <http://www.kluweronline.com/issn/0892-7553>.

Natalia von Ellenrieder, PhD

••••

Social insects

The fascination of social insects

The social insects represent one of evolution's most magnificent, successful, and instructive developments. Ever since the behavior of ants, bees, wasps, and termites was first recorded in antiquity, these insects have exerted a powerful hold on our imaginations.

Three characteristics of social insects account for this interest. The first is the very habit of living in social groups. To biologists, this way of life represents a fascinating evolutionary innovation, yet it also poses a basic dilemma. Contrary to what we may have commonly learned, the essential history of life on Earth has not been the story of the rise and fall of dominant groups of animals like the dinosaurs. Instead, life's history has been the succession of what evolutionary biologists J. Maynard Smith and E. Szathmáry termed "major transitions" in evolution. Without any one of these transitions, living things today would be fundamentally different.

Two of these major transitions are the evolution of sexually reproducing organisms from those that reproduce asexually, and the evolution of multicellular from single-celled organisms. These transitions share a radical reorganization of living matter. In particular, at each transition, existing units coalesced to form larger units. In the process, the original units lost some or all of their power of independent reproduction, and the way genetic information is transmitted was changed. The social insects are the prime examples of the next major transition, from solitary organisms to social organisms. As a result of this change, workers of social insects have largely lost the power of reproducing independently, and genetic information has come to be primarily transmitted from generation to generation via the reproduction of the queens (and kings in termites). Why these workers have foregone their power of reproduction is a basic dilemma that insect sociality poses for biologists.

The second characteristic of social insects, which explains why they command attention, is their overwhelming numerical and ecological dominance. Ants, for example, occur from the Arctic tundra to the lower tip of South America. They swarm in diverse habitats, ranging from desert to rainforest, and from the depths of the soil to the heights of the rainforest canopy. The abundance and ubiquity of social insects give them an ecological importance that is unmatched among land-dwelling invertebrates.

The third and final fascinating characteristic of social insects is the way in which they perform work. Clearly, social insects carry out cooperative tasks, such as building nests, with results that are stunningly complex and precise. Termites build nests that are towering hills of red clay, with intricate internal architecture designed to maintain a cool, stable interior. The nests of honey bees house waxen combs composed of thousands of exquisitely arrayed hexagonal cells. Social wasps construct nests in which parallel tiers of cells are enclosed within a delicate paper globe.

These nest forms point to the ability of social insects to achieve, as a collective, impressive feats of architectural complexity. However, biologists have long puzzled over how social insects complete these complex tasks as a group when individual workers frequently strike us as inept. "It seems to me that in the matter of intellect the ant must be a strangely overrated thing," wrote Mark Twain after watching the fumbling meanderings of a worker carrying a grasshopper's leg back to its nest. The ant's seemingly unnecessary scaling of an obstacle was, Twain went on to say, "as bright a thing to do as it would be for me to carry a sack of flour from Heidelberg to Paris by way of Strasburg steeple." In this chapter, we first review some basic biology of social insects, then consider the three principal characteristics of social insects in turn.

An outline of social insect biology

What are social insects?

The social insects are not a taxonomic group. Instead, the transition to living in societies has evolved independently in several lineages of insects. This happened in the very distant past. For example, ants are thought to have arisen by approximately 80 million years ago, and the termites even earlier. The societies, or colonies, of social insects are distinct from mere swarms or aggregations. Species of insect qualifying as truly social, or eusocial, exhibit three traits. The first, and most important, is a reproductive division of labor. Some individuals (the queens and kings) are fertile and reproduce, whereas others (workers, or neuters) are sterile or semisterile. The workers either do not reproduce, or reproduce to a far lesser extent. The second trait is that the workers perform tasks of benefit to the colony, and in particular, cooperatively

1820

A honeybee (*Apis mellifera*) worker with yellow pollen sacs performing a "wag-tail" dance to indicate the direction of food. (Photo by Kim Taylor. Bruce Coleman, Inc. Reproduced by permission.)

rear the young. The third trait is that the society has some permanence and, specifically, that more than one adult generation coexists in the colony.

The major groups of eusocial insects are to be found in the Hymenoptera (ants, bees, wasps, and relatives) and the Isoptera (termites). The ants (Formicidae) are all eusocial, and represent the most species-rich group of social insects, with over 9,000 described species. The majority of bees are not social, but the group includes such well-known eusocial groups as the honey bees (Apini), stingless bees (Meliponinae), and bumble bees (Bombini), as well as less-conspicuous eusocial groups such as the sweat bees (Halictinae). There are about 1,000 eusocial species of bees. Eusocial groups among the wasps (Vespidae) include the hover wasps (Stenogastrinae), the paper wasps (Polistinae), and the yellowjackets and hornets (Vespinae). There are around 1,200 eusocial species of wasps. The termites (Isoptera), like the ants, consist only of eusocial species, and number over 2,800 species. Eusocial representatives also occur in at least two other insect orders, the Hemiptera (bugs and aphids) and the Thysanoptera (thrips). All eusocial hemipterans and thrips live in plant galls and possess a defensive "soldier" morph. Eusocial hemipterans consist of 60 aphid species in the two subfamilies, Hormaphidinae and Pemphiginae; about six species of eusocial thrips have been described in the family Phlaeothripidae.

The structure and life cycle of insect societies

The typical social insect colony is a family. In ants, bees, and wasps, the workers are all females. Males die after mating with queens and so take no part in social life. The colony therefore consists of a queen, her adult worker daughters, and

the brood (eggs, larvae, and pupae) produced by the queen. The job of the workers is to rear the brood to produce new workers, new queens, and males. The new queens and males are collectively referred to as reproductives, or sexuals. The workers carry out their job by foraging for food, building or defending a nest, and feeding and tending the brood. Termite colonies have the same basic structure, except that workers are of both sexes and the queen is accompanied by a reproductive male, a king.

The life cycle of the typical social insect begins with the formation of a colony by a mated queen (in the ants, bees, wasps) or by a queen and king (in termites). The colony then enters a growth phase, with the founding individual or pair producing worker offspring that gather resources for the colony and thus allow it to expand. Once the colony is sufficiently populous, the reproductive brood of new queens and males is produced. As adults, these leave the colony, mate or pair, and disperse to found new colonies of their own. The mating and dispersal flights of ants can involve the sudden summertime emergence of thousands of winged queens and males ("flying ants"), which attract attention, and occasionally dismay, in the cities of northern Europe and North America.

There are many variations on this basic colony structure and life cycle. Colonies may be started by a single queen, as in the garden ant (*Lasius niger*), or multiple queens, as in the desert ant (*Messor pergandei*). Mature colonies may be headed by one queen, again as in the garden ant, or by more than one queen, as in the red ant (*Myrmica rubra*). Likewise, in termites, mature colonies may be headed by just a single royal pair or several such pairs.

Queens may mate with one male or many males. For example, queens of the red imported fire ant (*Solenopsis invicta*) always mate with one male; whereas queens of the domestic honey bee (*Apis mellifera*) mate on average with about 17 males, and queens of *A. nigrocincta* from Southeast Asia mate, on average, with the extraordinarily high number of 40 males.

Colonies may reproduce, not by releasing winged sexuals, but by splitting in two, as in the wood ants (*Formica* spp.). They may occupy a single nest site or, as in the acorn-nesting ant (*Leptothorax longispinosus*), several neighboring nests. Colonies may be annual or perennial. Those of bumble bees (*Bombus* spp.) are annual, founded by a single queen in spring. Before the year is over, she produces workers and reproductives and then dies. She leaves behind the new generation of mated queens to found fresh colonies in the following year. Colonies of honey bees are perennial. They are founded by a queen and a swarm of workers and, over the next two to three years, the queen produces reproductives each year before dying.

Queens of perennial social insects are probably the longest-lived of all insects, with the record being held by a captive garden ant queen that lived 29 years. Insect colonies also vary enormously in average size. At one extreme, some halictine bee colonies contain just a handful of workers. At the other, colonies of the army ants and driver ants of tropical America and Africa (*Eciton* and *Dorylus* spp.) have a workforce numbering hundreds of thousands or even millions.

Ants (*Crematogaster scutellaris*) caring for larvae. (Photo by Bartimeu Borrell. Bruce Coleman, Inc. Reproduced by permission.)

Social insects exhibit two additional important variations in their social structure. The first is social parasitism. Some species of ants, bees, and wasps do not produce workers, and queens subsist as parasites in the nests of other social insects, in the way that cuckoos parasitize the labor of nesting birds. One example of this practice is found in ants of the genus *Epimyrma*. After mating, a queen enters a colony of a related species and kills the existing queen by the macabre method of straddling her back and slowly throttling her. The workers, fatally to their interests, seem unaware of the death of their queen, and go on to rear eggs laid by the parasite.

The second important variation in social structure is unicoloniality. Populations of social insects typically exist as a set of distinct colonies that are tightly defended against all intruders, including workers of the same species. In unicolonial ants and termites, the social boundaries that maintain the distinctiveness of colonies have disappeared, and the population is a loose network of nests among which individuals appear to be interchangeable. In its introduced, European range, the Argentine ant (*Linepithema humile*) occurs in a vast unicolonial population, or "supercolony," which runs for over 3,728 mi (6,000 km) along the Mediterranean coast from Spain to Italy. Unicoloniality represents the emergence of another grade of sociality among insects. In fact, to the extent that unicolonial populations represent groupings of whole societies, we see in them the beginnings of a further major transition in evolution.

Caste in the social insects

The queens and workers of social insects are often referred to as castes. The distinction between castes may be behavioral or morphological. For example, in paper wasps (such as *Polistes* spp.), the queen and her workers do not differ morphologically. Instead, the queen maintains her position as the principal or only egg layer, by aggressively dominating the workers. By contrast, in vespine wasps (such as *Vespula* spp.), the queen is larger than the workers, differs from them morphologically, and is rarely aggressive. The most extreme caste dimorphism between queens and workers is found in ants and termites, especially in species with very large colonies. The queens of termites are among the largest. With a massively enlarged abdomen, a queen *Macrotermes* can reach 5.5 in (14 cm) long and 1.4 in (3.5 cm) wide, and be capable of laying 30,000 eggs per day. In general, queens, specialized for dispersal and egg laying, have a relatively large thorax and abdomen. The workers, specialized for labor, lack these features and are relatively small. In ants and termites, workers are always wingless, whereas young queens bear wings that they shed after their dispersal or mating flight.

Workers may themselves form separate subcastes. In the leaf cutter ants (*Atta* spp.), there are at least four distinct size-classes of worker in each colony, varying 200-fold in body weight. Each subcaste performs the tasks most suited to its body size. The largest, the major workers, cut green leaves from plants outside the nest. The smallest, the minor workers, who labor in the narrow tunnels deep within the nest, insert the chopped-up leaf pieces into the growing mass of the symbiotic fungus on which these ants feed. In other ants, major workers act as a defensive subcaste, the so-called soldiers. Only ants and termites exhibit morphological worker subcastes. However, even in species without them, worker subcastes exist in a temporal sense, since workers tend to change their jobs as they grow older. Typically, young workers perform tasks inside the nest such as feeding and cleaning the brood, and old workers perform external tasks such as defense and foraging.

The differences between the castes are, with few or even no exceptions, not based on genetic differences. Instead, whether a female egg develops into a queen, worker, or worker subcaste depends on how she is reared, and particularly on her nutrition. For example, in the honey bee, queens are reared in large cells and receive a special diet from workers, the so-called royal jelly. Workers are reared in small cells and do not receive royal jelly. At the genetic level, development as either caste must depend on the differential expression of genes present in all females. In other words, honey bee larvae in large cells that are fed royal jelly switch on one set of genes and develop as queens. Larvae treated otherwise fail to switch on these genes, or switch on an alternative set, and develop as workers. Seven of the genes involved in this process have been isolated and identified in honey bees, but very little is known about corresponding genes in other social insects.

The evolution of social behavior in insects

The origin of sociality

Explaining the origin of sociality in insects has been one of the great successes of evolutionary theory. Biologists generally agree that the likeliest pathway by which insect sociality arose was through offspring remaining in the nest as adults to help rear their mother's young. Many Hymenoptera exist (so-called solitary or nonsocial species) in which females feed their larvae themselves, unaided by workers. Some populations of halictine bees contain both nests of this sort and nests in which the parent is assisted in brood-rearing by her earliest-emerging daughters. We can therefore observe, in a single population, the likely state of affairs immediately before and immediately after the appearance of sociality. What, though, might be the evolutionary mechanism driving this process?

A very widely accepted answer is kin selection. Kin selection theory, devised by the outstanding evolutionary biologist W. D. Hamilton in the 1960s, assumes that social behavior (here the altruistic rearing of brood by sterile workers) is subject to genetic variation. In fact, genes that affect social behavior have been isolated and sequenced in the red imported

fire ant. Next, kin selection theory proposes that genes for altruism can be naturally selected if aid is directed at relatives (kin). This is because relatives of an altruist are, by definition, likely to share its genes for altruism. Therefore, by rearing enough relatives, workers can add sufficient genes for altruistic behavior to the next generation to make up for those they fail to add by not producing offspring themselves. This is why the basic social structure of a social insect colony is a simple family in which workers rear their mother's sexual offspring, which are the workers' siblings. As we have seen, in some species there are several queens per colony, and in others queens mate multiply. Their colonies are extended, rather than simple families. Such cases are assumed to be secondary developments in which the dilution in relatedness among individuals has been compensated for by greater efficiency among workers in rearing broods.

Hamilton's kin selection theory is equally applicable to the social Hymenoptera and the termites. In both groups the first societies are likely to have involved individuals remaining in the nest to help rear siblings. There has been a great deal of discussion about a secondary facet of Hamilton's theory, which highlighted the unusual genetic system exhibited by the Hymenoptera, termed haplodiploidy. In haplodiploid organisms, females are diploid (have two sets of chromosome per cell) and develop from fertilized eggs; males are haploid (have one set of chromosomes per cell) and develop from unfertilized eggs. This leads to an unusually high relatedness among sisters. Female offspring of the same mother and father can receive one of two genes from their mother at each locus (position on a chromosome). But they receive only one gene from their father, as being haploid, he has only one gene copy per locus. This raises their average chance of gene-sharing from the usual one-half (as found in wholly diploid organisms) to three-quarters.

Hamilton speculated that the especially high relatedness among haplodiploid sisters accounted for the concentration of independently evolved social lineages in the Hymenoptera and for the fact that workers in the Hymenoptera are always females. This has been termed the haplodiploidy, or three-quarters relatedness, hypothesis. The hypothesis has proved difficult to test and remains unproven. The reason is that features of the Hymenoptera other than the three-quarters relatedness of sisters may have led to the group's propensity to evolve societies based on female-only workers. Separating the contribution of these features from that of haplodiploidy itself requires a good knowledge of the evolutionary tree (phylogeny) of the nonsocial and social Hymenoptera and the traits associated with different points along the tree. As more is learned about these, it should become clearer whether the haplodiploidy hypothesis is upheld.

Conflict inside social insect colonies

A close observer of life inside a bumble bee colony, especially toward the end of the colony cycle, will notice that there is a great deal of friction among the members of the society. Workers jostle the queen, eat her eggs, and attempt to lay their own eggs. The queen retaliates by acting aggressively toward the workers and trying to eat the workers' eggs. Such a colony departs from the traditional picture of a social insect

A paperwasp (*Polistes hebraeus*) works on its nest. (Photo by J-C Carton. Bruce Coleman, Inc. Reproduced by permission.)

society in two ways. The first is that the workers are not completely sterile. This is possible because in bumble bees and many other social Hymenoptera, the workers possess ovaries but have lost the ability to mate. Due to haplodiploidy, this means the workers can lay eggs, but because they are unfertilized, these eggs always develop into males. The second way in which the colony is unlike a traditional insect colony is that it is clearly not an internally harmonious society. In some cases, aggressive, reproductive bumble bee workers even kill the queen, their own mother. This is the antithesis of how social insect workers are supposed to behave.

A great strength of the theory of kin selection is that it offers not just a reason for why a society evolves in the first place, but also an explanation of why conflict occurs within existing societies. The essential reason behind conflict is that different sets of individuals within the society are related to the reproductive brood to different degrees. Since natural selection leads them to act toward the reproductives according to their relatedness, conflict over reproductive decisions will therefore arise. In the bumble bees and other social Hymenoptera sharing their social structure, with a single, once-mated queen and workers capable of egg laying, a kin-selected conflict arises over male parentage. Queens are more closely related to the sons they produce and so favor these males being reared. Workers are more closely related to the sons *they* produce and so favor them being reared. The result is a prolonged squabble over egg laying. The conflict peaks toward the end of the colony cycle because, at this point, the colony moves from rearing workers (over whose production there is little disagreement, because both queen and workers favor the colony growing to be large and productive) to rearing sexuals (over which, for male sexuals, there is disagreement).

Workers killing their own mother, which permanently prevents the queen from interfering with the workers' reproductive attempts, is an extreme, but logical, outcome of this conflict over male parentage.

Another well-studied form of kin-selected conflict in the social Hymenoptera is conflict between the queen and workers over the sex ratio (ratio of new queens to males). This arises because, due to their comparatively high relatedness to females, workers favor rearing a more female-biased sex ratio than is favored by the queen. This conflict has different outcomes according to the practical power held by each party. In some ants, the queen can apparently control the caste of female eggs, presumably by adding to the eggs substances that affect differential gene expression during their subsequent development. In this way, a queen can force the workers to rear the sex ratio optimal for her. In other (perhaps most) ants, the queen lacks the ability to determine the caste of eggs. The workers achieve their optimal sex ratio either by selectively killing the extra males in the queen's brood or by channeling more female larvae into development as queens.

Several other types of kin-selected conflict are known from social insects. In fact, potential conflict is present whenever more than one individual is capable of laying eggs (conflict over male parentage being a special case of this). The ponerine ant (*Dinoponera quadriceps*) is a species that has lost its morphological queen caste. One worker (the alpha) dominates the others as the chief reproductive, and is capable of producing both female and male eggs because, unusually, she remains able to mate. She uses her sting to smear any high-ranking challenger with a chemical that induces the other workers to pin down the upstart for as long as several days. As a result, the immobilized challenger loses her high rank. Both the alpha and the other workers gain from this behavior, because each is less closely related to the brood that a successful challenger would produce than to the alpha's brood. Such collective suppression, or "policing," of selfish behavior can also be provoked by the conflict over male parentage. It is likely to be a general mechanism for maintaining reproductive harmony. Hence, social insect colonies are indeed the integrated organizations that they are popularly supposed to be. But close investigation shows that this state of affairs is conditional. It depends, first on a high degree of common interest within the society (due to relatedness) and, second on social mechanisms that keep costly conflicts in check.

The ecological success of social insects

The habit of social living has clearly made the social insects extremely successful. By insect standards, they are not exceptionally species-rich, although in some habitats their local species diversity may be relatively high. But social insects are exceptionally abundant. The renowned evolutionary biologist and conservationist E. O. Wilson estimates that, although making up approximately 2% of known insect species, social insects form as much as 50% of the world's insect biomass.

As a result of their abundance and their diversity of diets and habits, social insects have a huge impact on other species and on the structure and functioning of ecological communities. They are outstanding "ecological engineers," in that their activities alter the physical environment and affect the availability of resources to other organisms. Such activities include pollination of flowers (by bumble bees, stingless bees, and honey bees), the dispersal of seeds (by harvesting ants), the cropping of green vegetation (by leaf cutter ants), the turning of soil (by soil-dwelling ants and termites), the consumption of wood and grass (by termites), and the predation of other insects (by predatory ants and wasps). Biologists have highlighted a certain set of "ecosystem services" supplied by living organisms on which the world's ecological health, and hence the economic health of human society, depends. Of these services, social insects contribute to at least three, namely pollination, soil formation, and the biological control of harmful arthropods. For example, the annual economic value of honey bee pollination in the United States has been estimated to lie between $1.6 and $5.7 billion. There are many reasons for valuing and conserving social insects, but this fact alone highlights the dangers stemming from any loss of pollinating bees from ecological communities.

Social insects are also partners in a series of unique symbioses with a vast range of other organisms. These include bacteria (in the leaf-cutting ants), fungi (in the leaf-cutting ants and some termites), and flowering plants (some ants only live inside special structure on certain plant species, which they protect from the attacks of herbivorous insects). These symbioses also include those with other arthropods (many mite species live symbiotically inside ant nests) and other insects (lycaenid butterflies are reared inside ants' nests as mutualistic guests or, in some species, as unwelcome parasites).

Another aspect of the ecological success of social insects is their extraordinary ability to invade areas outside their normal range, to which they have been introduced by human activity. No fewer than seven of the 14 species recognized as the world's most persistent invasive insects are social insects. These seven are the yellow crazy ant (*Anoplolepis gracilipes*), the Argentine ant, the big-headed ant (*Pheidole megacephala*), the red imported fire ant, and the little fire ant (*Wasmannia auropunctata*), along with the European wasp (*Vespula vulgaris*) and the termite *Coptotermes formosanus*.

One key to the success of social Hymenoptera as invasive species is the ability of a single queen to found a colony whose many reproductives can then seed the invasion of a large island or even a continent. In ants, another key is undoubtedly unicoloniality. A unicolonial network of nests can carpet fresh territory without competition between conspecific nests that might otherwise limit the ants' spread. Many, if not all, of the successful invasive ant species are indeed unicolonial. Once established in new territory, even introduced social insects that are not unicolonial can spread rapidly and reach very high numbers. For example, the European wasp (*V. vulgaris*) increased its range in New Zealand by 22 mi (35 km) per year and has attained the extremely high density of 12 nests per acre (30 nests per hectare).

Self-organization in social insects

Part of the great ecological success of social insects can be attributed to their sophisticated means of communicating with one another. In addition, by weight of numbers, a colony can overpower enemies and monopolize food sources. It can also perform many tasks simultaneously, and one set of workers can finish tasks that another set has left incomplete. Social insects communicate with their nest mates principally by releasing pheromones, but one exception is the famous dance language of honeybees. Here, a returning scout bee directs its fellow foragers to a flower patch by means of a dance whose orientation and duration indicate the bearing of and distance to the patch.

Communication among workers may be used to coordinate tasks, but there is very little evidence that any one individual, even the queen, directs the overall work of the colony. Control of colony activity is decentralized. As we have seen, social insects as a group create complex structures and complete sophisticated tasks at the same time as individual workers seem largely rudderless in their activity. These complex structures include not only nests, but also more transient creations such as the elaborately branching raiding columns of army and driver ants. Social insects considered as working collectives therefore pose a paradox. How can the evident complexity of their works be achieved by the group in the absence of central control or a high degree of individual competence?

The answer to this paradox—self-organization—was established by biologists in the 1970s and 1980s. Self-organization is the process (verified by mathematical simulations and even by experiments with teams of robots) by which relatively unthinking and autonomous units, following a few simple rules, can achieve complex tasks when they interact collectively. Other entities made of many interacting parts, such as brains and computers, probably act in a self-organized way in some respects. The price that self-organized systems pay for their way of operating is a level of error and inconsistency among their individual subunits. Indeed, the errors made by social insect workers in their task performance may even form the raw material from which they discover new ways of collectively achieving goals, in the way that genetic errors (mutations) provide the raw material from which natural selection can craft complex adaptations. This is why social insects may individually seem inefficient and even stupid, but the colony as a whole can achieve something that appears to be the result of planned, intelligent design. Working in the emerging field known as "swarm intelligence", which owes much to observations of the way social insects do things, computer scientists have applied the principles of self-organization to running several types of complex systems required by human economic activity. These include telephone networks, Internet search systems, factory production lines, and even route-planning for fleets of commercial trucks. What is lost by tolerating a level of individual error is more than made up for by the robustness, flexibility, and economic efficiency of the system as a whole. In this way, we continue to learn from social insects in ways never imagined by earlier observers of these enthralling creatures.

Resources

Books

Abe, T., D. E. Bignell, and M. Higashi, eds. *Termites: Evolution, Sociality, Symbiosis, Ecology.* Dordrecht, The Netherlands: Kluwer Academic Publishers, 2000.

Agosti, D., J. D. Majer, L. E. Alonso, and T. R. Schultz, eds. *Ants: Standard Methods for Measuring and Monitoring Biodiversity.* Washington, DC: Smithsonian Institution Press, 2000.

Bolton, B. *A New General Catalogue of the Ants of the World.* Cambridge, MA: Harvard University Press, 1995.

Bourke, A. F. G. "Sociality and Kin Selection in Insects." In *Behavioural Ecology: An Evolutionary Approach,* edited by J. R. Krebs and N. B. Davies. Oxford: Blackwell, 1997.

Bourke, A. F. G., and N. R. Franks. *Social Evolution in Ants.* Princeton: Princeton University Press, 1995.

Camazine, S., J.-L. Deneubourg, N. R. Franks, J. Sneyd, G. Theraulaz, and E. Bonabeau. *Self-Organization in Biological Systems.* Princeton: Princeton University Press, 2001.

Choe, J. C., and B. J. Crespi, eds. *The Evolution of Social Behavior in Insects and Arachnids.* Cambridge: Cambridge University Press, 1997.

Crozier, R. H., and P. Pamilo. *Evolution of Social Insect Colonies.* Oxford: Oxford University Press, 1996.

Hamilton, W. D. *Narrow Roads of Gene Land.* Vol. 1, *The Evolution of Social Behaviour.* Oxford: W. H. Freeman/Spektrum, 1996.

Hölldobler, B., and E. O. Wilson. *The Ants.* Berlin: Springer-Verlag, 1990.

Keller, L., ed. *Queen Number and Sociality in Insects.* Oxford: Oxford University Press, 1993.

Maynard Smith, J., and E. Szathmáry. *The Major Transitions in Evolution.* Oxford: W. H. Freeman, 1995.

Michener, C. D. *The Bees of the World.* Baltimore: Johns Hopkins University Press, 2000.

Ross, K. G., and R. W. Matthews, eds. *The Social Biology of Wasps.* Ithaca, NY: Comstock Publishing Associates, 1991.

Seeley, T. D. *The Wisdom of the Hive.* Cambridge: Harvard University Press, 1995.

Williams, D. F., ed. *Exotic Ants: Biology, Impact, and Control of Introduced Species.* Boulder, CO: Westview Press, 1994.

Wilson, E. O. *Success and Dominance in Ecosystems: the Case of the Social Insects.* Oldendorf/Luhe, Germany: Ecology Institute, 1990.

Periodicals

Allen-Wardell, G., et al. "The Potential Consequences of Pollinator Declines on the Conservation of Biodiversity and Stability of Food Crop Yields." *Conservation Biology* 12 (1998): 8–17.

Bonabeau, E., M. Dorigo, and G. Theraulaz. "Inspiration for Optimization from Social insect Behaviour." *Nature* 406 (2000): 39–42.

Chapman, R. E., and A. F. G. Bourke. "The Influence of Sociality on the Conservation Biology of Social Insects." *Ecology Letters* 4 (2001): 650–662.

Crespi, B. J., D. A. Carmean, L. A. Mound, M. Worobey, and D. Morris. "Phylogenetics of Social Behavior in Australian Gall-Forming Thrips: Evidence from Mitochondrial DNA Sequence, Adult Morphology and Behavior, and Gall Morphology." *Molecular Phylogenetics and Evolution* 9 (1998): 163–180.

Evans, J. D., and D. E. Wheeler. "Differential Gene Expression Between Developing Queens and Workers in the Honey Bee, *Apis mellifera.*" *Proceedings of the National Academy of Sciences* 96 (1999): 5575–5580.

Giraud, T., J. S. Pedersen, and L. Keller. "Evolution of Supercolonies: the Argentine Ants of Southern Europe." *Proceedings of the National Academy of Sciences* 99 (2002): 6075–6079.

Hammond, R. L., M. W. Bruford, and A. F. G. Bourke. "Ant Workers Selfishly Bias Sex Ratios by Manipulating Female Development." *Proceedings of the Royal Society of London,* Series B. 269 (2002): 173–178.

Krieger, M. J. B., and K. G. Ross. "Identification of a Major Gene Regulating Complex Social Behavior." *Science* 295 (2002): 328–332.

Monnin, T., F. L. W. Ratnieks, G. R. Jones, and R. Beard. "Pretender Punishment Induced by Chemical Signalling in a Queenless Ant." *Nature* 419 (2002): 61–65.

Palmer, K., B. Oldroyd, P. Franck, and S. Hadisoesilo. "Very High Paternity Frequencies in *Apis nigrocincta.*" *Insectes Sociaux* 48 (2001): 327–332.

Passera, L., S. Aron, E. L. Vargo, and L. Keller. "Queen Control of Sex Ratio in Fire Ants." *Science* 293 (2001): 1308–1310.

Organizations

International Isoptera Society. Web site: <http://www.cals.cornell.edu/dept/bionb/isoptera/homepage.html>

International Union for the Study of Social Insects. Web site: <http://www.iussi.org>

Other

"Social Insects World Wide Web (SIWeb)" [cited April 1, 2003]. <http://research.amnh.org/entomology/social_insects/>.

Andrew F. G. Bourke, PhD

• • • • •

Insects and humans

The relationship between insects and humans is long and complex. Since antiquity, insects have infected us with disease, attacked our crops, infested our food stores, and pestered our animals. And although we derive considerable benefit from their services, including pollination, honey and wax production, and the biological control of pests and weeds, the pestiferous (troublesome) activities of a mere fraction of all insect species negatively shapes humankind's perception of them. Still, throughout history insects have managed to permeate our spiritual, cultural, and scientific endeavors.

Cultural entomology

The field of cultural entomology explores the manifestation of insects in human culture. Insects have been used around the world, particularly in ancient cultures, as symbols of gods or to portend good and evil events. Their images have appeared on murals, coins, and stamps. Insects have long inspired writers, musicians, and poets—all have used their images in story, song, and verse. Artists and craftsmen still use insects as models in painting, sculpture, jewelry, furniture, and toys.

Insects in mythology, religion, and folklore

Numerous tribes around the world have used insects as totems or as symbols to explain creation myths. In the mythology of a South American tribe, a beetle created the world, and from the grains of earth that were left over, created men and women. The bizarre giraffe-like *Lasiorhyncus barbicornis* is one of the most grotesquely shaped weevils in New Zealand. Because of its striking resemblance to the shape of their canoes, the Maori (native New Zealanders) dubbed the weevil *Tuwhaipapa*, the god of the newly made canoe. Insect symbolism became even more widespread in the ancient world, especially in the Middle East. For example, hornets were symbolic of the kingdom of the first Egyptian dynasty (around 3100 B.C.), due to their fierce and threatening nature. The sacred scarab played a significant role in the religious lives of the early Egyptians. The word *psyche* was not only the ancient Greek word for "butterfly," but also a metaphorical reference to the soul. The ancient Greeks also used beetles, fleas, and flies in their plays and fables, such as Aesop's well-known accounting of the ant and the grasshopper. Other insects were used by the Greeks to symbolize everything from industriousness to insignificance. Early Christian animal symbolism related insects to foulness and wickedness, but eastern religions emphasizing spiritual unity with nature considered insects to signify good luck. Depictions of bees were familiar on the shields of medieval knights. Grasshoppers, beetles, and moths found their way onto the coat of arms of numerous European families. Today, modern "clans" such as college and professional sports teams have insect names or use them as mascots.

Insects in literature, language, music, and art

Scenes of honey hunting first appeared in a Spanish cave some 6,000 years ago. Later, banquet scenes depicting skewered locusts graced the walls of an ancient Egyptian tomb. In ancient Greece, images of insects appeared on jewelry and coins. Sacred scarab ornaments and other insect jewelry appeared in Egypt as early as 1600 to 1100 B.C. and were commonly worn as amulets, necklaces, headdresses, and pendants. For centuries the Japanese have carved pendants, or *netsuke*, in a variety of insect forms. In the Middle Ages, insects were used to adorn decorative boxes, bowls, and writing sets. Medieval monks incorporated insects in their illuminations, the brightly painted borders surrounding important religious manuscripts. The increase of geographic exploration and scientific knowledge saw insect subjects migrating to center page as part of descriptions of newly discovered species. One of the best-known references to insects in literature is in Franz Kafka's *Metamorphosis*, in which the lead character is transformed into a beetlelike insect. Today insect jewelry is very popular, with beetles, butterflies, cicadas, and dragonflies carved from stone or fashioned from glass and metal. The durable bodies of the beetles are used as pendants and earrings, while butterfly-wing art is commonly sold to tourists in tropical countries. Many countries celebrate insects on their postage stamps. Insects have even inspired the names of rock bands and nicknames for automobiles. They also serve as the root for such derogatory words as lousy, nitpicker, grubby, and beetle-browed. Other insects have found their way onto the big screen, inspiring fear and loathing in B-movies or providing the basis for comical, even sympathetic, characters in major motion pictures and animated features.

Insects as pests

Approximately 1% of all insect species are considered to be of any negative economic importance. Yet these relatively small numbers of species are responsible for significant economic loss as a result of their feeding activities on timber, stored products, pastures, and crops. Insects become pests as a result of a complex set of circumstances created or enhanced by concentrations of plant or animal foods. Plant-feeding insects, particularly those feeding on legumes, tomatoes, potatoes, melons, gourds, and grains, are some of humankind's greatest competitors for food. One-third to one-half of all food grown for human consumption worldwide is lost to damage caused by insects. Blood-feeding insects, such as mosquitoes and other flies, are responsible for spreading numerous pathogens, resulting in millions of human illnesses and deaths annually.

Agricultural pests

The cultivation of plants for agricultural or horticultural purposes has greatly affected how we perceive insects. Initially, crops were raised in small and widely distributed plots, resulting in small and localized pest outbreaks. But the development of large monocultures of grain (corn, rice, wheat) and fiber sources (cotton) covering vast regions during the past 150 years has resulted in a host of new insect pests. Faced with abundant, predictable, and nutritious concentrations of food in the virtual absence of natural predators and parasites, plant-feeding insects soon achieved unimaginable numbers. Once established, these pests quickly multiplied and spread, threatening economic ruin for entire regions. Every crop grown as food (for humans and their animals), as well as those grown for recreational consumption, either legally (coffee, tea, cocoa, and tobacco) or illegally (marijuana, coca, poppies) has one or more insect pests.

More than 6,000 species of thrips, true bugs, homopterans, beetles, flies, butterflies and moths, ant, wasps, and even a few termites are considered serious agricultural pests throughout the world. In some regions, the catastrophic loss of food crops as a result of locust or caterpillar depredations not only causes enormous economic harm, but can also lead to human malnutrition, starvation, and death. Large numbers of insects reduce crop yields by feeding on vegetative structures or removing nutritious fluids. Species with piercing/sucking mouthparts (aphids, psyllids, leafhoppers, mealybugs, and plant bugs) transmit plant pathogens such as bacteria, fungi, and viruses, that reduce plant vigor. Beetles are implicated in the spread of viral diseases and pestiferous nematodes.

Forest pests

In temperate forests, adult and larval moths and beetles, and to a lesser extent sawfly and horntail wasps and a few flies, attack stands of living, dying, or dead trees. Although these insects form an essential part of the nutrient cycling system, their presence in trees managed for timber can quickly propel them to pest status. These insects bore into trunks and limbs, while others focus their attack on cones, seeds, foliage, buds, and roots. Living trees may suffer from reduced vitality, stunted growth, or death. In most cases, forest pests attack trees that are already under stress from lack of water, poor nutrition, or injured by disease or fire. With their impaired defense systems, trees often fall victim to plant diseases introduced by these insects.

A beekeeper examines a frame of honey produced by *Apis mellifera*. (Photo by ©Nigel Cattlin/Holt Studios International/Photo Researchers, Inc. Reproduced by permission.)

Forest pests not only cause stunted growth, reduced yield, and lowered values of lumber and other products, but are also responsible for the loss of habitats, damaged watersheds, and increased fire hazards. Habitat disruption also leads erosion and flooding. Pest outbreaks are more likely to occur in pure stands, old growth forests, and in plantations. Ice storms, hail storms, floods, high winds, drought, disease, and fire may trigger forest-pest outbreaks.

Structural pests

Termites, wood-boring beetles, and carpenter ants can and do infest dry and treated timber used to build homes and outbuildings. They damage wood by feeding or continually chewing tunnels until the wood is completely hollowed out, leaving only a deceptive outer shell. Termites are the most destructive structural pests and attack timber throughout the world. While termites tunnel secretly, leaving behind only the occasional pile of frass in door and window jams, successive generations of anobiid ("death watch") and bostrichid (false powderpost) beetles clearly mark their presence with shotlike exit holes covering wooden surfaces.

Household pests

Long before the appearance of humans, insects were nibbling on scattered seeds and grains, scavenging rotting fruit, hair, feathers, or decaying flesh. The moment humans began to store and process these materials for our own use, our relationship with these insects was forever cemented. The scavengers quickly adapted to living in human habitation, becoming nearly cosmopolitan (worldwide in distribution) as a result of commerce and other human activities.

The larvae of clothes moths and carpet beetles destroy woolen clothing, rugs, and hides. Powderpost beetles destroy finished wood products, damaging floorings, cabinetry, and furniture. The cigarette beetle (*Lasioderma serricorne*) is a serious pest of spices, legumes, grains, and cereal products. The drugstore beetle (*Stegobium paniceum*) attacks spices and legumes, as well as herbs, crackers, and candy. The omnivorous sawtoothed grain beetle (*Oryzaephilus surinamensis*) infests cereals, bread, pasta, nuts, cured meats, sugar, and dried fruits. The confused flour beetle (*Tribolium confusum*) is one of the most important pests of food stored in supermarkets and homes. The rice weevil (*Sitophilus oryzae*) infests stored cereals, especially rice; the closely related grain weevil (*S. granarius*) prefers wheat and barley products. Larvae of the Indian mealmoth (*Plodia interpunctella*) feed on coarsely ground grains, such as cornmeal (also known as "Indian" meal), as well as crackers, dried fruit, nuts, and pet food, leaving sheets of silk in their wake. The American (*Periplaneta americana*), Oriental (*Blatta orientalis*), German (*Blattella germanica*), and brown-banded (*Supella longipalpa*) cockroaches not only invade food stores, but also destroy paper products and clothing. Ants frequently invade homes in search of food and water.

Pests of domestic animals

Flies, fleas, and lice harass, weaken, infect, or kill livestock, poultry, and pets, and the damaged hides and reduction in meat, milk, egg and wool production result in economic loss. Some insect pests are quite specific when selecting their hosts, others attack many species of domestic and wild animals. Horn flies, mosquitoes, black flies, blowflies, and screwworms are all considered general blood-feeding pests. The irritating bites of female heel flies can cause entire herds of livestock to stampede. Blood-feeding horn flies and lice may feed in such numbers that their hosts suffer anemia from blood loss. Screwworm maggots live in the wounds of injured animals, where they feed on living tissues, sometimes causing death of the host. Horseflies, blowflies, and stable flies transmit anthrax and other pathogens that cause keratitis, and mastitis. Parasitic fleas and lice spend most, if not all, of their lives on the bodies of their bird and mammal hosts. Infestations of cattle lice may lead to terrible irritation and thus constant scratching against fences and posts, resulting in raw skin, hair, and blood loss. Infested animals lose their vigor and fail to gain weight. Hog lice carry swinepox and other infectious swine diseases. Flat, wingless sheep keds are parasitic flies that pierce the skin and suck the blood of sheep, causing them to rub, bite, and scratch. Some species, such as fleas, beetles, and cockroaches, are intermediary hosts of disease-causing organisms such as poultry tapeworm and heartworm, and may infect domestic animals and pets if ingested.

Vectors of human disease

Lice, flies, mosquitoes, fleas, and assassin bugs are important insect vectors of human disease. Until World War II, many more soldiers died as a result of infectious diseases spread by insects than as a result of injury on the battlefield. Houseflies are implicated in the mechanical spread of disease, and myasis, or the infection of the body with maggots, is a common medical phenomenon. Assassin bugs carry the flagellate protozoan that causes Chagas's disease.

Mosquitoes spread several viral diseases, including yellow fever and various encephalitides. They are also vectors of disease-causing nematodes and protozoans. Yellow fever and malaria infect 200 to 300 million people each year, killing approximately 2 million. The most notorious disease-spreading mosquitoes belong to the genera *Aedes* and *Anopheles*. *Aedes aegypti* is found worldwide in warmer regions and is the vector of yellow fever. Other species of *Aedes* are associated with dengue, eastern and western encephalitis, and Venezuelan equine encephalitis. Several species of the widely distributed *Anopheles* are vectors of malaria.

The tsetse are biting, blood-feeding, African flies and are efficient vectors of African sleeping sickness in humans and nagana in animals, diseases caused by a parasitic trypanosome. Although tsetse prefer other animals as hosts, they frequently bite humans. Historically, sleeping sickness dramatically retarded the exploration and settling of much of the African continent. Although sleeping sickness is usually under control in most regions, nagana remains an important and widespread disease.

Fleas ingest the plague-causing bacterium by feeding on infected rodents. As the rodents die from the disease, the fleas bite humans to feed, passing on the bacterium to their new host. Bubonic, pneumonic, and septicemic plague are all forms of the plague. Bubonic plague, or Black Death, was responsible historically for the deaths of millions of people in Europe, Asia, and Africa. Although plague is treatable today with antibiotics, future pandemics are still likely. With increased bacterial resistance to antibiotics, the specter of resistant strains of the plague-causing bacterium remains a deadly possibility. Oriental rat fleas also spread endemic typhus.

Outbreaks of the body louse can reach epidemic proportions in elementary schools and military posts. These lice are vectors for the rickettsial disease epidemic typhus. Unlike mosquitoes, the lice do not inject their host with the pathogen. Instead, human hosts infect themselves by crushing lice at the bite site, or by exposing their eyes, nose, and mouth to the feces of infected lice. Typhus is fatal to body lice. They pick up the disease from the blood of an infected host and soon die, but not before they can infect others.

Controlling insect pests

Today, many pests are effectively controlled by a system known as integrated pest management (IPM). IPM programs entail a combination of chemical, mechanical, cultural, physical, or biological controls. These methods are augmented by legislation mandating proper pest-control procedures, implementing quarantines of plant and animal hosts, and certifying pest-free imported and exported agricultural commodities.

Effective pest control, whether in the home, forest, or farm, is absolutely dependent upon the correct identification of the insect and a knowledge of its life cycle. Armed with this information, pest-control efforts can be directed at the pest's most vulnerable stages of development. Even closely related species can have slight differences in biology that can render useless methods designed to control their relatives. Knowledge of a pest's biology, coupled with thorough and regular survey and detection programs, frequently enable the implementation of preventive measures before pest populations cause serious economic damage.

Chemical controls can attract, repel, or poison pests. They are used in traps or applied as dusts, granules, sprays, aerosols, and fumigants. Systemics penetrate plant tissues, killing only the insects that feed on the plants. Chemical insecticides include inorganics, oils, botanicals, and synthetic organophosphates. These compounds act as physical poisons, killing the insects by asphyxiation or by abrasion that causes the loss of body fluids. Protoplasmic poisons precipitate proteins, while respiratory poisons deactivate respiratory enzymes. Many commonly used insecticides, including parathion, pyrethrins, and dichlorodiphenyltrichloroethane (DDT), are nerve poisons. Abuse of insecticides, especially those that persist in the environment such as DDT, can and do lead to catastrophic declines in predators that rely directly or indirectly on insects as food. Insect resistance to pesticides, which can occur in a matter of generations, may preclude the usefulness of chemical applications.

Mechanical and physical controls directly affect the insects or their environments. Labor-intensive methods such as hand picking, trapping, and barriers are often impractical on a large scale. Exposing storage containers to extreme temperatures will reduce insect populations in grain elevators, while soaking plant bulbs in warm-water baths kills some pests infesting nurseries.

Cultural controls involve the utilization of various agricultural practices to reduce insect pest populations and require extensive knowledge of the pest's life cycle. Crop rotation prevents the buildup of pest populations over time, as does the careful selection of mixed crops. Small, expendable plantings known as trap crops are planted to lure pests away from major crops. Once infested, these crops are treated with pesticides, plowed under, or both. Tillage reduces populations of soil insects by eliminating their host plants, exposing them to dry air and predators, or crushing them. Clean culture is the removal of all crop residues, leaving nothing behind for pests to lay their eggs on or to eat. Changing planting and harvesting times to reduce crop exposure to pests can also be effective. Genetically altered crops, although controversial, may be less attractive to pests or more tolerant of pest damage. Some possess insecticidal qualities or lack essential nutrients required by pest species.

Insect populations may be naturally or biologically controlled by a host of "checks and balances." When a species becomes established outside its natural range, it may achieve pest status because its normal population controls did not also become established. This same disruption may occur when a species encounters an unlimited food supply in the highly disturbed ecosystems typical of the monoculture practices of modern agriculture. Entomologists are then deployed to find the geographic origin of the pest and to identify its naturally occurring predators, parasites, and pathogens. These "biological control" organisms are then selected for the host specificity, effectiveness, ease of rearing, and ability to adapt to their host's new environment. Typically, successful natural or biological control is marked by closely synchronized fluctuations between prey and predator populations.

Pest outbreaks are followed by heavy predation followed by prey scarcity that depletes the predator population. The goal of biological control programs is not to eradicate the pest, but to keep populations below levels at which they are economically damaging.

Pest control of any kind has its risks and may adversely affect other species or their habitats. Biological control is based on the fact that the introduction of alien organisms will disrupt established populations. Tests are usually conducted to determine whether introduced organisms will adversely affect nontarget organisms, but extensive investigations of the delicate balance between pests, their natural enemies, and other insect species in the community are seldom practical and difficult to assess. In some cases, species introduced as biocontrols, such as the marine toad into Australia to control sugar cane beetles, have become part of the much larger problem of alien species invasions, a major factor in local and species-wide extinction.

Insect phobias

The dislike of humans for insects is understandable, as they bite, sting, invade our homes, infest our food, and ravage our gardens and crops. But for some, this dislike is replaced by an intense, irrational fear called entomophobia. Entomophobia in childhood usually develops in both sexes, but if developed in adulthood occurs mostly in women. Entomophobes not only fear insects, but also the places where they might occur, and avoiding contact with insects can be quite problematic given the animal's universality. As with those who suffer other phobias, entomophobes experience typical fright responses such as an increased heart rate, sweating, labored breathing, tremors, and dizziness, as well as temporary paralysis or running away. Entomophobia (and other phobias) may develop as a result of innate fears regarding threatening objects or situations, or genetic predisposition. Treatment may include psychoanalysis or modeling therapy, in which a therapist, friend, or relative touches an insect in the presence of the entomophobe.

Delusory parasitosis, or Ekbom's syndrome, is when a person is under the false illusion he or she is being attacked by insects or other parasites. Sufferers typically seek help from entomologists and pest-control operators, rather than qualified mental health care professionals, convinced that "bugs" are burrowing under their skin. Sufferers, often middle-aged men and menopausal women, subject themselves to caustic fluids and compounds to rid themselves of the imagined pests, or mutilate themselves with knives or razors in an effort to dig them out. Friends and relatives are sometimes so convinced of the veracity of the sufferer's condition that they develop sympathetic itching.

The scientific study of insects

Entomology is the science of insect study. Entomologists around the world observe, collect, rear, and experiment with insects, either for purely scientific or for practical reasons. Regardless of the approach, the basis for useful and accurate entomological work is insect classification. Taxonomy, the

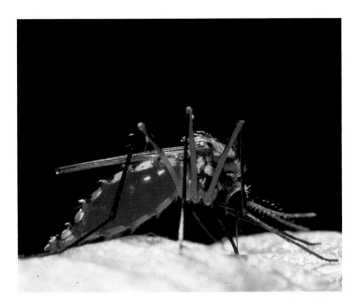

A blood-filled mosquito (*Aedes trisseriatus*) biting its human host. (Photo by Dwight Kuhn. Bruce Coleman, Inc. Reproduced by permission.)

science of naming things, is one of the basic elements of insect study. Unlike common names, which can vary considerably between regions, cultures, and languages, scientific names are universally recognized. Papers published around the world in different languages all use the same scientific names for the same species. The use of this universally recognized system greatly facilitates the storage and retrieval of biological information. Efforts to standardize common names are usually applied only to economically important insects.

Insect systematics is the study and ordering of the natural diversity of insects. The discipline blends taxonomy with information from other fields of biology, such as morphology, behavior, genetics, biogeography, phylogenetics, and DNA sequencing, to arrive at a classification that reflects the evolutionary paths of insects. Systematics also contributes information used in other insect studies, including faunistics, the study of some or all insects in a region; ecology, the study of insect interrelationships with their environment; and zoogeography, the study of insect habitats and distributions, past and present. Insect distributions based on detailed faunistic and taxonomic studies provide clues to the nature and extent of past climates. The fossil remains of insects whose species or genera are extant today, combined with information on their current habitat requirements, can be used with a considerable degree of confidence to infer ancient ecological conditions that prevailed tens or hundreds of thousands of years ago. The validity of these and other conclusions all depend on the quality of current systematic knowledge. Systematic data is also critical to our ability to conserve insects threatened with extinction and to preserve their habitats. Laws and regulations drafted by legislators and other administrative bodies to protect sensitive species are more likely to succeed if they are based on systematic information that accurately reflects the biology, distribution, and habitat requirements of the insect.

There are a number of entomological subdisciplines. Forensic entomologists examine insects found at crime scenes to determine such facts as time and place of death, and also investigate infestations of food and materials to determine liability. Forest entomologists track the depredations of insects attacking managed timber. Economic entomologists focus their attention on controlling pest species that attack gardens, crops, vineyards, and orchards. Medical and veterinary entomologists study insects as vectors of disease. As humans continue to modify and destroy habitats, we increase our contact with insect vectors, increasing the frequency and importance of insect-borne diseases that afflict both humans and animals. Battling diseases, such as the West Nile virus, that affect humans, domestic animals, and wildlife will require the close cooperation of researchers in the biomedical and veterinary sciences. Conservation medicine connects these two fields of endeavor to explore the links between wildlife species and the health of ecosystems and people. By pinpointing environmental sources of pathogens, scientists can begin to understand the ecological causes of changes in human and animal health, and entomologists will make significant contributions to the development of this field.

Beneficial insects

The human benefits derived from insects are enormous, not only as objects of pure scientific interest, but also for the services and products they provide. They not only pollinate crops, but also directly produce food, fiber, and other useful products. Insects are an important source of nutrition and income in many parts of the world. They also help control pest species, reducing our dependence upon chemical controls. Insects provide educators with a rich and diverse template from which they can introduce students to the natural world and provide scientists with sensitive ecological indicators.

Pollinators

Insects are the most important animal pollinators. Brightly colored flowers with highly attractive odors draw insects seeking nutritious pollen and nectar. Pollen is an excellent source of protein; nectar is high in carbohydrates. The relationship between flowers and insects is one of mutualism: angiosperms provide insect pollinators with food, while the insects are essential to the propagation of the angiosperms. Insects in at least 29 orders participate in pollination. Although many flower-visiting insects focus their attention on pollen and nectar, others eat part or all of the inflorescence.

The effectiveness of pollinators of agricultural crops, especially bees, is determined by their host-plant specificity, foraging distance, daily foraging period, prevailing temperature and humidity, number of flowers visited, and pollen-carrying capacity. Although those honey bees managed in association with extensive monoculture are effective pollinators, solitary native species or other social bees are frequently better adapted for pollinating orchards and other crops. Bumblebees are used to pollinate greenhouse tomatoes in North American and Europe and are the sole pollinators of red clover. New Zealand has imported "long-tongued" bumblebees to increase their red clover yields with great success. In Europe, flies are the most effective pollinators of special varieties of cabbage and other cruciferous crops.

In Japan, fruit growers used bundled hollow reeds as artificial nests to encourage the pollination activities of Japanese horn-faced bees (*Osmia cornifrons*) with spectacular success. This solitary species has since been imported into the United States for orchard-fruit pollination. The European alfalfa leaf-cutter bee (*Megachile rotundata*) has been imported to facilitate alfalfa pollination. Small colonies of the native social Alkali bee (*Nomia melanderi*) have also been used in some parts of the United States as pollinators. These and other solitary bees are not as easily manipulated as honey bees, but their use as pollinators in commercial crops is increasing.

Oil palms, natives of West Africa, were imported into Southeast Asia as a major plantation crop, but with limited success. Outside their natural range, oil palms had to be pollinated by hand. Entomologists searched oil palms in West Africa and soon discovered a complex of beetles, mostly weevils, responsible for their pollination. One species was imported to Malaysia, where its establishment has resulted in consistently higher yields of oil palms.

Apiculture

Apiculturists, or beekeepers, manage colonies of bees that have been selected for their docility and foraging capability. The bee genus *Apis* consists of four species in Europe and Asia, of which one, *A. mellifera*, is widely kept by humans for their honey and wax production. For centuries, honey was the only sweetener available to Europeans, and fermented honey was used to make wine such as mead in Europe and *tej* in Ethiopia. The antiseptic qualities of honey also made it useful as a dressing for wounds. Artists and sculptors have long used beeswax to cast and mold sculptures, construct masks, and to build models. Beeswax was also used in ointments, suppositories, cosmetics, candies, lubricants, adhesives, varnish removers, and furniture and shoe polishes, but has been supplanted by paraffin. The pollination services of honeybees are just as valuable as their production of honey and wax. Beekeepers transport millions of hives each year in an effort to market their bees' pollination services and to ensure a continuous supply of pollen and nectar for them throughout the season.

Early beehives in Greece were made of clay pottery, while those European and Africa were made of hollow logs. Later the Europeans began using dome-shaped beehives constructed of woven straw. Modern hives first appeared in or around 1850 in the United States and Europe. These bee enclosures consist of a stack of boxes, each with frames suspended inside that serve as a foundation for the waxen honey and brood comb. The brood capacity of each hive is expanded throughout the season by adding new boxes and frames. The bees usually store their brood in the deeper boxes below, while keeping honey in the shallower boxes and frames above. In temperate regions, the bulk of the honey, up to 222.5 lb (100 kg) per box is harvested annually. The bees are fed a sugar solution at regular intervals during the late summer and throughout the winter.

Other valued bee products include wax, royal jelly, new swarms, and propolis. The latter is a resinous material col-

An eastern lubber grasshopper (*Romalea guttata*) eating grass. (Photo by J. Charmichael. Bruce Coleman, Inc. Reproduced by permission.)

lected by foraging bees and is used to attach combs to the roof of the hive. It has antibiotic qualities and has been claimed to successfully treat some human ailments. Royal jelly is high in protein and vitamins. It is fed to all larvae for the first five days, but only those destined to be queens receive it until pupation. Pollen has many therapeutic properties and is reported to alleviate the symptoms of colds, influenza, asthma, hay fever, arthritis, rheumatism, prostate disorders, and menstruation and menopause, as well as to enhance virility. Pollen treatments often involve ingesting a mixture with unrefined honey or propolis. In Eastern Europe and Western Asia, areas with intensive apiculture, people frequently live to be 100 years or older. Unrefined honey, including bits of wax, pollen, and propolis, is a regular component of the local diet. However, the precise value of these components, and their relationship to longevity, remains unclear.

Sericulture

Commercial silk is harvested primarily from cocoons spun by the silkmoth (*Bombyx mori*), also known as the Oriental silkworm or the mulberry silkworm. The silk is produced by the salivary glands of the larvae, or silkworms. A native of

China, the silkworm is the world's only completely domesticated insect; none are known in the wild. In other parts of China, Africa, and India, cocoons of other moth species are first collected in the wild before the silk can be harvested.

The earliest records of sericulture date back to 2600 B.C. Raw silk became an important item of trade between China and Europe. The Silk Road, opened in A.D. 126, stretched westward nearly 6,000 miles across China, Turkestan, and Iraq, before ending at the northeastern shore of the Mediterranean Sea. China carefully guarded its silk industry for centuries, but about 150 B.C., eggs were smuggled out of the country and into India. In A.D. 300, the Chinese sent four women to Japan to instruct their royal court in the art of sericulture, establishing what would become the world's largest silk industry. In A.D. 555, two European monks smuggled silkworm eggs and mulberry seeds out of China, marking the beginning of sericulture in the western world.

Sericulture is widely practiced today. China and Japan remain the largest producers of silk, but thriving silk industries are found in South Korea, Indonesia, Brazil, Thailand, and Uzbekistan. China accounts for about 80% of the world's raw silk production. Once the largest producer of silk, Japan is now the world's largest consumer, its market driven by the manufacture of kimonos. Italy is the largest producer of silk in Europe, while the United States is the largest importer. The present value of silk production around is estimated at more than $1 billion.

Raising silkworms is incredibly labor intensive. About 1,700 cocoons are required to make one dress; and their caterpillars must consume 125 lb (56.7 kg) of mulberry leaves before pupation. Just one pound of caterpillars can consume 12 tons (10.9 t) of mulberry leaves before reaching the pupal stage. The larval host plant, mulberry, is easily and widely grown, but most caterpillars in Japan are reared on a completely synthetic diet. Silkworms are susceptible to various maladies. Interestingly, the investigation of one of these diseases ultimately led the French microbiologist Louis Pasteur to correctly deduce the microbial origin of diseases.

To harvest the silk, mature caterpillars are transferred to a rack where they can spin their cocoons before pupating inside. The cocoons are then boiled to kill the pupae within and to remove the sericin, a dull and chalky outer coating. The ends of the silk filaments from several cocoons are located, unraveled, wound together, and attached to a spool. The number of filaments attached to the spool determines the size of the thread. The spools of raw silk are then soaked and dyed before they are reeled onto skeins. The skeins are bundled into books, which are gathered into bales and shipped to mills for weaving.

Insect farming

Apiculture and sericulture were the earliest forms of insect farming. Today, other insects are reared in captivity for commercial, research, and biocontrol purposes. Insect farms may be small-scale operations or massive commercial enterprises. Cockroaches, grasshoppers, crickets, fly maggots, bloodworms, fruit flies, mealworms, and other insects are regularly reared throughout the world as live pet food, fish bait, re-

search subjects, and educational tools. Still others are mass reared to combat weeds and insect pests.

Butterflies and other insects are raised around the world and shipped live to insect zoos and butterfly houses, research facilities, or preserved for the dead-stock trade. Most specimens are used as ornaments or in decorative displays. Hobbyists and researchers requiring quality specimens also drive a significant portion of the dead-stock trade. The sale of preserved insects, mostly beetles and butterflies, amounts to tens of millions of dollars annually.

Butterfly farms, established in Central and South America, Malaysia, and Papua New Guinea, have been regarded as a benefit to butterfly conservation because they do not rely on specimens caught in the wild to supply the commercial demand for living and preserved specimens. Butterfly farmers enclose their breeding colonies with shade cloth to protect the stock from parasites and predators. Some pupae are shipped live, others are kept until they reach adulthood. After the butterflies emerge, some are released to enhance local wild populations, while the remainder is packaged for sale. Proceeds from the sales are used to support families, maintain the farm, and to preserve or enhance the environment that replenishes the breeding stock.

Insects as biological controls

Biological control, or biocontrol, is the augmentation of naturally occurring enemies of a pest to reduce or eliminate the reliance on expensive and potentially ecologically disastrous pesticides. Another form of biological control involves irradiation to produce large numbers of sterilized male insects, such as fruit flies and screwworms, for release. When combined with cultural practices, mechanical controls, and carefully managed use of insecticides, biological control is an important component of any IPM program.

Numerous predatory and parasitic insects are commercially reared for use in biological-control programs in greenhouses, fields, and orchards. Parasitic wasps are routinely released worldwide to control plant-feeding pests such as moth caterpillars, aphids, and scale insects. After some early and well-documented successes, the widespread introduction of ladybird beetles (or ladybugs) to control aphids and scales has achieved mixed results. Although most insect biocontrols target garden and agricultural pests, others attack medical and veterinary pests or weeds. For example, cactus-feeding moths, scales, and beetles have been imported into Australia and parts of Africa to control invasive cacti.

Insects as food

Insects are an excellent source of fat and protein and were no doubt a staple in the diet of early hunter/gatherers. Today, eating insects, or entomophagy, is widely practiced. Throughout tropical Africa, people invade termite mounds for the large queens and eat them on the spot. The queens' swollen, egg-filled abdomens (the size of a breakfast sausages) are an extremely rich source of fat. Another caterpillar, the mopane worm (*Imbrasia belina*) is regularly harvested in southern Africa, where it is dried and eaten like a cracker or soaked in tomato sauce. Large grasshoppers are commonly eaten in parts of Africa and Asia. In Thailand, locusts are roasted and eaten on satay sticks, while giant water bugs are roasted or soaked in brine and eaten like crackers. June beetles, weaver ants, and mole crickets are eaten in the Philippines, and adult and larval weevils and scarab beetles are consumed in parts of Asia and Australia. Australian aborigines eat a wide variety of insects, including the well-known cossid moth larva known as the *witjuti* or "wichety" grub. The aborigines also excavate nests of honeypot ants in search of the nectar-filled ants known as *repletes*. A different genus of honeypot ant is known from western North America, where some rural Mexicans also dig them up for food. Western European culture has largely ignored insects as food, although they appear regularly as a curiosity at entomological gatherings and some eateries. Given their nutritional value, abundance, and diversity, the commercial opportunities afforded by the development of insects as food are boundless.

Medicinal insects

The plethora of defensive chemicals, mating pheromones, toxins, and other compounds produced by insects undoubtedly have a therapeutic value and, as with botanicals, their early use as "folk medicines" should not lessen their prescription or discourage further research. The real or imagined benefit of using insects as medicine probably developed as a result of their consumption as food. Various insects were burned, roasted, and pulverized into numerous concoctions purported to have some curative effect. Early Europeans used powdered ladybird beetles to relieve toothache and to cure measles and colic. Whirligig and rhinoceros beetles were used in preparations to increase the libido, and one Javanese click beetle is the primary ingredient of a particularly sexually stimulating potion. Japanese folk medicine incorporated species from several families of beetles to treat conditions as varied as convulsions, cancer, hydrophobia, and hemorrhoids.

In Medieval Europe, concoctions derived from insects were thought to transfer their outstanding qualities to the patient, and thus conspicuously hairy bees and flies became ingredients in hair tonics and baldness cures. Singing crickets and katydids were recommended to treat disorders of the ears and throat. Earwigs were prescribed for earache, perhaps because of their fan- or ear-shaped hind wings. Wasp galls were pressed into service as treatment for hemorrhoids; simply carrying one in a pocket was all that was required. Even today in parts of Latin America and Asia, the ingestion of large, powerful horned beetles is thought to impart the same robust qualities to the consumer. As far as is known, any therapeutic effects achieved by these treatments were attained entirely through the power of suggestion!

Insect defenses have been adapted and exploited for pharmaceutical use. Bee stings have long been used to treat patients with rheumatism, as the venom acts to increase blood flow to the afflicted areas while simultaneously stopping pain. Several commercial preparations using bee venom are sold today. Cantharidin, a blistering defensive compound found in the bodily secretions of blister beetles, renders them distasteful to predators. Cantharidin is highly toxic to humans if ingested; 30 mg is lethal. In smaller doses, cantharidin can cause severe gastroenteritis and irreversible kidney damage.

In spite of its toxicity, cantharidin extracted from the European Spanish fly (*Lytta vesicatoria*) and other blister beetles has been used to treat epilepsy, sterility, asthma, rabies, and lesions resulting from gonorrhea. *Lytta vesicatoria* is best known for its purported qualities as an aphrodisiac, which were first noted in the sixteenth century. Today pharmaceutical catalogues mention cantharidin for use in some ointments and plasters, the preparation of tinctures in veterinary medicine, and an ingredient in some hair-restorers. In Peru, warts are scarified, then covered with a pulp made from blister beetles. A blister forms over the wart, and after several days of treatment, the wart is destroyed.

During the Napoleonic Wars and the U. S. Civil War, military surgeons noted that untreated soldiers with deep wounds infested with maggots healed more quickly and in greater numbers than their treated comrades. During World War I, it was discovered that some wound-infesting maggots consumed only dead tissue. As they fed, the maggots excreted large quantities of the nitrogenous substance called allantoin that acted as a sterilizing agent. For years surgeons have employed maggots carefully raised under aseptic conditions to clean deep wounds, especially those filled with pus or associated with bone fractures. However, synthesized allantoin and the use of antibiotics have rendered maggot therapy all but obsolete, except in the most extreme of cases.

Ants and beetles with large, piercing mandibles have been used as crude sutures to close wounds. The insects are held up to the edge of the wound and allowed to bite the skin, bringing the two sides together. The heads are then separated from the body and left in place until the wound has healed. As mentioned earlier, antiseptic honey has been used to dress wounds, while sterile and inert beeswax is used in some orthopedic surgeries to fill spaces in bones.

Other products produced by insects

Several species of scale insects and their relatives have been used for centuries to produce varnishes, dyes, and inks. Shellac is a sticky, resinous material produced as a shell-like coating by the lac insect (*Laccifer lacca*), native to India. These insects live in small groups, encrusting twigs of acacia, soapberry, and fig. Twice each year, the twigs are gathered and the insects scraped off. Bright red lac dye is then extracted from the body tissues of the female with hot water, while the resinous "shellac" is melted and filtered. The shell-lac eventually cools, forming the flaky texture of commercial shellac. Because of its elasticity, resistance to solvents, adhesive properties, and insulative qualities, shellac is used in furniture and shoe polishes, printing inks, electrical insulators and sealants, clothes buttons, and a variety of hair-care products. Before compact discs and vinyl albums, shellac-based records were the standard in the recording industry. India exports 50.7 million lb. (23 million kg) per year, mostly to Europe and the United States.

The red dye cochineal is produced from the dried and pulverized bodies of the cactus mealybug, or cochineal insect (*Dactylopus coccus*). The pigment is repugnant to the insect's predators and probably deters parasites. Feeding exclusively on cactus, the cochineal insect is distributed from southwestern United States to South America. The Pima Indians of Arizona and the Aztecs of Mexico cultivated and used these insects as dye. The Spanish explorer Hernando Cortez was the first westerner to report the use of cochineal, and the intense vermilion dye soon became an important item in European commerce. Spain's monopoly of cochineal was eventually broken, and imported scales and their cacti were soon raised throughout the Mediterranean region and in India and Australia. However, despite the efforts in these various locales, the insects managed only to establish themselves in the Canary Islands. To produce cochineal, the insects are scraped off the cactus and killed in boiling water, which also removes their waxy coating. The bodies are dried in the sun and then ground into a powder. Approximately 70,000 insects are needed to produce 3 lb (1.36 kg) of powder, or 1 lb (0.45 kg) of cochineal. The importance of cochineal has declined with the increased use of synthetic dyes, although it still enjoys some popularity as a natural food coloring in Latin America and in some soft drinks. Its use in cosmetics is complicated by the fact that Orthodox Jews do not consider cochineal insects kosher. Today Peru and the Canary Islands are the principal producers of cochineal.

Insects as bioindicators

Researchers use insects as indicator species to measure environmental disturbances, the effects of habitat fragmentation, and changes in biodiversity. The responses of some sensitive insects to habitat disturbances are rapid, predictable, and easily analyzed and measured. For example, aquatic beetles and the larvae of mayflies, stoneflies, dragonflies, and damselflies are especially sensitive to even subtle changes in water temperature, chemistry, and turbidity. The presence or absence of a particular species may indicate habitats polluted as a result of illegal chemical dumping, agricultural and mining runoff, effluent from power plants and sewage treatment, or increased erosion as a result of logging. Careful monitoring of aquatic insect populations can signal problems long before pollutants manifest themselves in plant and vertebrate populations. Changes in some terrestrial insect populations, especially those species with high ecological fidelity, are used to measure changes in plant communities, such as those affected by urban and agricultural development. Soil-dwellers have long been used to indicate soil fertility and levels of pollution. The presence of easily identifiable, taxonomically well-known, and intensively studied species, such as butterflies and tiger beetles, are also used to identify significant ecological habitats worthy of protection.

Insect conservation

Insect populations are subject to decline, extirpation, and extinction, primarily as a result of habitat loss. Urbanization, agricultural development, water pollution and impoundments, wetland modification, exotic introductions, and pesticides have all been linked to the decline of some species. Even lesser-known, although no less significant, phenomena such acid rain, electric lights, and off-road vehicles exact a toll on insect populations. Collecting, although touted by some as a significant threat to localized species, has never been shown to result in the extinction or extirpation of populations. Nonetheless, collecting large numbers

Oriental cockroaches (*Blatta orientalis*) gather at night on an uncleared café table. (Photo by Kim Taylor. Bruce Coleman, Inc. Reproduced by permission.)

of specimens from a single population is considered unethical by many entomological societies, and is stated as such in their guidelines. The Xerces Society, based in the United States, is the only international organization dedicated solely to the conservation of all invertebrates, including insects. Several European countries publish catalogues of sensitive, threatened, or endangered species known as Red Lists, and other national organizations sponsor insect-conservation programs that include the formulation of plans for habitat management and restoration and captive breeding, and reintroduction programs.

Most legislative efforts to conserve insects have been directed at butterflies and beetles, although other insect orders receive some recognition. Internationally, the Convention on International Trade in Endangered Species (CITES) monitors the importation and exportation of listed species. The Endangered Species Act (ESA) of the United States is a comprehensive legislative effort, not only to list and protect threatened and endangered species, but also to provide for the preservation of their habitat. In Papua New Guinea, laws protect some species, regulate the sale of others, and preserve important habitats. The collection and sale of Papua New Guinea insects may be sanctioned only by a government marketing board known as the Insect Farming and Trading Agency. National parks and wildlife reserves may have laws protecting all wildlife in their jurisdiction, including insects. Other local, state, or provincial agencies also protect insects. Many species are afforded protection because they occur in small populations living in highly restricted or sensitive habitats, such as caves or sand dunes.

Insects as educational tools

Insects afford primary, secondary, and college students an inexhaustible resource that stimulates curiosity and provides important insights into the biological world. Processes found in all animals, such as growth and reproduction, are easily observed in insects. Their relatively short life cycles and dramatic physical transformations during the growth process make them ideal subjects for classroom and laboratory study. They are readily available, inexpensive, easy to maintain and handle, and their use seldom raises the ethical issues associated with using vertebrate animals. Outdoor environmental education with insects provides students with opportunities to study and observe food webs, especially pollination and predator-prey relationships. In addition to numerous field guides, there are numerous instructional materials available, particularly at the primary and secondary levels.

High schools and universities dissect large numbers of cockroaches and locusts to study morphology and physiology.

Fruit flies have been used in genetic research for decades and are excellent models for students studying inheritance, as the giant chromosomes found within their larval salivary glands are easily observed. Adult features, such as eye color and wing length, are easily recognized and used in simple breeding experiments. Ants and crickets demonstrate pheromone and acoustic communication systems, respectively. Dissections of mud dauber wasp nests provide opportunities to observe such ecological interactions as parasitism and nutrient recycling. Mealworms (*Tenebrio molitor*) and tobacco hornworms (*Manduca sexta*) provide insight into complex behaviors resulting from competition and overcrowding. Madagascan hissing cockroaches (*Gromphadorhina portentosa*) make excellent research subjects for studies of communication, dominance hierarchies, and learning, while mantids offer fascinating opportunities to observe growth, development, and predatory behavior.

Resources

Books

Adams, J., ed. *Insect Potpourri: Adventures in Entomology.* Gainesville, FL: Sandhill Crane Press, Inc., 1992.

Akre, R. D., G. S. Paulson, and E. P. Catts. *Insects Did It First.* Fairfield, WA: Ye Galleon Press, 1992.

Berenbaum, M. R. *Bugs in the System: Insects and Their Impact on Human Affairs.* Reading, MA: Addison-Wesley Publishing Company, 1995.

Buchman, S. L., and G. P. Nabhan. *The Forgotten Pollinators.* Washington, DC: Island Press, 1996.

Collins, N. M., and J. A. Thomas, eds. *The Conservation of Insects and Their Habitats: 15th Symposium of the Royal Entomological Society of London.* London: Academic Press, 1991.

Elias, S. A. *Quaternary Insects and Their Environments.* Washington, DC: Smithsonian Institution Press, 1994.

Gordon, D. G. *The Eat-A-Bug Cookbook: 33 Ways to Cook Grasshoppers, Ants, Water Bugs, Spiders, Centipedes, and Their Kin.* Berkeley, CA: Ten Speed Press, 1998.

Grissell, E. *Insects and Gardens.* Portland, OR: Timber Press, 2001.

Hamel, D. R. *Atlas of Insects on Stamps of the World.* Falls Church, VA: Tico Press, 1991.

Klausnitzer, B. *Insects: Their Biology and Cultural History.* New York: Universe Books, 1987.

Menzel, P., and F. D'Aluisio. *Man Eating Bugs: The Art and Science of Eating Insects.* Berkeley, CA: Ten Speed Press, 1998.

Samways, M. J. *Insect Conservation Biology.* London: Chapman and Hall, 1994.

Periodicals

Black, S. H., M. Shepard, and M. M. Allen. "Endangered Invertebrates: The Case for Greater Attention to Invertebrate Conservation." *Endangered Species Update* 18, no. 2 (2001): 42–50.

Goodman, W. G., R. Jeanne, and P. Sutherland. "Teaching About Behavior with the Tobacco Hornworm." *The American Biology Teacher* 63, no. 4 (2001): 258–261.

Hogue, C. L. "Cultural Entomology." *Annual Review of Entomology* 32 (1987): 181–199.

Kellert, S. R. "Values and Perceptions of Invertebrates." *Conservation Biology* 7, no. 4 (1993): 845–855.

Matthews, R. W. "Teaching Ecological Interactions with Mud Dauber Nests." *The American Biology Teacher* 59, no. 3 (1997): 152–158.

Matthews, R. W., L. R. Flage, and J. R. Matthews. "Insects As Teaching Tools in Primary and Secondary Education." *Annual Review of Entomology* 42 (1997): 269–289.

Pyle, R., M. Bentzien, and P. Opler. "Insect Conservation." *Annual Review of Entomology* (1981) 26: 233–258.

Sherman, R. A., J. Sherman, L. Gilead, M. Lipo, and K. Y. Mumcuoglu. "Maggot Debridement Therapy in Outpatients." *Archives of Physical Medicine and Rehabilitation* 82 (2001): 1226–1229.

Other

"Cultural Entomology Digest." [cited 29 Apr. 2003] <http://www.insects.org/ced/index.html>.

"Food Insects Newsletter." [cited 29 Apr. 2003] <http://www.hollowtop.com/finl_html/finl.html>.

Arthur V. Evans, DSc

Conservation

Conservation ethics

Imagine a world without the buzz and beauty of insects. As with so many other animals, insects are under pressure from human population growth and economic development. Crickets are being silenced as lawns are manicured, and butterflies are being lost as cultivated flowers replace wild ones. Above all, it is the devastation to tropical forests, where most insects live, that is taking an unprecedented toll on the rich tapestry of insect life. In response to this insect impoverishment, there has been an upwelling of activity to secure their future. Before asking how to ensure insect survival, we must ask why we should do it. This is the realm of conservation ethics, and it underpins all that we do.

This conservation awareness derives from two fundamentally different philosophical approaches. The first is more basic, ethically speaking, and values nature according to its usefulness to us as humans. This is the utilitarian approach. Whether we consider ecosystems, species, or the actual products of nature, all are human survival tools. The term *sustainable utilization*, which is used widely in conservation circles, means essentially the harvesting of nature in a way that does not compromise future human generations. The second approach is considered ethically deeper and views nature as having a right to exist according to its intrinsic value and not on its merit for human survival. All forms of nature, from fox to fish to fly, have their innate worth and deserve a place on this planet, even if they do not necessarily benefit us directly.

In practical terms, these two approaches are not necessarily mutually exclusive. A reserve may be set up to conserve the tiger, which also gives space and place to a whole range of insect species. This does not mean, however, that the insects were part of the planning process, although conservationists know that the conservation of the tiger, which needs vast natural areas for survival, will embrace many, but not necessarily all, insect habitats.

When we talk of utilitarian value, we also may be including aesthetics. Utilitarian merit does not reside merely in the fact that we eat or wear parts of nature. Birdwing butterflies, being immensely beautiful and for the most part very rare, are exploited (and protected) because we prize their beauty. In reality, and taking a deeper philosophical approach, the rights pertaining to the butterfly and the physiological way a butterfly works are essentially no different from these aspects of a flea. This human perspective on nature colors all that we do in conservation. Some have argued that it is not species that have rights at all, but only individuals. If that is the case, it means that most people would sanction and act upon looking after a panda rather than an earwig. Many of our everyday expressions underscore these views. A despised person may be described as a "cockroach," while an exceptionally gentle person is "not hurting a fly."

"Coarse-filter" and "fine-filter" approaches to insect conservation

A salient feature of insect conservation is that overall we do not know exactly what we are conserving. This is so because most insect species are not scientifically described, and among those that are, there is generally little information available on their biological makeup for meaningful conservation decisions. This dearth of knowledge about the species that we aim to conserve is known as the *taxonomic challenge*.

This state of affairs, in turn, leads to a dichotomy in the way we approach insect conservation. There has to be conservation and management of whole landscapes so as to embrace as many habitats and microhabitats as possible. This "coarse-filter" approach allows all the activities in which insects normally engage to continue, which ensures the insects' long-term survival; it comes about after a process of biogeographical prioritization. This prioritization is a global and regional methodology that ascertains the relative importance of landscapes that are special or what is termed *irreplaceable*, that is, found nowhere else. Such irreplaceable landscapes are given high priority in conservation planning. In addition, representative landscapes also are selected, because they are home to more widespread species. Formerly widespread landscapes and species, such as those of the prairies, have suffered enormous impoverishment. Such remnant areas of naturalness also are given high priority in conservation planning.

The other branch in the dichotomous approach to insect conservation is the targeted one, where specific species are the subject of conservation. This goes hand in hand with conservation of their habitats. Usually these are large, charismatic species whose existence is known to be seriously threatened. This is the "fine-filter," or "species," approach and often complements the landscape level, or coarse-filter, one. Focal species in this approach often are red-listed globally. After quantitative assessment of their increasing rarity and threat levels and the outlook for their future, certain species are categorized according to criteria outlined by the IUCN (World Conservation Union). After ratification by independent experts, a submission is made to the IUCN Species Survival Commission for inclusion on the Red List. This is an immensely important document that is held in high regard worldwide. It plays a major role in conservation planning of all types across the globe. It is also important from an insect point of view, because any one insect species has as much entry space on the Red List as, say, a whelk, wombat, or whale. In other words, philosophically, the Red List is based on the intrinsic value of species and not necessarily on how important they are to humans.

In the case of insects, however, we often do not have adequate data on which to make a sound assessment of their threat status or of exactly how to improve their conservation management. In such cases, the species that are strongly suspected of being seriously threatened are flagged as Data Deficient, a designation that leads to further research, which then fine-tunes their conservation status. In some cases, a species may not be as threatened as was thought and may be removed from the Red List. In other instances, a species may turn out to be under grave threat, and conservation recommendations will be proposed, so that immediate action to ensure its survival can be put in place.

The species on the IUCN Red List are globally threatened and usually confined to one country and often to one specific area within that country. If they become extinct in the local area, they are lost completely. Many countries, and even states, provinces, and municipalities, also have their own list of threatened species. These species are included in national or local red data books and may or may not be on the IUCN Red List. Nevertheless, it is important to distinguish between the global Red List and the various red data books, because some species on local lists may be threatened in that area and yet be very common elsewhere. The wart biter katydid (*Decticus verrucivorus*), a large, bright green insect that lives in grassland, is included in the British Red Data Book (Insects) because it is so threatened in Britain, yet in many parts of Europe and Asia, it is very common and so is not entered onto the IUCN Red List.

Sometimes a particular subspecies or morph is threatened, and it may be cited on the Red List or in a red data book, yet its genetic relatives, which are not threatened, may not be. These different forms are known as evolutionary significant units (ESUs). The gypsy moth (*Lymantria dispar*) has several forms that differ genetically in only small ways, yet that difference can be highly significant in terms of conservation. The British ESU of the gypsy moth became extinct early in the twentieth century, but the Asian form, which has invaded the country, is such a forest pest that its presence must be reported to the agricultural authorities.

Dung beetle conservation efforts include asking motorists to avoid running over these beetles in the road. (Photo by Ann & Steve Toon Wildlife Photography. Reproduced by permission.)

Conserving insect function

In most people's eyes, insects do not have the glamour of large vertebrates, so their conservation requires particular promotional approaches. These perceptions change over time. During the nineteenth century and earlier, insects were revered. In the Middle Ages lice, which at that time were not known to be vectors of the deadly disease typhus, were considered "Pearls of God" and were viewed as a sign of saintliness. As the body of Archbishop Thomas à Becket lay in Cantebury Cathedral on the night of his murder, the cooling of his body caused the lice to crawl out from under his robes. It is said that the vermin that left his body "boiled over like water simmering in a cauldron, and the onlookers burst into alternate fits of weeping and laughter, between the sorrow of having lost such a head and the joy of having found such a saint."

Today we view such so-called vermin and their habitat as an ecosystem, and our focus often is on conservation of these systems and all the interactions that they contain. The task of knowing all the interactions that take place in even a small

A biologist with a bumble bee (*Bombus* sp.) that has been tagged with a tiny radar antenna. The radar dishes shown in the background will track the bee as it forages. (Photo by ©James King-Holmes/Science Photo Library/Photo Researchers, Inc. Reproduced by permission.)

ecosystem is formidable. A small temperate pond may have as many as 1,000 species, which, in turn, generate 0.5 million interactions. Furthermore, these interactions are dynamic and vary in strength from one moment to the next. To conserve these interactions, we have to take a black-box approach, that is, conserve what we do not know. This leads us back to the coarse-filter approach mentioned earlier.

The reason that this approach is so important in terms of insect conservation is that in many terrestrial ecosystems insects dominate or are very important in determining the species composition and the functioning of those systems. If suddenly we took away all the insects, we probably would immediately see a radical transformation of the system, which would become unrecognizable very soon. Insects pollinate, suck, and chew plants and serve as a vector for disease. Insects are the derminants of the way most terrestrial ecosystems work. This view recognizes the fact that insects are small and most have high reproductive rates. Springtails (Collembola) can reach densities of 9,300 per square foot (100,000 per square meter) in leaf litter. A single gravid aphid, were its reproductive potential to be realized unchecked, would give rise to such an abundance of individuals after one year that the world would be more than 8.7 mi (14 km) deep in aphids.

The point is that insect conservation is tied intimately tied to conservation of all aspects of the natural systems around us. Insect conservation is a major, indisputable component of biodiversity conservation.

Threats to insects

The process of habitat loss usually is characterized by a series of landscape transformations. First, roads dissect the indigenous ecosystems, allowing people to enter the area and develop infrastructures, whether a factory, housing, or agricultural plots. This activity leads to the development of holes in the indigenous landscape and perforation of the once continuous naturalness. As the process of perforation continues to expand, the natural areas become completely divided, and only patches are left, isolated from one another. This is termed fragmentation. Human pressures around the edges of these patches, such as partial logging and slash-and-burn agriculture, cause these patches to shrink, with high-quality habitat remaining only in the centers of the patches. This final phase is termed *attrition*.

While the patch may look more or less intact, it has suffered an increase in edge and a decrease in core conditions.

A monarch butterfly (*Danaus plexippus*) with lepidopterist's migration tracking tag. (Photo by ©M. H. Sharp/Photo Researchers, Inc. Reproduced by permission.)

This comes about because exterior conditions impinge on the edge. Such conditions may be, for example, hot, dry, sunny impacts that dry out the soil and render it unsuitable for many shallow-rooted tropical plants. Such landscape attrition may not have an immediate effect upon insect population. Only over time, and because of shrinkage of quality habitat and vulnerability during times of adverse conditions, as well as increased genetic risks, may the loss of insect populations and species become apparent. The subsequent loss of species from a patch is known as ecological relaxation and is one of the most insidious and least-understood processes threatening insects.

Most insects live in the tropics, and it is there where the greatest pressures upon them exist. In the second half of the twentieth century, 10 million square miles (16 million square kilometers) of forest were cleared, and in the early twenty-first century tropical forests were being lost at the rate of 80,778 square miles (130,000 square kilometers) or more per year. These forests, in a constantly moist and warm climate and with complex vegetational architecture, are extremely vulnerable. Tropical forest removal and transformation have devastated insect populations across the continents, so much so that it has been estimated that as many as 30 insect species are becoming extinct every week. Most of these extinctions

are of scientifically nameless species. In other words, we do not know what we have been losing. This loss of unknown species has been termed "Centinelan extinction," from Centinela Ridge in Ecuador, where botanists tried to go to describe new species of trees, only to find that the trees had been chopped down and lost to science and the world forever.

Logging of tropical forests usually is the fastest way for timber corporations and their investors to make lots of money quickly. The number of tropical forest tracts allocated for logging is at least eight to ten times higher than the limited number of areas set aside for parks and reserves. These remnant reserves, in turn, become more vulnerable, because their edges are exposed. Although selective logging of tropical forests in Indonesia, for example, would seem to be sustainable utilization, there nevertheless has been a clear decrease in butterfly diversity. The loss becomes greater as the intensity of logging grows. There is evidence that some taxonomic groups of insects are more susceptible than others—moths are more sensitive than beetles, for instance. Changes in one insect group after logging has taken place do not necessarily correlate with changes of another. Certain groups, such as the arboreal dung beetles in Borneo and the ants in Ghana, can survive in agricultural areas even after the forest has been removed. This contrasts, however, with the fact that many insects, especially such large ones as morpho and birdwing butterflies, need massive areas of primary forest.

The situation in cool forests, such as those in the boreal zone, often mirrors that in the tropics, with some species tolerating little disturbance and others greater disturbance. Nevertheless, it is critical in all climatic zones to maintain large areas of virgin forest for the large-sized, ecologically specialized, reluctant-to-disperse endemic species. These areas often have particular soil conditions, litter depths, or even large logs that specialist moths, beetles, and other insects need. It is not necessarily any particular taxonomic group that is under threat any more than another group but rather insects with a particular way of life and restricted conditions in which they can survive. This is precisely why the situation in the tropics is so tragic. Many insects there, as far as we can tell, have very restricted host plant (and even host insect) preferences and often are confined to a small geographical area. This inevitably means that widespread loss of tropical forest results in large-scale insect diversity loss. Such loss is particularly acute in Indonesia, South America, and West Africa, with some estimates suggesting that intact rainforest in Madagascar will be gone by 2025. It is perhaps in the Congo basin where tropical insects are safest, because years of human conflict and the impact of AIDS have left considerable areas intact.

Island faunas, particularly on tropical islands, are especially vulnerable because of their small size and very high levels of pressure from humans. Their sensitivity may not necessarily be because certain species do not have the behavioral mechanisms to cope with change but because loss of even a few individuals from these small populations may put the species at a demographic risk. The same applies to insects of special habitats, such as caves. These insects are at risk because over millennia they have adapted to unusually stable environmental conditions, which are readily perturbed by humans.

It is not just in the tropics that insects are at risk but also across all ecosystems. Mediterranean-type ecosystems, because of their suitability for humans and high-value agriculture, have been especially susceptible. The same may be said of grasslands, which in prehistoric times were so extensive, yet now have been largely cultivated. The transformation of landscapes and consequent habitat loss have not affected all insect species equally. Among those that have suffered greatly are larger, specialist species, such as dung beetles, ant-dependent blue butterflies, wetland species, and old-growth forest species. Certain other species, in contrast, have benefited enormously from human activity. Transformation of landscapes has enabled many opportunistic species to expand beyond the confines of where they would normally live. Many other species have been translocated accidentally or deliberately to foreign lands and have proliferated there, often having a devastating effect on the indigenous fauna. This is evident in Hawaii, which before human settlement had no ants. Today, invasive foreign ants are having a major adverse impact on the native Hawaiian insects.

This proportionate change in abundance is seen when ecosystems are polluted. While some species are extremely vulnerable and are readily lost from the system, others can proliferate. In freshwater systems, stoneflies and mayflies may be the first to succumb, whereas blackflies (*Simulium* spp.) may multiply. Alien invasive plants also may change an indigenous ecosystem radically and affect the native insect fauna. This may come about because human modification of the landscape encourages the spread of weeds, sometimes to such an extent that native vegetation is overwhelmed and thus becomes unavailable to the former insect fauna.

The important point, and it underscores all aspects of human pressures upon insects, is that the various threats can be synergistic with each other—one threat compounds the next. This is evident, for example, in some South African streams where the highly endemic dragonfly and damselfly fauna is being threatened on all sides. Alien invasive trees shade the stream banks, while cattle trample the edges, breaking up the vegetation and silting the water. Extraction of too much water for the decidous fruit and wine industries retards water flow and causes a drop in oxygen levels. Introduced rainbow trout compound the situation through predation. Pollution may further affect the populations. The result has been a radical retraction of geographical ranges, so that many species today are found only in preserved mountain catchments. Three species may even be extinct, not having been rediscovered for several decades.

This synergism can build to the point where much of the insect fauna can no longer withstand all the pressures, leading to a major shift in its character or ecological integrity. Such a major, relatively sudden shift is termed a *discontinuity*. The concern is that many such discontinuities are imminent, especially with the potential widespread impact of global warming; if they become far-reaching, the original faunal composition may not return to its natural state. It has been calculated that if these discontinuities continue at the same rate worldwide, it may take millions of years for the original diversity to return. Global climatic change and the random walk of speciation, however, mean that insect diversity in the future will not be the same as it is today.

The conservation process

Taking the coarse-filter approach, the initial step in the conservation process is to select which geographical areas are necessary to conserve so as to maintain both irreplaceable and typical species. This prioritization process has identified, for example, at the global scale those countries with unusually rich and irreplaceable biotas. These global hot spots then become a top priority for further research and conservation management. Among those countries are Indonesia, Brazil, South Africa, Mexico, and Madagascar. This global prioritization process is based principally on plants and vertebrates. Insects (especially butterflies, dragonflies, and some beetle groups), however, are beginning to play a greater role in the process rather than falling under the umbrella of the bigger animals and plants. Nevertheless, from the perspective of practicality some conspicuous insect groups have to serve as surrogates for other less well known and difficult-to-sample

Butterfly houses, such as this one in Kuranda, Queensland, Australia, provide one way of breeding rare or endangered butterflies while allowing the public to see them alive. (Photo by Rosser W. Garrison. Reproduced by permission.)

groups. This is a taxonomic nested approach. In using the better known groups, we hope that we have done our best to include most of the other species.

Prioritization is one thing, while translating science into practice is quite another. Often it is in those countries where there is widest insect diversity and greatest threat that the least is being done to conserve insects. There are exceptions. Costa Rica, for example, has committed itself to conserving its biodiversity, including its insects, as a major national asset. Although it covers only 0.2% of the nation, Santa Rosa Park in northwestern Costa Rica contains breeding populations of 55% of the country's 135 species of sphingid moth.

In northern Europe, many insect species on the Red List or in red data books are protected. Germany has gone one step further by giving protection to and banning collection of all dragonfly species. The British organization Butterfly Conservation, which has a membership of several thousand, has become an immensely influential organization, devoting itself to conserving British butterflies and moths and achieving considerable success in cooperation with land owners and government organizations. Amateur membership and activity have played a major role in identifying populations of threatened species needing protection.

Because conservation is a crisis science, we also have to prioritize in terms of what we can and should do. A particular area of the world may not be the hottest hot spot, but it may be under severe threat by, say, logging of a certain tropical forest. Urgent attention and resources must be directed to saving as much as possible of that area through appropriate management. This is termed *triage* and directs attention to where it is most effective rather than to areas that are relatively safe or where the damage is so great that any new efforts would see little benefit.

Prioritization can take place at various spatial scales, from the finer scale of habitat or landscape to the coarser scale of whole reserves. Although biogeography rarely obeys political boundaries, conservation, in contrast, is organized essentially along political lines, which impinges on the prioritization process. Once biological prioritization has identified areas for conservation, however, this information becomes available for policymakers to make more informed decisions. This is evident today in the creation of Transfrontier Parks, wildlife reserves that cross national boundaries. One example is the wilderness area of the Drakensberg Mountains across Lesotho and South Africa.

In terms of the fine-filter approach, once species have been identified as being in need of conservation management and have been included on the IUCN Red List or in a local red data book, they must be protected. This requires political muscle, and each country has its suite of laws to address species conservation. In the United States, it is the Endangered Species Act. This act has engendered considerable discussion and debate, because different groups may have different interests and motives. The national or provincial laws coincide with international laws when a country has officially ratified a particular international convention. The Convention on International Trade in Endangered Species (CITES) monitors closely, at international and national levels, trade in various species so as to protect them from poaching. Certain birdwing butterflies, for example, may not be collected in the wild and traded internationally.

Conservation management

After deciding where and what to conserve, it is necessary to address the how of conservation. This process calls for deploying the most ergonomic measures to maximize the chance of survival of habitats and species in the long term. A baseline hope is that we have first preserved all the parts, that is, all the species and their interactions. This is termed the *precautionary principle*. Management involves maintaining the remnant patches with minimal pressures on their edges, so as to avoid loss of species through ecological relaxation. These patches require maximum connectedness. Wherever possible, corridors are encouraged or created to link the patches. For these corridors to be effective, they need to be of an ecological nature similar to the patches; in this way, insects can move between them without having to overcome barriers that are adverse to the species. A forest patch needs to be connected to a like patch with a forest corridor and not to a grassland (unless, of course, the patches are grassland). These corridors give the insects greater opportunities in times of environmental crisis and also encourage the maintenance of genetic diversity.

The patches may be in need of management to simulate former natural conditions that no longer apply as the patches become small. For example, fire or grazing may be excluded and may have to be reintroduced deliberately. Such management must be carefully carried out so as to approximate the natural situation closely. This often involves rotational management, where, for example, different plots are burned or grazed at different times. Although this may be done to encourage general insect diversity, special procedures also may be carried out to help a particular species. Butterfly larva food plants, for example, may be supplemented.

Some management entails minimized human impact. Old trees, with much dead wood, may be left to encourage specialist wood-inhabiting insects. Field margins may be left unsprayed with pesticide and herbicides so as to create more area for indigenous insects. These minimal input margins are known as *conservation headlands* or *conservation corridors*, and they have benefit for both the overall insect fauna and agriculture. The reason is that the margins encourage the maintenance of the natural food webs and, in doing so, provide a home for such predatory insects as ground and ladybird beetles that move onto the adjacent crop after pesticide residues

have weathered. They eat and thus reduce the pest populations, whether they are caterpillars or aphids.

All management actions are carried out after the conservation question is clearly framed. Is our aim, for example, to encourage a wide range of indigenous butterfly species in an area, or is it to pull out all the stops to prevent extinction of a critically endangered specific species of butterfly? These two goals may not be mutually exclusive. Indeed, the reality of insect conservation is embedded in overall biodiversity conservation and also in the conservation of whole landscapes. After prioritization and selection of good indigenous habitat for protection, it often happens that certain of those quality remnants support individual noteworthy insects of conservation concern.

In South Africa, the KwaZulu-Natal mist-belt grasslands have been converted in large part to agricultural land, with less than 1% remaining. These remnants have been a top priority for conservation, because they are composed of unique plant assemblages as well as many endemic insects. Four remnants are the last habitat patches for populations of the lycaenid Karkloof blue butterfly (*Orachrysops ariadne*). This is an ant-dependent habitat specialist favoring cool slopes with a particular creeping variety of its host plant. Conservation management involves rotational burning of the habitat, which benefits not only the butterfly but also the plant community as a whole, because human-induced ignition simulates the original lightning strikes that prevent the plant community from becoming senescent. In addition, there is restoration of the periphery of the site with the removal of planted pine trees. Moreover, as with many conservation management programs, there is maintenance of the site, entailing removal of alien invasive plants, especially bramble, that continually threaten to reduce quality habitat.

In principle, these conservation approaches also can be carried out in suburban gardens and amenity areas. A first approach, and one that has become very popular, is to plant nectar-producing vegetation to attract butterflies. This butterfly gardening, at its simplest level, may involve only the planting of alien flowers and shrubs as nectar sources for adults. An ecologically deeper approach is to restore areas, albeit small ones, with indigenous vegetation, to cater to all aspects of the life cycle. Natural grassland, with its indigenous flowers, shrubs, and trees, is left to grow serendipitously or is planted deliberately, to encourage a wide range of insect species, from bees to butterflies. This method has become very important among dedicated enthusiasts in the United States and northern Europe, where so much natural habitat has been lost. Some plant and seed merchants even specialize in indigenous vegetation and, as such, are contributing enormously to insect conservation.

Ecological landscaping may involve the creation of ponds, especially for dragonflies. This has deep cultural roots in Japan, where old rice paddies have been converted to hugely popular dragonfly reserves. Small ponds can be constructed by anyone with a will to conserve pond insects. Such a pond need not be deeper than 3.3 ft (1 m), have sloping sides, and be well enclosed with water plants of all types and with rocks both submerged and emergent. Water levels also must be kept constant. The water should be well aerated (usually with abundant water weed), and bushes or trees should overhang part of the pond to provide shelter, perches, and hunting grounds for dragonflies chasing midges. These gardening approaches generally contribute greatly to increasing the local abundance of formerly more widespread species. They are a worthwhile "at home" activity, but they should not detract from lobbying for large tracts of natural and unique landscapes, which support specialized and endemic species that, if lost from one locale, may well be lost forever. In some cases it may even be necessary to breed certain insect species in captivity for reintroduction into the wild once field conditions are made secure. This is being done for several species, including the well-known European field cricket, *Gryllus campestris*, at the Invertebrate Conservation Centre at the London Zoo.

Viewed in the greater scheme of things, insect conservation takes many approaches. While there are general principles to consider, often it is necessary to tailor conservation management to specific local conditions and aims—to think globally and act locally. The realm of insect conservation is exciting and immensely important, and it is crucial for future generations. This area is by no means just for professionals. Everyone can play a key role, whether by participating in an official project or by encouraging indigenous flowers and trees in one's own garden. Whether farmer or financier, everybody can place a conservation brick into the wall of our heritage, so that we leave the world as rich a mosaic of insects as it was when we came into it.

Resources

Books

Collins, N. M., and J. A. Thomas, eds. *The Conservation of Insects and Their Habitats.* London: Academic Press, 1991.

Fry, R., and D. Lonsdale. *Habitat Conservation for Insects: A Neglected Green Issue.* Middlesex, U.K.: The Amateur Entomologists' Society, 1991.

Kirby, P. *Habitat Management for Invertebrates: A Practical Handbook.* Sandy, Bedfordshire, U.K.: Royal Society for the Protection of Birds, 1992.

New, T. R. *An Introduction to Invertebrate Conservation Biology.* Oxford, U.K.: Oxford University Press, 1995.

———. *Butterfly Conservation*, 2nd edition. Melbourne, Australia: Oxford University Press, 1997.

Samways, M. J. *Insect Conservation Biology.* London: Chapman and Hall, 1994.

The Xerces Society. *Butterfly Gardening.* San Francisco: Sierra Club Books, 1998.

Michael J. Samways, PhD

Protura
(Proturans)

Class Entognatha
Order Protura
Number of families 4

Photo: Front view of a member of the family Acerentomidae, collected from rainforest litter in Queensland, Australia. (Scanning micrograph by D. E. Walter. Reproduced by permission.)

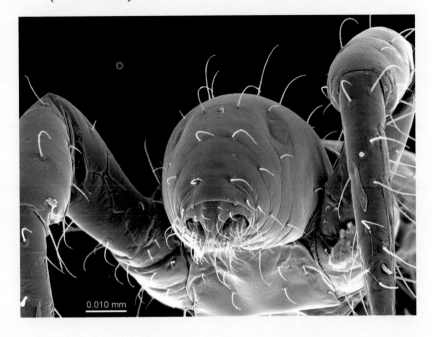

0.010 mm

Evolution and systematics

No fossil proturan is known, but the order is believed to date from the Devonian. Most authors include Protura together with Collembola and Diplura in the class Entognatha within the subphylum Hexapoda, because their mouthparts are enclosed in gnathal pouches formed by the ventral overgrowth of the lateral portions of the head. Some authors consider proturans to be related most closely to collembolans, and join these two orders in a separate class called Ellipura. The order Protura is subdivided into two suborders: Eosentomoidea and Acerentomoidea, each including two families—Eosentomidae and Sinentomidae in the first and Protentomidae and Acerentomidae the second.

Physical characteristics

Proturans are cryptic, pale, and minute wingless hexapods ranging from 0.024 to 0.06 in (0.6–1.5 mm) in length. The head is conical and prognathous, and only the tips of the styliform mouthparts can be seen externally. A pair of "pseudoculi," with an olfactory or chemosensory function, is found on the head. Eyes and antennae are wanting, and the sensory function of the latter is taken over by the forelegs, which are longer than the middle and hind legs and are provided with numerous sensillae. All legs have five segments and end in a single claw; they are longer in proturans living near the soil surface and shorter in those living deeper in the soil. The adult abdomen has 11 segments plus the telson; the first three segments each bear a pair of small, segmented abdominal appendages on the posterior corners, called styli. Cerci are lacking.

Larvae look like adults, except for the number of abdominal segments, which increases from eight in the first larval instar to the definitive 11 of the adult (a phenomenon called anamorphosis). The third instar ("larva II") has nine segments, the fourth instar ("maturus junius") has 11 segments but incomplete chaetotaxy (bristle arrangement) and no genital armature, and the fifth instar ("preimago," present only in Acerentomidae) has complete chaetotaxy but incomplete genital armature; there is no instar with 10 segments. It is unknown whether proturans keep molting during adult life. Only some of the proturans (Eosentomidae) possess a tracheal system for respiration, the remaining groups breathe through the integument.

Distribution

Present in every zoogeographic region, with the exception of the Arctic and Antarctic.

Habitat

Protura are found primarily in leaf litter, rotten logs, soil, and moss in wooded areas but also inhabit agricultural soils, meadows, and turf. Most proturans live close to the soil surface, although some small species can be found as deep as 10 in (25 cm). They prefer moist organic soils that are not too acidic.

Behavior

Proturans move slowly; the forelegs are held forward in front of the head and the middle and hind legs are used for walking. They have been observed to curve the telson over the head and discharge a sticky secretion on enemies. Proturans sometimes form large conspicuous aggregations.

Dorsal view of a member of the family Acerentomidae, collected from rainforest litter in Queensland, Australia. (Scanning micrograph by D. E. Walter. Reproduced by permission.)

Feeding ecology and diet

Some species have been seen sucking the contents of free hyphae of fungi (subterranean network of fungi consisting of threadlike tubular filaments that feed, grow, and ultimately may produce a mushroom or some other kind of reproductive structure) associated with oak and hornbeam roots, but nothing else is known about the trophic relationships of proturans.

Reproductive biology

Males deposit spermatophores on the ground, and unattended females collect them. Life cycles are mostly unknown.

In the known European Acerentomidae, eggs are laid in early spring and adults overwinter; among the Eosentomidae there may be more than one generation per year, because larvae are found throughout the year. Proturans living near the soil surface tend to have one generation per year in temperate zones, whereas those that live deeper in the soil are apt to be reproductively active throughout the year. There are also species that spend the summer near the surface and migrate down into the soil stratum for the winter.

Conservation status

About 500 species have been described, none of which is listed by the IUCN.

Significance to humans

Proturans are so small and cryptic that people rarely notice them. They can be seen in soil samples of Berlese-type funnels. (A Berlese-type funnel is an insect collecting device that consists of a large funnel containing a piece of screen suspended over a container; soil is placed in the funnel, and heat from a light placed above the funnel forces the hidden insects down the funnel into the container.) They are decomposers, helping in the breakdown and recycling of organic nutrients. No proturan is considered a pest.

1. *Sinentomon yoroi*; 2. *Berberentulus huisunensis*; 3. *Neocondeellum japonicum*; 4. *Eosentomon palustre*. (Illustration by John Megahan)

Species accounts

No common name
Berberentulus huisunensis

FAMILY
Acerentomidae

TAXONOMY
Berberentulus huisunensis Chao and Cheng, 1999, Nantou, Huisun, Taiwan.

OTHER COMMON NAMES
None known.

PHYSICAL CHARACTERISTICS
Slender body and elliptical head; mouthparts are slender and elongated. Foretarsal claw with one tooth near its base; *squama genitalis* with pointed acrostylus.

DISTRIBUTION
Central Taiwan.

HABITAT
Nothing is known.

BEHAVIOR
Nothing is known.

FEEDING ECOLOGY AND DIET
Nothing is known.

REPRODUCTIVE BIOLOGY
Nothing is known.

CONSERVATION STATUS
Not listed by the IUCN.

SIGNIFICANCE TO HUMANS
None known. ◆

No common name
Eosentomon palustre

FAMILY
Eosentomidae

TAXONOMY
Eosentomon palustre Szeptycki and Slawska, 2000, Potoczek reserve southeast of Kêpice, Poland.

OTHER COMMON NAMES
None known.

PHYSICAL CHARACTERISTICS
Elongate forms with small head and broad, sturdy mouthparts. Middle and hind claws knife-shaped. Female squama genitalis are short, with a long "beak" perpendicular to the median line and with a rounded, distinctly sclerotized apex of the stylus.

DISTRIBUTION
Restricted area in northern Poland.

☐ *Eosentomon palustre*

HABITAT
Hydrophilic (water-loving) species, associated with peat bogs and other swampy areas in soil, litter, and mosses of humid pine forests or on their borders.

BEHAVIOR
Nothing is known.

FEEDING ECOLOGY AND DIET
Nothing is known.

REPRODUCTIVE BIOLOGY
Nothing is known.

CONSERVATION STATUS
Not listed by the IUCN.

SIGNIFICANCE TO HUMANS
None known. ◆

No common name
Neocondeellum japonicum

FAMILY
Protentomidae

TAXONOMY
Neocondeellum japonicum Nakamura, 1990, Chichibu-shi, Saitama Prefecture, Japan, 656 ft (200 m).

OTHER COMMON NAMES
None known.

Sinentomon yoroi

Berberentulus huisunensis

Neocondeellum japonicum

PHYSICAL CHARACTERISTICS
Stout body; sturdy mouthparts; abdominal styli bearing terminal vesicles.

DISTRIBUTION
Japan.

HABITAT
Deciduous forest of *Zelkova serrata*.

BEHAVIOR
Nothing is known.

FEEDING ECOLOGY AND DIET
Nothing is known.

REPRODUCTIVE BIOLOGY
Nothing is known.

CONSERVATION STATUS
Nothing is known.

SIGNIFICANCE TO HUMANS
None known. ◆

No common name
Sinentomon yoroi

FAMILY
Sinentomidae

TAXONOMY
Sinentomon yoroi Imadaté, 1977, Sengodani, Takefu City, Fukui Prefecture, Japan, 656 ft (200 m).

OTHER COMMON NAMES
None known.

PHYSICAL CHARACTERISTICS
The body is ornamented with spines. All immature stages are known, except for the prelarva stage. Larva I has one row of dorsal setae on the first through seventh abdominal segments, with accessory setae shifted posteriorly, giving the appearance of two rows. Larva II has anterior rows of setae on the second to seventh terga, giving the appearance of three rows; "matures junior" (fourth instar larva) has two setae on the eleventh sternum, and the adult has six setae.

DISTRIBUTION
Japan.

HABITAT
Areas with vegetation of moso bamboo, *Phyllostacys pubescens*, mixed with *Cryptomeria japonica*; steep slanting ground with scarce litter, where tightly intertwined bamboo roots occupy upper layer of soil to a depth of about 5.9 in (15 cm).

BEHAVIOR
Nothing is known.

FEEDING ECOLOGY AND DIET
Nothing is known.

REPRODUCTIVE BIOLOGY
Nothing is known.

CONSERVATION STATUS
Nothing is known.

SIGNIFICANCE TO HUMANS
None known. ◆

Resources

Books

Bernard, E. C., and S. L. Tuxen. "Class and Order Protura." In *Immature Insects*, edited by F. W. Stehr. Vol. 1. Dubuque, IA: Kendall/Hunt, 1987.

Copeland, T. P., and G. Imadaté. "Insecta: Protura." In *Soil Biology Guide*, edited by D. Dindal. New York: John Wiley and Sons, 1990.

Imadaté, G. *Fauna Japonica Protura (Insecta)*. Tokyo: Keigaku, 1974.

———. "Protura." In *The Insects of Australia: A Textbook for Students and Research Workers*, Vol. 1, edited by CSIRO. 2nd edition. Carlton, Australia: Melbourne University Press, 1991.

Janetschek, Heinz. *Handbuch der Zoologie eine Naturgeschichte der Staemme des Tierreiches*. 2nd edition. Vol. 4, *Arthropoda. Half, Insecta*. Part 2/3, Special, Protura. Berlin: Walter de Gruyter, 1970.

Nosek, Josef. *The European Protura: Their Taxonomy, Ecology and Distribution, with Keys for Determination*. Geneva, Switzerland: Museum d'Histoire Naturelle, 1973.

Tuxen, S. L. *The Protura: A Revision of the Species of the World with Keys for Determination.* Paris: Hermann, 1964.

———. *Fauna of New Zealand.* Number 9, Protura. New Zealand: Science Information Publishing Centre, 1986.

Wooten, Anthony. *Insects of the World.* New York: Blanford Press, 1984.

Periodicals

Berlese, A. "Monografia dei Myrientomata." *Redia* 6 (1909): 1–182.

Ewing, H. E. "The Protura of North America." *Annals of the Entomological Society of America* 33 (1940): 495–551.

Houston, W. W. K., and P. Greenslade. "Protura, Collembola and Diplura." *Zoological Catalogue of Australia* 22 (1994): 1–188.

Nosek, J., and D. Keith M. Kevan. "Key and Diagnoses of Proturan Genera of the World." *Annotationes zoologicae et botanicae (Bratislava)* 122 (1978): 1–54.

Tuxen, S. L. "Monographie der Proturen. 1. Morphologie nebst Bemerkungen über Systematik and Ökologie." *Zeitschrift für Morphologie und Oekologie der Tiere* 22 (1931): 671–720.

Natalia von Ellenrieder, PhD

•
Collembola
(Springtails)

Class Entognatha
Order Collembola
Number of families 20

Photo: A springtail (order Collembola) on a mossy branch near a pond. (Photo by Kim Taylor. Bruce Coleman, Inc. Reproduced by permission.)

Evolution and systematics

Rhyniella praecursor, a springtail from the Devonian (about 400 million years ago [mya]), represents the oldest known fossil of terrestrial animals. Springtails constitute an extremely ancient group that branched off very early during the evolution of the line that led to the more advanced hexapods, but their exact position within Arthropoda has not been resolved. Springtails traditionally were considered members of the subclass Apterygota within the class Insecta. Some authors now join springtails with proturans and diplurans, with which they share mouthparts enclosed in a pouch formed by elongated lateral portions of the head, in a taxon called Entognatha within the Hexapoda. Other authors include them together with Proturans in a taxon called Ellipura within the Hexapoda, and still others exclude them from the Hexapoda in a separate class (Collembola) and suggest that the hexapod characteristics (development of a head with five segments, thorax with three locomotory segments, and reduction of appendages on remaining body segments) evolved independently at least four times.

Collembola are classified into three suborders, Arthropleona, Neelipleona, and Symphypleona. Arthropleona comprises species that are longer than they are wide and in which the divisions between the thoracic and abdominal segments are easily visible. In contrast, the Neelipleona and Symphypleona are spherical, with poorly defined intersegmental boundaries on the body. Neelipleona differs in that the body is formed largely by expansion of the thoracic rather than the abdominal segments, as is the case in Symphypleona.

Physical characteristics

Springtails are small, wingless arthropods, usually 0.04–0.12 in (1–3 mm) long. They range from a little more than 0.008

in (0.2 mm), the smallest hexapods in the world (some species of Neelidae), to 0.4 in (10 mm) for some members of the tropical subfamily Uchidanurinae. Species living in caves or deep in leaf litter or soil tend to be white or gray, whereas species living in more open environments often are clothed with colored scales. Compound eyes are absent or reduced to a cluster of not more than eight ommatidia, and antennae are four-segmented (segments sometimes are subdivided, giving the appearance of more than four segments). The name Collembola, derived from the Greek *coll*, meaning "glue," and *embol*, meaning a "wedge," refers to the *collophore*, a peg-shaped structure on the venter of the first abdominal segment. The collophore once was thought to function as an adhesive organ, but it is involved mainly in maintaining the water balance, absorbing moisture from the environment. The common name springtails refers to a forked jumping organ, the furcula, found on the underside of the fourth abdominal segment. The furcula is retracted against the venter of the abdomen and held there by a structure on the third abdominal segment, the retinaculum. The genital opening is found on the fifth abdominal segment. There are only six abdominal segments (not 11 as in higher hexapods), and caudal appendages, or cerci, are lacking.

Distribution

Springtails occur worldwide, including the frozen Antarctic, where they reach the most southerly distribution of any hexapod (84°47′ south latitude) and survive at temperatures lower than −76°F (−60°C).

Habitat

Springtails are most abundant in warm, damp places, and many live in leaf litter and soil. Some live in caves or in the

A springtail, also known as a snow flea (family *Isotomidae*), in snow in Michigan, USA. (Photo by L. West. Bruce Coleman, Inc. Reproduced by permission.)

burrows of small mammals. A few species live in nests of social insects. They also occur in moss, under stones, in intertidal zones along the coast, on lake and pond surfaces, on snowfields, in the canopy of tropical rainforests, and in the deserts of Australia. Common in grassy or wooded habitats worldwide, they can be extremely abundant in certain habitats; in some grassland communities, population densities exceed 300 million individuals per acre (750 million per hectare).

Behavior

The "jumping" behavior characteristic of springtails occurs when the retinaculum releases the furcula, causing it to snap down against the substrate and flip the organism some distance (up to 8 in, 20 cm) through the air. This device, present in all but a few genera, seems to be an effective adaptation to avoid predation. Some Collembola use the collophore not only for water balance but also for adhering to smooth surfaces. Others extend a pair of eversible vesicles from the collophore's tip, move their front legs across the tip, and use the vesicles for grooming themselves.

Feeding ecology and diet

Springtails are primarily soil or litter dwellers, and most species feed on the fungi and bacteria found in rotting organic matter. Many arboreal and soil species also feed on algae, and some eat other plant materials. A few other species are carnivorous, feeding on nematodes and other springtails and their eggs. Mouthparts are styliform for sucking or have molar plates for grinding.

Reproductive biology

Like other non-insect hexapods, these insects continue to molt after they reach sexual maturity; unlike other hexapods, reproductive activity occurs only during alternate instars. Each reproductive stage is followed by a molt, a short period of feeding, and another molt. Reproduction usually is bisexual, although there are some parthenogenetic species. In many species males deposit a stalked spermatophore on the substrate for the females to find. Spermatophores can be deposited anywhere. Alternatively, the males may wait to find a receptive female and then deposit spermatophores nearby, or they may use the third pair of legs to transfer a drop of sperm to the female's genital opening. In some species competing males eat one another's spermatophores before setting up their own in the same place. Many symphypleonids display elaborate courtship behavior: the male dances and butts heads with the female to entice her to take up a spermatophore deposited on the substrate. In *Sminthurides aquaticus*, the male holds a potential mate with specialized antennal claspers and is carried by her until she becomes sexually receptive.

Eggs are deposited singly by some species or laid in large masses by several females. Some female sminthurids cover their eggs with a mixture of eaten soil and fecal material, which protects them from dehydration and fungal attack. Springtails lack metamorphosis; the eggs hatch into young that are similar in appearance but smaller than the adults and lacking reproductive organs. They usually molt four to five times before reaching sexual maturity and continue to molt periodically (up to 50 times) throughout the rest of their adult lives. Some species have many generations in a single year (multivoltine), particularly in the tropics, whereas many others have only one (univoltine); some species in the Antarctic may take up to four years per generation.

Springtails (order Collembola) massing in the spring. (Photo by Kim Taylor. Bruce Coleman, Inc. Reproduced by permission.)

Conservation status

There are more than 6,000 species worldwide, plus an estimated 25% unknown species. No springtails are listed by the IUCN. This is a highly adaptable and resistant group. Endemic species within endangered areas could be candidates for conservation programs.

Significance to humans

Springtails are part of the community of decomposers that break down and recycle organic waste, and, in this respect, they play a significant role in energy flow for many ecosystems. Most people see springtails when they lift stones in a garden or turn over compost. Swarms on snow are called "snow fleas."

Springtails are considered pests in houses, but they are harmless uninvited guests. A few species feeding on living plants constitute pests: *Bourletiella hortensis* (the garden springtail) may damage seedlings in early spring, *Sminthurus viridis* (the lucerne flea) is a pest of legume pastures, and *Hypogastrura armata* is a frequent pest of commercial mushrooms.

1. Lucerne flea (*Sminthurus viridis*); 2. Varied springtail (*Isotoma viridis*); 3. Water springtail (*Podura aquatica*). (Illustration by Amanda Humphrey)

Species accounts

Varied springtail
Isotoma viridis

FAMILY
Entomobryidae

TAXONOMY
Isotoma viridis Bourlet, 1839, Europe.

OTHER COMMON NAMES
English: Green springtail, snow flea.

PHYSICAL CHARACTERISTICS
Body is clothed with short hairs. Grows to 0.08–0.16 in (2–4 mm) in length. Colors may be dark green, greenish yellow, lilac, blackish blue, reddish, purple, or dark brown, usually with small, pale dorsal spots. Well-developed furcula.

DISTRIBUTION
Palearctic region.

HABITAT
This species dwells in surface litter and is common in gardens in soil, grass, and snow.

BEHAVIOR
They run actively and have a strong springing movement.

FEEDING ECOLOGY AND DIET
Fungal hyphae, spores, decaying leaf matter, and algae constitute their normal diet. When such food items are in short supply, they may feed on nematodes or exhibit cannibalism.

REPRODUCTIVE BIOLOGY
Spermatophores are deposited on the substrate by the male and subsequently picked up by the female; mature females lay clutches of 27–54 eggs.

CONSERVATION STATUS
Not listed by the IUCN.

SIGNIFICANCE TO HUMANS
None known. ◆

Water springtail
Podura aquatica

FAMILY
Poduridae

TAXONOMY
Podura aquatica Linnaeus, 1758, Europe.

OTHER COMMON NAMES
German: Wasserfloh; Italian: Pulce d'acqua.

PHYSICAL CHARACTERISTICS
This species is 0.08 in (2 mm) long and dark blue to reddish brown in color; it has short legs and antennae.

DISTRIBUTION
Northern Hemisphere.

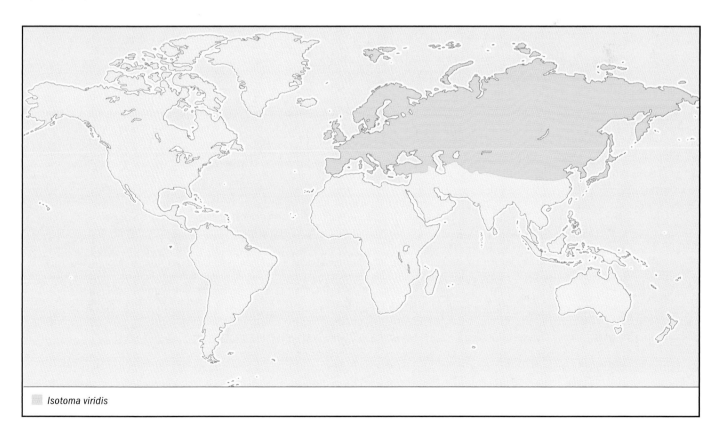

Isotoma viridis

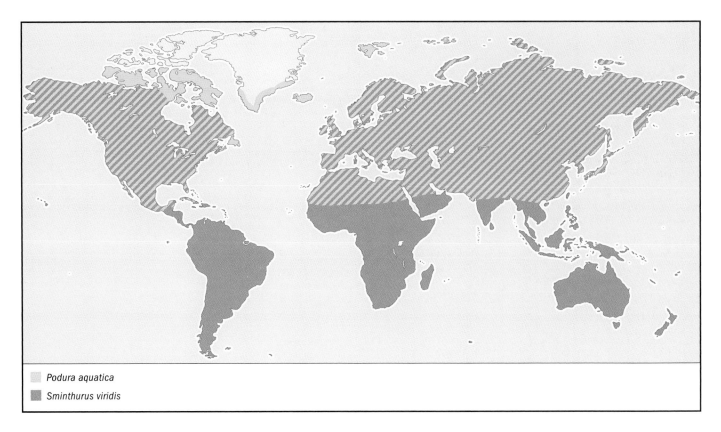

Podura aquatica
Sminthurus viridis

HABITAT
Semiaquatic. They live and feed on the surface of standing water, but they do not lay their eggs in the water.

BEHAVIOR
Nothing is known.

FEEDING ECOLOGY AND DIET
They are scavengers, feeding on decaying animal and vegetable matter.

REPRODUCTIVE BIOLOGY
Nothing is known.

CONSERVATION STATUS
Not listed by the IUCN.

SIGNIFICANCE TO HUMANS
None known. ◆

Lucerne flea
Sminthurus viridis

FAMILY
Sminthuridae

TAXONOMY
Podura viridis Linnaeus, 1758, "Europe."

OTHER COMMON NAMES
English: Clover springtail, alfalfa springtail; German: Luzerne-floh.

PHYSICAL CHARACTERISTICS
They have a distinct and well-developed furcula and a globular shape. Grow to 0.1 in (2.5 mm) in length. Long, elbowed antennae. An irregular pattern of pigment (green, brown, yellow) over the body.

DISTRIBUTION
Cosmopolitan; originally from Europe, now spread around the world through commerce.

HABITAT
Occur in areas where temperature and rainfall are suitable (more than 9.8 in, or 250 mm, of rain in the growing season).

BEHAVIOR
When disturbed, is able to jump as far as 12 in (30 cm).

FEEDING ECOLOGY AND DIET
Biting mouthparts; the young eat patches of leaves, and adults skeletonize leaves.

REPRODUCTIVE BIOLOGY
The male attaches a stalked spermatophore to the soil or low vegetation. The female places it into her genital opening. Females lay clusters of 40 eggs in the soil during winter; there are three generations each winter. In spring drought and temperature-resistant eggs are produced, which do not hatch until the following autumn.

CONSERVATION STATUS
Not listed by IUCN.

SIGNIFICANCE TO HUMANS
Considered a pest of legume pastures (lupines, lentils, beans, and field peas). The predatory mites *Bdellodes lapidaria* and *Neomolgus capillatus* keep it under biological control. ◆

Resources

Books

Christiansen, Kenneth A., and Peter F. Bellinger. *The Collembola of North America North of the Rio Grande: A Taxonomic Analysis.* Grinnell, IA: Grinnell College, 1981.

Coleman, David C., and D. A. Crossley. *Fundamentals of Soil Ecology.* San Diego: Acedemic Press, 1996.

Maynard, E. A. *A Monograph of the Collembola or Springtail Insects of New York State.* Ithaca, NY: Comstock Publishing Co, Inc., 1951.

Hopkin, S. P. *Biology of the Springtails (Insecta: Collembola).* Oxford, U.K.: Oxford University Press, 1997.

Lubbock, J. B. *Monograph of the Collembola and Thysanura.* London: Ray Society, 1873.

Salmon, J. T. *An Index to the Collembola.* Vol. 1. Bulletin no. 7. Wellington, New Zealand: Royal Society of New Zealand, 1964.

Handschin, E. *Die Tierwelt Deutschlands und der angrenzenden Meeresteile.* Vol. 16: *Urinsekten oder Aperygota (Protura, Collembola, Diplura und Thysanura)*, edited by F. Dahl. Jena, Germany: [n.p.], 1929.

Other

Bellinger, P. F., K. A. Christiansen, and F. Janssens. "Checklist of the Collembola, 1996–2003." (11 Feb. 2003). <http://www.collembola.org>.

"Collembola." Feb. 1997 (11 Feb. 2003). <http://www.missouri.edu/bioscish/coll.html>.

Natalia von Ellenrieder, PhD

Diplura
(Diplurans)

Class Entognatha
Order Diplura
Number of families 9

Photo: Campodeids typically are white, or nearly white, have no eyes, and have long, slender antennae. (Photo by ©David R. Maddison, 2003. Reproduced by permission.)

Evolution and systematics

Diplura is one of three orders grouped in the Entognatha, a lineage of six-legged arthropods separate from the true insects. The Entognatha share a common ancestor with true insects, but developed along their own distinct evolutionary path. Diplura means "two tails," and classified within the order are all Entognatha with paired tail appendages. This definition is rather superficial, and possibly delimits an artificial group. Multiple evolutionary units may be present within the order as currently defined, and new orders may eventually be created to separate them.

The fossil record of Diplura is sparse, owing to their soft-bodied nature. The oldest fossil commonly referred to this order lived in the Upper Carboniferous. Presently, the order contains approximately 800 species grouped into three suborders and nine families.

Physical characteristics

Diplurans are long and slender, cylindrical or dorsoventrally flattened in cross-section, 0.12–1.97 in (3–50 mm) in length. Body color is normally white or pale yellow, and the cuticle is often semitransparent. Wings are absent in all stages of development. The body is usually covered with hairlike structures called setae, infrequently with scales. The head bears no eyes, but supports a pair of long antennae bristling with complex sensory hairs. Each antennal segment contains its own musculature (a primitive feature), and some families possess unique telescoping antennae. Diplurans have chewing mouthparts concealed within a pouch formed by the front of the head, the defining feature of the Entognatha. The mandibles have one point of articulation with the head, another primitive, ancestral character.

The abdomen has ten segments. A pair of tail appendages, called cerci, is located on the last abdominal segment, and takes several forms in the various families. Cerci are either long and multisegmented, as in the family Campodeidae, or forcepslike, as in the family Japygidae. Pairs of small leglike structures called styli are present on the underside of most abdominal segments, and these act to support the long abdomen. Pairs of eversible sacs are also found on the underside of the abdomen, and these most likely function to maintain water balance.

Distribution

The order Diplura is widely distributed throughout the world, although a few families have restricted distributions.

Habitat

Diplurans are a component of the soil fauna. On the surface, they can be found in moist microhabitats, such as under rocks, logs, leaf litter, and tree bark. Some species live part or all of their lives within caves.

Behavior

Many methods of burrowing are found within the order. Members of the family Campodeidae prefer loose soil, burrowing with wormlike movements of their streamlined bodies. Campodeids are also capable runners on the surface. Species in the family Japygidae push into preexisting soil cavities with their strong legs, which are useless for running. Burrowing is aided by the mouthparts, and by telescoping antennae in species that have them.

Species with long, flexible tail appendages disengage them readily if grasped by a predator, allowing the rest of the animal to escape. Species with the tail appendages modified into forceps use them effectively for defense. Several accounts describe forceps being used to capture prey, but this is not believed to be their primary use.

Feeding ecology and diet

Diplurans feed on live prey, dead soil animals, fungi, living plants, and decaying vegetation. Most species are omnivorous, combining several food sources. Diplurans locate prey with their antennae; they stalk to attain striking distance, and rush to capture it. Prey items include small, soft-bodied mites,

Japygidae species are found in decaying vegetation and damp soils. (Photo by Roger D. Akre/M. T. James Entomolgical Collection.)

worms, myriapods, and various adult insects and insect larvae, including smaller diplurans.

Reproductive biology

Male diplurans attach stalked sperm packets to the soil at random. Females with mature eggs search for these sperm packets and collect them with the genital opening. The eggs are also laid on stalks, and are placed in small clutches in leaf litter or in soil cavities. Parental care exists in several families, involving the female parent curling around her egg clutches, and remaining with the first and second instar larvae until they are capable of surviving independently.

The first two instars are prelarval stages, incompletely developed and incapable of feeding. Third instar larvae attain the typical adult form. Diplurans gradually increase in size and attain more setae with successive molts. No distinct change marks sexual maturity, the only indication being a full complement of body setae. Adults continue to molt, enabling regeneration of broken appendages. Up to 30 molts have been recorded, and lifespan may exceed two years.

Conservation status

No species of Diplura is listed by the IUCN, but populations of Diplura are little known. There are a large number of endemic species in the order with very restricted distributions, and habitat destruction could easily threaten these more locally distributed species. Some species appear to be coping well with the presence of man; for example, considerable diversity has been recorded in urban areas such as Vienna, Austria.

Significance to humans

A few species are known to cause minor damage to garden vegetables. Most do not affect humans in any way.

1. *Holjapyx diversiunguis*; 2. *Campodea fragilis*; 3. *Heterojapyx gallardi*. (Illustration by John Megahan)

Species accounts

No common name
Campodea fragilis

FAMILY
Campodeidae

TAXONOMY
Campodea fragilis Meinert, 1865, Copenhagen, Denmark.

OTHER COMMON NAMES
None known.

PHYSICAL CHARACTERISTICS
Length 0.12–0.19 in (3–5 mm). Body thin, flexible, translucent white to pale yellow. Tail appendages multisegmented, long, and antennalike. Legs and abdominal styli well developed. Precise identification of this species requires examination of the setae.

DISTRIBUTION
All continents except Antarctica.

HABITAT
Damp, loamy soil.

BEHAVIOR
Gregarious; retreats from light when exposed.

FEEDING ECOLOGY AND DIET
Feeds on plant matter and dead insects.

REPRODUCTIVE BIOLOGY
Males produce large numbers of sperm packets, as many as 200 per week. Females lay eggs within natural soil spaces, avoiding direct contact between eggs and soil. Eggs hatch in 12–13 days.

CONSERVATION STATUS
Not threatened.

SIGNIFICANCE TO HUMANS
None known. ◆

No common name
Heterojapyx gallardi

FAMILY
Heterojapygidae

TAXONOMY
Heterojapyx gallardi Tillyard, 1924, Epping, New South Wales, Australia.

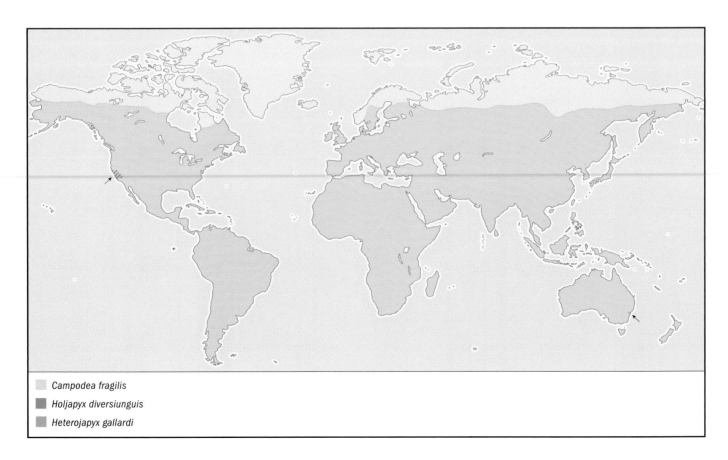

Campodea fragilis
Holjapyx diversiunguis
Heterojapyx gallardi

OTHER COMMON NAMES
None known.

PHYSICAL CHARACTERISTICS
Length 1.2–2 in (30–50 mm); a giant of the order. Females larger than males. Body mostly white. Abdominal segment eight tinged with light brown, segments 9, 10, and forceps dark brown and shiny. Forceps have asymmetrically serrated inner margins.

DISTRIBUTION
Coastal mountains of New South Wales, Australia, in the vicinity of Sydney.

HABITAT
Lives in soil. Found under large rocks and logs when on the surface.

BEHAVIOR
Burrows to depths below 18 in (45 cm). Found deeper during dry weather, emerging onto the surface when the soil is wet. Antennae continually scan the surroundings while on the surface.

FEEDING ECOLOGY AND DIET
Predaceous. Waits just below the surface of the soil with forceps exposed to capture prey. Once prey is grasped, it is pulled beneath the soil and fed upon.

REPRODUCTIVE BIOLOGY
Females apparently offer parental care, having been observed with early instar larvae. Females probably guard their eggs as well.

CONSERVATION STATUS
Not threatened.

SIGNIFICANCE TO HUMANS
None known. ◆

No common name
Holjapyx diversiunguis

FAMILY
Japygidae

TAXONOMY
Japyx diversiunguis Silvestri, 1911, Yosemite National Park, California, United States.

OTHER COMMON NAMES
None known.

PHYSICAL CHARACTERISTICS
Length 0.2–0.3 in (6–8 mm). Body pale yellow, tenth abdominal segment and forceps dark brown. Antennae have 26 segments. Forceps each with one prominent tooth. Species identification accomplished by examination of the setae and abdomen.

DISTRIBUTION
Central one-third of California, United States, from the coast east to the crest of the Sierra Nevadas.

HABITAT
In soil and on the surface, under rocks and damp leaves.

BEHAVIOR
A deep burrower, descending to depths around 30 in (76 cm).

FEEDING ECOLOGY AND DIET
Nothing is known.

REPRODUCTIVE BIOLOGY
Report of female with eight small larvae suggests parental care. Females probably also guard their small clutch of eggs.

CONSERVATION STATUS
Not threatened.

SIGNIFICANCE TO HUMANS
None known. ◆

Resources

Books

Condé, B., and Pagés, J. "Diplura." In *The Insects of Australia: A Textbook for Students and Research Workers.* 2nd edition, edited by CSIRO. Carlton, Australia: Melbourne University Press, 1991.

Ferguson, L. M. "Insecta: Diplura." In *Soil Biology Guide,* edited by D. Dindal. New York: John Wiley & Sons, 1990.

Other

"Diplura" *Tree of Life Web Project* [March 31, 2003]. <http://tolweb.org/tree?group=Diplura&contgroup;=Hexapoda>.

Jeffrey A. Cole, BS

Microcoryphia
(Bristletails)

Class Insecta
Order Microcoryphia
Number of families 2

Photo: Bristletails (*Petrobius maritimus*) climb on rocks in Wales. (Photo by Adrian Davies. Bruce Coleman, Inc. Reproduced by permission.)

Evolution and systematics

Members of the order Microcoryphia, or Archaeognatha, are superficially similar to common silverfish (order Thysanura). Both have life cycles without metamorphosis and never develop wings. Microcoryphia are true insects, but they diverged from the main developmental path leading to other insect orders very early in evolutionary time. The order is structurally primitive, retaining many features believed to have been possessed by the ancestral insect. The earliest known fossil is from the Lower Devonian period. There are two modern families, Machilidae and Meinertellidae, containing approximately 450 species.

Physical characteristics

Microcoryphia range in size (excluding appendages) from 0.3–0.8 in (7–20 mm). Microcoryphia are cylindrical, with enlarged dorsal thoracic plates, making them appear teardrop shaped or humpbacked. The body is covered with scales. The name Microcoryphia means "small head." The head bears a pair of long, threadlike antennae, along with a pair of enlarged maxillary palps that resemble a fourth pair of legs. The eyes are large, meeting on top of the head, and can apparently detect movement and form images. The 10-segmented abdomen bears three long, antennalike tail appendages, and ventrally, segments two through nine possess pairs of styli, which support the abdomen and enable it to glide smoothly across the substrate. Pairs of eversible sacs are frequently found on abdominal segments one through seven.

Common silverfish, in contrast, have flattened bodies and smaller, widely separated eyes. Microcoryphia have mandibles with only one point of articulation, similar in structure to those of noninsect hexapod groups such as the springtails (order Collembola). All other true insects, including silverfish, have mandibles with two points of articulation.

Distribution

Microcoryphia are known from all continents except Antarctica, and are distributed from sea level to 15,750 ft (4,800 m) in the Himalayas.

Habitat

Most Microcoryphia live near the ground, on the surface of the soil, in leaf litter, and on or under rocks. Species inhabiting tropical rainforests are often partially or entirely arboreal.

Behavior

Activity in a large number of species is crepuscular or nocturnal. Many species are known to aggregate, and these aggregations may consist of members of multiple species and genera from the same family, but not from different families. Individuals seem to follow certain paths and routes. Chemical signals may be important in trail forming and aggregating behavior.

Microcoryphia jump by bending and suddenly releasing the tip of the abdomen, striking it against the ground. These insects are also excellent climbers, using their abdominal styli to anchor themselves while they move like an inchworm up vertical surfaces.

Feeding ecology and diet

Microcoryphia are herbivores, feeding on algae, fungi, lichens, and leaf litter.

Reproductive biology

Elaborate tactile courtship behavior often occurs between pairs, involving special sensory structures. There are three common methods of sperm transfer. In the first, the carrier thread method, males spin a thread between the ground and the elevated tip of the abdomen. Sperm droplets are placed on the thread, which females collect with the ovipositor. In the second method, the male attaches a sperm packet to the ground. He then pushes his head and thorax below the female, "pick-a-back" style, and maneuvers her over the sperm packet, whereupon she takes it up with her ovipositor. In the third method, the male and female elevate the tips of their abdomens and bring them into contact, aligning themselves at a 45° angle. The male deposits a sperm droplet onto the ovipositor of the female.

Females lay eggs in protected places, usually cementing them to the substrate. Development generally progresses through six to eight instars. Adults continue to molt, an ancestral characteristic enabling regeneration of lost appendages. Individuals may live up to three years, and molt as many as 60 times.

Conservation status

No species of Microcoryphia is listed by the IUCN.

Significance to humans

No significance to humans has been noted for Microcoryphia.

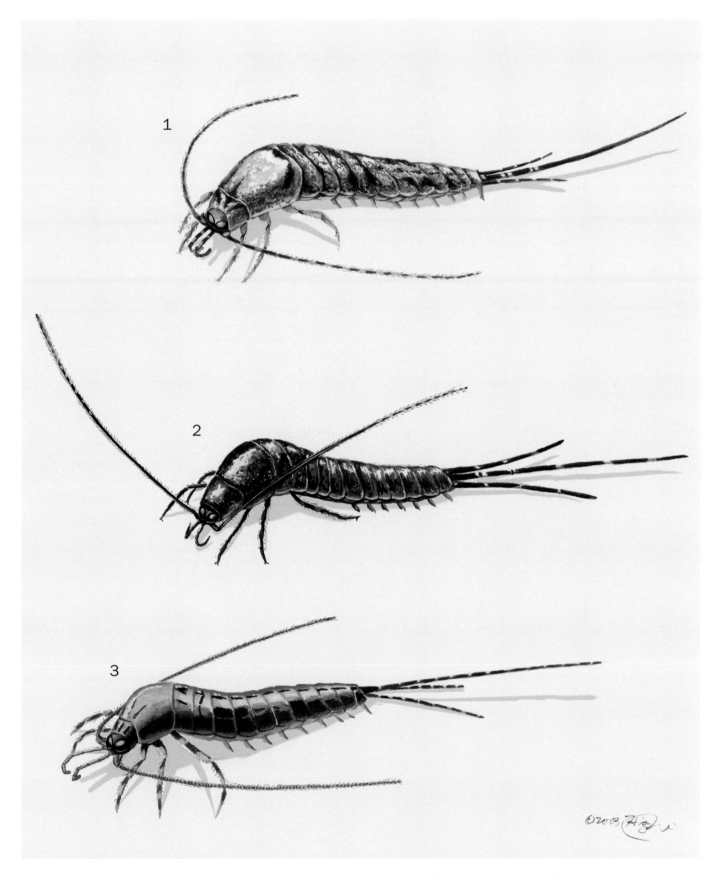

1. *Trigoniophthalmus alternatus*; 2. *Neomachilellus scandens*; 3. *Petrobius brevistylis*. (Illustration by Jonathan Higgins)

Species accounts

No common name
Neomachilellus scandens

FAMILY
Meinertellidae

TAXONOMY
Neomachilellus scandens Wygodzinsky, 1978, vicinity of Manaus, Amazonas, Brazil.

OTHER COMMON NAMES
None known.

PHYSICAL CHARACTERISTICS
Length of body is 0.47 in (12 mm), tail is 0.51 in (13 mm), and antennae are 0.67 in (17 mm). Body pigment yellowish white, mostly obscured by a thick covering of darker scales. Diagnostic features of the species include large spots on the eyes and a single dark ring on the first antennal segment.

DISTRIBUTION
A circle with a radius of 62 mi (100 km), centered on Manaus, Amazonas, Brazil.

HABITAT
Inhabits both primary and secondary growth dryland and inundation forests of the central Amazon. Believed to have originated in dryland areas, colonizing inundation forests via waterways in the remarkably water-resistant egg stage.

BEHAVIOR
Almost entirely arboreal in dryland forests. In inundation forests, individuals inhabit the forest floor during the dry season and migrate to the canopy at the onset of rains. Has a very well-developed vertical jumping ability, an adaptation to life in trees.

FEEDING ECOLOGY AND DIET
Grazes in leaf litter, consuming fungi, algae, and decaying leaves.

REPRODUCTIVE BIOLOGY
Copulates in the "pick-a-back" posture. The male attaches a sperm packet to the ground. He then pushes his head and thorax below the female and maneuvers her over the sperm packet, whereupon she takes it up with her ovipositor. Females deposit eggs in leaf litter. The generations are continual in dryland forests, but an annual life cycle has developed in inundation forests. Eggs remain submerged for prolonged periods, surviving five to six months of inundation in the wet season. A dry forest floor induces hatching, and the larvae develop rapidly, requiring only three months to reach maturity.

CONSERVATION STATUS
Not threatened.

SIGNIFICANCE TO HUMANS
None known. ◆

□ *Neomachilellus scandens*
■ *Petrobius brevistylis*
■ *Trigoniophthalmus alternatus*

No common name
Petrobius brevistylis

FAMILY
Machilidae

TAXONOMY
Petrobius brevistylis Carpenter, 1913, Portraine, Dublin County, Ireland.

OTHER COMMON NAMES
None known.

PHYSICAL CHARACTERISTICS
Body length is 0.43 in (11 mm) excluding antennae and tail filaments, which are both about as long as the body. Body scaling silvery gray, contrastingly marked with irregular patches of dark scales. Antennae completely covered with dark scales.

DISTRIBUTION
Northern Europe. Introduced into northeastern North America, perhaps on rocks used for ballast.

HABITAT
Cliffs and boulders along rocky seacoasts. Prefers sheer surfaces with few cracks and little loose sediment. Most abundant above the high tide line. In Europe, also inhabits building walls and masonry.

BEHAVIOR
Individuals move about on the surface of cliffs and on rocks throughout the littoral zone. Rests in cracks in rock or under stones.

FEEDING ECOLOGY AND DIET
Grazes on algae, lichens, and mosses growing on rocks.

REPRODUCTIVE BIOLOGY
Fertilization approaches an internal mode. The male and female elevate the tips of their abdomens and bring them into contact, aligning themselves at a 45° angle. The male deposits a sperm droplet onto the ovipositor of the female. Tactile contact is minimal. Some populations consist almost entirely of females and are apparently parthenogenetic. Groups of malleable eggs are laid in narrow crevices in rocks, and conform to the shape of the crevice containing them. Larval development requires three and a half months. Adults may live up to three years.

CONSERVATION STATUS
Not threatened.

SIGNIFICANCE TO HUMANS
None known. ◆

No common name
Trigoniophthalmus alternatus

FAMILY
Machilidae

TAXONOMY
Trigoniophthalmus alternatus Silvestri, 1904, Italy.

OTHER COMMON NAMES
None known.

PHYSICAL CHARACTERISTICS
Body scaling largely pale, with dark scales forming a patchwork pattern. Antennae, palps, and tail appendages covered with dark scales and numerous, narrow rings of lighter scales. Simple eyes triangular in shape.

DISTRIBUTION
Europe except Portugal and Scandinavia; also recorded from Iran. Introduced into eastern United States, first recorded in 1911.

HABITAT
Disturbed, rocky areas. Frequently found in rural and semiurban settings, inhabiting rock walls, cement building walls, and quarries. A historical account of this species (Stach, 1939, as listed in Wygodzinsky and Schmidt, 1980) living on bricks at a steamboat landing offers an interesting possible explanation as to how these insects become introduced.

BEHAVIOR
Crepuscular, active only briefly, just before and just after twilight. Gregarious, with groups sometimes numbering in the hundreds. When inactive, individuals conceal themselves close to the ground, such as beneath rocks or building foundations.

FEEDING ECOLOGY AND DIET
Grazes on green algae.

REPRODUCTIVE BIOLOGY
Some populations are parthenogenetic, consisting entirely of females that do not require males for reproduction. Parthenogenesis permits a single individual to establish a population. Eggs are laid on the underside of rocks, avoiding soil contact.

CONSERVATION STATUS
Not threatened.

SIGNIFICANCE TO HUMANS
None known. ◆

Resources

Books
Ferguson, L. M. "Insecta: Microcoryphia & Thysanura" In *Soil Biology Guide* edited by D. Dindal. New York: John Wiley and Sons, 1990.

Periodicals
Smith, E. L. "Biology and Structure of California Bristletails and Silverfish (Apterygota, Microcoryphia, Thysanura)." *The Pan Pacific Entomologist* 46, no. 3 (1970): 212–225.

Wygodzinsky, P., and K. Schmidt. "Survey of the Microcoryphia (Insecta) of the Northeastern United States and Adjacent Provinces of Canada." *American Museum Novitates* 2701 (1980): 1–17.

Other
"Archaeognatha (=Microcoryphia), Bristletails" *Tree of Life Web Project* [March 29, 2003]. <http://tolweb.org/tree?group= Archaeognatha&contgroup;=Insecta>.

"Jumping Bristletails" [31 March 2003]. <http://www.ent3 .orst.edu/moldenka/taxons/Petrobius.html>.

Jeffrey A. Cole, BS

Thysanura
(Silverfish and fire brats)

Class Insecta
Order Thysanura
Number of families 4

Photo: A silverfish (*Lepisma saccharina*) seen in leaf matter, southern California, USA. (Photo by ©Mark Smith/Photo Researchers, Inc. Reproduced by permission.)

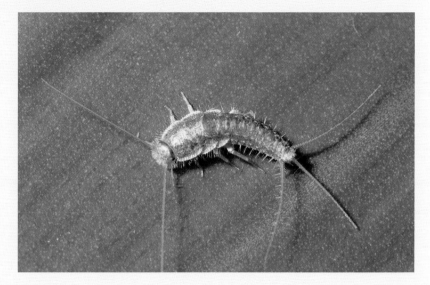

Evolution and systematics

Fossils of thysanurans are known from the upper Carboniferous period, including *Ramsdelepidion schusteri*, a very large silverfish 2.36 in (60 mm) long found in the state of Illinois in the United States.

The order Thysanura is considered the sister group of the Pterygota (winged insects), with which it shares a double articulation between mandibles and head and a well-sclerotized thoracic pleuron.

The Thysanura, or Zygentoma, includes four extant families: Lepidotrichidae, with only one extant species; Nicoletiidae, with soil, subterranean, and myrmecophilous (associated with ants) species; Lepismatidae, including the common domestic species, and Maindroniidae, including species of restricted distribution that live under rocks.

Physical characteristics

Thysanurans are primitively wingless insects. Most species are covered with overlapping silvery-gray scales although some lack scales. Members of the order are usually 0.4–0.8 in (10–20 mm) long, but may range from 0.04–1.9 in (1–50 mm); the bodies are dorsoventrally flattened. The compound eyes are small or absent; there are long filiform antennae; and the external mouthparts include mandibles with two points of articulation to the head. Thysanurans have short styli on abdominal segments two through seven, and two cerci and a median caudal filament at the tip of the abdomen. Females have a jointed ovipositor. The eggs are elliptical, about 0.04 in (1 mm) long. They are soft and white when first laid, but after several hours turn yellow and eventually brown. The larvae resemble small adults.

Distribution

There are about 370 species of thysanurans worldwide.

Habitat

Thysanurans are found in humid locations; under bark, rocks, rotting logs, and leaf litter; in caves; in ant and termite nests; and in synanthropic situations (those associated with human habitation). A few species live in sandy deserts.

Behavior

Silverfish hide under stones or leaves during the day and emerge after dark to search for food. All are fast running. They are nocturnal, even the early larval instars.

Feeding ecology and diet

Thysanurans are omnivorous, feeding on decaying or dried vegetable material and animal remains. Domestic species feed on starchy material such as paper, binding, and artificial silk. Some species associated with ants are cleptobiotic, robbing food from the ants. Species that live in deserts are able to absorb water through the rectum.

Reproductive biology

Thysanurans have ametabolous development; the larvae of silverfish resemble the adults, but are generally smaller in size. Silverfish continue to molt throughout their lives and individuals may live for up to six years. Most silverfish reproduce sexually, with the male depositing a sperm packet on the substrate beneath a silken thread, which is picked up by the female. Some species are parthenogenetic. Females use their ovipositor to insert the oval, whitish eggs into cracks and soil litter.

Conservation status

No species in Thysanura is listed by the IUCN.

Significance to humans

Domestic species of thysanurans are common household pests, causing extensive damage to household goods by feeding on wallpaper paste, book bindings, cardboard, and other paper products, and starch sizing of some textiles.

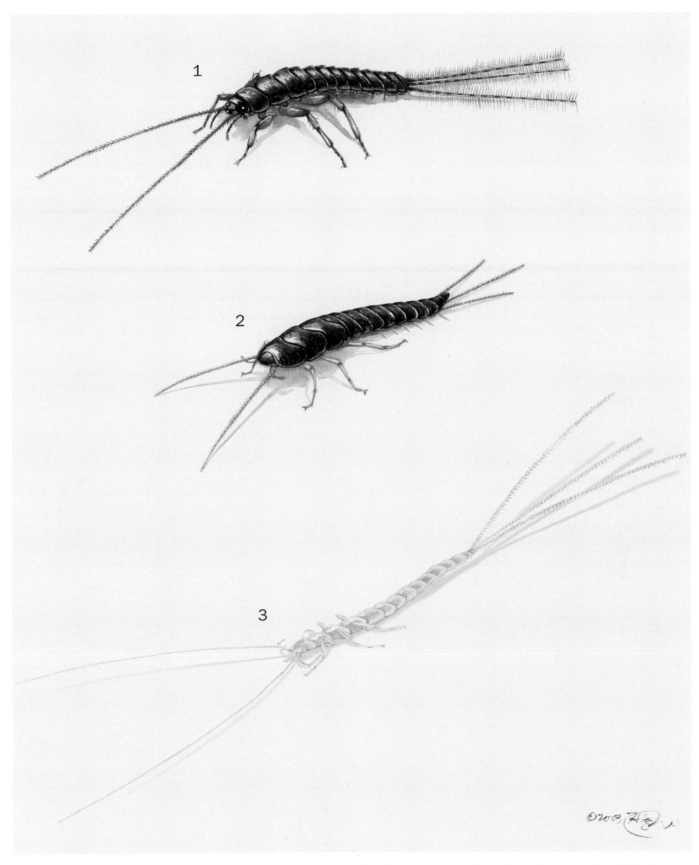

1. Relic silverfish (*Tricholepidion gertschi*); 2. Silverfish (*Lepisma saccharina*); 3. *Cubacubana spelaea*. (Illustration by Jonathan Higgins)

Species accounts

Relic silverfish
Tricholepidion gertschi

FAMILY
Lepidotrichidae

TAXONOMY
Tricholepidion gertschi Wygodzinsky, 1961, 2 mi (3.2 km) north of Piercy, Mendocino County, California, United States.

OTHER COMMON NAMES
English: Venerable silverfish.

PHYSICAL CHARACTERISTICS
Length 0.47 in (12 mm); body elongated, lacks scales. Compound eyes and three ocelli; caudal appendages longer than rest of body. Color reddish brown to yellowish.

DISTRIBUTION
Restricted to the redwood-mixed conifer forest of the coastal range of northern California.

HABITAT
Lives under rotten bark and decaying wood of Douglas fir (*Pseudotsuga menziesii*).

BEHAVIOR
When uncovered, tries to run away at high speed or displays warning posture elevating the body above the substrate, standing on the tips of the claws and moving the tip of the abdomen and caudal appendages laterally.

FEEDING ECOLOGY AND DIET
Feeds on vegetable detritus and terrestrial algae.

REPRODUCTIVE BIOLOGY
After male and female encounter one another, foreplay takes place, in which the male waves his caudal appendages intensively and for short periods in rapid succession, and the female follows the male in "tandem-walks," touching the male appendages with her antennae. The male then rotates and faces the female, runs past the female several times, and finally bends the abdomen dorsally and laterally, at the same time walking slowly and rotating while secreting threads of silk, upon which he ultimately place a spermatophore. The female then walks toward the male, and the male grasps her with his caudal appendages until she picks up the spermatophore with her ovipositor.

CONSERVATION STATUS
Not listed by the IUCN. This species lives in association with fallen Douglas firs in an advanced state of decay, and is also myrmecophilous, depending on the presence of a species of carpenter ants (*Camponotus* spp.). It will not be at risk as long as forest management practices leave sufficient fallen trees in a suitable stage of decay, and do not alter the thermal or moisture conditions of the forest understory.

SIGNIFICANCE TO HUMANS
The family Lepidotrichidae was known only from Oligocene fossils until living specimens were discovered in northwestern California in 1959, and is represented today only by *Tricholepidion gertschi*. The family is thus of interest to scientists because it is considered the most primitive of the thysanurans and the link between primitive wingless and winged insects. ◆

Silverfish
Lepisma saccharina

FAMILY
Lepismatidae

TAXONOMY
Lepisma saccharina Linnaeus, 1758, America, Europe, Sweden.

OTHER COMMON NAMES
English: Bristletail, common silverfish, fishmoth, fringetail, furniture bug, paper moth, shiner, silver witch, slicker, sugar fish, sugarlouse, tasseltail, wood fish; French: Poisson d'argent; German: Gemeines Silberfischchen; Spanish: Pescadito de plata.

PHYSICAL CHARACTERISTICS
Length 0.4 in (10 mm); body covered with silver scales (modified setae).

DISTRIBUTION
Probably native to tropical Asia; has been spread by humans around the world.

HABITAT
Domestic, found in warm, damp places, such as basements, closets, bookcases, shelves, and baseboards.

BEHAVIOR
Nocturnal.

FEEDING ECOLOGY AND DIET
Immature and adult stages are fond of flour and starch and are sometimes found in cereal; they can also feed on muslin, starched collars and cuffs, lace, carpets, fur, and leather. They are also cannibalistic, feeding on molted skins and dead and injured individuals.

REPRODUCTIVE BIOLOGY
Male spins a silk thread and deposits a spermatophore (packet of sperm) underneath; female picks it up and introduces it in her genital chamber. Eggs are laid singly or in batches of two to three, and are deposited in crevices or under objects. Under optimum conditions, an adult female lays an average of 100 eggs during her life span. Larvae have no scales up to their third molt. After 10 molts they reach sexual maturity, and the adults, with a life span of two through eight years, keep molting about four times per year.

CONSERVATION STATUS
Not threatened.

SIGNIFICANCE TO HUMANS
Household pest; does not transmit any disease. ◆

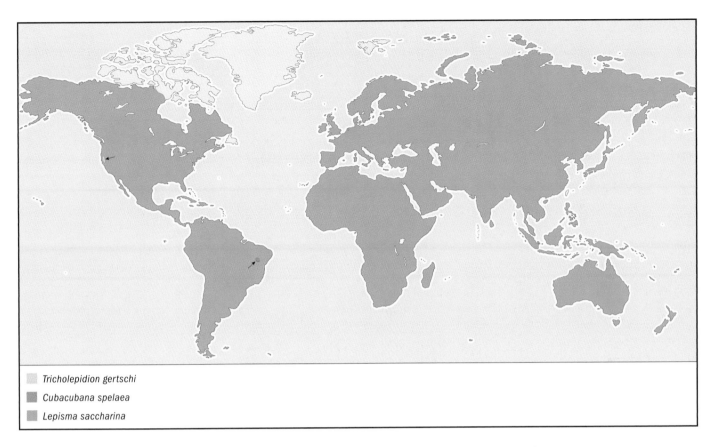

Tricholepidion gertschi

Cubacubana spelaea

Lepisma saccharina

No common name
Cubacubana spelaea

FAMILY
Nicoletiidae

TAXONOMY
Cubacubana spelaea Galan, 2000, Toca da Boa Vista, Bahia, Brazil.

OTHER COMMON NAMES
None known.

PHYSICAL CHARACTERISTICS
Length without appendages 0.47 in (12 mm), total length 1.41 in (36 mm); body elongated. Blind; antennae as long as one and a half times body length, cerci slightly longer than body length; lacks scales and pigmentation (white to entirely transparent).

DISTRIBUTION
Cave system Toca da Boa Vista in the north of Bahia, Brazil.

HABITAT
Deepest and moistest areas of the cave close to water ponds.

BEHAVIOR
Fast running on rocks and stalagmites.

FEEDING ECOLOGY AND DIET
Probably feeds on dry leaves and other vegetable detritus transported into the cave by bats, as well as on the paper wrapping of topographic bases placed by speleologists in the cave years ago.

REPRODUCTIVE BIOLOGY
Males still unknown, which could indicate parthenogenesis.

CONSERVATION STATUS
Not threatened.

SIGNIFICANCE TO HUMANS
None known. ◆

Resources

Books
Quintero, D., and A. Aiello, eds. *Insects of Panama and Mesoamerica*. New York: Oxford University Press, 1992.

Smith, G. B., and J. A. L. Watson. "Thysanura." In *The Insects of Australia. A Textbook for Students and Research Workers*, Vol. 2 (CSIRO), 2nd edition. Carlton, Australia: Melbourne University Press, 1991.

Wygodzinsky, P. "Order Thysanura" In *Immature Insects*, Vol. 2, edited by F. W. Stehr. Dubuque, IA: Kendall/Hunt Publishing, 1987.

Periodicals
Sturm, H. "The Mating Behaviour of *Tricholepidion gertschi* Wygodzinski, 1961 (Lepidotrichidae, Zygentoma) and Its

Resources

Comparison with the Behaviour of Other 'Apterygota.'" *Pedobiologia* 41, nos. 1–3 (1997): 44–49.

Wygodzinsky, P. "On the Surviving Representative of the Lepidothrichidae (Thysanura)." *Annals of the Entomological Society of America* 54 (1961): 621–627.

———. "A Review of the Silverfish (Lepismatidae, Thysanura) of the United States and Caribbean Area." *American Museum Novitates* 2,481 (1972): 1–26.

Natalia von Ellenrieder, PhD

Ephemeroptera
(Mayflies)

Class Insecta

Order Ephemeroptera

Number of families 37–40

Photo: A giant mayfly sitting on a leaf on a forest trail in Thailand. (Photo by Rosser W. Garrison. Reproduced by permission.)

Evolution and systematics

The scientific name of the order refers to the short life of the adult, or winged state. The English common name is associated with the month of May, when adults emerge more frequently in the Northern Hemisphere. The members of this order appear to be the most primitive of the extant flying insects. Fossil representatives that can be assigned to Ephemeroptera are known from the Upper Carboniferous, with some of the extinct groups (*Bojophlebia prokopi*) reaching gigantic sizes (a wingspan of up to 18 in, or 45 cm). Permian species had aquatic larvae, but unlike extant species, adults had functional mouthparts and pairs of wings of almost equal size.

The relationships of Ephemeroptera with other insect orders are still the subject of controversy. While some authors maintain that this order is related more closely to the dragonflies and damselflies (Odonata), forming the group Palaeoptera, others consider the Ephemeroptera basal to the other winged insect orders. The higher classification of the order is also questionable, with differing classifications at the suborder and superfamily levels. There are about 40 recognized fossil and extant families.

Physical characteristics

Body size ranges from very small (0.04 in, or 1 mm) to large (3.2 in, or 80 mm). Adults are of varying colors—white, yellow, pink, or black—generally with hyaline wings. Males generally are svelte, with the abdomen almost completely filled with air. The forelegs are always functional and very long. In females the body is more robust and stout, with the abdomen and part of the thorax filled with eggs. The hind wings can be absent, but when they are present, they always are smaller than the forewings. In some groups (e.g., Poly-

mitarcyidae) the adult legs are nonfunctional with the exception of the first pair on the male; some leg segments may even be lost (e.g., the American genus *Campsurus*). The caudal filaments (two or three) are always present and are longer in males. While the overall appearance of the adults is remarkably standard, the larvae possess a wide variety of body shapes, ranging from depressed and shield-like among the swirl current dwellers to long and cylindrical among secluded tunnel inhabitants or spindle-like in the free-swimming forms.

A unique characteristic of Ephemeroptera is the presence of an extra winged stage, called the "subimago." It is similar to the adult stage, or imago, but the coloration is duller, the body more robust, and the wings usually smoky. This extra stage allows for the dramatic elongation of some parts of the body (male legs and genitalia). They are covered with water-repellent microhairs, which are important in allowing them to pass from the aquatic medium to the air.

Distribution

Cosmopolitan, except Antarctica, the extreme Arctic, and a few small oceanic islands.

Habitat

The larvae inhabit a wide variety of bodies of water, from swiftly running streams high in the mountains to lowland rivers, ponds, and lakes. Many live under rocks or debris, whereas others burrow in living or decaying plant tissues or the clay bottoms of rivers and lakes. Others swim freely in small ponds, and a few live in the film formed by water seepage. While most species are sensitive to pollution, some can tolerate a certain level of contamination of their habitat. For example, some species of the genera *Callibaetis* and *Baetis*

(Baetidae) can tolerate high contents of organic matter and relatively low oxygen, while some species of *Caenis* (Caenidae) can stand certain high levels of suspended sediments. The adults, which normally are weak fliers, stay close to the water, where immature insects live.

Behavior

Most larvae hide during the day to avoid predation, but some swim freely at this time. Some species are gregarious, especially those that live in restricted patchy environments, but no social structure is known. Adults spend most of the day resting in the shade of vegetation close to the water, until swarming time, at which point they fly actively.

Feeding ecology and diet

Mayflies feed only as larvae; mouthparts and the digestive system are atrophied in adults. Most larvae are detritivores or herbivores, but a few are predators. Some modifications of mouthparts or forelegs or both occur in some species of Oligoneuriidae and Polymitarcyidae that filter food. Larvae

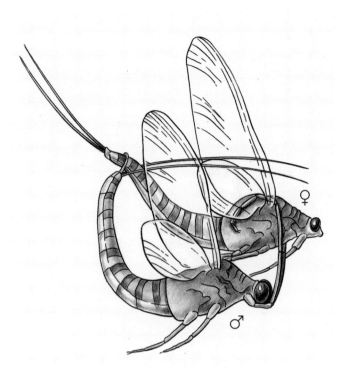

Copulating pair of *Rhithrogena semicolorata*. (Illustration by Patricia Ferrer)

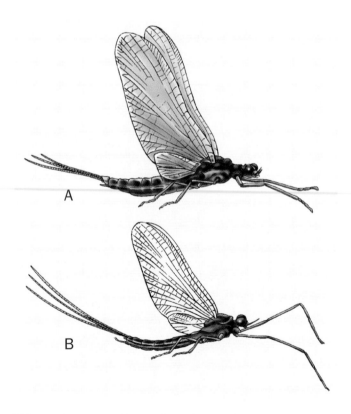

Difference between subimago (A) and imago (B). The subimago has duller coloration, a more robust body, and shorter legs and caudal filaments. (Illustration by Patricia Ferrer)

A pair of mayflies (*Ephemera danica*) on a dandelion stem. (Photo by John Shaw. Bruce Coleman, Inc. Reproduced by permission.)

Mayflies (*Hexagenia* sp.) emerge from water and swarm to tree branches. (Photo by ©James H. Robinson/Photo Researchers, Inc. Reproduced by permission.)

of species living in tunnels filter food from the water current that is produced by the active movement of the abdominal gills. Mayflies are important components of the food chain; fish, birds, and invertebrates eat both adults and larvae. These insects are popular with fly fishermen, who tie flies to imitate certain species of larvae and adults and time their fishing to emergence patterns for best results.

Reproductive biology

The mating flight, or "swarm," of the male imagoes is one of the distinctive features of the order. The swarm time, location, flight pattern, and number of individuals participating are specific to each species. Some swarms are composed of a few males and others by hundreds of thousands. The female flies above the swarm and is taken by a male from below, using his long forelegs. Shortly after copulation the female deposits eggs in the water. The female can carry 100 to 12,000 eggs. Egg development ranges from one week to a year and the larval period from three weeks to three years.

Conservation status

Some species have a restricted distribution and strict ecological tolerance and are more endangered than are the widely distributed species. The main reasons for population decline are habitat degradation by pollution, deforestation of river margins, dam construction, and introduction of exotic fish. Two mayfly species are listed as Extinct and one as Vulnerable by the IUCN.

Significance to humans

The larvae, as primary consumers, filter and remove large amounts of nutrients from the water and are important as food for other aquatic organisms. Owing to their sensitivity to contaminants, mayflies are used as bioindicators of water quality. The winged stage facilitates nutrient and energy exportation from aquatic to terrestrial ecosystems. A few larvae burrow in wooden structures, and when some species that form massive swarms are attracted to town lights, they cause allergy-related problems for some people.

1. *Chiloporter eatoni* nymph stage; 2. *Ephemera vulgata*; 3. *Stenonema vicarium*. (Illustration by Patricia Ferrer)

Species accounts

No common name
Chiloporter eatoni

FAMILY
Ameletopsidae

TAXONOMY
Chiloporter eatoni Lestage, 1931, Chile.

OTHER COMMON NAMES
None known.

PHYSICAL CHARACTERISTICS
Adult body size 0.6–0.9 in (15–22 mm); body and wings yellowish. Larvae have a big head and almost circular gills covering the abdomen. Gill color varies from pale yellow or pink to violet.

DISTRIBUTION
Southern Argentina and Chile.

HABITAT
Cold, well-aerated Patagonian creeks, streams, and lake margins.

BEHAVIOR
Larvae can hide under rocks and debris, although they normally are agile crawlers and fast swimmers, using the gills to propel themselves. Adult behavior is unknown.

FEEDING ECOLOGY AND DIET
Larvae are active predators, feeding mainly on other aquatic insect larvae.

REPRODUCTIVE BIOLOGY
Not known.

CONSERVATION STATUS
Distribution very scattered. Populations probably are endangered by introduction of trout, owing to their large size, brilliant coloring, and conspicuous behavior.

SIGNIFICANCE TO HUMANS
Small, but important in fish diet. ◆

Brown mayfly
Ephemera vulgata

FAMILY
Ephemeridae

TAXONOMY
Ephemera vulgata Linnaeus, 1758, Europe.

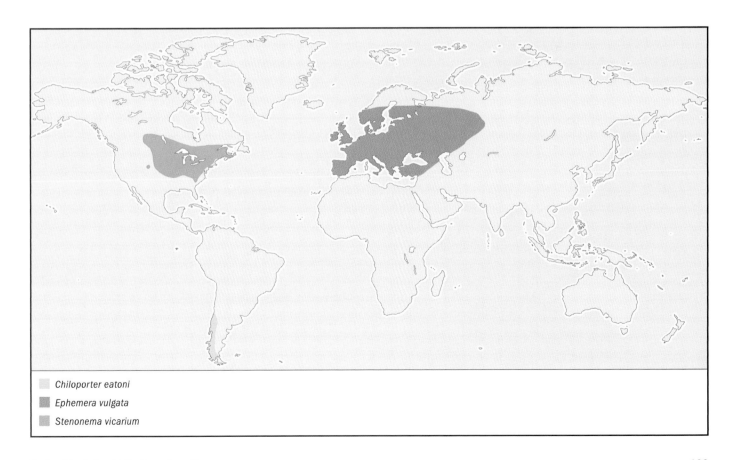

☐ *Chiloporter eatoni*
■ *Ephemera vulgata*
▨ *Stenonema vicarium*

OTHER COMMON NAMES
English: Brown drake.

PHYSICAL CHARACTERISTICS
Adult grow to 0.5–0.9 in (14–22 mm) and have spotted wings. Larvae have mandibular tusks for burrowing and big abdominal fringed gills (directed dorsally) that produce a water current inside the burrow.

DISTRIBUTION
Europe, including Great Britain and Scandinavia, south of the Arctic Circle and eastward to central Siberia.

HABITAT
Mostly in still waters that are not too cold (ponds, lakes, and riverine estuaries) and slow-flowing lowland rivers.

BEHAVIOR
Larvae burrow tunnels in silty and sandy substrates or in fine gravel. Imagoes fly in masses in the evening. Subimagoes hatch on the water surface.

FEEDING ECOLOGY AND DIET
Larvae are active filterers, feeding on small organic particles.

REPRODUCTIVE BIOLOGY
Males form small to large swarms along shores. Females deposit their eggs while drifting on the water surface. The life cycle is completed within two (rarely three) years, depending on water temperature.

CONSERVATION STATUS
Not threatened.

SIGNIFICANCE TO HUMANS
Small, but highly valued by fly fishermen, who tie flies that imitate larvae and adults. Fishing can provide high rental fees for water owners. ◆

No common name
Stenonema vicarium

FAMILY
Heptageniidae

TAXONOMY
Baetis vicaria Walker, 1853, Saint Lawrence River.

OTHER COMMON NAMES
None known.

PHYSICAL CHARACTERISTICS
Adult body size is 0.4–0.6 in (10–14 mm); wings are hyaline with dark brown veins. Larvae are flattened, with broad head and spreading legs; caudal filaments longer than the body.

DISTRIBUTION
Eastern United States and southern Canada.

HABITAT
Variety of water habitats, generally in moderately to rapidly flowing streams of varying sizes.

BEHAVIOR
Larvae are strongly thigmotactic (contact-loving), clinging to some substrates. They crawl rapidly but seldom swim, which they do poorly. The full-grown larva floats to the water surface, from where the subimago hatches.

FEEDING ECOLOGY AND DIET
Larvae are collectors (gatherers), feeding on particulate organic matter.

REPRODUCTIVE BIOLOGY
Males form small swarms of 10 to 20 individuals. After mating, the female rests on the water, releasing eggs slowly or depositing a few at a time while touching the water with the tip of the abdomen. The life cycle lasts from a few months to one year.

CONSERVATION STATUS
Not threatened.

SIGNIFICANCE TO HUMANS
Small, but important in fish diet. ◆

Resources

Books
Berner, Lewis, and Manuel L. Pescador. *The Mayflies of Florida.* Rev. ed. Gainesville: University Presses of Florida, 1988.

Domínguez, Eduardo, ed. *Trends in Research in Ephemeroptera and Plecoptera.* New York and London: Kluwer Academic/Plenum Publishers, 2001.

Landolt, Peter, and Michel Sartori, eds. *Ephemeroptera and Plecoptera: Biology, Ecology, Systematics.* Fribourg, Switzerland: Mauron, Tinguely, and Lachat SA, 1997.

Needham, James G., Jay R. Traver, and Yin-Chi Hsu. *The Biology of Mayflies, with a Systematic Account of North American Species.* Ithaca, NY: Comstock Publishing, 1935.

Periodicals
Allan, J. Dave, and Alexander S. Flecker. "The Mating Biology of a Mass-Swarming Mayfly." *Animal Behavior* 37, no. 3 (1989): 361–371.

Brittain, John E. "Biology of Mayflies." *Annual Review of Entomology* 27 (1982): 119–147.

Edmunds, George F. Jr., and W. Patrick McCafferty. "The Mayfly Subimago." *Annual Review of Entomology* 33 (1988): 509–529.

Peters, William L., and Janice G. Peters. "In the Predawn Mass Mating of Sand-Burrowing Mayflies, Timing Is Everything." *Natural History* (1988): 8–14.

Resources

Ruffieux, Laurence, Jean-Marc Elouard, and Michel Sartori. "Flightlessness in Mayflies and Its Relevance to Hypotheses on the Origin of Insect Flight." *Proceedings of the Royal Society London Series B.* 265, no. 1410 (1998): 2135–2140.

Other

Hubbard, Michael D. "Ephemeroptera Galactica." 13 Feb. 2002. [2 Apr. 2003] <http://www.famu.org/mayfly/>.

McCafferty, Patrick W. "Mayfly Central." [2 Apr. 2003] <http://www.entm.purdue.edu/entomology/research/mayfly/mayfly.html>.

Eduardo Domínguez, PhD

Odonata

(Dragonflies and damselflies)

Class Insecta
Order Odonata
Number of families 28

Evolution and systematics

Odonates appeared by the early Permian period, and lineages corresponding to the three extant suborders flourished in the Mesozoic—Zygoptera and Anisozygoptera in the Triassic and Anisoptera in the Jurassic. Debate exists about the relationship of the three suborders; conventional classification fails to match the true evolutionary relationships of the groups, with "Anisozygoptera" being an artificial grouping. Zygoptera (damselflies) comprise four superfamilies and 18 families, the so-called Anisozygoptera has only one family and genus, and Anisoptera (dragonflies) contains three superfamilies and four families.

Physical characteristics

Wingspans range from 6.5 in (162 mm) in the Australian dragonfly, *Petalura ingentissima*, to 0.8 in (20 mm) in the Southeast Asian damselfly, *Agriocnemis femina*. They have large compound eyes and chewing mouthparts. The two posterior segments of the thorax are fused together. The legs are well developed for seizing prey and for perching; locomotion is almost solely by flight. The large, strong, multiveined wings usually have an opaque pterostigma near the wing tip. The ten-segmented abdomen is long and slender. In males unique secondary genitalia evolved on the underside of the second and third abdominal segments, separated from the actual genital opening near the abdomen tip. Damselfly and several dragonfly females have well-developed ovipositors used to insert eggs into plant tissue; in some dragonflies the ovipositor valves are reduced, and eggs are dropped into water. Both sexes have caudal appendages at the tip of the abdomen, which in males work like claspers to grasp the female during mating. Larvae are aquatic and have a unique lower jaw specialized for grasping prey. Damselfly larvae are long and narrow and have three caudal lamellae used for breathing. Dragonfly larvae have broad bodies and breathe through tracheal gills located in the rectum.

Distribution

Dragonflies are found worldwide, except in frozen polar areas. Their greatest diversity is in the tropics.

Habitat

Larvae are mostly aquatic and are found under stones, buried in mud or detritus, or clinging to vegetation in stagnant and running freshwater. A few inhabit small water reservoirs in plants; others live in moist terrestrial burrows in forests. Adults occur over almost any kind of freshwater, where they mate and oviposit (lay eggs).

Behavior

Odonates regulate their body temperature by assuming different postures and selecting perching sites. In cool weather

A newly emerged dragonfly (*Rhionaesetina variegata*) with larval skin. (Photo by Rosser W. Garrison. Reproduced by permission.)

they engage in wing whirring and land on sun-facing perches, whereas in hot weather they avoid overheating by assuming an "obelisk" position, with the abdomen exposing the least possible area to the sun. Many males are territorial and patrol an area of water, chasing rival males from it. Females cruise through territories, attracted to possible egg-laying sites. In some species, males perform threatening displays for other males or courtship displays for females by exposing color patches on the head, legs, abdomen, or wings. After emergence, some species undertake long-distance migrations; others disperse short distances when mature, searching for suitable sites to oviposit.

Feeding ecology and diet

Larvae and adults are active or ambush predators. Adults capture and eat insects on the wing, and larvae eat mosquito larvae, other aquatic invertebrates, and even fish and tadpoles. Adult adaptations for feeding include large eyes, which allow them to see in virtually all directions; legs forming a "basket" to scoop up prey; and strong wings, providing amazing flight maneuverability. Larvae capture prey by rapidly extending the

labium and seizing prey between the two movable hooks at its tip.

Reproductive biology

Mating is unique. The male caudal appendages grasp a female at the back of the head (dragonflies) or anterior part of the thorax (damselflies), forming the "tandem position." Before copulation, the male arches his abdomen, transferring sperm from near the tip to the secondary genitalia at the base. Copulation ensues when the female arches her abdomen to bring her genital opening into contact with the accessory male genitalia, forming the "wheel position." After copulation, the female oviposits either alone or guarded by the male, who continues to hold her in the tandem position or flies near her. Eggs are laid in aquatic plant tissue, mud, or water. The growing larva sheds its skin several times before metamorphosing into an adult. Larvae live from six months to five years, depending on water temperature and food supply. The adult is the dispersal stage and lives from one to two months in temperate areas to a full year in the tropics.

Conservation status

Of the more than 5,500 known species of odonates, 137 are included on the IUCN Red List: two as Extinct; 13 as Critically Endangered; 55 as Endangered; 39 as Vulnerable; 17 as Lower Risk/Near Threatened; and 11 as Data Defi-

Dragonfly mating wheel. (Illustration by Kristen Workman)

A damselfly (*Archilestes californica*) female (below) ovipositing while guarded by a male (above) in tandem position. (Photo by Rosser W. Garrison. Reproduced by permission.)

cient. Little is known about distribution and habitat preferences for most species, and continuing habitat destruction precludes obtaining essential information necessary to ensure conservation of sensitive species. Nature preserves to conserve dragonflies exist in Japan, Europe, and the United States. Valuable conservation efforts are under way in local areas and for specific problems in South Africa, Australia, and India, but many areas with high numbers of unique species are in tropical regions without habitat protection. Japan,

where artificial habitats are created and managed for propagating dragonflies, represents one of the best examples of dragonfly conservation.

Significance to humans

Despite menacing common names (e.g., "devil's darning needles" or "horse-stingers"), odonates are harmless; they have no sting. They consume large numbers of harmful insects (including disease-transmitting mosquitoes) and also are excellent indicators of freshwater quality. For the Navaho Indians they symbolize pure water. Traditionally known as the "invincible insect," the dragonfly was a favorite symbol of strength among Japanese warriors, and the old name for the island of Japan (Akitsushima) means "Island of the Dragonfly."

Dragonfly oblisking behavior. (Illustration by Kristen Workman)

1. Living fossil (*Epiophlebia laidlawi*); 2. Forest giant (*Megaloprepus caerulatus*); 3. Wandering glider (*Pantala flavescens*). (Illustration by Jacqueline Mahannah)

Species accounts

Living fossil
Epiophlebia laidlawi

FAMILY
Epiophlebidae

TAXONOMY
Epiophlebia laidlawi Tillyard, 1921, Himalayas.

OTHER COMMON NAMES
None known.

PHYSICAL CHARACTERISTICS
Black with bright yellow stripes on the thorax and abdomen. One of the only two extant species of Anisozygoptera, it shares some characteristics with damselflies (forewings and hind wings similar in shape and venation and a well-developed ovipositor) and others with dragonflies (eyes separated by less distance than their width, a pair of superior caudal appendages and a single inferior one in the male, and broad-bodied larva with rectal breathing).

DISTRIBUTION
Confined to the eastern Himalayas in Nepal and India.

HABITAT
Breeds in streams between 6,000 and 11,500 ft (1,800–3,500 m). Adults fly in clearings within dense bamboo forests.

BEHAVIOR
Larvae stridulate when disturbed. During the maturation period adults fly high above breeding areas. When mature, males fly slowly, low down and close to the stream; females skulk at the water's edge in the vegetation.

Epiophlebia laidlawi

FEEDING ECOLOGY AND DIET
Nothing is known.

REPRODUCTIVE BIOLOGY
Males grasp females by the back of the head to form the tandem position. A solitary female inserts eggs into the stems of plants growing at stream margins. Larval period lasts from six to nine years, the longest known for any odonate.

CONSERVATION STATUS
Until 1980 considered Endangered but since then discovered at several new sites, appearing to be widespread and common. Given the necessary habitat protection, it can be considered safe.

SIGNIFICANCE TO HUMANS
None known. ◆

Wandering glider
Pantala flavescens

FAMILY
Libellulidae

TAXONOMY
Libellula flavescens Fabricius, 1798, India.

OTHER COMMON NAMES
English: Rainpool glider, globe skimmer.

PHYSICAL CHARACTERISTICS
Yellowish-red in color. The base of the hind wing is noticeably broadened, with a small, diffuse yellowish patch at the base. Pterostigma of the forewing longer than that of the hind wing. Strongly tapering abdomen, with a black mid-dorsal stripe.

DISTRIBUTION
A cosmopolitan species, found in tropical and temperate regions around the world. Common in the tropics but rarely seen in Europe.

HABITAT
Breeds in small, shallow, often temporary pools. Adults frequently are observed far from water.

BEHAVIOR
Strong, high-gliding flight, rarely settling. The species is gregarious and may form large feeding and migratory swarms. Feeding flights may continue beyond dusk. They have been seen far out at sea, flying even at night, when they frequently are attracted to the lights of ships.

FEEDING ECOLOGY AND DIET
A study of the gut contents of adults feeding over rice fields in Bangladesh showed that their diet consisted mainly of mosquitoes.

REPRODUCTIVE BIOLOGY
Males patrol territories about 30–150 ft (9–45 m) in length. After mating, the male remains in tandem while the female lays her eggs. Females oviposit by tapping the surface of the water

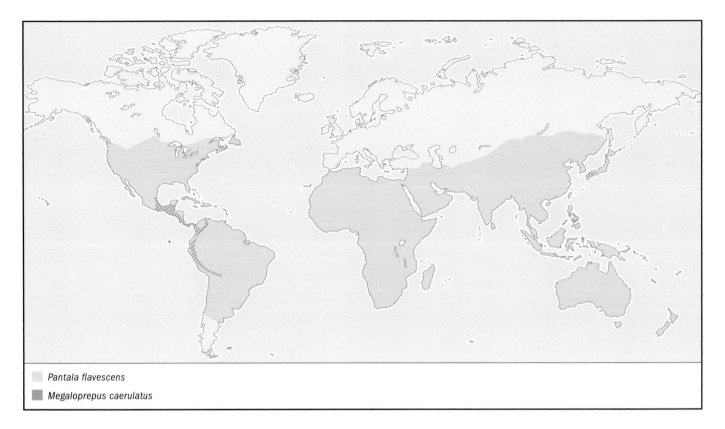

Pantala flavescens

Megaloprepus caerulatus

with the tip of the abdomen. Larval development is rapid, an adaptation that allows for the use of temporary pools (including swimming pools) as breeding sites.

CONSERVATION STATUS
Not listed by the IUCN.

SIGNIFICANCE TO HUMANS
None known. ◆

Forest giant
Megaloprepus caerulatus

FAMILY
Pseudostigmatidae

TAXONOMY
Libellula caerulata Drury, 1782, Bay of Honduras.

OTHER COMMON NAMES
Spanish: Helicóptero.

PHYSICAL CHARACTERISTICS
Largest damselfly, with a wingspan of 6.4 in (160 mm) and a very long abdomen, at 4 in (100 mm). Wings lack pterostigma and have a wide, dark blue band. Sexually dimorphic (males and females look different); males are larger, with a white patch before the blue band and a hyaline wing tip, while females are shorter, with only white on the wing tip.

DISTRIBUTION
Rainforests of Central and South America, from Mexico to Bolivia.

HABITAT
Larvae breed in water-filled plant containers. Adults frequent sunlit gaps or small clearings in the forest.

BEHAVIOR
Because of their particular breeding sites scattered throughout the forest, this forest giant is never found in great numbers.

FEEDING ECOLOGY AND DIET
Adults are specialist foragers; they detect nonmoving prey and pluck small web-building spiders, occasionally also taking wrapped prey from webs. Larvae feed on mosquito and fly larvae, microcrustaceans and tadpoles, and conspecifics sharing the same tree hole.

REPRODUCTIVE BIOLOGY
A territorial male uses a slow wing beat frequency to appear as a pulsating blue and white beacon to both potential mates and competing males in an open forest gap. He can aggressively defend a particular tree hole for up to three months. After copulating, the female uses her long abdomen to lay her eggs inside tree holes with water.

CONSERVATION STATUS
Not listed by the IUCN.

SIGNIFICANCE TO HUMANS
None known. ◆

Resources

Books

Corbet, P. S. *Dragonflies: Behavior and Ecology of Odonata.* New York: Cornell University Press, 1999.

Dunkle, S. W. *Dragonflies Through Binoculars: A Field Guide to Dragonflies of North America.* New York: Oxford University Press, 2000.

Needham, James C., Minter J. Westfall, and Michael L. May. *Dragonflies of North America.* Revised edition. Gainesville, FL: Scientific Publishers, 2000.

Silsby, J. *Dragonflies of the World.* Collingwood, Australia: CSIRO Publishing, 2001.

Westfall, Minter J. Jr., and Michael L. May. *Damselflies of North America.* Gainesville, FL: Scientific Publishers, 1996.

Organizations

British Dragonfly Society. Membership Office, 53 Rownhams Road, Maybush, Southampton, SO16 5DX United Kingdom. Web site: <http://www.dragonflysoc.org.uk>

Dragonfly Society of the Americas. 2091 Partridge Lane, Binghamton, NY 13903 United States. Web site: <http://www.afn.org/iori/dsaintro.html>

Gesellschaft Deutschsprachiger Odonatologen. Web site: <http://www.libellula.org>

International Odonata Research Institute. E-mail: iori@afn.org Web site: <http://www.afn.org/iori>

Societas Internationalis Odonatologica. P.O. Box 256, Bilthoven, NL-3720 AG Netherlands.

Worldwide Dragonfly Association. P.O. Box 321, Leiden, 2300 AH Netherlands. Web site: <http://powell.colgate.edu/wda/dragonfly.htm>

Other

"Dragonfly (Odonata) Biodiversity." [22 Dec. 2002]. <http://www.ups.edu/biology/museum/UPSdragonflies.html>.

Natalia von Ellenrieder, PhD

Plecoptera
(Stoneflies)

Class Insecta

Order Plecoptera

Number of families 16

Photo: A stonefly (*Leuctra fusca*) on a balsam flower in Europe. (Photo by Kim Taylor. Bruce Coleman, Inc. Reproduced by permission.)

Evolution and systematics

Stonefly fossils date from the early Permian, about 258 to 263 million years ago. The fossil record of just under 200 species is most diverse in the Jurassic and is considered fragmentary in comparison with other aquatic insects, probably due to stoneflies' preference for running waters that are not conducive to burial and fossilization. Stoneflies comprise a hemimetabolous (i.e., undergoing complete metamorphosis) order divided into two suborders: Arctoperlaria, containing 12 families (Capniidae, Chloroperlidae, Leuctridae, Nemouridae, Notonemouridae, Paltoperlidae, Perlidae, Pteronarcyidae, Scopuridae, Styloperlidae, and Taeniopterygidae); and Antarctoperlaria, containing four families (Austroperlidae, Diamphipnoidae, Eustheniidae, and Gripopterygidae). The order includes five superfamilies and more than 2,000 species.

Physical characteristics

Adult stoneflies vary in body length from about 0.19 to 1.97 in (5 to 50 mm) and in color from brown or black to green or yellow, usually marked with distinctive light or dark patterns.

They are typically winged, except, for example, a wingless aquatic adult of the family Capniidae known from the depths of Lake Tahoe, Nevada. Wings are typically fully winged (macropterous), but the wings of one or both sexes of some species or high altitude or latitude populations of a particular species are shortened (brachypterous) and are not functional. The ordinal name (Plecoptera = "folded wings") describes the hind wings that typically have an expanded posterior lobe that folds under the main wing. Adults have 10 abdominal segments, a three-segmented tarsus, and a pair of terminal, usually multisegmented, cerci. The multisegmented larval cerci become reduced to fewer segments in some taxa and to a single segment in males of the families Leuctridae, Nemouridae, and some Taeniopterygidae. Males have distinctive genitalia consisting of various modifications of the ninth and tenth segments into paired hooks, lobes (paraprocts), sclerotized paired stylets, or a median terminal probe (epiproct). The aedeagus is housed inside the abdomen and extruded during copulation from behind the ninth sternum. The external female genitalia consists of a lobe-like plate usually on the eighth abdominal sternum covering the genital opening. Larvae may or may not

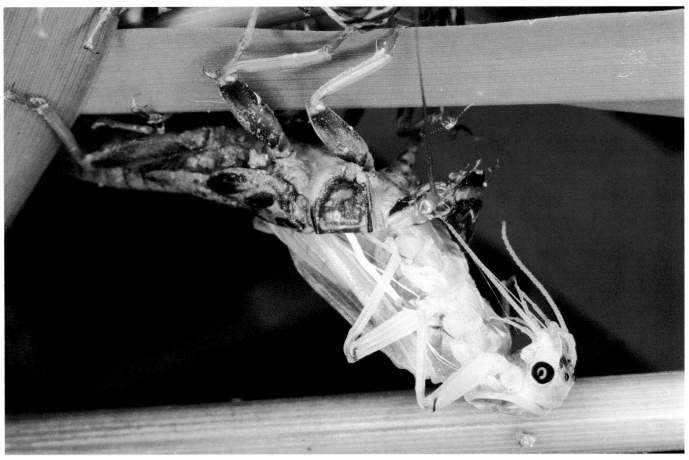

A stonefly (*Leuctra fusca*) nymph sheds its exoskeleton. (Photo by Roy Morsch. Bruce Coleman, Inc. Reproduced by permission.)

resemble their adult forms, and those of particular families or genera vary from being gill-less to having simple or branched gills, diagnostically located and structured, arising from parts of the body such as near mouthparts, thorax, coxae, or abdomen. Larvae always have a pair of multisegmented cerci, and the adults of gilled taxa usually retain stubs or vestiges of the larval gills that aid in their identification.

Distribution

This order is distributed worldwide, on all continents except Antarctica, and on most major islands except Cuba, Fiji, Hawaii, and New Caledonia. Species of the suborder Arctoperlaria are generally distributed in the Northern Hemisphere; exceptions are the family Notonemouridae, which occurs only in southern South America, southern Africa, Madagascar, Australia, Tasmania, and New Zealand, and the two genera of the family Perlidae, *Anarcroneuria* and *Neoperla*, that have moved south across the Equator in recent times. The suborder Antarctoperlaria is restricted to the Southern Hemisphere.

Habitat

Stoneflies are almost exclusively inhabitants of streams, where their larvae inhabit organic or mineral substrates.

Adults occur in streamside microhabitats, including on or under rocks, moss, debris, leaf packs on the bank or projecting above the water surface, and in riparian vegetation. The larvae of a few species occur in wave-swept substrates of cold alpine or boreal lakes or in intermittent streams.

Behavior

The typical mating system of Northern Hemisphere stoneflies involves aggregation of males and females at encounter sites near streams, and vibrational communication for mate finding. Males call for females with species-specific drumming signals while performing a ranging search, and receptive stationary females answer with a simple drumming signal. These signals and the accompanying positioning allow males to minimize their search and triangulate on a particular female. Copulation occurs immediately after location with no specialized display. Males are polygamous, but females mate only once and then will no longer answer male drumming calls.

Feeding ecology and diet

Stonefly adults of particular taxa are either nonfeeders (e.g., members of the group Systellognatha of Arctoperlaria) or are mainly herbivorous, feeding on pollen, nectar, or other plant parts (e.g., some species of the group Euholognatha of Arctoperlaria). Larvae are either primarily herbivore-detritivores, insectivores, or omnivores, and in some cases their diet shifts among these categories as they develop through 10 to 25 instars. Herbivore-detrivores have molariform mandibles and are either scrapers, grazers, collector-gatherers, shredders, or gougers. Predators have sharp-cusped mandibles and toothed laciniae for grasping and holding their prey, which they actively seek in the interspaces of their leaf-pack or on mineral substrates in streams.

Reproductive biology

The mating of stoneflies involves a male mounting a female, curving his abdomen around her left or right side, engaging and pulling down the subgenital plate with his external genitalia, and typically inserting his aedeagus into her bursa. Sperm is therefore typically conveyed into the female by the intromittent aedeagus, but in some species sperm is deposited in a pocket beneath the subgenital plate and then aspirated internally by the female, while in other species sperm is conveyed through a hollow male epiproct. Eggs vary considerably in size, shape, and chorionic (egg shell) ornamentation and sculpturing. Penetration by sperm through a micropyle and fertilization, as in most insects, is delayed until just before oviposition (laying of eggs). Eggs are deposited in pellets or masses and released in one of two ways: 1) directly into water by the female splashing into the surface or dropping eggs from the air during an oviposition flight; or 2) being washed off in shallow water. Eggs may hatch within three to four weeks or enter a diapause (a period of arrested development) lasting from three months to one or more years. Generation time for larvae varies from four months to three or four years, depending on the taxa and on environmental conditions. The short-lived adults provide no parental care.

Conservation status

Stoneflies generally require unpolluted streams for continued population health and are therefore important biological indicators of stream water quality. They are important components of one of the major biomonitoring measures, the EPT (Ephemeroptera-Plecoptera-Trichoptera) Index. Some species are endemic to particular stream watersheds or rare and restricted in distribution to small geographic regions, and are therefore listed on various regional or national endangered species lists. Stonefly larvae can generally be thought of as similar in relative size, space requirements, and clean water physical and chemical tolerance to trout and other fish juveniles. Management practices of stream fisheries are therefore conducive also to stonefly management, although habitat or species management practices specifically for stoneflies have rarely been proposed or practiced.

A giant stonefly (*Pteronarcys* sp.) completing metamorphosis. (Photo by Joe McDonald. Bruce Coleman, Inc. Reproduced by permission.)

The 2002 IUCN Red List includes four stonefly species. *Alloperla roberti* is categorized as Extinct; *Leptoperla cacuminis* and *Riekoperla darlingtoni* as Vulnerable; and *Eusthenia nothofagi* as Data Deficient.

Significance to humans

Stoneflies are entirely beneficial to humans as integral and important components of stream food webs and biological indicators of good water quality. Most adults have vestigial mouthparts and cannot bite. Their importance as fish food makes them, along with mayflies, caddisflies, and midges, of much interest to fly fishermen.

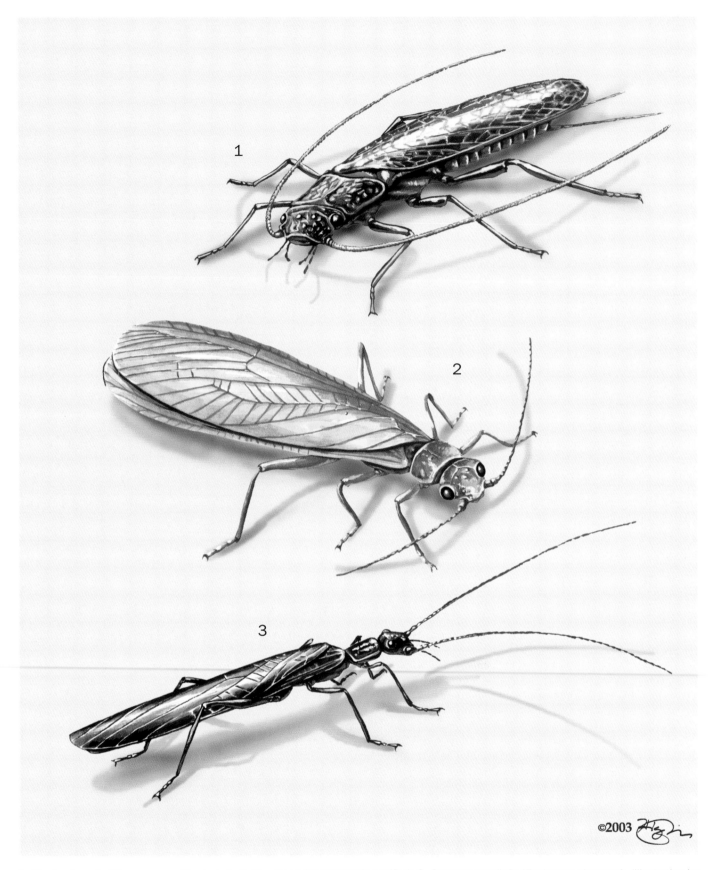

1. Giant salmonfly (*Pteronarcys californica*); 2. Golden stone (*Hesperoperla pacifica*); 3. Common needlefly (*Zealeuctra claasseni*). (Illustration by Jonathan Higgins)

Species accounts

Common needlefly
Zealeuctra claasseni

FAMILY
Leuctridae

TAXONOMY
Zealeuctra claasseni (Frison), 1929, Herod, Illinois, United States.

PHYSICAL CHARACTERISTICS
Adults and larvae with typical elongate body of the family Leuctridae; males have a single cercal segment. Larvae are gill-less.

DISTRIBUTION
Midwestern United States.

HABITAT
Small, often intermittent streams.

BEHAVIOR
Nothing is known.

FEEDING ECOLOGY AND DIET
Herbivore-detritivore.

REPRODUCTIVE BIOLOGY
A winter-emerging species (November–February). Males call females on leaf mats and plants with a drumming signal of about 20 beats, and females answer with a similar signal. Drumming and mating occur within a day after emergence, and oviposition follows the same day or a day later.

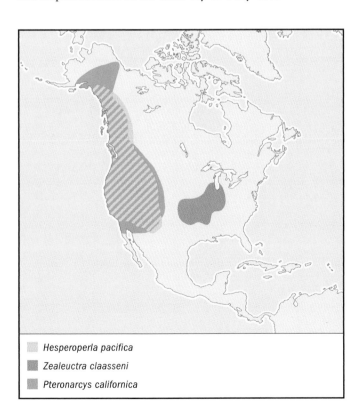

Hesperoperla pacifica
Zealeuctra claasseni
Pteronarcys californica

CONSERVATION STATUS
Not threatened. A hardy stonefly, able to survive for several drought years as diapausing eggs in intermittent streams.

SIGNIFICANCE TO HUMANS
None known. ◆

Giant salmonfly
Pteronarcys californica

FAMILY
Pteronarcyidae

TAXONOMY
Pteronarcys californica Newport, 1848, California, United States.

PHYSICAL CHARACTERISTICS
Large (1.18–1.97 in [30–50 mm]), dark brown stoneflies. Larvae have profuse, filamentous thoracic and abdominal gills, retained as stubs in adults.

DISTRIBUTION
Widespread in western North America.

HABITAT
Cobble substrates of large streams and rivers.

BEHAVIOR
Adults aggregate on riparian vegetation for drumming and mating.

FEEDING ECOLOGY AND DIET
Major shredders of decomposing leaves in streams.

REPRODUCTIVE BIOLOGY
Males call females on debris and plant substrates with heavy 6-beat signals, and females answer with similar signals. Spring-emerging species. Larvae require two to three years for development.

CONSERVATION STATUS
Not threatened.

SIGNIFICANCE TO HUMANS
Important stream food web components and major trout food. Very important to fly fishermen, who create artificial fly lure that resembles both adults and larvae. ◆

Golden stone
Hesperoperla pacifica

FAMILY
Perlidae

TAXONOMY
Hesperoperla pacifica (Banks), 1900, Olympia, Washington, United States.

PHYSICAL CHARACTERISTICS
Yellow-orange colored stoneflies, 0.71–1.18 in (18–30 mm) body length. Larvae have branched, filamentous gills arising from the ventral thorax and tenth abdominal segment, retained as stubs in adults.

DISTRIBUTION
Widespread in western North America.

HABITAT
Coarse gravel and cobble substrates of small and large streams and rivers.

BEHAVIOR
Adults aggregate on streamside rocks, debris, and vegetation for drumming and mating.

FEEDING ECOLOGY AND DIET
An insectivorous predator that feeds mainly on bloodworm (Chironomidae) larvae and mayfly and caddisfly larvae.

REPRODUCTIVE BIOLOGY
Males call females with monophasic, 12-beat drumming signals, females answer with 16-beat signals, then males reply with 22-beat signals. A spring-emerging species. Hatchling larvae require more than one year for development.

CONSERVATION STATUS
Not threatened.

SIGNIFICANCE TO HUMANS
An important trout food insect. Fly fishermen model their lure after it. ◆

Resources

Books
Resh, V. H., and D. M. Rosenberg. *The Ecology of Aquatic Insects.* New York: Praeger Publishers, 1984.

Sinitschenkova, N. D. "Paleontology of Stoneflies." In *Ephemeroptera and Plecoptera*, edited by Peter Landolt and Michel Sartori. Fribourg, Switzerland: Mauron and Tinguely; Lachat SA, 1995.

Stark, B. P., S. W. Szczytko, and C. R. Nelson. *American Stoneflies: A Photographic Guide to the Plecoptera.* Columbus, OH: Caddis Press, 1998.

Stewart, K. W. "Vibrational Communication (Drumming) and Mate-searching Behavior of Stoneflies (Plecoptera); Evolutionary Considerations." In *Trends in Research in Ephemeroptera and Plecoptera*, edited by Eduardo Dominguez. New York: Kluwer Academic/Plenum Publishers, 2001.

Stewart, K. W., and P. P. Harper. "Plecoptera." In *An Introduction to the Aquatic Insects of North America*, 3rd ed., edited by R. W. Merritt and K. W. Cummins. Dubuque, IA: Kendall/Hunt Publishing, 1996.

Stewart, K. W., and B. P. Stark. *Nymphs of North American Stonefly Genera (Plecoptera)*, 2nd ed. Columbus, OH: Caddis Press, 2002.

Periodicals
Shepard, W. D., and K. W. Stewart. "Comparative Study of Nymphal Gills in North American Stonefly (Plecoptera) Genera and a New, Proposed Paradigm of Plecoptera Gill Evolution." *Miscellaneous Publications Entomological Society of America* 55 (1983): 1–57.

Stewart, K. W. "Theoretical Considerations of Mate-finding and Other Adult Behaviors of Plecoptera." *Aquatic Insects* 16 (1993): 95–104.

———. "Vibrational Communication in Insects: Epidomy in the Language of Stoneflies." *American Entomologist* 43 (1997): 81–91.

Zwick, P. "Phylogenetic System and Zoogeography of the Plecoptera." *Annual Review of Entomology* 45 (2000): 709–746.

Kenneth W. Stewart, PhD

Blattodea
(Cockroaches)

Class Insecta

Order Blattodea

Number of families ca. 7

Photo: A German roach (*Blattella germanica*) resting on a leaf in Florida, USA. (Photo by Bob Gossington. Bruce Coleman, Inc. Reproduced by permission.)

Evolution and systematics

Fossil cockroaches or cockroach-like species are so numerous in the coal seams of the Upper Carboniferous (350 million to 400 million years ago [mya]) that the period is known as the "Age of Cockroaches." Only about 5% of fossil cockroaches represent entire insects; 90% consist of wings or wing fragments, with the remaining 5% consisting of other body parts. For the most part, these fossils appear to be very similar to the forms living today. However, the females of many fossils from the Carboniferous and Permian deposits (270 to 225 mya) of Europe, Asia, and North America, as well as from the Triassic (225 to 180 mya) and Jurassic deposits (135 to 100 mya) of Russia, had long, external ovipositor valves. This indicates that they probably laid single eggs in soil or soft plant tissue, rather than a number of eggs in an ootheca, which is characteristic of today's forms. During the Jurassic period the ovipositor valves decreased in length and gradually evolved into those found in all present-day species.

The oldest fossil genus in the Blattellidae (one of the largest families), *Pinablattella*, dates to the Cretaceous (135 mya) of Siberia. The coexistence of fossil cockroaches and fossil plants in the same geological stratum suggests that there was a close association between them during this early period, and it is speculated that the cockroaches fed on these plants. Cockroaches are related closely to termites and mantids, with which they often have been grouped in the order Dictyoptera. This chapter treats cockroaches as comprising the order Blattodea, which does not include termites and mantids. Cockroaches can be divided into two large superfamilies, the Blattoidea and the Blaberoidea. These two groups evolved from oviparous ances-

tors; the Blattoidea remained oviparous, and the Blaberoidea eventually evolved ovoviviparity and viviparity.

Physical characteristics

The chewing mouthparts are directed downward. The antennae consist of numerous segmented annuli and usually are longer than the body. Compound eyes typically are present, but they may be reduced in size or absent, especially in cavernicolous (cave-dwelling) species. The pronotum is large and often shieldlike and covers the head. When present, the forewings typically are modified into hardened tegmina, which may be abbreviated or absent; hind wings may be reduced in size or absent, if present, they are membranous, with well-developed veins. Legs are adapted for running or sometimes digging. The coxae are adpressed against the body. The tarsi have five segments, often with pulvilli, which may be reduced, or absent. Tarsal claws almost always are present, with or without an arolium between them.

Sizes vary widely, the smallest being about 0.8 in (2 mm) long, the size of a mosquito (e.g., *Nocticola* species). The largest species generally are in the Blaberidae; the wingless *Macropanesthia rhinoceros* (Panesthiinae) probably is the bulkiest cockroach known, and it may reach 2.6 in (65 mm) in length and weigh 0.53–1.1 oz (15–30 g) or more. Another wingless blaberid, *Gromphadorhina portentosa*, may reach a length of 3.1 in (80 mm). Some fully winged neotropical blaberids may have a body length of 3 in (75 mm) (*Archimandrita tessellata*) or 3.1 in (80 mm) (*Blaberus giganteus*).

Numerous species are uniformly dark (black, brown, or reddish brown), but many are distinctively marked. A large

An oriental cockroach (*Blatta orientalis*) cleaning its antennae. (Photo by Adrian Davies. Bruce Coleman, Inc. Reproduced by permission.)

number (especially Panesthiinae) are aposematic (brightly colored as a warning signal to predators), with white or yellow markings that warn potential predators that they are poisonous or dangerous. While many species do indeed secrete distasteful chemical compounds for defense, others may only mimic this adaptation with their coloration.

Distribution

Cockroaches are worldwide in distribution, although some genera are endemic to certain countries. The greatest number of species occurs in the tropics. Some pest species, if not controlled, may build up huge populations in homes, businesses, and other buildings and in sewers.

Habitat

Feral cockroaches inhabit almost every conceivable habitat, with the exception of extremely cold regions, although *Eupolyphaga everestinia* (Polyphagidae) was collected on Mount Everest at 18,500 ft (5,639 m); pest species (e.g., the German cockroach) can survive indoors in extreme cold. Cockroaches are found in caves, mines, animal burrows, bird nests, ant and termite nests, deserts, and water (subaquatic). Most cockroaches live outdoors and, during the day, usually are found near the ground and hiding under bark, dead leaves, soil, logs, or stones. Numerous species have adapted to human beings and live in man-made structures (homes, restaurants, food stores, hospitals, and sewers) where temperatures and levels of humidity are relatively stable and the cockroaches are protected from adverse climatic conditions.

Many species are found in association with plants, but the significance of the relationships is obscure. Undoubtedly,

most of the associations are accidental, but some damage plants. Not all cockroach/plant associations are harmful. There are some tropical species that live in the canopy and are involved in pollinating certain plants. Numerous species in at least six ovoviviparous genera (Blaberidae: Epilamprinae), are amphibious or semiaquatic. They usually live on land at the edges of streams or pools, but they may spend brief periods of time in the water. About 25 genera and about 62 species of cockroaches have been found in bromeliads, where water collects in the leaf bases. Many of these species are not restricted to this habitat, but some may be truly bromeliadicolous.

In the United States, desert species of *Arenivaga* (Polyphagidae) migrate vertically on plants during the day and avoid the heat of spring, summer, and autumn. Between November and March nocturnal temperatures are lower, and the insects burrow into the sand among roots of shrubs. During the winter, soon after darkness, they become active during peaks of nighttime surface temperatures.

Behavior

There are three types of social behavior and communication: gregarious (aggregation), subsocial (adults care for larvae), and solitary; communication (especially during courtship behavior) is accomplished with pheromones. Vision apparently plays little or no significant role in sexual recognition, courtship, and copulation, in spite of the fact that many species have large, well-developed, pigmented eyes.

There are many records of predation on cockroaches. Among the arthropods, ants and spiders are the most important predators in the tropics. Cockroach remains have been found in the stomachs of fish, salamanders, toads, frogs, turtles, geckos, and lizards. Several species of birds eat these insects, and a Peruvian wren ("cucarachero") apparently specializes in cockroaches. Among mammalian predators on cockroaches are opossums, porcupines, monkeys, rodents, and cats.

Cockroaches are subject to parasites, for example, viruses, bacteria, protozoa, fungi, and helminths. Among the parasite arthropods are wasps, flies, and beetles. One whole wasp family, the Evaniidae, favors only cockroach eggs. Larvae and adult cockroaches are eaten by species in three families of Hymenoptera, two of Diptera, and one of beetles. Species of *Ampulex* and *Dolichurus* (Hymenoptera: Ampulicidae) provision their nests only with cockroaches.

Many species burrow into the substrate when disturbed or during periods of inactivity. Some conceal themselves in folded dead leaves, in caves they burrow into guano, or in crevices. Some species run rapidly or fly or become immobile at the approach of an attacker. Cockroaches have mechanoreceptors, especially on their cerci, that respond to the slight acceleration of air that signals the approach of a predator. Some species that become immobile when attacked by ants cling so tightly to the substrate that their vulnerable undersurfaces cannot be harmed. Many species have defensive glands that produce a variety of irritating chemicals, the most common being trans-2-hexenal, which is sprayed forcibly on the attacker. Many immature cockroaches and some adults secrete a sticky proteinaceous substance on their terminal abdominal segments and cerci that gums up ant or beetle attackers.

Some tropical cockroaches have warning coloration that involves mimicry, and they resemble lampyrid beetles (fireflies), coccinellid beetles, or wasps. Virtually nothing is known about the relationships of the mimic (the cockroach) and model; in fact, in many instances the models are unknown.

Cockroaches produce sound in many ways, but two methods, namely, stridulation and expulsion of air through the second abdominal spiracles, have evolved in ovoviviparous Blaberidae: Oxyhaloinae. The stridulating structures consist of parallel striae (thickenings) on the ventral lateroposterior margins of the pronotum and on the dorsoproximal regions of the costal veins of the tegmina. When disturbed, the insect rubs the pronotum sideways against the modified region of the costal veins.

Under normal periods of light and dark, many pest species are nocturnal and increase their activity just as it begins to get dark; activity ceases after five or six hours, and they remain quiet throughout the following day. Some feral Australian Polyzosteriinae that live exposed are diurnal and are active during the day. Domestic cockroaches may aggregate in large numbers during the day and at night move out and migrate to obtain food and water.

Feeding ecology and diet

Cockroaches seem to eat almost anything, from plants to animals. They may exhibit preferences and discriminate when given a choice. When deprived of food and water, they can live from five days (*Blattella vaga*) to 42 days (female *Periplaneta americana*). When given dry food but no water, they live for about the same period of time as insects that are starved; if they are provided with water, most live longer. Some species can live for two to three months on water alone. In bat-inhabiting caves they feed on guano, and in sewers they consume human feces. Some species live in dead trees and feed on wood.

Reproductive biology

Courtship precedes mating and is controlled by pheromones. Some females assume a calling position by raising the wings, expose intertergal membranes, expand their genital chambers, and release a pheromone to attract the male. In many species the newly emerged females, while still white and teneral, attract and mate with older males (e.g., *Diploptera punctata*).

Some species show little courtship behavior. The female does not mount the male, nor does she feed on or palpate his tergites; he either mounts or backs into her and makes the connection. Many cockroaches have stridulating organs, but in most species their role in courtship, if any, has not been determined. In some wingless species (e.g., *Gromphadorhina* and *Macropanesthia*) males hiss during courtship and simply

The death's head cockroach (*Blaberus giganteus*) lives in the Amazon rainforest in Peru. (Photo by George D. Dodge. Bruce Coleman, Inc. Reproduced by permission.)

back into and seize the female genitalia, and the pair remains joined, with their heads facing opposite directions. In all species, while the two are joined, the male forms and transfers a spermatophore to the female's bursa copulatrix.

The eggs of almost all cockroaches are enclosed in a tan proteinaceous capsule called an "ootheca," which may have as few as four or five eggs or as many as 243. When the ovaries mature, the eggs are extruded through a blob of soft, rubbery, colleterial gland secretion, and, with the help of the ovipositor valves, are lined up vertically and alternately on one side and then the other; when completed, there are two rows of eggs covered by a capsule. The eggs are extruded with the micropylar ends (the region where sperm enter the egg) and oothecal keel (the area that contains the air chambers for egg respiration) upward. The position in which the ootheca is carried at the time it is deposited (i.e., whether or not it is rotated 90°) is significant taxonomically and played an important role in the evolution of blaberid and blattellid ovoviviparity.

There are four types of reproduction: oviparity, false ovoviviparity, true ovoviviparity, and viviparity. In oviparity (all families except Blaberidae), the ootheca is dropped shortly after it is completed and while it still is in the vertical position (all Blattidae and Blattellidae: Pseudophyllodromiinae) or after it is rotated 90° to the right or left (Blattellidae, Nyctiborinae, and Ectobiinae). After the female deposits the ootheca, she leaves and has nothing more to do with the eggs. Initially, these eggs have enough yolk to complete development, but water is obtained from the substrate.

In false ovoviviparity (almost all Blaberidae and four genera of Blattellidae), the ootheca, after it is formed, is retracted into a uterus or brood sac, where it remains during gestation (e.g., *Nauphoeta cinerea*). The oothecal membrane is greatly reduced in the Blaberidae and less so in the Blattellidae. When first laid, the eggs have enough yolk to complete development and obtain water from the mother during gestation.

An American cockroach (*Periplaneta americana*) female with egg case. (Photo by Kim Taylor. Bruce Coleman, Inc. Reproduced by permission.)

In most cockroaches that carry the ootheca externally, the female deposits it soon after its formation; some females dig a hole and cover it with debris or conceal it in some other way. They then leave and have nothing more to do with the eggs. Those species that carry the ootheca externally during gestation probably provide some protection from parasites or predators. The Blaberidae protect their eggs by retracting them into a brood sac. Additional protection is afforded after the ovoviviparous species give birth and the newborn larvae aggregate, usually for relatively short times, under the female until the cuticle hardens. This is true of *Byrsotria fumigata*, *Gromphadorhina portentosa*, and *Rhyparobia maderae*. In *Phlebonotus*, *Thorax*, and *Phoraspis*, the female's tegmina are large and arched; the wings may be reduced and the abdominal tergites depressed, forming a chamber in which the larvae hide, covered by the tegmina.

Conservation status

The greatest threat to feral cockroaches is the destruction of habitats, particularly in the tropics. Those species that are restricted to a particular niche are especially vulnerable. At least one cockroach, *Ectobius duskei*, the Russian steppe cockroach, has been eliminated as a result of the cultivation of wheat in virgin steppes. Other than conservationists and students of biodiversity, few people probably would object if a cockroach, especially a pest species, were eliminated. Today, support is given for the search for medically and economically important substances that are derived from animals and plants, and it is conceivable that some compounds they produce will be found to be of practical value. No species is listed by the IUCN.

Significance to humans

The biology of domesticated species that are pests of humans has been investigated thoroughly, often with the aim of controlling or eradicating them, which is virtually impossible. Some of these species are used in biology classes or in commercial or government laboratories to study physiological problems. University researchers are developing "robotic" cockroaches, using the large, wingless *Gromphadorhina portentosa*. Engineers and neurologists have developed a wristwatch-sized sensory and video package to control, start, and steer the "Biobot" cockroach. One of the aims for these robots is to make possible remote measurements of environmental conditions where humans cannot easily or safely go (e.g., bombed buildings).

Cockroaches have been used as food by humans and are said to taste like shrimp. The Aborigines of Australia and the Lao Hill tribe of Thailand collect and eat raw cockroaches. The Laos in Korat eat cockroaches, and the children collect the oothecae for frying. In southern China and in Chinatown in New York City, dried specimens of *Opisthoplatia orientalis* are sold for medicinal purposes. Large nondomiciliary cockroaches are cultured or grown readily, and many have been maintained as pets. Probably one of the most popular is the slow-moving Madagascan hissing cockroach, *Gromphadorhina portentosa*.

True ovoviviparity (only the blaberid Geoscapheinae, specifically, four Australian genera: *Geoscapheus*, *Macropanesthia*, *Neogeoscapheus*, and *Parapanesthia*) differs from false ovoviviparity in that an ootheca is not formed. The eggs pass directly from the oviduct into the uterus or atrium, where they lie in a jumbled mass until parturition. The eggs have enough yolk initially to complete development but obtain water from the mother as needed.

With viviparity (only one species, that is, *Diploptera punctata*, Blaberidae: Diplopterinae), the very small ootheca is rotated and retracted into the uterus. It has only about a dozen very small eggs enclosed in an incomplete membrane; the larvae are quite large when born. At first, the eggs lack sufficient yolk and water to complete development, but during gestation the embryos drink water and dissolved proteins and carbohydrates synthesized and transported by the mother's uterus.

Many species that live underground or in dead trees burrow into soil or wood and form a chamber where the insects live and emerge to feed or carry down food (e.g., dead leaves) to the chamber. All the species of *Cryptocercus* (Cryptocercidae) live in and feed on decaying wood; the cockroaches have special protozoa that digest cellulose. Another group of cockroaches, the Panesthiinae, also eat wood, but their cellulose is digested by bacteria.

The panesthiine cockroaches *Geoscapheus robustus* and *Macropanesthia rhinoceros* burrow in sandy areas in Queensland, especially where the cypress pine grows. The adults make a nest of grass, roots, and dead leaves about 2 ft (0.6 m) below the surface and probably play an important role in litter turnover.

The fact that domesticated cockroaches live in homes, sewers, privies, and hospitals and feed on human foods and feces and cadavers makes it possible for them to be vectors of a variety of human diseases by carrying infectious organisms internally and externally. They have been known to harbor viruses, bacteria, fungi, protozoa, and helminths. At least 18 bacteria that cause human diseases have been found naturally in domesticated pest species, notably the American and German cockroaches. *Blattella germanica* collected from hospitals and residential areas have been found to carry various species of five genera of fungi of medical importance. The evidence for cockroaches' transmitting human disease organisms is essentially circumstantial, but they have the potential to transmit pathogens indirectly when foods or utensils used to prepare foods are contaminated. Although their importance as vectors of disease is circumstantial, there is no doubt that some people are allergic to cockroaches, and hypersensitive individuals may experience asthma when they are exposed to cockroach allergens. Allergies are common among laboratory workers who study cockroaches. The culprits are commonly the German, American, and Oriental cockroaches, but other species (e.g., *G. portentosa*) also have been implicated. Many species of cockroaches produce defensive secretions that may cause burning sensations, vertigo, or nausea. Normal cockroach integumental secretions cause dermatitis and conjunctival edema in entomologists who frequently study or work with these insects.

1. Oriental cockroach (*Blatta orientalis*); 2. Madeira cockroach (*Rhyparobia maderae*); 3. Cinereous cockroach (*Nauphoeta cinerea*); 4. German cockroach (*Blattella germanica*); 5. Asian cockroach (*Blattella asahinai*); 6. American cockroach (*Periplaneta americana*); 7. Brownbanded cockroach (*Supella longipalpa*); 8. Suriname cockroach (*Blatta surinamensis*). (Illustration by Amanda Smith)

Species accounts

Suriname cockroach
Blatta surinamensis

FAMILY
Blaberidae

TAXONOMY
Pycnoscelus surinamensis (Linnaeus), 1758, Surinam.

OTHER COMMON NAMES
None known.

PHYSICAL CHARACTERISTICS
Medium size, 0.71–0.94 in (18–24 mm) long. Shiny brown to black in color. Tegmina often do not extend beyond the end of the abdomen. Black pronotum, with yellow area along the anterolateral margins.

DISTRIBUTION
Originated in Indo-Malayan region and spread by trade throughout humid and subhumid tropics. In the United States it has been found in Alabama, Florida, Texas, Louisiana, and Iowa.

HABITAT
In northern states can become established in heated greenhouses and zoos. In the wild frequently burrows into loose soil or sand but also occurs under stones, manure, wood, rubbish, or chicken feces.

BEHAVIOR
As usually only females are produced, courtship rarely occurs. When it does, the nonfunctional male mounts the female and grasps her genitalia from above.

FEEDING ECOLOGY AND DIET
Destroys palms and ferns by eating the hearts. In greenhouses, girdles rose bushes and eats aerial roots and orchid petals.

REPRODUCTIVE BIOLOGY
Parthenogenetic, producing females. Males are produced rarely; when they occur, they are nonfunctional. At about 86°F (30°C), the virgin female deposits her first ootheca in the uterus seven days after emergence and gives birth 34 days later. During an average life span of 307 days she produces three litters, each consisting of about 21 young. Larvae undergo eight to 10 molts and mature in 140 days when reared in groups. The fecundity of clones from different parts of the world varies considerably.

CONSERVATION STATUS
Not threatened.

SIGNIFICANCE TO HUMANS
Household pest in the southern United States, Philippines, Tanzania, and Trinidad. Intermediate host of the nematode *Oxyspirura mansoni*, which causes worm infections and blindness in poultry. Spread simplified by parthenogenesis, because the species can become established through the introduction of a single larva or unfertilized adult female into a suitable habitat. ◆

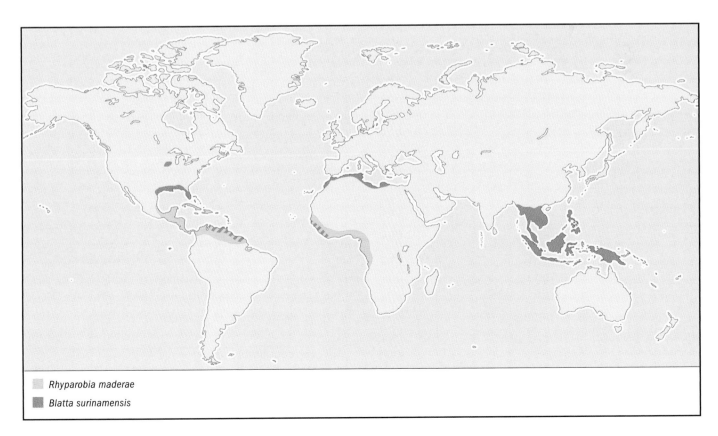

☐ *Rhyparobia maderae*
■ *Blatta surinamensis*

Cinereous cockroach
Nauphoeta cinerea

FAMILY
Blaberidae

TAXONOMY
Nauphoeta cinerea Olivier, 1789, Isle of France, Mauritius.

OTHER COMMON NAMES
English: Lobster cockroach.

PHYSICAL CHARACTERISTICS
Large, reaching 0.98–1.14 in (25–29 mm) in length. The sexes are similar, except that male tegmina and wings reach about the fifth abdominal tergite and are slightly longer in the female. Ashy colored. Pronotum with a lobster-like design.

DISTRIBUTION
Apparently a native of East Africa, the species has become distributed widely in the tropics by commerce.

HABITAT
Found in native huts in the Sudan, in outhouses and stores, and in hospitals in Australia.

BEHAVIOR
Males stridulate when courting nonreceptive females. Newly hatched larvae crawl under the female or under her wings, remaining there until about an hour after hatching.

FEEDING ECOLOGY AND DIET
The cinereous cockroach infests mills producing animal feeds in Florida and poultry food sheds in Honolulu. Fond of feeds containing fish oil; known to kill and eat the cypress cockroach, *Diploptera punctata*.

REPRODUCTIVE BIOLOGY
At 86–96.8°F (30–36°C), the mated female produces the first ootheca, containing about 33 eggs, in 13 days; gestation lasts 31 days. Larvae undergo seven or eight molts and are reared in groups. Males mature in 72 days and females in 85 days. The adult male lives 365 days, and the female lives 344 days, during which time she produces six litters.

CONSERVATION STATUS
Not threatened.

SIGNIFICANCE TO HUMANS
Was implicated in an outbreak of *Salmonella* poisoning in an Australian hospital. ◆

Madeira cockroach
Rhyparobia maderae

FAMILY
Blaberidae

TAXONOMY
Blatta maderae Fabricius, 1781, Madeira.

OTHER COMMON NAMES
English: Fatula cockroach, knocker cockroach.

PHYSICAL CHARACTERISTICS
Reaches 1.57–1.97 in (40–50 mm) in length. Sexes are similar in habitus. Pale brown to tan. Tegmina and wings fully devel-

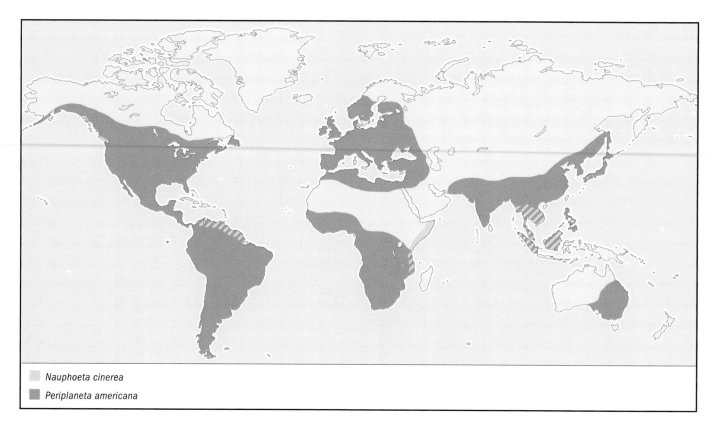

Nauphoeta cinerea
Periplaneta americana

oped, with two black lines in basal region and mottled posterior zones.

DISTRIBUTION
Native of West Africa; introduced into West Indies more than a hundred years ago and through commerce became circumtropical. Abundant in countries that border the Caribbean Sea south of the United States.

HABITAT
Especially infests food stores. In 1950 became established in basements of buildings in New York City occupied by people from Puerto Rico; spread little since originally reported. Outdoor species in tropics found in sugar cane fields and associated with palms, guava, and bananas. Spread from banana ships and intercepted at ports by quarantine inspectors.

BEHAVIOR
Gregarious; builds up huge colonies. The male may tap the substrate with his thorax, which may be a method of attracting the female. She feeds on secretions on his second abdominal tergite during courtship. The pair remain attached for 20–30 minutes. Young larvae forage with the mother.

FEEDING ECOLOGY AND DIET
Probably omnivorous. Bananas and grapes are favorite foods.

REPRODUCTIVE BIOLOGY
At 86–96.8°F (30–36°C), the mated female produces first ootheca, usually containing up to about 40 eggs, 20 days after becoming adult. Eggs develop within the uterus in 58 days. Larvae undergo seven or eight molts and are reared in groups. Males mature in 121 days; females require 150 days. Life expectancy is up to 2.5 years.

CONSERVATION STATUS
Not threatened.

SIGNIFICANCE TO HUMANS
Important pest in some areas. Used extensively as an experimental laboratory animal. ◆

Asian cockroach
Blattella asahinai

FAMILY
Blattellidae

TAXONOMY
Blattella asahinai Mizukubo, 1981, Okinawa.

OTHER COMMON NAMES
None known.

PHYSICAL CHARACTERISTICS
The sexes are similar; fully winged. Reaches 0.31–0.47 in (8–12 mm) in length. Yellowish or yellowish brown. Adult pronotum has a pair of light to dark brown longitudinal stripes. This species is almost impossible to distinguish from the German cockroach but differs markedly in behavior.

DISTRIBUTION
This species is the most recent introduction into the United States. Although it is not distributed as widely as the German cockroach, it has been found (as its synonym *Blattella beybienkoi*) in Sri Lanka, Andaman Islands, Myanmar, Chagos Archipelago, China, India, Thailand, and Okinawa.

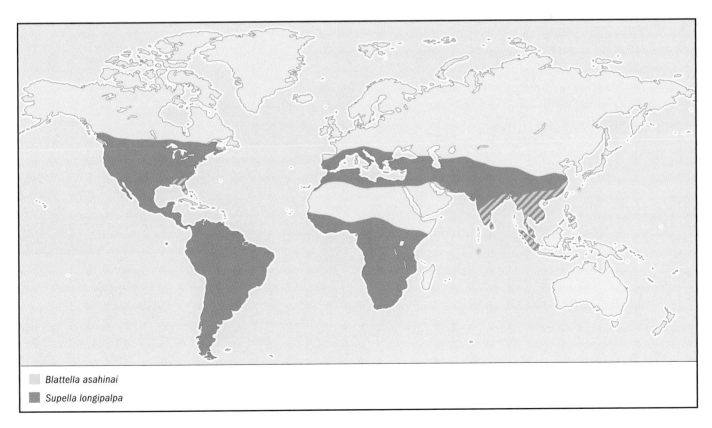

☐ *Blattella asahinai*
■ *Supella longipalpa*

HABITAT

In Florida, where it was introduced and became established and spread, it occurs outdoors on lawns, bushes, and trees.

BEHAVIOR

Flies readily. Active at sunset. Adults are attracted to white walls and illuminated buildings. Adults and larvae are active in grass and mulch. In houses adults fly to lights and sit on walls, tables, and dishes. Winged adults are half an inch (12.7 mm) long and have a pair of longitudinal stripes on the thorax; the larvae lack wings but also are striped.

FEEDING ECOLOGY AND DIET

In Florida it often feeds on aphid honeydew.

REPRODUCTIVE BIOLOGY

In the laboratory Asian cockroach males cross with German cockroach females and produce offspring. Crosses between male German cockroaches and Asian females did not produce offspring.

CONSERVATION STATUS

Not threatened.

SIGNIFICANCE TO HUMANS

Considered a pest on lawns in Florida, sometimes making outdoor activities almost impossible. ◆

German cockroach
Blattella germanica

FAMILY

Blattellidae

TAXONOMY

Blatta germanica Linnaeus, 1767, Denmark; type Brünnich, 1763 [suppressed by the International Commission on Zoological Nomenclature].

OTHER COMMON NAMES

English: Croton bug, steamfly, steambug, Yankee; German: Russische Schabe; Russian: Prussak.

PHYSICAL CHARACTERISTICS

The sexes are similar; fully winged. Relatively small at 0.4–0.5 in (10–13 mm) in length. Pale yellowish-brown to tawny. Pronotum has distinct dark, parallel, longitudinal bands.

DISTRIBUTION

Cosmopolitan; found in and around human habitations throughout the world. Present in Greenland, Iceland, and the Canadian High Arctic, where it survives indoors.

HABITAT

Preferring a warm, moist habitat, the German cockroach is a common pest of kitchens, larders, and restaurants. In temperate climates it can be found outdoors, under houses without basements, under rubbish and date palms, and in city dumps. It also has been taken in gold mines and caves in South Africa. It is one of the most prevalent cockroaches in the galleys and storerooms of ships.

BEHAVIOR

Gregarious and may build up huge populations: a four-room apartment in Texas harbored 50,000–100,000 mostly German cockroaches. It is a poor flier despite having fully developed wings, which it uses to glide downward from a high resting place. The sexually active female assumes a calling position by raising the wings and emitting a pheromone produced in the tenth abdominal tergite that attracts the male. On contact with

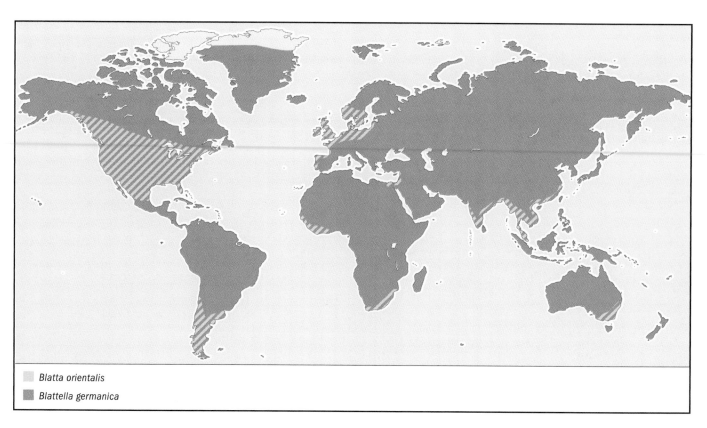

Blatta orientalis
Blattella germanica

the female, the male raises the wings, exposing a pair of glands on his back. The female mounts the male, apparently feeding on a glandular secretion. The male clasps the female genitalia, and the pair remain joined until the male inserts the sperm-filled spermatophore.

FEEDING ECOLOGY AND DIET
Omnivorous.

REPRODUCTIVE BIOLOGY
At 86°F (30°C) the mated female produces first ootheca, eight days after adult emergence. Average number of eggs per egg case is 37; the incubation period is 17 days. The male lives 128 days; the female lives 153 days, during which time she produces, at intervals of 22 days, seven oothecae, each of which she carries externally, protruding from the end of the abdomen, until or shortly before eggs hatch. Larvae undergo five to seven molts; they are reared in groups and mature in 40—41 days.

CONSERVATION STATUS
Not threatened.

SIGNIFICANCE TO HUMANS
The species causes asthmatic attacks and is known to carry or is suspected of transmitting various bacterial disease organisms. ◆

Oriental cockroach
Blatta orientalis

FAMILY
Blattidae

TAXONOMY
Blatta orientalis Linnaeus, 1758, America, Oriente.

OTHER COMMON NAMES
English: Schad roach, black beetle.

PHYSICAL CHARACTERISTICS
This dimorphic, shiny, blackish-brown species is 0.71–0.94 in (18–24 mm) long. The tegmina and wings of the male are reduced but cover about two-thirds of the abdomen. The tegmina of the female are small lateral pads that extend only to the middle of the metanotum; hind wings are absent.

DISTRIBUTION
Found in port cities throughout the world and occurs in almost every state of the United States. It also is found in England, northern Europe, Israel, southern Australia, and southern South America.

HABITAT
Usually found in basements and cellars, service ducts, crawl spaces, toilets, and behind baths, sinks, radiators, ovens, and hot-water pipes; they may congregate in large numbers around sources of water. In north-central states of the United States, they often are found outdoors around homes during the summer and in garbage and trash dumps.

BEHAVIOR
Often occurs outdoors. In buildings, usually occur below or on ground floor, but small numbers may be found up to the fifth floor.

FEEDING ECOLOGY AND DIET
Probably omnivorous.

REPRODUCTIVE BIOLOGY
At 86–96.8°F (30–36°C), the first ootheca is deposited, 12 days after the female becomes an adult; the eggs take 44 days to hatch. The larvae undergo eight to 10 molts. When raised in groups, males require 146 days and females 165 days to mature. The average number of eggs in an ootheca is 16, and the female averages only 2.5 oothecae during a brief average adult life span of 44 days. The life cycle of this species is seasonal; in some areas adults appear in May and June, but some adults can be found in almost all months. Having considerable resistance to cold, they have been found breeding out of doors in England and southern Russia.

CONSERVATION STATUS
Not threatened.

SIGNIFICANCE TO HUMANS
A significant household pest. ◆

American cockroach
Periplaneta americana

FAMILY
Blattidae

TAXONOMY
Blatta americana Linnaeus, 1758, America.

OTHER COMMON NAMES
English: Bombay canary, ship cockroach; Dutch: Kakerlac.

PHYSICAL CHARACTERISTICS
Grows to 1.1–1.7 in (28–44 mm) in length. Tegmina and wings developed in both sexes. Reddish-brown, with pale yellow zone around the edge of the pronotum.

DISTRIBUTION
An important cosmopolitan pest in tropical and subtropical areas, having been distributed by commerce to many regions of the world.

HABITAT
Its original home was Africa, where it is found commonly both inside and outside human dwellings. In the United States it probably is found in all urban areas. This species prefers a warm, moist habitat and in tropical and subtropical America is common outdoors and may be found in dumps, woodpiles, sewers, and cesspools.

BEHAVIOR
May fly short distances on occasion. Gregarious and may build up huge populations. After spraying the walls of a trickling filter plant in Florida, over 2.5 cu yd (2.3 m^3) of American cockroaches were collected, and several times that quantity died in the adjacent woods. Females produce a pheromone that can attract the male from as far away as 98 ft (30 m).

FEEDING ECOLOGY AND DIET
Probably omnivorous; known to feed on human feces.

REPRODUCTIVE BIOLOGY
Five-year life-history study under room conditions in Lafayette, Indiana (64.4–80.6°F, or 18–27°C, during winter, with a maximum summer temperature 95°F, or 35°C), found average duration of larval development to be 468 days for females and 551 days for males. Females lived, on average, 441

days and produced 58 oothecae (maximum of 90). Oothecae contained 16 eggs, and the incubation period was 53 days. In another study conducted at 86–96.8°F (30–36°C), larvae went through nine to 13 molts. When reared in groups, females required 161 days and males 171 days to mature.

CONSERVATION STATUS
Not threatened.

SIGNIFICANCE TO HUMANS
Together with the German cockroach, probably the most important cockroach pest. Many bacteria, viruses, fungi, and helminths, a number of them pathogenic to man, have been found in the American cockroach. ◆

Brownbanded cockroach
Supella longipalpa

FAMILY
Blattellidae

TAXONOMY
Supella longipalpa Fabricius, 1798, India. Until recently known as *Supella supellectilium* (Serville).

OTHER COMMON NAMES
English: Furniture cockroach, TV cockroach.

PHYSICAL CHARACTERISTICS
A small cockroach, reaching only 0.39–0.57 in (10–14.5 mm) in length. Male tegmina completely cover the abdomen; in the female they rarely reach the tip of the abdomen. Color varies widely. Dark pronotum, often with a pale area in the center; chestnut tegmina, with pale areas at base and in middle.

DISTRIBUTION
Probably originated in Africa but, as a domiciliary pest, spread by commerce throughout warm temperate countries; may occur outdoors in subtropical countries. Found in almost every state of the United States.

HABITAT
In dwellings, it can be found in almost every room: on kitchen chairs; in cupboards and pantries; underneath tables and shelves in closets; behind pictures and picture molding, in TV sets, bookshelves, drawers, and shower stalls.

BEHAVIOR
Tends to fly when disturbed. The female emits a pheromone that has been identified and attracts the male from a distance.

FEEDING ECOLOGY AND DIET
Often feeds on gum sizing of books and on paste behind wallpaper, stamps, and gummed labels. Visits kitchens when searching for food.

REPRODUCTIVE BIOLOGY
Oothecae may be attached throughout the house to walls and ceilings and about kitchen sinks, desks, tables, and bedding. At about 86°F (30°C), larvae undergo six to eight molts. When reared in groups, both sexes mature in about 55 days. Males live 115 days and females 90 days. The female deposits her first ootheca, which usually contains 16 eggs, 10 days after emergence. During her lifetime she produces an average of 11 oothecae, which are deposited at six-day intervals. Eggs hatch in 40 days.

CONSERVATION STATUS
Not threatened.

SIGNIFICANCE TO HUMANS
Important household pest that spreads throughout infested homes. ◆

Resources

Books
Asahina, S. *Blattaria of Japan*. Tokyo: Nakayama-Shoten, 1991. (Japanese, with parts in English).

Bell, W. J., and Adiyodi, K. G., eds. *The American Cockroach*. London: Chapman and Hall, 1982.

Bey-Bienko, G. Y. "Fauna of the U.S.S.R." In *Insects*. Moscow: Institute of Zoology, Academy of Sciences of the USSR, 1950.

Carpenter, F. M. *Treatise on Invertebrate Paleontology*. Part R, *Arthropoda*. Vol. 3, *Superclass Hexapoda*. Boulder, CO: Geological Society of America, 1992.

Cornwell, P. B. "The Cockroach." In *A Laboratory Insect and an Industrial Pest*, vol. 1. London: Hutchinson, 1968.

Gordon, David George. *The Compleat Cockroach: A Comprehensive Guide to the Most Despised (and Least Understood) Creature on Earth*. Berkeley, CA: Ten Speed Press, 1996.

Roth, L. M., and Alsop, D. W. "Toxins of Blattaria." In *Arthropod Venoms*. Edited by S. Bettini. Handbook of Experimental Pharmacology, vol. 48. New York: Springer-Verlag, 1978.

Taylor, R. L. *Butterflies in My Stomach; or, Insects in Human Nutrition*. Santa Barbara, CA: Woodridge Press, 1975.

Vishniakova, V. N. "Mesozoic Blattids with External Ovipositors and Details of Their Reproduction." In *Jurassic Insects of Karatau*. Edited by B. B. Rohdendorf. Moscow: Akademiya Nauk SSSR, Ordelenide Obschei Biolgii, 1968. (Russian).

Periodicals
Carpenter, F. M. "A Review of Our Present Knowledge of the Geological History of Insects." *Psyche* 37 (1930): 15–34.

Grandcolas, P. "El Origen de la Diversidad en las Cucarachas: Perspectiva Filogentica de su Gregarismo, Reproducion, Comunicacion y Ecologia." *Boletin de la Sociedad Entomologica Aragonesa* 26 (1999): 397–414 (English translation, 415–420).

McKittrick, F. A. "Evolutionary Studies of Cockroaches." *Memoir of the Cornell University Agricultural Experimental Station* 389 (1964): 1–197.

Moore, Thomas E., Seldon B. Crary, Daniel E. Koditschek, and Todd A. Conklin. "Directed Locomotion in Cockroaches: 'Biobots' *Acta Entomologica Slovenica* 6, no. 2 (1998): 71–78.

Resources

Princis, K. "Zur Systematik der Blattarien." *Eos: Revista Española de Entomología* 36, no. 4 (1960): 427–449.

Rehn, J. A. G. "Man's Uninvited Fellow Traveler: The Cockroach." *Scientific Monthly* 61 (1945): 265–276.

Roth, L. M. "Evolutionary Significance of Rotation of the Ootheca in the Blattaria." *Psyche* 74 (1967): 85–103.

———. "The Evolution of Male Tergal Glands in the Blattaria." *Annals of the Entomological Society of America* 62 (1969): 176–208.

———. "Evolution and Taxonomic Significance of Reproduction in Blattaria." *Annual Review of Entomology* 15 (1970): 75–96.

———. "A Taxonomic Revision of the Panesthiinae of the World." I. "The Panesthiinae of Australia (Dictyoptera: Blattaria: Blaberidae)." *Australian Journal of Zoology* Supplement series 48 (1977): 1–112.

———. "The Mother-Offspring Relationship of Some Blaberid Cockroaches (Dictyoptera: Blattaria: Blaberidae)." *Proceedings of the Entomological Society of Washington* 83, no. 3 (1981): 390–398.

———. "A Taxonomic Revision of the Genus *Blattella Caudell* (Dictyoptera, Blattaria: Blattellidae)." *Entomologica Scandinavica* Supplement no. 22 (1985): 1–221.

———. "*Blattella asahinai* Introduced into Florida (Blattaria: Blattellidae)." *Psyche* 93 (1986): 371–374.

———. "New Cockroach Species, Redescriptions and Records, Mostly from Australia, and a Description of *Metanocticola christmasensis* gen. nov., sp. nov., from Christmas Island (Blattaria)." *Records of the Western Australian Museum* 19 (1999): 327–364.

———. "Systematics and Phylogeny of Cockroaches (Dictyoptera: Blattaria)." *Oriental Insects* 37 (2003): 1–139.

Roth, L. M., and G. P. Dateo. "Uric Acid Storage and Excretion by Accessory Sex Glands of Male Cockroaches." *Journal of Insect Physiology* 11 (1965): 1023–1029.

Roth, L. M., and T. Eisner. "Chemical Defenses of Arthropods." *Annual Review of Entomology* 7 (1962): 107–136.

Roth, L. M., and H. B. Hartman. "Sound Production and Its Evolutionary Significance in the Blattaria." *Annals of the Entomological Society of America* 60 (1967): 740–752.

Roth, L. M., and E. R. Willis. "A Study of Cockroach Behavior." *American Midland Naturalist* 47, no. 1 (1952): 66–129.

———. "The Medical and Veterinary Importance of Cockroaches." *Smithsonian Miscellaneous Collections* 134, no. 10 (1957): 1–147.

———. "An Analysis of Oviparity and Viviparity in the Blattaria." *Transactions of the American Entomological Society* 83 (1958): 221–238.

———. "The Biotic Associations of Cockroaches." *Smithsonian Miscellaneous Collections* 141 (1960): 1–470.

Rugg, D., and H. A. Rose. "Biology of *Macropanesthia rhinoceros* Saussure (Dictyoptera: Blaberidae)." *Entomological Society of America* 84, no. 6 (1991): 575–582.

Schal, C., J. Y. Gautier, and W. J. Bell. "Behavioural Ecology of Cockroaches." *Biological Reviews* 59 (1984): 209–254.

Shelford, R. "Mimicry Amongst the Blattidae; with a Revision of the Genus *Prosoplecta* Sauss., and the Description of a New Genus." *Proceedings of the Zoological Society of London* (1912): 358–376.

Sreng, L. "Cockroach Mating Behaviors, Sex Pheromones, and Abdominal Glands (Dictyoptera: Blaberidae)." *Journal of Insect Behavior* 6, no. 6 (1993): 715–735.

Willis, E. R., G. R. Riser, and L. M. Roth. "Observations on Reproduction and Development in Cockroaches." *Annals of the Entomological Society of America* 51 (1958): 53–69.

Other
Driscoll, T. M. *Insect Pet Care: The Madagascar Hissing Cockroach*. 1999. (Video; includes a booklet with directions for housing, and biological information).

Louis M. Roth, PhD

Isoptera
(Termites)

Class Insecta
Order Isoptera
Number of families 7

Photo: Nasute termites congregate on a tree in the rainforest of Costa Rica. (Photo by Peter Ward. Bruce Coleman, Inc. Reproduced by permission.)

Evolution and systematics

Termites make up the order Isoptera, comprising approximately 2,800 described species. The number of described species is thought to be no more than one-fourth to one-half the actual number, but this depends in part on the species concept applied. Some newer phylogenetic species concepts recognize every diagnosable geographically distinct population as a species, whereas the traditional biological species concept includes many geographic forms as subspecies within a smaller number of described species. Irrespective of theoretical issues of species limits, it is widely recognized that certain portions of the termite fauna, such as the New World soldierless termites, have been neglected by systematists, and many parts of the world remain to be thoroughly explored in terms of their termite biodiversity.

Termites are among the most primitive of the terrestrial Neoptera, which also includes the Blattodea (cockroaches) and Mantodea (mantids). These are the most basal of extant orders of winged insects to lay eggs on land, rather than in the water,

and all share a similar adaptation for protection of the eggs from desiccation, laying the eggs as a mass encased in a protective secretion called an ootheca. Because of this shared character, these orders comprise a group formerly called the Oothecaria, but now more commonly known as the Dictyoptera.

Recent phylogenetic studies have suggested that the mantids were the earliest offshoot of the Dictyoptera, leaving the cockroaches and termites as sister groups. The most primitive family of the cockroaches, the Cryptocercidae, are the cockroaches most like termites: they eat wood, live in tunnel systems inside logs, have symbiotic flagellates in their hindgut, and live in small subsocial family groups in which the parents share their burrows with one brood of offspring they feed with proctodeal fluids. Likewise, termites in the most primitive family, the Mastotermitidae, are the termites most like cockroaches. Mastotermitidae is represented by only one living species, *Mastotermes darwiniensis*, the only termite that still lays eggs in an ootheca. It is also the only

Termites at their nest entrance in the Amazon rainforest of Peru. (Photo by Michael Fogden. Bruce Coleman, Inc. Reproduced by permission.)

termite that has five-segmented tarsi and an anal lobe in the hind wing, characters it shares with cockroaches.

Because of these similarities, it could be argued that termites are actually social cockroaches. However, termites lack the many distinctive features of cockroaches, a wide flattened body with the pronotum expanded over the head as a shield, shortened and thickened anterior wings that barely project beyond the tip of the abdomen, and very spiny legs, and therefore termites seem morphologically simpler and more primitive than cockroaches. The split in their lineages must have occurred very early, so that all extant members of each group form monophyletic clades of equal ordinal rank. Termites are more like stoneflies in that they retain a more primitive cylindrical body form and elongate flying wings. Cockroaches developed a more robust body form suited for a free-living, detrivorous lifestyle; termites pursued a life of tunneling inside wood or soil substrates and developed a more advanced level of family integration. Most systematists prefer to regard cockroaches as having split very early from a common protorthopteroid ancestor, which possessed a mix of characters predefinitive of either order. Hence, cockroaches are not asocial termites, and neither are termites social cockroaches.

The traditional hypothesis of termite origin is that they descended from a common ancestor of the primitive cock-

roach family Cryptocercidae, which is comprised of the single living genus *Cryptocercus*, of Holarctic distribution. Cryptotcercids digest wood and live in tunnel systems inside rotten logs. They also display a primitive subsociality in which a reproductive pair of adults share a tunnel system, usually with a single brood of up to about 20 offspring. The mandibles and intestines of cryptocercids are also very similar to those of primitive rottenwood termites in the family Termopsidae. A further similarity is that several genera of symbiotic protozoa are common to the hindguts of these Cryptocercids and primitive termites. However, some studies suggested that Cryptocercids may not be basal within the phylogeny of the cockroach order Blattaria, and therefore, if termites have a shared ancestry with Cryptocercidae, then the termite order Isoptera is cladistically encompassed within the Blattodea, rather than a sister group of the cockroaches. Consequently, termites must be considered a type of derivative social cockroach. The conventional view, reflected by current taxonomy, is that the three orders branched from common ancestors now long extinct, and form distinct and valid independent clades of equal ordinal rank.

The oothecarian orders belong to a larger assemblage called the Orthopteroid insects, which all undergo gradual metamorphosis and possess chewing mouthparts and include the orders Orthoptera (grasshoppers and crickets), Grylloblatta (rockcrawlers), Phasmida (walking sticks), Dermaptera (earwigs), Embioptera (webspinners), Zoraptera (zorapterans), and Psocoptera (psocids).

Physical characteristics

Termites have gradual metamorphosis and branching developmental pathways leading to three or more different terminal castes, hence are polymorphic. In all other insects with gradual metamorphosis, the hatchling's mouthparts harden so that they are able to feed themselves. However, in termites, the eggs hatch to a form called a "larvae," or "white larvae," so called because their mandibles do not harden and darken, thus remaining white. The white larvae are kept together near the eggs in specific areas called nurseries. They are incapable of feeding on wood and are fed by the workers, both on stomodeal regurgitant and the proctodeal liquid which contains the protozoan fauna needed to establish their own stomach colony of digestive symbionts. The larval stage usually lasts for two instars.

The white larvae are a developmental character state that distinguishes the life cycle of all termites from that of all other insects. In most other insects with gradual metamorphosis, the larvae are independent after hatching. By laying small eggs that do not have enough yolk to hatch as independently functional individuals, the queen economizes. The white larvae hatch prematurely, essentially in an embryonic state. The advantage of this for the termite colony is that the queen's investment of yolk per egg is minimized, thereby making it possible for her to maximize the number of eggs she produces. The burden of feeding these tiny white larvae falls to the workers, thus relieving the queen of the physiological cost in yolk investment per offspring. The delegation of care to the workers also helps to promote the queen's fecundity and the colony's fitness. The care of the queen by her offspring, and the care of younger offspring by their older siblings, are hallmarks of termite sociality. While parental care characterizes subsociality, this filial and sibling care, along with the physical expression of these relationships in a polymorphic caste system, underlies eusociality.

In the second developmental instar, young termites face a decisive developmental juncture, at which they are determined either as neuters or as nymphs. Which form they become is visibly expressed in their body form by the third instar. Neuters have a slightly larger head than nymphs, and nymphs have tiny wing buds on the corners of their thoracic segments. Also by the third instar, their mandibles harden and they start to feed themselves, thus acquiring the characteristic abdominal color of their food, which is visible through the semi-transparent abdominal cuticle. In the course of several more molts, neuters become either workers or soldiers and spend the rest of their lives in their home nest. Nymphs, in contrast, grow longer wing pads and accumulate fat with each molt. In their final molt, they grow wings, develop large eyes, and become sexually mature. They then fly from the nest to establish new colonies.

Among soil-inhabiting termites, which all have a true worker caste, the neuter/nymph juncture is irrevocable. The factors influencing this developmental switch between neuters and nymphs appear to be seasonal cues and nutritional factors that influence hormones, and may be mediated by feedback pheromones. In the worker line, they may undergo several more molts before attaining their mature worker size. Male and female workers may all be the same size, or either males or females may be larger, thus forming major and minor workers of different sexes. In addition, workers of certain stages are able to molt via a presoldier stage to a final soldier stage. Soldiers are usually all of one size, but there may be two or three soldier subcastes. Soldier differentiation results from an increase in juvenile hormone released from small paired glands behind the brain, the corpora allata. Soldiers produce an inhibitory pheromone that regulates the worker to soldier ratio.

Instead of developing in the direction of neuter workers and soldiers, larva can develop in the direction of nymphs. In the first several instars, nymphs look about the same as small workers, the only differences are that their heads are slightly smaller and their thoracic segments are pointed at the posterior corners, exhibiting rudimentary wing-bud development. The wingbuds grow larger with each instar, so that by the last instar, large wing pads lay flat over the thorax. Nymphs do not develop until colonies are about five years old, and from that time a cohort of nymphs will develop annually. Nymphs lead relatively pampered lives compared to workers. They rarely, if ever, engage in the tunneling, building, and feeding behaviors that occupy the ever-busy workers. They may circulate in the population with workers, but do not generally participate in work. They feed themselves and also receive food from workers. As their wing pads lengthen, their bodies gradually change color to a milky white, reflecting the tremendous deposition of fatty tissue in their bodies. Their thorax and abdomen grow longer than those of workers.

Distribution

Termites are predominantly tropical and subtropical insects. Their diversity falls off sharply in temperate regions, and they are absent altogether in boreal and arctic regions. They tend to be most conspicuous in savannas (tropical grasslands), where their mounds often predominate in the landscape. They are more diverse on continents than on islands, and their diversity usually diminishes the more distant the islands are from continents. They are frequently quite diverse in deserts, but in this habitat do not build conspicuous aboveground epigeal nests or mounds.

Habitat

Termites have been called "dwellers in darkness," a reference to their cryptic lifestyle; they always stay inside substrates (either wood or soil), or tunnels, shelter tubes, and nests they construct from soil, feces, and saliva. Even when they forage, they usually do so through tunnels in the soil or under the cover of mud shelter tubes they construct. Other wood-boring insects are generally free living as adults. Indeed, this lifestyle is one of the many important and unique features of termite biology and ecology. This lifestyle partially accounts for the reason why so many people are unfamiliar with termites, and also for why they are so feared as pests.

Only a few termite species have evolved the ability to venture into the open to forage. The only such termites in North

America are the narrow-nosed nasute termites of the genus *Tenuirostritermes*. In this genus, the workers and soldiers are dark bodied, a defense against ultraviolet radiation that they share with other free-foraging species. The only other time termites emerge from the hidden confines of their galleries, tunnels, and foraging tubes, is when they stage their brief annual alate flights.

Most primitive termites are xylophagous, that is they feed on dead wood. However, more advanced termites forage on a much wider array of foods, including fine woody debris, plant litter, leaf litter, dead grass, organic layers of soil, humus, highly decomposed wood, live wood, live herbaceous plants and grass, dung, fungi, fungus gardens, and lichens. Most termites are adapted for hot, warm, and humid climates, and therefore tend to be most abundant at low elevations and in coastal regions.

Behavior

All termites are eusocial. They have a caste system with soldiers, primary reproductives, secondary reproductives, and workerlike pseudergates or true workers, and live in cooperative colonies with long-lived queens and kings. In fact, termites are the only order that is entirely eusocial. The other major cluster of eusocial insects is found in the order Hymenoptera in the suborder Apocita, which includes the stinging wasps, ants, and bees. Within this group of stinging wasps, adult female societies have evolved a dozen or more times.

Two families of termites are socially more primitive than the others, possessing pseudergates or false workers instead of true workers, and may be designated as "pro-social." These are the rotten wood termites (family Termopsidae) and the dampwood and drywood termites (family Kalotermitidae). In these families, instead of an early developmental bifurcation between neuter and nymphal developmental lines, all individuals follow the nymphal line of development. To prevent too many offspring from maturing and leaving the colony as alates, the king and queen physically manipulate some of them by nibbling on the wing pads or hind legs. These minor physical injuries trigger regressive molts, causing such individuals to regress their wing pads and thus delay development for at least one season. In this condition, they are known as pseudergates. They effectively serve as workers, although this developmental condition is not permanent and they can later resume nymphal development and eventually become alates. Because colonies of pro-eusocial termites are confined to individual items of wood, their populations often number only a few thousand individuals, rarely exceeding 10,000.

The next level of sociality exhibited by termites may be termed meso-eusocial. In these termites (most species in the families Mastotermitidae, Rhinotermitidae, and Serritermitidae), a true worker caste is present. In addition, because such termites always have subterranean habits, they can access a larger base of food resources and attain much larger populations. Populations in mature colonies may number from tens of thousands to millions. However, their nest remains a primitive and poorly defined collection of galleries and cells inside a log, stump, or tree trunk. Such meso-eusocial species have frequently evolved a franchising approach to resource acquisition, in which neotenic supplementary reproductives are generated to occupy any suitable new nesting resource, and colonies spread by a creeping process of colony budding.

Because pro-eusocial termites live inside their food source (dead wood), they have no need to recruit, or lay a pheromone trail, to some distant food source. The only recruitment that occurs is that of workers and soldiers to the breach points, so that when galleries break open they are quickly defended by soldiers and repaired by workers. The sternal gland pheromone used for this "breach recruitment" in pro-eusocial termites is hexanoic acid. In contrast, meso-eusocial termites must travel through soil tunnels to find and link their diverse foraging resources, and the sternal glands have evolved a different pheromone, dodecatrienol, for this new function.

In the pro-eusocial termites, the developmental dichotomy between pseudergates and nymphs occurs late, following wing pad injury, and the differentiation in terms of size and fat body accumulation is negligible. Both pseudergates and nymphs feed for themselves and accumulate fat reserves at about the same rate, and all retain approximately an equal chance of future reproduction. With the transition to meso-eusociality and the early, developmentally controlled bifurcation of neuter and nymphal lines, a true worker caste is defined. Coincident with this is a distinctive differentiation of the workers and nymphs. True workers develop slightly enlarged heads, mainly to house their enlarged mandibular muscles. They also develop enlarged salivary glands to meet the increased demand placed on them for feeding the dependent nymphs.

Nymphs of meso-eusocial termites also undergo a new and distinctive differentiation, which mainly involves a pronounced tendency to accumulate fat and for the abdomen and thorax to grow longer to accommodate these new reserves. Nymphs now become truly nutritional "bank accounts" for the colony. The worker-nymph dichotomy at the pro-eusocial stage is merely one of wing pads, but at the meso-eusocial stage is accentuated by a dramatic differential social/nutritional investment. This is profoundly important, for now the entire termite colony population has split along somatic/generative lines. At the pro-eusocial stage, the vast majority of the population, all but the soldiers, continue to express their development and behavior in a manner which leaves open the possibility of alate differentiation and personal reproduction (= direct fitness). But at the meso-eusocial stage, a large portion of the population, the workers, now pursue a course of development that forecloses the possibility of alate development. At the same time, they express altruistic feeding behaviors toward siblings, thus sacrificing their own reproductive potential to promote the future reproductive potential of their larval siblings. Workers therefore engage in a strategy in which *indirect* fitness benefits (proxy reproduction by close kin) exceed personal reproduction efforts. Meso-sociality is seen in most species in the families Mastotermitidae, Rhinotermitidae, and Serritermitidae.

The next stage in termite social evolution, meta-eusociality, is characterized by the construction of an organized nest. Two families of termites, the Hodotermitidae and the Termitidae, have achieved meta-eusociality. The nest is usually in the soil and separate from any feeding source. It may be entirely below the surface (hypogeal), or on and above the surface (epigeal). In desert environments, hypogeal nests are often found below large surface rocks (sublithic). Nests may also be attached to the trunks or limbs of trees (arboreal). In wood- or cellulose-feeding species, nests are generally composed of a hardened, pulpy, fecal plaster material called carton, which varies in color from light tan to dark brown. In humus and soil-feeding species, nests are made of a darker "stercoral" fecal material, presumably high in lignins and polyphenols. Other nests are essentially excavations of tiers of galleries in the soil, containing chambers lined with fecal plastering.

With the development of an organized nest, the social life of the colony is transformed. Workers and soldiers must evolve a division of functions between nest tasks and foraging tasks. In all cases, the younger stages are occupied with in-nest tasks, and the older workers and soldiers assume the more perilous tasks associated with foraging at greater distances from the nest. With this age-related polytheism, there is often a division of workers into sex-based size subcastes. In most cases, males become the smaller, minor workers, and females the larger, major workers, but these roles are reversed in the Macrotermitidae. While both sexes may become workers, the soldier caste frequently becomes restricted to one or the other sex, and may become di- or trimorphic. Such nests have a distinctive growth pattern of their own, starting as a single chamber (the nuptial cell, or copularium), and growing by the addition and/or reorganization of more cells and diverse tunnels and tiers over time.

The population size of mature nest-building termites is sometimes quite modest and tends to overlap, rather than exceed, the population ranges of meso-eusocial termites. There can be as few as several thousand up to a few million termites, mainly reflecting ecosystem differences in the resource productivity that can be harvested by a population with a central nest. Some meta-eusocial species have evolved interconnected multiple-nest systems (polycalism). These systems are analogous to those of meso-eusocial termites whose colonies spread by budding, and allow meta-eusocial species to achieve populations of several millions.

The final stage in termite social evolution is ultra-eusociality. Ultra-eusocial termites usually have multiple worker and soldier subcastes and very large nests called mounds. Mound-builders occur in at least one species of *Coptotermes*, *Coptotermes acinaciformis* of Australia; the fungus-growing Macrotermitinae in the genera *Odontotermes* and *Macrotermes* of Africa and Asia; in the Nasutitermitinae in the genera *Syntermes* and *Cornitermes* of South America; some *Nasutitermes*, including the cathedral-mound building termite (*Nasutitermes triodiae*) of northern Australia; some Amitermitinae, such as *Amitermes laurensis*, *A. meridionalis*, and *A. vitosus*, of northern Australia; and in *Amitermes medius* of Panama.

Mounds differ from nests not only in size, but also in having a massive outer wall of soil, usually clay. Because the mounds are so large and conspicuous, the termites must invest much more in fortifying them for defense. Sometimes the larger size also involves the addition of ventilation shafts

A termite (*Macrotermes*) queen with her attending workers in Kenya. (Photo by Wolfgang Bayer. Bruce Coleman, Inc. Reproduced by permission.)

and chimneys. In addition, mound builders often store large quantities of food in peripheral storage pits or in storage chambers in the mound itself. Mounds are always long-lived structures, lasting for many decades, if not for centuries. Female reproductives of mound-building termites attain a prodigious size, due to the expansion of their abdomens. They attain a length of 2–4 in (5–10 cm), and have an egg-laying rate and duration exceeding that of all other animals. The larger nests of some arboreal nasutes, such as *Nasutitermes rippertii* of the Bahamas, could qualify those species as ultra-eusocial. But even relaxing the definition, only a handful of species on each continent and probably no more than 50 to 100 in total can qualify as ultra-eusocial.

The ultra-eusocial grade reflects not only group-level selection leading to an adaptive demography of the population and an efficient subdivision of behavioral tasks among castes, but a quantum advance in the level of internal homeostasis. Ultra-eusocial termites distinctively exceed the most advanced hymenopteran societies, the honeybees and leafcutter ants, in population size and nest scale. Their mound populations typically number into the millions, and their biomass is substan-

mainly lignocellulosic matter. This dead plant biomass is high in caloric value, but low in nutritive value.

Lignocellulosic matter is the most abundant material produced annually in the biosphere by photosynthesis. Fungi are the main organisms that consume it, and they are able to do so because their threadlike hyphae have appropriate enzymes, and because the microscopic threads that make up wood-rotting fungi are extremely economical in their nutrient requirements. Vertebrate animals in general cannot derive sufficient nutriment from lignocellulosic matter, and hence almost none, except ruminant ungulates (hoofed animals with complex stomachs such as cows), utilize this abundant matter as food. However, a few groups of insects have evolved as successful detritivores, in all cases by coevolving symbiotic relationships with microbial organisms such as bacteria, protozoa, or fungi (these more primitive organisms have very diverse and useful metabolic machinery).

The main groups of insect detritivores are cockroaches, termites, crickets, flies, and beetles. Termites are the only insect detritivores that are social, leading to a higher level of coordinated foraging and increased foraging reach, and therefore to an extraordinary level of lignocellulosic processing power. Termites also comprise an entire insect order of considerable antiquity (dating perhaps to the late Paleozoic or early Mesozoic periods), and thus their symbioses are coevolved to a high level of integration and efficiency. Furthermore, they have ecologically radiated so that specialist genera exist for virtually every type of plant detritus, wood, humus, and dung, in every stage of decay, in virtually every type of temperate, subtropical, and tropical habitat.

In termites, three main patterns of endosymbiotic relationships may be outlined. First, all lower termite families have a complex paunch community including flagellate protozoa; second, the fungus-growing Macrotermitinae have lost the flagellate protozoans and instead cultivate a basidiomycete fungus, *Termitomyces*, as fist-sized lumps called fungus combs; and third, the remaining higher termite subfamilies have also lost the flagellate protozoans (although some harbor amoebae or ciliates) and have instead various segments, chambers, and diverticula of the hindgut in which bacterial cultures are involved in various ways in digestive metabolism.

Among the lower termites harboring flagellate protozoan, three subpatterns may be outlined. First, in the family Mastotermitidae, small gastric ceca (presumably containing bacteria) are present in the anterior midgut and there are blattobacteria in fat body bacteriocytes. A hindgut community of flagellate protozoa, spirochetes, and bacteria is also present. Second, in the families Termopsidae and Hodotermitidae, the blattobacteria are lost, but gastric ceca are present as well as hindgut flagellates. Third, in the families Kalotermitidae, Rhinotermitidae, and Serritermitidae, both the blattobacteria and gastric ceca are lost, and digestion symbiosis is relegated entirely to the community of flagellates, spirochetes, and bacteria in the paunch.

In the Macrotermitinae, digestion is accomplished through an initial rapid pass of the food through the intestines of minor workers, after which it is deposited as "primary feces" on the fungus comb. It is then further degraded by the action of

Vigorous development of a floodplain termite mound of *Amitermes* sp. has enveloped an abondoned tire resting on a stump beside an outback road, Gulf County, Queensland, Australia. (Photo by ©Wayne Lawler/Photo Researchers, Inc. Reproduced by permission.)

tial due to their larger bodies. But beyond these aspects of scale and biomass, such societies have a permanence and impact on the ecosystems that is exceptional. In the terrestrial landscape, only the relatively recently evolved human civilizations rival the majestic organizational complexity and domineering ecological impact of the ultra-eusocial termites.

Feeding ecology and diet

Termites are detritivores, in other words, they eat dead plant material. Dead plant material contains very little cytoplasm for two reasons: first, because much of the biomass of a plant consists of hollow, water transporting, vascular cells; and second, because when plants or plant parts die slowly, the plant retranslocates its cytoplasmic nutrient reserves to living tissue. Thus, dead wood and withered leaves and grass are mostly composed of plant cell-wall material, which is primarily made up of two types of plant carbohydrate polymers, cellulose and lignin. Plant detritus, therefore, whether large, coarse, and woody or small twigs, leaves, and plant debris, is

the cultivated fungus, which produces nutrient-enriched conidiophores. The conidiophores are eaten by the termites and fed to the dependent castes, while older workers feed on spent fungus comb. The intestine of termites in the Macrotermitinae does not contain symbionts, instead the midgut is substantially enlarged and the hindgut is short and simple.

The other subfamilies of the higher termites (excluding the Macrotermitinae) have evolved a completely modified hindgut with several new segments, chambers, and diverticula for housing new types of symbionts, mostly bacterial floras of various types. In one such modification, the junction of the midgut and hindgut is a unique intestinal innovation called the mixed segment. This region of the intestine posterior to the midgut is composed of a lobe of mesenteric tissue that forms an elongate diverticulum adjacent to the anterior-most section of the hindgut. The Malpighian tubules, which are involved in nitrogen excretion, form a network of convolutions over the mesenteric lobe of the mixed segment. The exact function of the mixed segment is a subject of much interest and conjecture. It seems likely that it has many functions, including roles in nitrogen excretion, osmotic regulation, and hindgut irrigation, as well as housing bacteria that may be involved in nitrogen recycling.

Another gut modification is the elongation and enlargement of the first chamber of the hindgut as a "secondary crop." Posterior to this is the enteric valve, which, in humivorous species, has evolved a complex armature involving spines, forks, and featherlike structures that projects into the succeeding bacterial pouch and rakes bacteria into the food bolus as it passes through the narrow valve. The bacterial pouch is the modified anterior-most portion of the paunch. In the *Cubitermes* group of genera from Africa, there is an additional bacterial diverticulum on the paunch. The anterior colon is made up of modified chambers which may bear internal cuticular processes, to which are attached various bacteria of unknown function. Thus, although the structural details vary considerably, in almost every case there are three or more distinct, sequential zones of the hindgut in which the food bolus is sequentially processed by different bacterial communities.

Reproductive biology

In general, termites may be characterized as monogamous. The timing of alate flights within a species is synchronized so that termites from different colonies can meet and form outbreeding pairs. After flight, the termites shed their wings, a process called dealation, and the newly dealated individuals seek a mate. The females in some species call by releasing a pheromone from their sternal or tergal glands. When the male finds the female, there usually ensues a brief chaotic zigzagging run, with the female leading and the male following in close tandem. This tandem run apparently constitutes their courtship. If a female fails to put on a good run, or if a male fails to keep pace, the suitor is rejected. This courtship evaluation must be made within minutes, because termites are extremely vulnerable to predation during this brief period of life outside the colony.

After courtship, pairs seek an appropriate place in the soil or in a crevice, crack, or hole in wood to serve as a copularium. Only after they are sealed inside, and therefore committed irrevocably to each other, do they mate. The copularium is the starting point for the growth of the termite family and the expansion of a new gallery system of the colony. Pairing, and then delaying mating until after the copularium is sealed, ensure lifelong monogamy and mate fidelity. In only a few species, and even then only rarely, do more than one male and female join in forming a copularium. In these species, polygamous associations may be adaptively advantageous in getting the growth rate of the colony off to a faster start. Although much attention has been paid to these exceptional cases of polygamy, they should not overshadow the fact that virtually all species in most cases form monogamous unions.

Monogamy is extremely important because it ensures that the growing population is comprised of full siblings. In a monogamous, outbred family, each offspring has exactly 50% of the genome of each parent, and shares with each sibling a 50% chance of having a particular gene in common. In other words, an exact equivalence of relatedness exists among all individuals in the colony. The importance of this is that colony-mates (siblings) are of equal genetic value as potential offspring. Therefore, if a termite has the opportunity, through a redirection of effort, to trade offspring production for sibling production, then it is an even tradeoff. The endo-substrate mode of life tends to enforce monogamous mating, and in consequence yields conditions of high intrafamilial relatedness that are conducive to a re-allocation of effort, away from risky potential offspring toward a secure investment in additional siblings.

Because of these conditions, immature termites in the early stages of social evolution may be thought of as willing helpers for their parents. They were able to trade direct reproduction for equally valuable indirect reproduction, simply by choosing to feed their parents and siblings, in other words, by evolving alloparenting. However, while pursuing this indirect strategy a further selfish reproductive benefit exists, the possibility of inheriting the parental nest when a pseudergate transforms into a neotenic replacement reproductive. That selfish interests still count is evident in the siblicidal battles that occur among replacement reproductives in lower, proeusocial termites. In the family Termopsidae, replacement reproductives soldiers may develop in orphaned colonies, and in this context their hypertrophied mandibles may be turned toward the nasty business of siblicidal neotenic combat, rather than the customary and more noble altruistic function of soldiers as colony defenders. The peculiar occurrence of reproductive soldiers among the ecologically most-primitive rottenwood termites has led to the suggestion that soldiers actually originated, not as colony defenders, but as a selfish adaptation, or "Cainism," an interesting parallel of the Biblical account in which one brother kills the other.

In pro-eusocial termites, the potential for neotenic transformation is present in workers, nymphs, and even presoldiers in some species. This provides great reproductive flexibility for individuals to express their own best situational strategy. In this respect, the pro-eusocial termites are exceedingly interesting, and far from being fully understood. In meso-eusocial termites, neotenic tendencies become more limited and are usually restricted to either the nymphal or worker line. In the Termitidae, where meta-eusociality is the rule, this individual reproductive flexibility is greatly restricted or completely lost. For example, termites in the subfamilies Macrotermitinae and Apicotermitinae do not have neotenics, and replacement reproduction is only possible by unflown alates. Thus, while individual reproductive strategies and a mixed portfolio of indirect and direct reproductive benefits were important in the early stages of termite social evolution, once critical social grades were reached, individual developmental options became rigidly channeled and the reproductive prerogative restricted. Eventually, this resulted in stronger domination by the primary reproductive pair, the king and queen, for whom the colony became merely a somatic extension, serving the reproductive output of their gonads.

Conservation status

No species of Isoptera is known to be threatened. Nevertheless, because of the extensive clearing of tropical forests for timber production and the conversion of much land to intensive agricultural production, it is likely that many termite species have been ecologically marginalized. Ancient termite-mound dominated landscapes are easily obliterated by bulldozers. Because so many insectivorous animals depend on termite populations for food, termite conservation is important and should be given greater attention.

Significance to humans

Termites have traditionally been regarded as pests. Even the ancients recognized termites as the original terminators and destroyers of man's wooden constructs. The name termite derives from the Latin root word "term -es, -in, -it," as in terminal: the end; and terminate: to bring to an end or destroy. Until the 1940s, the only defense against their depredations was good construction and vigilance. But with the advent of synthetic pesticides, an arsenal of poison was unleashed. To destroy drywood termites, chemical fumigants, including the ozone-depleting chemical methyl bromide, were used. For subterranean termites, chemicals such as pentachlorophenol and CCA have been used as wood preservatives, and lindane, aldrin, dieldrin, chlordane, and chlorpyrifos (most of which have now been banned) were used as soil termiticides. Despite some regulatory progress in removing the more hazardous and environmentally persistent synthetic pesticides, many equally toxic compounds remain on the market in both developed and developing countries.

However, there are hopeful signs of progress in termite pest management. These include heating, liquid nitrogen, microwaves, and electrical treatments for drywood termites; baiting and trap-treat-release systems for subterranean termites; and borates as wood preservatives. These newer control techniques offer the promise of effective control with greatly reduced hazard of human exposure and environmental contamination with synthetic pesticides. So far, even with these controls, urban pest termites appear to be holding their own, and even if completely eradicated from the urban environment, will likely persist in natural environments.

With so much emphasis on termite elimination, termite conservation has been given relatively little attention, even though numerous studies document that termite biodiversity is negatively impacted by intensive agriculture and forest clearing. In tropical areas, termites have been estimated to constitute up to 75% of total insect biomass and 10% of total animal biomass. No other taxon, except perhaps earthworms, represents such an major component of tropical ecosystems. Therefore, their conservation is vital to the operational integrity of ecosystems, including trophic relationships, soil microstructure and processes, and major flows of energy and nutrients in biogeochemical cycles. The danger is that, with effective baits, "control measures" might be implemented against termites as management tools in forestry, agriculture, or range management. This could be devastating, not only for termites, but for the ecosystems in which termites are keystone organisms.

Since termites are major converters of lignocellulosic matter to animal biomass, it would seem that they have potential for improving human utilization of lignocellulosic residues and wastes. Examples of such materials include scrap lumber and sawdust from saw mills; agricultural residues such as straw, bean pods, and sugar cane pulp; and animal dung from dairies and feed lots. Termites might be cultivated on such wastes and then harvested as feed for aquaculture or poultry production. The chemical energy in lignocellulosic wastes is usually dissipated to carbon dioxide by microbial degradation. By feeding such waste materials to termites, vast amounts of biochemical energy could be channeled into food production. Because termites have symbiosis with nitrogen fixing bacteria, they possess the rare metabolic machinery needed to convert plant cell wall matter into animal biomass. The substantial flow of plant biomass through termite intestines represents a significant pathway in the terrestrial carbon cycle that humankind has yet to productively tap into. Termites could be important organisms for humans to learn how to cultivate. The promise of large-scale termiticulture, however, will require much more research on termite physiology, nutrition, and respiration before many technical obstacles are overcome.

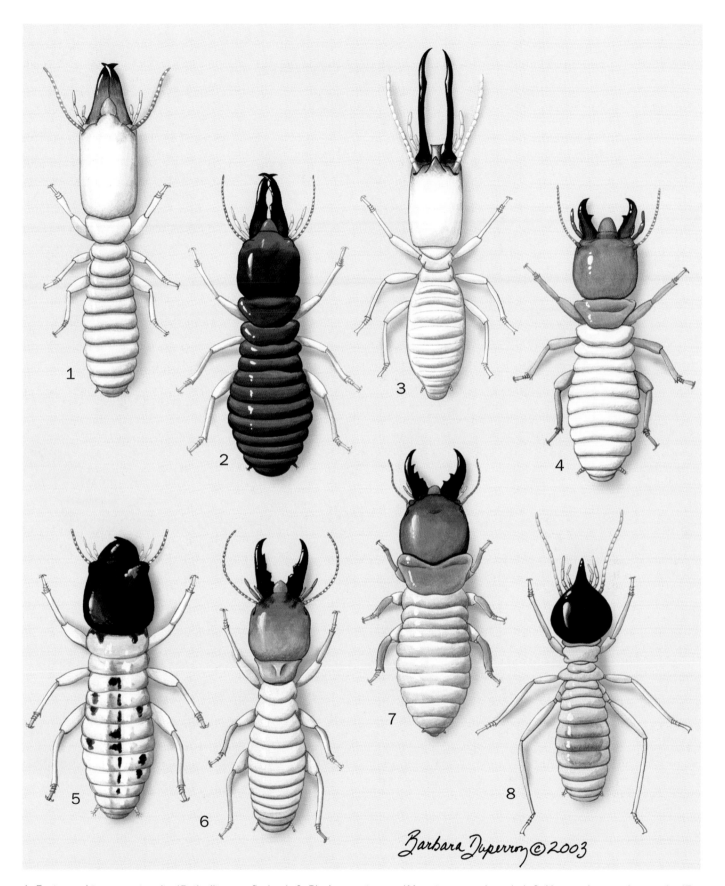

1. Eastern subterranean termite (*Reticulitermes flavipes*); 2. Black macrotermes (*Macrotermes carbonarius*); 3. Linnaeus's snapping termite (*Termes fatalis*); 4. Giant Australian termite (*Mastotermes darwiniensis*); 5. West Indian powderpost drywood termite (*Cryptotermes brevis*); 6. Wide-headed rottenwood termite (*Zootermopsis laticeps*); 7. Giant Sonoran drywood termite (*Pterotermes occidentis*); 8. Black-headed nasute termite (*Nasutitermes nigriceps*). (Illustration by Barbara Duperron)

Species accounts

West Indian powderpost drywood termite
Cryptotermes brevis

FAMILY
Kalotermitidae

TAXONOMY
Termes brevis Walker, 1853, Jamaica.

OTHER COMMON NAMES
English: Furniture termite.

PHYSICAL CHARACTERISTICS
Alates medium size, 0.4–0.5 in (10–12 mm) from head to tip of wings; median vein curving forward to meet anterior margin in outer one-third of wing; yellow brown. Soldiers have distinctive phragmotic (pluglike), deeply wrinkled head with high frontal flange; short mandibles.

DISTRIBUTION
The most widespread termite species. Regarded as a "tramp" species because easily spread in any item of wood furniture, in the wooden spars, masts, and planking of ships, in wooden pallets, and dunnage (packing material in ships). Native to the West Indies, widely distributed on wood sailing ships after discovery of New World. Now established in most oceanic archipelagos, including the Canary Islands, New Caledonia, the Hawaiian Islands, Bermuda, the Azores, Brazil, Australia, South Africa, and most of the Gulf Coast cities of the United States, especially peninsular Florida.

HABITAT
Usually found in urban areas in structural timbers of houses, in furniture, and in boats. Rarely occurs in natural settings, prefers drier wood found in human habitation. Also requires humid air and is usually only found in coastal and island localities.

BEHAVIOR
Colonies typically small with only a few thousand individuals, but infested structures may have numerous colonies. Each colony occupies galleries extending a few meters in length. Fecal pellets distinctively shaped, short, six-sided cylinders. Infestation usually detected by piles of fecal pellets which termites dump out of galleries through small round "kick holes" quickly sealed with fecal plaster.

FEEDING ECOLOGY AND DIET
Xylophagous, survives in variety of woods, particularly sound hardwood and softwood timbers, needs very dry and sound wood.

REPRODUCTIVE BIOLOGY
Colonies pro-eusocial, headed by primary reproductives or secondary neotenic replacement reproductives. Neotenic reproductives develop quickly when primaries are removed and engage in lethal fighting until single reproductive pair is

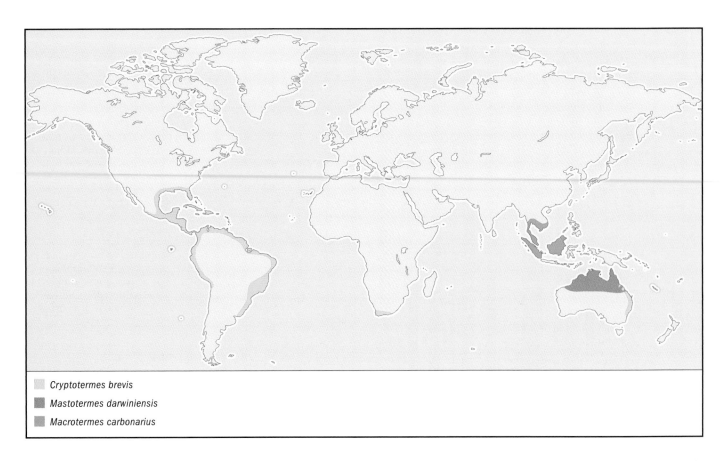

☐ *Cryptotermes brevis*
■ *Mastotermes darwiniensis*
■ *Macrotermes carbonarius*

reestablished. In Florida, alates fly between dusk and dawn from April to June. After short dispersal flight, wings are broken off and pairs search for holes and crevices in which to form copularium. Mating does not occur until pair seal themselves in.

CONSERVATION STATUS
Not threatened; expanding and prospering due to human activity.

SIGNIFICANCE TO HUMANS
Destructive pest of human-made wooden structures, particularly houses and historical buildings. Sometimes referred to as the furniture termite because of its unusual ability to form colonies in relatively small, moveable, wooden items and furnishings. Also known to attack books and archived documents. ◆

Giant Sonoran drywood termite
Pterotermes occidentis

FAMILY
Kalotermitidae

TAXONOMY
Termes occidentis Walker, 1853, Baja California, Mexico.

OTHER COMMON NAMES
None known.

PHYSICAL CHARACTERISTICS
Large, alates 0.7–0.78 in (18–20 mm) from head to wing tips; antennae with 20 or 21 segments, reddish brown. Soldiers heavy bodied, 0.5–0.6 in (14–15 mm); toothed mandibles, round heads with black compound eyes, broad pronotum, thorax with wing pads, brilliant orange and yellow. Pseudergates large with large compound eyes.

DISTRIBUTION
Sonoran Desert, including Baja California and Sonora, Mexico, and southwestern Arizona, United States.

HABITAT
Dead standing branches of paloverde trees of the genus *Cercidium*.

BEHAVIOR
Colonies pro-eusocial, rarely exceed 3,000 individuals; most mature colonies have standing population of 1,000 to 1,500. All individuals develop as nymphs. About 5% of nymphs have wing pad scars resulting in development as pseudergates, comprising about 10 to 15% of population. Pseudergates may molt to presoldiers and then to soldiers. Soldiers comprise about 2% of population, guard breaches of the galleries. Nymphs molt to alates in July, small numbers fly from nest at night over flight season from late July through September. Dispersal flights last at least four minutes but not more than one hour. Alates seek beetle emergence holes on dead paloverde trees, which male and female pairs enter and seal off copularium. Primary reproductives always bite off the distal halves of their antenna for reasons that remain unknown.

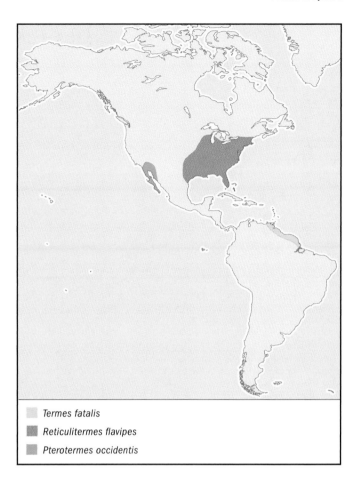

☐ *Termes fatalis*
■ *Reticulitermes flavipes*
▨ *Pterotermes occidentis*

FEEDING ECOLOGY AND DIET
Xylophagous; colonies excavate wide, meandering galleries inside dead paloverde branches. Occasionally found in saguaro cactus skeletons.

REPRODUCTIVE BIOLOGY
Most colonies headed by pair of primary reproductives. However, if one or both are removed, pseudergates start to molt within weeks to become replacement reproductives. When excess numbers molt following death of primaries, lethal fighting follows among neotenic replacements until one male and one female reproductive again become established and suppress, presumably by inhibitory pheromones, further neotenic molting.

CONSERVATION STATUS
Not threatened. Critical habitat (dead standing branches of paloverde trees) not utilized by humans except sometimes as firewood. However, as the species is monophagous on one tree species and standing branches on tree are limited, could be subject to local extirpation if intensively collected near urban areas.

SIGNIFICANCE TO HUMANS
Excellent study organism because of large size and ease of maintenance; one of the most well-studied North American drywood termites. Not a pest; no known economic value. ◆

Giant Australian termite
Mastotermes darwiniensis

FAMILY
Mastotermitidae

TAXONOMY
Mastotermes darwiniensis Froggatt, 1896, Port Darwin, West Australia.

OTHER COMMON NAMES
None known.

PHYSICAL CHARACTERISTICS
Large: alates up to 1.4 in (35 mm) with wings, 2 in (50 mm) wingspan; soldiers 0.45–0.5 in (11.5–13 mm); workers 0.4 in–0.45 in (10–11.5 mm). Only termite whose winged alates possess anal lobe in hind wing; tarsi have five segments, females have short blattoidlike ovipositor with ventral valves, long enough to overlap the dorsal valves; eggs laid in ootheca. Soldiers have round reddish heads with relatively stout short mandibles, long apical teeth; right mandible has two well-defined marginal teeth, but in left mandible only first marginal tooth is well defined, second and third are indistinct. Soldiers and workers have unique coxal armature, or flange, on front legs, and rows of small opposable teeth on femora and tibia, unique leg characters in this family.

DISTRIBUTION
Once cosmopolitan (as indicated by fossils), now confined to tropical northern Australia and nearby islands.

HABITAT
Xylophagous and subterranean, feeds on logs, dead standing trees, and surface wood. Also known to girdle live trees, including commercial tree plantings, then feed on killed trees.

BEHAVIOR
Meso-eusocial. Little is known aside from feeding ecology and reproductive biology.

FEEDING ECOLOGY AND DIET
Independently evolved subterranean foraging habits and ability to build mud shelter tubes to access wood above ground level. Simple nest chambers of several tiers of thin-walled carton cells usually in bole of tree or stump, near or below ground level. Colonies may exceed 1 million individuals and forage over 328 ft (100 m).

REPRODUCTIVE BIOLOGY
Workers frequently transform into ergatoid reproductives that replace primary reproductives when colonies are orphaned, but more typically develop as supplementary reproductives and form new colonies by budding off from parent colonies. Most reproduction probably by neotenic ergatoid reproductives.

CONSERVATION STATUS
Not threatened.

SIGNIFICANCE TO HUMANS
Of phylogenetic interest because of status as a monotypic family, exhibits unique morphological characters that place it as most basal termite and possibly most basal extant species of Dictyopterid. Serious structural, forest, and agricultural pest; most destructive termite in tropical northern Australia; damaging timber in buildings, bridges, poles, fence posts, railway sleepers, living trees, and crops. Also feeds on many host tree species, including live trees, and attacks plantations and crops

such as sugar cane. However, its large body, large colony size, and wide diet make it an excellent candidate for beneficial termiticulture on cellulosic wastes. ◆

Eastern subterranean termite
Reticulitermes flavipes

FAMILY
Rhinotermitidae

TAXONOMY
Termes flavipes Kollar, 1837, Vienna, Austria (where it was introduced).

OTHER COMMON NAMES
None known.

PHYSICAL CHARACTERISTICS
Alates small, 0.4 in (10 mm) from head to wing tips; black except for yellow tibiae. Soldiers with elongate, parallel-sided, subrectangular, yellow heads, and stout, black, toothless mandibles, strongly curved inward at tips. Top of soldier head with minute opening of frontal gland, or fontanelle. Workers about 0.2 in (6 mm) long; creamy white.

DISTRIBUTION
Native to eastern forests of United States from Florida to Maine, and in the west from Texas to Minnesota; introduced into Canada in southern Ontario, including the Toronto area.

HABITAT
Deciduous hardwood forests.

BEHAVIOR
Colonies are meso-eusocial and forage through shallow, narrow tunnels in soil that connect dead wood items, including stumps, logs, and roots. Also forages up trees on dead limbs and into rotten boles of trees with heart rot or butt rot. In urban environment, feeds on wood landscaping items such as fence posts, edging boards, firewood piles, wood-chip mulch, scrap lumber, and flower planter boxes. Once established in these items, explores foundation cracks and crevices and often finds access to structural wood framing, recruiting more workers and establishing numerous access points into the structure. Since feeding is from inside and most wood framing is concealed, may cause extensive damage over many years before being discovered.

FEEDING ECOLOGY AND DIET
Xylophagous, consumes hardwoods and softwoods, prefers sapwood to heartwood. Preferentially feeds on more porous spring wood of annual rings, leaving harder summer wood, thus damaged wood may have laminated appearance. Consumes both sound wood and partially decayed wood. Galleries often have small parchmentlike partitions of fecal paste and light tan specks of fecal plastering. When working above ground, in structures, or when foraging up trees, builds protective shelter tubes of fine soil particles and saliva, lined with fecal plaster. Shelter tubes are the most conspicuous sign of presence and activity, as no fecal pellets are produced.

REPRODUCTIVE BIOLOGY
No definitive nest, establishes reproductive chambers in logs, stumps, or other large moist wood items. Since structural wood is usually dry, reproduction rarely occurs in structural timbers. Moves deeper into ground in winter, staying beneath frost line,

probably occupying roots of old stumps. Generates large number of nymphoid neotenics from nymphs, which serve as supplementary reproductives and allow seeding of all potential reproductive resources within foraging territory with reproductives; thus, colonies expand continuously as resources are colonized. Most reproduction is by supplementary reproductives. Older colonies have no central nest or foraging territory headed by single primary or replacement pair, but instead extensive, loosely connected gallery systems with an extended family structure. Foraging territories are discovered, exploited to exhaustion, then abandoned as new foraging areas are expanded. This foraging-reproductive strategy is appropriate and well adapted, matched to the pattern of wood production and dispersion in the temperate zone.

CONSERVATION STATUS
Not threatened despite intensive control efforts.

SIGNIFICANCE TO HUMANS
Major subterranean termite pest in eastern North America. Responsible for hundreds of millions of dollars of damage and control annually; this damage and collateral expenses of inspection, control, and renovation make it one of the most destructive urban insect pests. ◆

Black macrotermes
Macrotermes carbonarius

FAMILY
Termitidae

TAXONOMY
Termes carbonarius Hagen, 1858, Malay Peninsula.

OTHER COMMON NAMES
None known.

PHYSICAL CHARACTERISTICS
Very large (largest in Southeast Asia); alates about 1.2 in (30 mm) from head to wing tips, wing span at least 2 in (50 mm). As suggested by species name *carbonarius*, workers and soldiers very dark, nearly black, a feature often found in free-foraging termites exposed to sunlight. Workers dimorphic, males larger than females. Soldiers also dimorphic but all females; with fleshy lobe at end of labrum; subrectangular heads; razor sharp, saberlike mandibles.

DISTRIBUTION
Southeast Asia, including Thailand, Cambodia, Malaysia, and Borneo.

HABITAT
Occurs in flat lowlands, uncommon in hilly terrain. Found in plantations such as rubber, coconut, durian, and teak, especially common in coastal dipterocarp forests.

BEHAVIOR
Colonies are ultra-eusocial. Builds large mounds up to 13 ft (4 m) high and 16 ft (5 m) wide at base. Fungus combs usually in large chambers around periphery of mound. Minor workers specialized for nest work, including tending king and queen, feeding larvae and nymphs, and nest repair. Major workers mainly forage.

FEEDING ECOLOGY AND DIET
Gramivorous, eating mainly dead grass, twigs, and surface debris cut into short pieces. Only species of subfamily Macrotermitinae to forage above ground; usually at night. Foraging parties involve major and minor soldiers and major workers, but not minor workers. Foraging area changes daily. After opening foraging hole, major workers build pavement trackways to foraging area. More workers then join, fanning out to collect dead grass and twigs; continuous cordon of major and minor soldiers guards outskirts of foraging area. Collected forage is carried below ground and passed to minor workers, who chew it and pass it rapidly through their digestive tract as "primary" feces, then deposit it on fungus comb upon which species of basidiomycete fungus (genus *Termitomyces*) grow.

REPRODUCTIVE BIOLOGY
Colonies usually headed by royal pair in thick-walled royal cell. Queen's abdomen unfolds and expands to gross dimensions (physogastry). Subfamily cannot produce neotenic reproductives; when primary reproductives die, colony may also die unless alates are there as replacements.

CONSERVATION STATUS
Not threatened.

SIGNIFICANCE TO HUMANS
Plantation pest. Alates sometimes harvested and eaten or used as chicken feed. ◆

Black-headed nasute termite
Nasutitermes nigriceps

FAMILY
Termitidae

TAXONOMY
Termes nigriceps Haldeman, 1858, western Mexico.

OTHER COMMON NAMES
English: Haldeman's black nasute

PHYSICAL CHARACTERISTICS
Queen physogastric; rusty yellow except for costal margins of wing scales which are dark brown, 0.7 in (18.5 mm) without wings. Soldiers with very dark heads; nasus wide and dark reddish; dense erect setae over head capsule. Workers dimorphic, rectangular heads; darkly pigmented.

DISTRIBUTION
Widely distributed from western Mexico as far north as Mazatlan, south to Panama and northern South America.

HABITAT
Coastal plains from sea level to about 3,280 ft (1,000 m).

BEHAVIOR
Colonies are meta-eusocial. Builds large conspicuous arboreal carton nests in trees, on fence posts and poles. May have more than one nest per colony.

FEEDING ECOLOGY AND DIET
Xylophagous; feeds mainly above ground via an extensive network of wide shelter tubes attached usually to the lower sides of tree branches.

Zootermopsis laticeps

Nasutitermes nigriceps

REPRODUCTIVE BIOLOGY
Colonies headed by primary reproductives.

CONSERVATION STATUS
Not threatened.

SIGNIFICANCE TO HUMANS
Occasional structural pest. ◆

Linnaeus's snapping termite
Termes fatalis

FAMILY
Termitidae

TAXONOMY
Termes fatalis Linnaeus, 1758, Para-Maribo, Suriname.

OTHER COMMON NAMES
None known.

PHYSICAL CHARACTERISTICS
Small, alate about 0.3 in (8.5 mm) with wings; brown, with large apical teeth. Soldiers monomorphic, with pale yellow, elongate, parallel-sided heads with hornlike tubercle projecting forward. Mandibles slender, elongate and rodlike; apices cupped together. Soldier labrum is narrow and rectangular with short points on anterior corners.

DISTRIBUTION
Northeastern South America, in Guyana, Suriname, Trinidad, and Amazonia, Brazil.

HABITAT
Rainforest.

BEHAVIOR
Colonies are meta-eusocial. Soldiers capable of violently snapping their mandibles, forcing them to cross downward, pushing the pointed head upward into ceiling of tunnel or nest chamber opening. Presumed function of snapping behavior is to lock soldiers' head into tunnel to block the advance of predators such as ants or other termites. Species of the subfamily Termitinae often build short, turretlike nests of hard dark fecal material. Nests composed of numerous interconnected cells. Little known about nesting behavior, but some species in the subfamily build nests inside mounds and nests of other termites.

FEEDING ECOLOGY AND DIET
Humivorous. Large apical teeth and molar without grinding ridges suggest diet of very soft decayed wood or humus.

REPRODUCTIVE BIOLOGY
Nothing known.

CONSERVATION STATUS
Nothing known.

SIGNIFICANCE TO HUMANS
First species of termite formally named by Linnaeus in 1758, thus has taxonomic significance as ordinal type. ◆

Wide-headed rottenwood termite
Zootermopsis laticeps

FAMILY
Termopsidae

TAXONOMY
Termopsis laticeps Banks, 1906, Florence and Douglas, Arizona, United States.

OTHER COMMON NAMES
None known.

PHYSICAL CHARACTERISTICS
Largest and most primitive North American termite. Winged alates 1–1.2 in (26–30 mm) from head to wing tips with wing span 1.8–1.9 in (46–48 mm); antennae with 26 segments; cerci with 5 segments; tarsi with 5 segments; reniform compound eyes, simple eyes or ocelli absent; body dark yellowish. Soldiers up to 0.6–0.9 in (16–23 mm) long with spectacularly long and jaggedly toothed mandibles; flattened heads; widest posteriorly; pronotum with anterior corners pointed; large spines on tibiae. Pseudergates develop from nymphs after wing-pad abscission or wing-pad biting. Pseudergates develop different pattern of hair over body and may molt several times, enlarging in size each time and developing large, wide head. Functionally reproductive replacement soldiers sometimes develop in orphaned colonies.

DISTRIBUTION
Central and southeastern Arizona to southern New Mexico, West Texas, United States, and Chihuahua and Sonora, Mexico; within altitudinal range 1,500–5,500 ft (457–1,676 m) above sea level.

HABITAT
Occurs in canyons and river valleys in rotten cores of boles and large branches of living riparian trees such as willow, cotton-

wood, sycamore, oak, alder, ash, walnut, hackberry, and other hardwoods. Not recorded from conifers or dead, rotten logs. In the relatively arid region it inhabits, only live trees can provide the moist conditions it requires.

BEHAVIOR
Colonies are pro-eusocial. Gallery excavations extend several feet (meters) with concentric, wandering, open chambers in rotten wood. Excavated central area becomes filled with caked mass of fecal pellets. Galleries typically originate from knot hole plugged with a mass of hard fecal pellets; galleries often damp or wet inside. Soldiers agile, defend openings against ants or other predatory intruders.

FEEDING ECOLOGY AND DIET
Mycetoxylophagous, feeding only on rotten hardwoods. Feeding probably helps advance development of heart rot in infested trees.

REPRODUCTIVE BIOLOGY
Alates fly in middle of night from late June through early August. Colonies initiated by alate pairs in tree scars, knot holes,

or small rot pockets where tree previously damaged by wind or beetles. Most field colonies headed by primary reproductive pairs, but replacement reproductives may develop from pseudergates or nymphs. Functional reproductive soldiers with heads smaller than typical soldiers have also been found as replacement reproductives in field colonies. Colonies rarely exceed 1,000 individuals.

CONSERVATION STATUS
Not threatened, but could be affected by agricultural or urban development in riparian habitats.

SIGNIFICANCE TO HUMANS
Attacks live trees and extends rot from dead to live portions of trees, hastening collapse or breakage of trees. However, most attacked tree species are not of economic importance, so not considered a pest. Could be pest in mature orchard crops such as pecan and pistachio, but this has not been reported. Could be used for physiological studies of termites because of large size, but collecting colonies difficult, requiring bucksaws or chainsaws, wedges, and sledgehammers. ◆

Resources

Books

Abe, T., D. E. Bignell and M. Higashi, eds. *Termites: Evolution, Sociality, Symbiosis, Ecology.* Dordrecht, The Netherlands: Kluwer Academic, 2000.

Choe, J. C., and B. J. Crespi, eds. *The Evolution of Social Behavior in Insects and Arachnids.* Cambridge, U.K.: Cambridge University Press, 1997.

Grassé, P.-P. *Termitologia,* 3 vols. Paris: Masson, 1982–1986.

Kofoid, C. A., et al., eds. *Termites and Termite Control.* Berkeley: University of California Press, 1934.

Krishna, K., and F. M. Weesner, eds. *Biology of Termites,* 2 vols. New York: Academic Press, 1969–1970.

Myles, T. G. "Resource Inheritance in Social Evolution from Termites to Man." In *The Ecology of Social Behavior,* edited by C. N. Slobodchikoff. New York: Academic Press, 1988.

Sands, W. A. *The Identification of Worker Castes of Termite Genera from Soils of Africa and the Middle East.* London: CAB International, 1988.

Uys, V. *A Guide to the Termite Genera of Southern Africa.* Plant Protection Research Handbook No. 15. Pretoria, South Africa: Agricultural Research Council, 2002.

Wilson, E. O. *The Insect Societies.* Cambridge: Belknap Press of Harvard University Press, 1971.

———. *Sociobiology: The Abridged Edition.* Cambridge: Belknap Press of Harvard University Press, 1980.

Periodicals

Myles, T. G. "Evidence of Parental and/or Sibling Manipulation in Three Species of Termites in Hawaii." *Proceedings of the Hawaiian Entomological Society* 27 (1986): 129–136.

———. "Reproductive Soldiers in the Termopsidae (Isoptera)." *Pan-Pacific Entomologist* 62, no. 4 (1986): 293–299.

———. "Review of Secondary Reproduction in Termites (Insecta: Isoptera) with Comments on Its Role in Termite Ecology and Social Evolution." *Sociobiology* 33 (1999): 1–91.

———. "Termite Eusocial Evolution: A Re-Examination of Bartz's Hypothesis and Assumptions." *The Quarterly Review of Biology* 63 (1988): 1–23.

Timothy George Myles, PhD

Mantodea
(Mantids)

Class Insecta
Order Mantodea
Number of families 15

Photo: A praying mantis (*Mantis religiosa*) on a prickly pear in Spain. (Photo by J-C Carton. Bruce Coleman, Inc. Reproduced by permission.)

Evolution and systematics

Praying mantids evolved during the early Cenozoic era (65 million years ago) from the cockroaches. Although it is widely accepted that mantids are related most closely to the cockroaches, the higher classification of mantids continues to be debated. Some entomologists contend that mantids should be grouped with the cockroaches in a single order, Dictyoptera, whereas others believe that the two groups should be placed in their own orders, Mantodea, the praying mantids, and Blattodea, the cockroaches. The latter approach is followed in this chapter.

The order Mantodea comprises 15 families that contain 434 genera and 2,300 species. The neotropical families Chaeteessidae and Mantoididae and the Old World family Metallyticidae all contain a single genus each and are considered by experts to be the most primitive mantid families. The Amorphoscelidae and Eremiaphilidae are two small families diversified throughout Africa and Asia. These five families generally have a thorax (pronotum) that is more or less square, as contrasted with the remaining families, whose pronotums are relatively elongated. The vast majority of mantids are grouped into these remaining families: Mantidae, Hymenopodidae, Acanthopidae, Liturgusidae, Tarachodidae, Thespidae, Iridopterygidae, Toxoderidae, Sibyllidae, and Empusidae. Many

A Japanese praying mantis (*Paratenodera aridifolia*) larva changing skin. (Photo by Kim Taylor. Bruce Coleman, Inc. Reproduced by permission.)

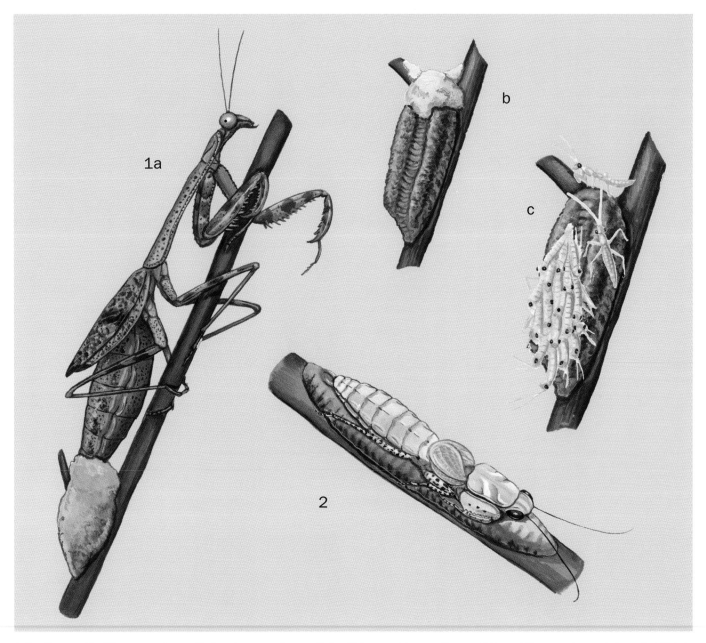

A Carolina mantis (*Stagmomantis carolina*) makes an oothecea (1a), which overwinters on its own (1b), and the young emerge in spring (1c). In contrast, *Tarachodula pantherina* (2) stands guard over her eggcase until the young emerge. (Illustration by Gillian Harris)

of the latter families were once subfamilies grouped under the Mantidae, but they have been elevated to family status.

Physical characteristics

Mantids generally are large, ranging in size from just under 0.4 in (1 cm) to more than 6.7 in (17 cm). Females usually are larger than males, sometimes twice their size. Their coloration depends primarily on where they live. Mantids that are found in savannas and meadows are straw-colored or light green. Those that inhabit leaf litter tend to be dark brown. Mantids that frequent flowers in search of prey typically are yellow, white, pink or light green.

All mantids are perhaps best known for their raptorial forelegs, which they use to capture live prey. Two rows of spines on the femur and an opposing row on the tibia (except in the family Amorphoscelidae, which has only a single femoral spine) enable them to impale prey. The number and arrangement of foreleg spines are important characteristics used by entomologists to classify mantids.

Flexible neck muscles allow mantids to turn their heads a full 180 degrees. The head is triangular in shape, except in some species, such as *Gongylus gongylodes*, that have an elongated vertex between the eyes. All mantids have two large, compound eyes and three small, simple eyes (ocelli). The antennae of most mantids are narrow and long and have

1. *Polyspilota aeruginosa* as she usually appears, cryptically colored dorsally to blend in with the foliage. 2. When attacked, her deimatic display reveals the colorful ventral surfaces and black spots on her legs. (Illustration by Gillian Harris)

many segments, but some species have antennae that are feathery, as in male *G. gongylodes.*

The thorax (pronotum) is longer than it is wide in most species and sometimes is expanded laterally, as in the subfamilies Deroplatyinae and Choeradodinae. Mantids generally have two pairs of wings, the forewings (tegmina) and the hind wings, but wing structure varies widely among mantid families from those having no wings (apterous) to those having two pairs of wings that are fully developed. The abdomen has 10 segments and terminates with the genitalia and a pair of multisegmented cerci.

Distribution

The majority of species are tropical and are concentrated in the rainforests of South America, Africa, and Southeast Asia. Mantid diversity decreases in temperate regions, and they are not found in boreal and tundra climates. Although most of these insects are limited in their distributions, a few species are widespread and found on more than one continent, such as the Chinese mantid, *Tenodera aridifolia sinensis,* and the European mantid, *Mantis religiosa.* These two species have become widespread since humans began transporting nursery stock with attached egg cases (ootheca) around the world.

A female mantid (*Stagmomantis limbata*) eating a male. (Photo by Bob Jensen. Bruce Coleman, Inc. Reproduced by permission.)

Habitat

Mantids are entirely terrestrial and inhabit rainforests, dry forests, primary and secondary forests, grasslands, and deserts. In temperate regions, mantids complete one entire life cycle per season, whereas in the tropics, mantids can have overlapping generations.

Behavior

Numerous interesting types of behavior have been studied among the praying mantids. They are known to groom themselves frequently. Using their forelegs, they wipe their eyes and heads and then systematically clean their forelegs with their mouths. They do the same with their antennae and middle and hind legs.

Mantids have evolved certain means of defending themselves against potential predators. When faced with danger, most species attempt to run or fly away. If this does not deter the predator, the mantid may react with a startle display that includes thrusting out the forelegs, flashing out the wings, and opening the mouth. In many species the undersides of the forelegs and wings are brightly colored, so the sudden flash of these colors produces a startling effect. If these kinds of defensive behavior fail, mantids may resort to playing dead (thanatosis) or biting and pinching.

Feeding ecology and diet

Ecologically, mantids are considered top arthropod predators in the food chain. They are generalist feeders and can catch and consume arthropods primarily of equal or smaller size. Rarely, large mantids have been known to ensnare small mice, lizards, frogs, and birds. Hatchling mantids typically feed on aphids and other insects of similar size, whereas adults prey upon larger insects, including butterflies. Most mantids are opportunistic feeders and perch motionless, awaiting suitable prey; a few species actually chase down prey. Known for being cannibals, mantids consume each other if the opportunity arises. Large eyes and extraordinarily quick foreleg strikes enable these insects visually hunting predators to capture prey in 1/20 of a second.

Reproductive biology

Perhaps the best-known myth that pertains to mantids is that females always decapitate males during copulation. Although this does occur occasionally, it is not commonplace. It is true, however, that if the female cannibalizes the male during copulation, he continues to mate with her even without his head. Once the male has inserted the sperm packet into the female's abdomen, she uses his sperm to fertilize her eggs. One to several egg cases containing 10–200 eggs (depending on species) are laid over the next several weeks. Eggs are encased in a frothy liquid that hardens and protects the eggs until they hatch.

Conservation status

Although there are perhaps a few rare extant mantid species (one species is known only from a few islands in the Galapagos archipelago), there are little data regarding the overall status of mantid populations. Global warming, habitat destruction, and misuse of pesticides, however, have a detrimental effect on species. One species is listed in the IUCN Red Book: *Apteromantis aptera*, found in localized areas of Spain and categorized as Lower Risk/Near Threatened.

Significance to humans

Humans have both feared and revered praying mantids for more than a thousand years. Mantids have been frequent subjects of art and literature. Their prayerful pose was once thought to help travelers find a way home. The Chinese used mantids in fighting games. Indeed, there is a style of kung fu that mimics the movements of mantids.

In more practical terms, the Chinese mantid, *T. a. sinensis*, has been used extensively as a means of biological control for plant pests. People purchase egg cases in the winter and place them in their gardens for hatching the following spring, the goal being to have mantids consume plant pests, such as aphids. Most people do not realize, however, that mantids are generalist predators; they consume plant pests in addition to any arthropod, including other mantids and beneficial insects.

Praying mantids have become common in the pet trade. Such species as *Deroplatys lobata*, *Hymenopus coronatus*, and *T. a. sinensis* are especially popular among hobbyists.

Chinese mantids (*Tenodera aridifolia sinensis*) hatching. (Photo by E. R. Degginger. Bruce Coleman, Inc. Reproduced by permission.)

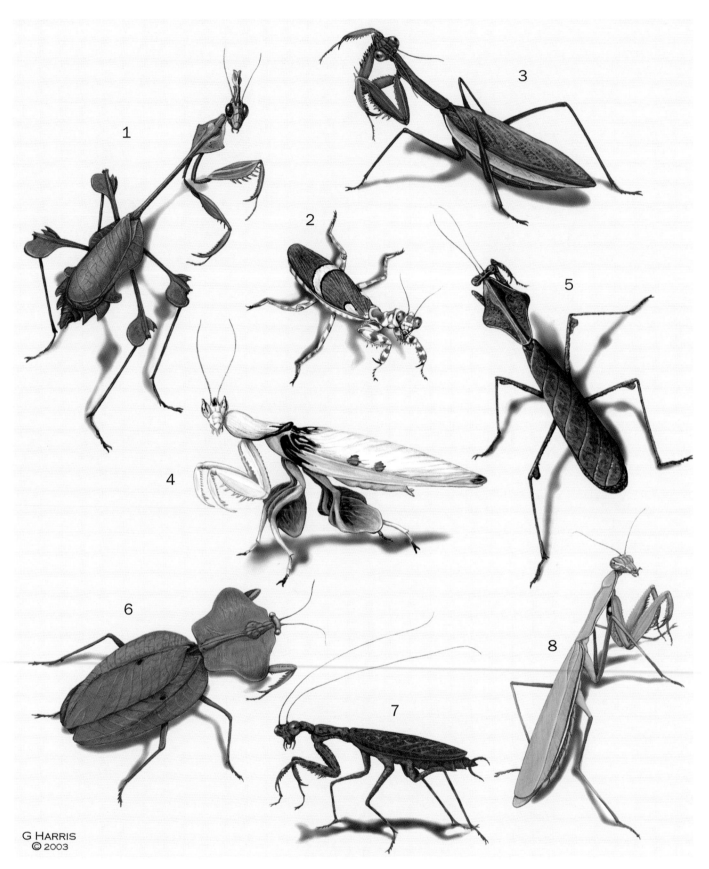

1. Wandering violin mantid (*Gongylus gongylodes*); 2. Boxer mantid (*Theopropus elegans*); 3. Chinese mantid (*Tenodera aridifolia sinensis*); 4. Orchid mantid (*Hymenopus coronatus*); 5. Dead-leaf mantid (*Deroplatys lobata*); 6. *Choeradodis rhomboidea*; 7. *Liturgusa charpentieri*; 8. European mantid (*Mantis religiosa*). (Illustration by Gillian Harris)

Species accounts

Wandering violin mantid

Gongylus gongylodes

FAMILY
Empusidae

TAXONOMY
Gongylus gongylodes Linnaeus, 1758, India.

OTHER COMMON NAMES
English: Rose mantid.

PHYSICAL CHARACTERISTICS
Varies from light to dark brown in color. The head has a conical extension at the vertex. The anterior portion of the extremely thin thorax is expanded laterally in a diamond shape. All legs have leaflike extensions. Males are 2.8–3.1 in (7–8 cm) long; females, 3.1–3.5 in (8–9 cm). Male antennae are feathery, whereas female antennae are threadlike.

DISTRIBUTION
Southern India, Sri Lanka, Thailand, and eastern Java.

HABITAT
Primary and secondary rainforests.

BEHAVIOR
Physical characteristics allow them to blend in well in leaf litter and shrubbery.

FEEDING ECOLOGY AND DIET
Preys chiefly upon flying insects that it can catch.

REPRODUCTIVE BIOLOGY
Oothecae containing 50–100 eggs are deposited on woody stems and hatch after several weeks.

CONSERVATION STATUS
Not listed by the IUCN, but is probably threatened by habitat destruction due to overpopulation.

SIGNIFICANCE TO HUMANS
None known. ◆

Orchid mantid

Hymenopus coronatus

FAMILY
Hymenopodidae

TAXONOMY
Hymenopus coronatus Olivier, 1792, Java.

OTHER COMMON NAMES
English: Flower mantid.

PHYSICAL CHARACTERISTICS
Adults are white with pink patches on the head, the anterior and middle portions of the forewings, and the legs. Females are twice as long as males, averaging 2 in (5 cm) and 1 in (2.5 cm), respectively). The eyes are conical and rise above the dorsal edge of the head. Legs have leaflike projections.

DISTRIBUTION
Southeast Asia.

HABITAT
Primary and secondary rainforests.

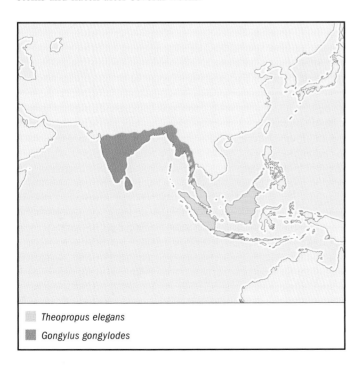

☐ *Theopropus elegans*
■ *Gongylus gongylodes*

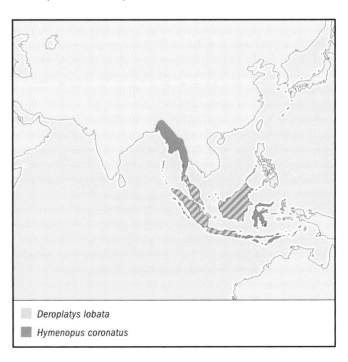

☐ *Deroplatys lobata*
■ *Hymenopus coronatus*

BEHAVIOR
Commonly found in or near flowers, awaiting prey. During courtship males tap their antennae and forelegs against the wings of the females, presumably to signal a willingness to mate.

FEEDING ECOLOGY AND DIET
Predators of small arthropods.

REPRODUCTIVE BIOLOGY
Long, narrow oothecae (2 in, or 5 cm) are laid on stems and branches of plants and shrubs. The first-stage larvae have red bodies and black heads, resembling ants. This ant mimicry is thought to protect them from predators.

CONSERVATION STATUS
Not threatened.

SIGNIFICANCE TO HUMANS
This is a popular species among hobbyists. ◆

No common name
Liturgusa charpentieri

FAMILY
Liturgusidae

TAXONOMY
Liturgusa charpentieri Giglio-Tos, 1927, Brazil.

OTHER COMMON NAMES
None known.

PHYSICAL CHARACTERISTICS
Mottled, lichen-colored elytra and legs camouflage this mantid well against tree trunks. The body is dorsoventrally flattened.

The head is held with the mouthparts facing forward (prognathous). Adult females are 2.5 in (6 cm) long; males, 1.6 in (4 cm).

DISTRIBUTION
Yucatán peninsula south through the Amazon basin.

HABITAT
Primary and secondary rainforests.

BEHAVIOR
Often found facing upside down on large tree trunks. When startled, they scamper around to the opposite side of the tree very quickly and then remain motionless to avoid detection.

FEEDING ECOLOGY AND DIET
Opportunistic carnivores of smaller arthropods.

REPRODUCTIVE BIOLOGY
Females attach oothecae inside bark fissures or under leaves. The exterior shell of the egg case is amber colored and opaque, allowing the eggs to be seen inside the case.

CONSERVATION STATUS
Not threatened.

SIGNIFICANCE TO HUMANS
None known. ◆

No common name
Choeradodis rhomboidea

FAMILY
Mantidae

TAXONOMY
Choeradodis rhomboidea Stoll, 1813, Surinam.

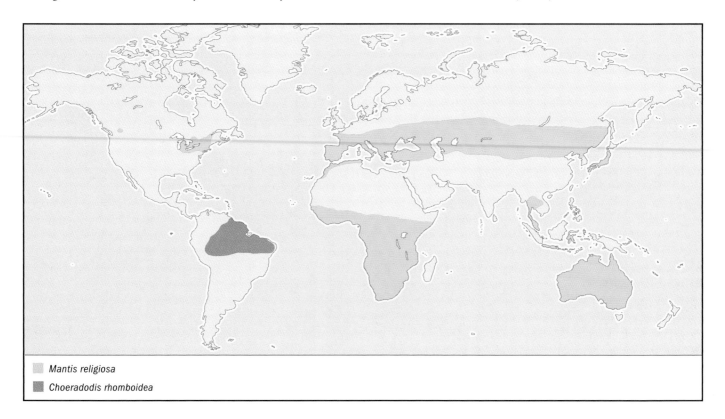

Mantis religiosa
Choeradodis rhomboidea

OTHER COMMON NAMES
None known.

PHYSICAL CHARACTERISTICS
Their uniformly leafy green color makes them difficult to find in the forest. The pronotum, or shield, is laterally flattened and leaflike. Males and females attain lengths up to 3 in (7.5 cm)

DISTRIBUTION
Amazon basin of South America.

HABITAT
Primary and secondary rainforests.

BEHAVIOR
Perch on leaves and remain motionless, completely blending into their verdant environment as they await passing prey.

FEEDING ECOLOGY AND DIET
Opportunistic feeders of smaller arthropods.

REPRODUCTIVE BIOLOGY
After mating, females deposit 50–100 eggs inside an ootheca attached to the underside of a leaf or branch.

CONSERVATION STATUS
Not threatened.

SIGNIFICANCE TO HUMANS
None known. ◆

Dead-leaf mantid
Deroplatys lobata

FAMILY
Mantidae

TAXONOMY
Deroplatys lobata Guérin-Méneville, 1838, type locality not known.

OTHER COMMON NAMES
None known.

PHYSICAL CHARACTERISTICS
These mantids are extraordinary mimics of dry, dead leaves. Their coloration is mottled light gray to dark brown. The leaflike pronotum is expanded laterally. Middle and hind legs have leaflike expansions, further adding to their camouflage. Adult males, at 2.5 in (6 cm) are approximately two-thirds the length of adult females, at 2.8 (7 cm).

DISTRIBUTION
Southeast Asia.

HABITAT
Primary and secondary rainforests.

BEHAVIOR
Dwells in leaf litter and shrubs. When threatened, these insects assume a startle posture by exposing their brightly colored forelegs as well as eyespots on the ventral sides of their forewings.

FEEDING ECOLOGY AND DIET
Carnivores of small arthropods.

REPRODUCTIVE BIOLOGY
Females lay oothecae on twigs. Hatchling mantids (50–100) emerge 30–50 days later.

CONSERVATION STATUS
Widespread, but habitat destruction threatens them. Not listed by the IUCN.

SIGNIFICANCE TO HUMANS
None known. ◆

European mantid
Mantis religiosa

FAMILY
Mantidae

TAXONOMY
Mantis religiosa Linnaeus, 1758, Africa.

OTHER COMMON NAMES
None known.

PHYSICAL CHARACTERISTICS
Varies from light green to brown in coloration, blending in with surrounding vegetation. Has a distinguishing bull's-eye spot on the inner forelegs. Adult males vary in length from 2–2.5 in (5–6 cm); females, from 2.5–3.2 in (6–8 cm).

DISTRIBUTION
Widely distributed in southern Europe, Africa, temperate Asia, Australia, the northeastern United States, and Canada.

HABITAT
Open fields and meadows.

BEHAVIOR
These mantids fly well; then fly toward bright lights at night.

FEEDING ECOLOGY AND DIET
Predators of small arthropods.

REPRODUCTIVE BIOLOGY
Egg cases (50–100 eggs each) are laid in autumn, primarily on low grass stems; they also can be found on rocks or buildings. Larvae hatch in spring.

CONSERVATION STATUS
Very widespread. Introductions into the United States have expanded its range. Not listed by the IUCN.

SIGNIFICANCE TO HUMANS
None known. ◆

Chinese mantid
Tenodera aridifolia sinensis

FAMILY
Mantidae

TAXONOMY
Tenodera aridifolia sinensis Saussure, 1871, type locality not known.

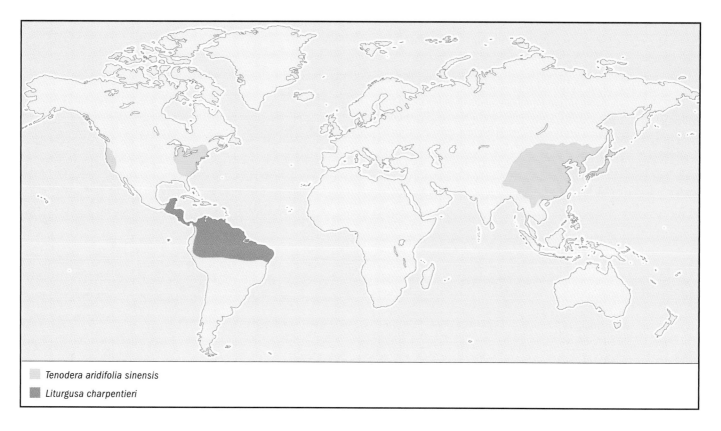

Tenodera aridifolia sinensis

Liturgusa charpentieri

OTHER COMMON NAMES
None known.

PHYSICAL CHARACTERISTICS
One of the largest mantids, females can attain lengths of 4 in (10 cm) or more. Chinese mantids are marbled green, brown, and gray, with a distinct pale green border on the anterior edge of the first pair of wings.

DISTRIBUTION
Found in temperate eastern Asia, the eastern United States, and California. Deliberately introduced to the United States in 1896.

HABITAT
Meadows and open fields.

BEHAVIOR
Seen on herbaceous plants and woody shrubs. Sometimes found adjacent to flowers, awaiting potential prey.

FEEDING ECOLOGY AND DIET
Carnivores. Devours any arthropod it can ensnare.

REPRODUCTIVE BIOLOGY
Females produce 100–200 eggs inside a spongy ootheca and attach it almost anywhere, including leaves, branches, buildings, and vehicles. Eggs overwinter in the ootheca, and the larvae hatch in spring.

CONSERVATION STATUS
Widespread in its range. Global warming eventually may restrict this species to a more limited temperate range. Not listed by the IUCN.

SIGNIFICANCE TO HUMANS
Often kept as pets. ◆

Boxer mantid
Theopropus elegans

FAMILY
Mantidae

TAXONOMY
Theopropus elegans Westwood, 1832, type locality not known.

OTHER COMMON NAMES
English: Banded mantid.

PHYSICAL CHARACTERISTICS
Hatchling larvae are red-and-black ant mimics until their first molt, after which they become green and white. Adult females are 1.6–2 in (4–5 cm) long, and males are 0.8–1.2 in (2–3 cm). Both sexes are spotted green and white with a large white transverse strip on the forewings. Hind wings are bright orange.

DISTRIBUTION
Southeast Asia.

HABITAT
Primary and secondary rainforests.

BEHAVIOR
Hides among flowers to catch prey. When encountering another mantid of the same species, they thrust out their forelegs in an apparent boxing motion—hence their common name.

FEEDING ECOLOGY AND DIET
Captures small arthropods for consumption.

REPRODUCTIVE BIOLOGY
Females lay oothecae on small branches or on the undersides of leaves. Larvae (30–75) hatch 30–50 days later.

CONSERVATION STATUS

Not listed by the IUCN, but probably threatened by habitat destruction.

SIGNIFICANCE TO HUMANS

None known. ◆

Resources

Books

Ehrmann, Reinhard. *Mantodea: Gottesanbeterinnen der Welt.* Münster, Germany: NTV, 2002.

Giglio-Tos, E. *Das Tierreich* Vol. 50, *Mantidae.* Berlin: Walter de Gruyter, 1927.

Helfer, Jacques R. *How to Know the Grasshoppers, Crickets, Cockroaches and Their Allies.* New York: Dover Publications, 1987.

Preston-Mafham, K. *Grasshoppers and Mantids of the World.* London: Blanford Press, 1990.

Prete, F. R., H. Wells, P. H. Wells, and L.E. Hurd, eds. *The Praying Mantids.* Baltimore: Johns Hopkins University Press, 1999.

Cynthia L. Mazer, MS

Grylloblattodea
(*Rock-crawlers*)

Class Insecta
Order Grylloblattodea
Number of families 1

Photo: Grylloblattidae are found in cold regions of China, Japan, Korea, Siberia, and western North America. They have been known to feed on insects found frozen in the snow. (Photo by Roger D. Akre/ M. T. James Entomolgical Collection.)

Evolution and systematics

About 300 species of fossil Grylloblattodea are known from the late Carboniferous to the early Cretaceous and were among the most abundant and diverse insects during the Permian. The order Grylloblattodea is considered ancestral to all other related orders (stoneflies, webspinners, and earwigs). There is only one living family, Grylloblattidae, which belongs to the suborder Grylloblattina.

Physical characteristics

The extant family Grylloblattidae is characterized by an elongate, subcylindrical, slightly flattened body. The head is short and prognathous, with small or absent eyes. The antennae are filiform, with 28–50 segments. All three thoracic segments are similar—more or less flattened. The legs are similar in appearance, with five segmented tarsi, and are adapted for running. Wings are entirely absent. The abdomen has 10 segments, and the cerci are long, with seven to 12 segments. The female has a long ovipositor consisting of six valves. The asymmetrical genitalia of the male are located in the ninth abdominal segment. Adult rock-crawlers range from 0.6 to 1.4 in (15–35 mm) in length. Adults are brown, with light brown legs and ventral surface of the abdomen. Nymphs are ivory white or yellowish; sometimes a completely black (melanistic) form occurs.

Distribution

The extant family Grylloblattidae includes five genera and 27 species from Siberia, northeastern China, Korea, Japan, United States, and Canada.

Habitat

Modern rock-crawlers live on and in soil, in caves, beneath stones, and in crevices of mountainous regions. Grilloblattidae also occur in mixed forests and near the snowfields at elevations of 656–10,499 ft (200–3,200 m). Rock-crawlers are adapted to cool-temperate habitats. For example, the optimum temperature for *Grylloblatta campodeiformis* is 38.7°F (3.7°C) to 59.9°F (15.5°C). Individuals increase activity until death occurs at 82°F (27.8°C). Cooling from 38.7°F (3.7°C) to 21.9°F (−5.6°C) results in decreased activity; prostration and death occur at 20.8°F (−6.2°C).

Behavior

Ice-crawlers typically are found singly or in sexual pairs. Occasionally, the female may suddenly eat the male. Individual populations occupy areas from 984 ft (300 m) to approximately 3,280 ft (1 km) in diameter. Rock-crawlers avoid light and are nocturnal. Because they are small, wingless insects, grylloblattids are limited in their migratory movements. They may be able to migrate for a few hundred meters at most; in general, populations occupy areas from 984 to 3,280 ft (300–1,000 m) in diameter.

Feeding ecology and diet

Both adults and nymphs are carnivorous and consume the soft tissues of insects and spiders. The nymphs also eat parts of plants or other organic matter in the soil. Grylloblattids first detect prey with their maxillae and bite them with their mandibles. The North American *Grylloblatta* moves at night to the surface of the snow to forage. Nymphs can be active without food for three to six months.

Reproductive biology

Courtship and mating of grylloblattids take place under stones. The male and female face each other, touching each other with their antennae. The male chases the female and grasps her right hind leg or cercus with his mandibles. He then seizes the posterior edge of the female's prothorax with his mandibles. The male mounts the female and twists his abdomen underneath hers to copulate. Owing to the asymmetry of the male external genitalia, the male always positions himself to the right of the female during copulation. Copulation lasts from about 30 minutes to four hours. In *Galloisiana nipponensis*, oviposition (the laying of eggs) occurs 10–50 days after copulation, throughout the year. The females lay 60–150 eggs in or on the soil, under stones and fallen leaves, or in decayed wood. The eggs are black and oval and measure 0.06–0.1 in (1.6–2.5 mm) in length. The first instar nymph hatches about 150 days after oviposition, at 50–53.6°F (10–12°C). Viable eggs may remain dormant from one to three years. There are eight or so nymphal instars, three during the first year, the next four at one-year intervals, and the last about six weeks after the preceding one. Both nymphs and adults are observed throughout the year.

Conservation status

No species of grylloblattids are listed by the IUCN.

Significance to humans

Rock-crawlers are important objects of study in terms of the physiological adaptation of insects to low temperatures and in the study of Pleistocene zoogeography.

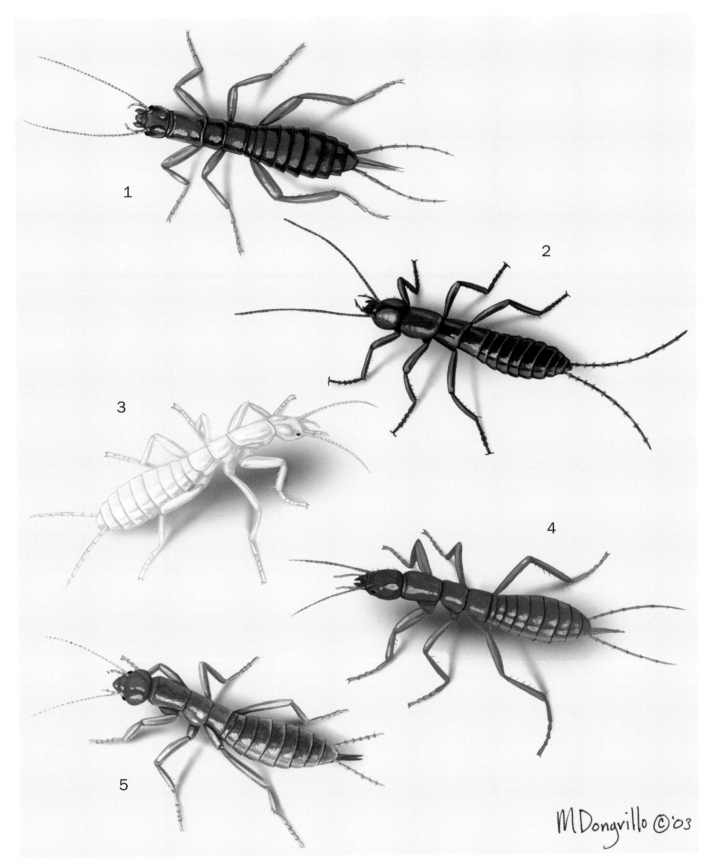

1. Pravdin's rock-crawler (*Grylloblattella pravdini*); 2. Biryong rock-crawler (*Namkungia biryongensis*); 3. Japanese rock-crawler (*Galloisiana nipponensis*); 4. Djakonov's rock-crawler (*Grylloblattina djakonovi*); 5. Northern rock-crawler (*Grylloblatta campodeiformis*). (Illustration by Marguette Dongvillo)

Species accounts

Japanese rock-crawler
Galloisiana nipponensis

FAMILY
Grylloblattidae

TAXONOMY
Galloisia nipponensis Caudell and King, 1924, Chuzenji near Nikko, Honshu, Japan.

OTHER COMMON NAMES
None known.

PHYSICAL CHARACTERISTICS
Body length of adult is 0.73–.0.87 in (18.5–22 mm).

DISTRIBUTION
Numerous local populations throughout Honshu and Shikoku, Japan.

HABITAT
This species lives in mixed forests near mountain streams at elevations of 656–3,281 ft (200–1,000 m) but sometimes occupies limestone caves.

BEHAVIOR
Nocturnal. Typically found singly or in sexual pairs.

FEEDING ECOLOGY AND DIET
Carnivorous. Nymphs also eat plant parts.

REPRODUCTIVE BIOLOGY
After copulation, females lay 60–150 black eggs, which hatch after about 150 days.

CONSERVATION STATUS
Not listed by IUCN.

SIGNIFICANCE TO HUMANS
Studied because of their adaptation to low temperatures. ◆

Northern rock-crawler
Grylloblatta campodeiformis

FAMILY
Grylloblattidae

TAXONOMY
Grylloblatta campodeiformis Walker, 1914, Sulphur Mountain, Banff, Canada.

OTHER COMMON NAMES
French: Grylloblatte du nord.

PHYSICAL CHARACTERISTICS
Body length of adult is 0.98–1.06 in (25–27 mm).

DISTRIBUTION
Divided into three subspecies. *G. c. campodeiformis* is known from southeastern British Columbia and southwestern Alberta, Canada, and northern Idaho, western Washington, and western and southern Montana in the United States. *G. c. athapaska* is distributed in southwestern British Columbia and *G. c. nahanni* in northeastern British Columbia.

HABITAT
This species usually is found near the timberline on mountains, along the margins of glacial bogs, and buried up to 3.3 ft (1 m) in rock scree.

BEHAVIOR
Nocturnal and negatively phototropic. Moves at night to snow surface for foraging.

Galloisiana nipponensis
Namkungia biryongensis
Grylloblattina djakonovi

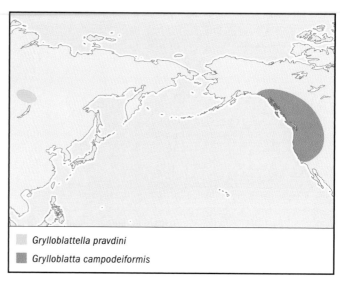

Grylloblattella pravdini
Grylloblatta campodeiformis

FEEDING ECOLOGY AND DIET
Scavenges dead insects and spiders.

REPRODUCTIVE BIOLOGY
The total cycle has been calculated to take about seven years.

CONSERVATION STATUS
Not listed by IUCN.

SIGNIFICANCE TO HUMANS
Important for scientific study for its adaptation to surviving at low temperatures. This species is also the emblem of the Entomological Society of Canada. ◆

Pravdin's rock-crawler
Grylloblattella pravdini

FAMILY
Grylloblattidae

TAXONOMY
Galloisiana pravdini Storozhenko and Oliger, 1984, Teletskoe Lake, Altai, Russia.

OTHER COMMON NAMES
None known.

PHYSICAL CHARACTERISTICS
Body length of adult is 0.64–0.69 in (16.3–17.5 mm).

DISTRIBUTION
The Altai mountains of Russia.

HABITAT
This species is found beneath stones in mixed forest at elevations of 1,312–2,297 ft (400–700 m).

BEHAVIOR
Not known.

FEEDING ECOLOGY AND DIET
Not known.

REPRODUCTIVE BIOLOGY
Not known.

CONSERVATION STATUS
Not listed by IUCN.

SIGNIFICANCE TO HUMANS
None known. ◆

Djakonov's rock-crawler
Grylloblattina djakonovi

FAMILY
Grylloblattidae

TAXONOMY
Grylloblattina djakonovi Bey-Bienko, 1951, Petrov Island, Primorye, Russia. Two subspecies are recognized.

OTHER COMMON NAMES
None known.

PHYSICAL CHARACTERISTICS
Body length of adult is 0.77–0.89 in (19.5–22.5 mm).

DISTRIBUTION
Southwestern part of the Sikhote-Alin mountain range and the northwestern part of the Chanbaishan mountains, Russia. The two subspecies are found on opposite sides of the Partizanskaya River: *G. d. djakonovi* is distributed in the Partizanskii Range eastward to the Partizanskaya River, and *G. d. kurentzovi* is found in the Livadiiskii Range westward to the Partizanskaya River.

HABITAT
Usually at elevations of 656–3,281 ft (200–1,000 m), but the population from Petrov Island occurs under stones in mixed forest at about 32.8–65.6 ft (10–20 m) from the seaside.

BEHAVIOR
Nocturnal. Typically found singly or in sexual pairs.

FEEDING ECOLOGY AND DIET
Carnivorous. Nymphs also eat plant parts.

REPRODUCTIVE BIOLOGY
After copulation, females lay 60–150 black eggs, which hatch after about 150 days.

CONSERVATION STATUS
Not listed by IUCN.

SIGNIFICANCE TO HUMANS
None known. ◆

Biryong rock-crawler
Namkungia biryongensis

FAMILY
Grylloblattidae

TAXONOMY
Namkungia biryongensis Namkung, 1974, Yongtanri, Jeongseonmyeon, Gangwon Province, Korea.

OTHER COMMON NAMES
Korean: Biryong-galleuwa beolle.

PHYSICAL CHARACTERISTICS
The largest living species of rock-crawler, with an adult body length of 1.34–1.38 in (34–35 mm).

DISTRIBUTION
Gangwon Province of Korea, 311 mi (500 km) northeast of Seoul.

HABITAT
This species is found under gravel about 65.6–98.4 ft (20–30 m) inside the entrances of the limestone caves Biryong donggul and Baekryeong dong-gul.

BEHAVIOR
Not known.

FEEDING ECOLOGY AND DIET
Not known.

REPRODUCTIVE BIOLOGY
Not known.

CONSERVATION STATUS
Not listed by IUCN.

SIGNIFICANCE TO HUMANS
None known. ◆

Resources

Books

Ando, H., and T. Nagashima. "A Preliminary Note on the Embryogenesis of *Galloisiana nipponensis* (Caudell et King)." In *Biology of the Notoptera*, edited by Hiroshi Ando. Nagano, Japan: Kashiyo-Insatsu, 1982.

Nagashima, T., H. Ando, and G. Fukushma. "Life History of *Galloisiana nipponensis* (Caudell et King)." In *Biology of the Notoptera*, edited by Hiroshi Ando. Nagano, Japan: Kashiyo-Insatsu, 1982.

Rentz, D. C. F. "A Review of the Systematics, Distribution and Bionomics of the North American Grylloblattidae." In *Biology of the Notoptera*, edited by Hiroshi Ando. Nagano, Japan: Kashiyo-Insatsu, 1982.

Storozhenko, S. Y. "Fossil History and Phylogeny of Orthopteroid Insects." In *The Bionomics of Grasshoppers, Katydids, and Their Kin*, edited by S. K. Gangwere, M. C. Muralirangan, and Meera Muralirangan. New York: Cambridge University Press, 1997.

———. *Systematics, Phylogeny and Evolution of the Grylloblattidan Insects (Insecta: Grylloblattida)*. Valdivostok, Russia: Dalnauka, 1998. (In Russian.)

———. "Order Grylloblattida Walker, 1914." In *History of Insects*, edited by A. P. Rasnitsyn, D. L. J. Quicke. Boston: Kluwer Academic Publishers, 2002.

Vickery, V. R., and D. K. M. Kevan. *A Monograph of the Orthopteroid Insects of Canada and Adjacent Regions*. Vol. 1. Ste. Anne de Bellevue, Canada: Lyman Entomological Museum and Research Laboratory, 1983.

Visscher, S. N., M. Francis, P. Martinson., and S. Baril. "Laboratory Studies on *Grylloblatta campodeiformis* Walker." In *Biology of the Notoptera*, edited by Hiroshi Ando. Nagano, Japan: Kashiyo-Insatsu, 1982.

Periodicals

Rasnitsyn, A. P. "Grylloblattide Are the Living Members of the Order Protoblattodea (Insecta)." *Doklady Biological Sciences* 228 (1976): 273–275.

Storozhenko, S. Y., and J. K. Park. "A New Genus of the Ice Crawlers (Grylloblattida: Grylloblattidae) from Korea." *Far Eastern Entomologist* 114 (2002): 18–20.

Vrsansky, P., S Y. Storozhenko, C. C. Labandeira, and P. Ihringova. "*Galloisiana olgae* sp. nov. (Grylloblattodea: Grylloblattidae) and the Paleobiology of a Relict Order of Insects." *Annals of the Entomological Society of America* 94, no. 2 (2001): 179–184.

S. Y. Storozhenko

Dermaptera
(*Earwigs*)

Class Insecta
Order Dermaptera
Number of families 28

Photo: Earwigs are harmless to humans and do not, despite their name, enter the ear. (Photo by ©J. H. Robinson/Photo Researchers, Inc. Reproduced by permission.)

Evolution and systematics

The oldest-known fossils of Dermaptera comprise about 70 specimens from the Jurassic, about 208 million years ago. Earwigs are considered orthopteroid insects, closely related to Orthoptera and Phasmatodea, and they are divided into four suborders. Suborder Archidermaptera, represented by only 10 fossil species from the Jurassic, had segmented adult cerci and four to five segmented tarsi. Forficulina, with about 1,800 species in 180 genera, is the largest suborder. Adult cerci are unsegmented and forceps-like; larval cerci also are unsegmented, except in two primitive groups. Arixeniina comprise five species in two genera, and Hemimerina consist of 10 species in one genus; they are wingless and have filamentous cerci. According to some phylogenetic studies, Archidermaptera constitutes the sister group of the remaining suborders. No fossil Hemimerina and Arixeniina earwigs are known. Some authors consider Hemimerina to be a separate order.

Physical characteristics

Dermaptera are brown or black, sometimes with a light brown or yellow pattern; a few are metallic green. The head is prognathous with chewing mouthparts. The antennae are long, thin, and filiform; ocelli are lacking, and the compound eyes are well developed, except in the blind Hemimerina and almost blind Arixeniina. The thorax bears two pairs of wings, of which the first, called "tegmina," is small and leathery, giving origin to the ordinal name (*derma*, meaning "skin," and *ptera*, meaning "wings"). Tegmina are short, covering the top of only the first segments of the abdomen and leaving the posterior part of the abdomen exposed. Hemimerina, Arixeniina, and some Forficulina are secondarily wingless; in the remaining Forfi-

A female earwig (*Forficula auricularia*). (Photo by John Markham. Bruce Coleman, Inc. Reproduced by permission.)

A female earwig brooding her young. (Photo by R. N. Mariscal. Bruce Coleman, Inc. Reproduced by permission.)

culina, the second pair of wings is membranous, large, almost semicircular, and complexly folded under the tegmina at rest. The abdomen is highly movable, with pair of unsegmented, usually pincer-like cerci at the posterior end (filamentous in Hemimerina and Arixeniina). Cerci usually are dimorphic: straight in females and curved or asymmetrical in males. Forficulina earwigs are elongate and slender, reaching 0.16–3.2 in (4–78 mm) in length (including the cerci). Hemimerina are about 0.4 in (10 mm) long, excluding cerci, and they have short, stout legs and a streamlined, smooth body for rapid movement through the fur of their hosts. Arixeniina have long and slender legs. Larvae of earwigs resemble adults except for the absence of wings; larvae of wingless species often are difficult to distinguish from adults. Larval cerci are simple and almost straight, and they resemble those of the female.

Distribution

Dermaptera are cosmopolitan (except polar regions), with the greatest diversity in the tropics and subtropics.

Habitat

Forficulina earwigs frequent humid crevices of all kinds; they can be found under bark, between leaves, and under stones. Hemimerina live on the bodies of giant rats in trop-ical Africa and Arixeniina live on bats in the Malayan-Philippine region.

Behavior

Earwigs prefer to hide in dark crevices during the day and become active at night. Cerci are used to open the wings, for grooming, and for defense. Some earwigs have defensive glands on the second or third abdominal segment that release a foul-smelling liquid, and they can squirt this fluid up to 4 in (100 mm).

Feeding ecology and diet

Most earwigs are omnivorous, but there are some species that are predominantly herbivorous, predacious (on chinch bugs, mole crickets, mites, scales, aphids, and caterpillars), or scavengers. Hemimerina feed on scurf and fungi growing on the skin of giant rats. Arixeniina feed on the skin-gland secretions of bats and occasionally on dead insects.

Reproductive biology

Hemimerina and Arixeniina are viviparous, and Forficulina are generally oviparous. In temperate climates Forficulina adults overwinter in soil, and in spring females, sometimes assisted by males, build a brood chamber or nest in the ground, within rotting vegetation, or under a rock. After mating and laying eggs, females chase males out of the nest. Females tend their eggs, turning them around and licking them to prevent the growth of fungi until they hatch. Females then forage for food, which they feed to their young larvae. Larvae stay in the nest until the second or third instar, completing up to four or five instars in all. If larvae do not leave the burrow after one or two molts, mothers may eat them.

Conservation status

Of the more than 1,800 species of earwigs known, only one, the St. Helena earwig (*Labidura herculeana*) is on the IUCN Red List; it is categorized as Endangered.

Significance to humans

The name "earwig" derives from the mistaken belief that this insect enters the ear and bores into the brains of sleeping people. Thus their common name in different languages often refers to the ear (Danish, Dutch, English, French, German, Russian, and Swedish) or to the forceps (Italian, Finnish, Portuguese, and Spanish). Most earwigs have little or no economic importance. They may do some damage in gardens by feeding on ornamental plants, but they also may be beneficial by eating other insects. A few species, if abundant, may damage blossoms of ornamental plants by chewing the stamens or petals.

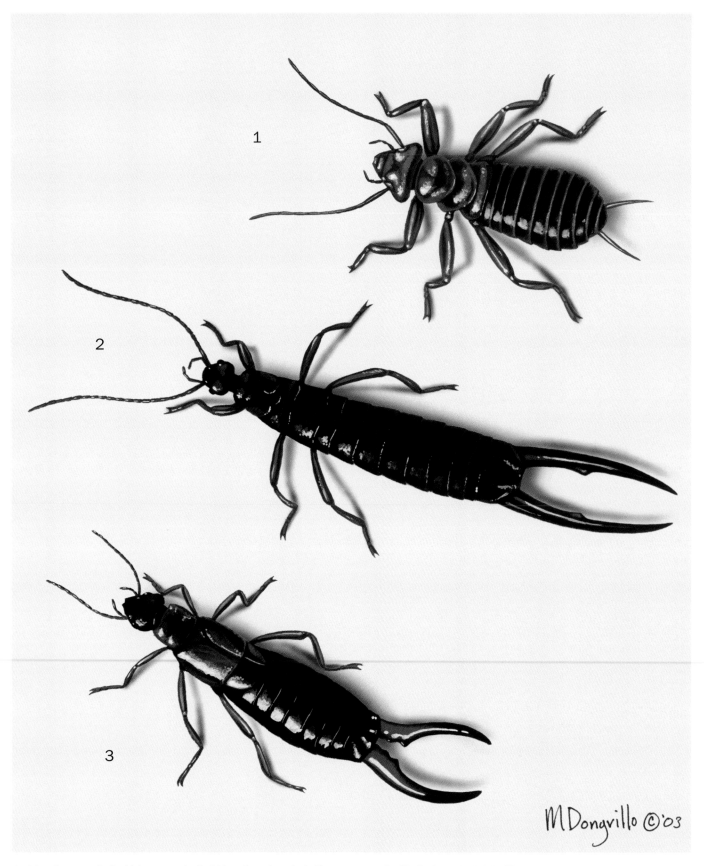

1. *Arixenia esau*; 2. St. Helena earwig (*Labidura herculeana*); 3. European earwig (*Forficula auricularia*). (Illustration by Marguette Dongvillo)

Species accounts

No common name
Arixenia esau

FAMILY
Arixeniidae

TAXONOMY
Arixenia esau Jordan, 1909.

OTHER COMMON NAMES
None known.

PHYSICAL CHARACTERISTICS
Robust, hairy, apterous, and almost blind, with long legs and rodlike cerci.

DISTRIBUTION
Malaysia, Indonesia, and the Philippines.

HABITAT
On the naked bat *Cheiromeles torquatus* (Molossidae) and its roosts.

BEHAVIOR
Obligatory associated with host; in other words, unable to survive away from the naked bat.

FEEDING ECOLOGY AND DIET
Skin-gland secretions of their hosts and occasionally dead insects.

REPRODUCTIVE BIOLOGY
Viviparous.

CONSERVATION STATUS
The host bat *Cheiromeles torquatus* is listed as Lower Risk/Near Threatened by the IUCN; maintenance of adequate populations of the host bat are necessary to preserve this earwig.

SIGNIFICANCE TO HUMANS
The world's first earwig reserve was created in Niah Great Cave, Sarawak, to protect this rare species. ◆

European earwig
Forficula auricularia

FAMILY
Forficulidae

TAXONOMY
Forficula auricularia Linnaeus, 1758, Europe.

OTHER COMMON NAMES
English: Common earwig; French: Perce-oreilles; German: Gemeiner Ohrwurm; Spanish: Tijereta europea; Italian: Forficola.

Arixenia esau
Labidura herculeana
Forficula auricularia

PHYSICAL CHARACTERISTICS

Reddish-brown. Tegmina and legs are yellow brown. Adults are winged, with a length of 0.47–0.59 in (12–15 mm). Male forceps are broadened and crenulated basally and 0.16–0.31 in (4–8 mm) long. Female forceps is 0.12 in (3 mm). Larvae look like adults but are wingless and smaller.

DISTRIBUTION

Originally known from the Palearctic region (Europe, western Asia, and North Africa), the European earwig has been introduced into East Africa, North America, the East Indies, Australia, New Zealand, Chile, and Argentina.

HABITAT

It hides among petals or leaves of garden plants or inside fruit, shrubbery, fences, woodpiles, bases of trees, and behind loose boards on buildings.

BEHAVIOR

Nocturnal, hiding during the day and roaming at night to find food and water.

FEEDING ECOLOGY AND DIET

Polyphagous, feeding on plants, ripe fruit, lichens, fungi, and other insects.

REPRODUCTIVE BIOLOGY

Clusters of 50–90 eggs are laid in chambers in moist soil from November to January; a second batch containing fewer eggs is deposited in March or April. Egg development takes from 10 to 90 days, depending on the temperature. The larval stage lasts 40–50 days, including four instars. Females guard the eggs and first instar larvae, excluding males from the nest. Univoltine (having one generation per year); all stages are present except in autumn, when there are only adults.

CONSERVATION STATUS

Not listed by the IUCN.

SIGNIFICANCE TO HUMANS

Not considered of great economic importance in Europe, but a serious pest in parts of the United States on flower crops, butterfly bush, hollyhock, lettuce, strawberry, celery, potatoes, sweet corn, roses, seedling beans and beets, and tender grass shoots and roots. They are sometimes beneficial, feeding on pests such as aphids, even on the above-mentioned crops. ◆

St. Helena earwig

Labidura herculeana

FAMILY

Labiduridae

TAXONOMY

Labidura herculeana Fabricius, 1798; St. Helena.

OTHER COMMON NAMES

None known.

PHYSICAL CHARACTERISTICS

Largest earwig, with a body length of 1.44–21.6 in (36–54 mm) and forceps of 0.6–0.96 in (15–24 mm), reaching up to 31.2 in (78 mm). Females are shorter than males. Black body, reddish legs, short tegmina, and no hind wings.

DISTRIBUTION

Horse Point Plain in the extreme northeast of the mid-Atlantic island of St. Helena.

HABITAT

Dry and barren, with stony soil, bushes, and tufts of grass.

BEHAVIOR

Living specimens have been found under stones or near burrows in the soil. They are nocturnal and active during summer rains.

FEEDING ECOLOGY AND DIET

Nothing is known.

REPRODUCTIVE BIOLOGY

Mating takes place between December and February, and females with eggs have been seen in March.

CONSERVATION STATUS

Rediscovered in 1965 for the first time since the original description. Dead remains seem to indicate a shrinking range, due to unknown causes. No living specimens have been found recently, and the species could be extinct. Study is needed to assess the status of the species. If it is extant, a detailed study on its ecology is necessary to design an appropriate protection and management program.

SIGNIFICANCE TO HUMANS

This species is of scientific interest because it is the largest earwig in the world. ◆

Resources

Books

Carpenter, F. M. "Superclass Hexapoda." In *Treatise on Invertebrate Paleontology*, Part R Arthropoda 4, edited by R. L. Kaesler. Boulder, CO: Geological Society of America, 1992.

Chopard, L. "Ordre de Dermaptères." In *Traite de Zoologie.* Vol. 9, edited by P. P. Grassé. Paris: Masson and Cie, 1949.

Rentz, D. C. F., and D. K. M. Kevan. "Dermaptera." In *The Insects of Australia (CSIRO)*, edited by D. F. Waterhouse, P. B. Carne, and I. D. Naumann. Ithaca, NY: Cornell University Press, 1991.

Sakai, S. *Dermapterorum Catalogus Praeliminaris.* 4 parts. Tokyo: Department of Biology and Chemistry, Daito Bunka University, 1970–1973.

Sakai, S. "A New Proposed Classification of the Dermaptera with Special Reference to the Check List of the Dermaptera of the World." Addition and Errata. *Dermapterorum Catalogus* 14 (1982): 1–108.

Periodicals

Giles, E. T. "The Comparative External Morphology and Affinities of the Dermaptera." *Transactions of the Royal Entomological Society of London* 115 (1963): 95–164.

Haas, F. "The Phylogeny of the Forficulina, a Suborder of the Dermaptera." *Systematic Entomology* 20 (1995): 85–98.

Natalia von Ellenrieder, PhD

Orthoptera
(Grasshoppers, crickets, and katydids)

Class Insecta
Order Orthoptera
Number of families 43

Photo: An oak brush cricket (*Meconema thalassinum*) leaping. (Photo by Kim Taylor. Bruce Coleman, Inc. Reproduced by permission.)

Evolution and systematics

Until the 1950s and 1960s the definition of the Orthoptera was very inclusive, and many entomologists placed within this order such groups as cockroaches (Blattodea), preying mantids (Mantodea), walking sticks (Phasmodea), and several others. Those representatives of the order that possessed jumping hind legs, such as crickets and grasshoppers, were termed Saltatoria. Currently, most taxonomists consider the order Orthoptera a monophyletic lineage, restricted in its composition to grasshoppers, crickets, katydids, and their closest relatives. Some distinguish between Orthoptera *sensu stricto*, an order that includes only short-horned grasshoppers and their relatives, and Gryllodea, an order that includes only long-horned grasshoppers, such as crickets and katydids. A division of the Orthoptera into 14 separate orders of insects was also suggested, but this classification scheme never found acceptance among entomologists. For the purpose of this chapter, a classification system that recognizes the order Orthoptera as comprising the suborders Ensifera (long-horned grasshoppers) and Caelifera (short-horned grasshoppers) is adopted. The sister group to the Orthoptera is still not clear; some authors consider them to be most closely related to Phasmodea, but recent molecular data indicate that they may be a sister group to a clade consisting of Phasmodea and Embiidina (web spinners).

The Orthoptera is one of the oldest lineages of insects; the oldest fossils attributed to this order are from the Carboniferous period. The two currently recognized suborders of Orthoptera, Ensifera and Caelifera, probably separated by the late Carboniferous. Most modern families of Ensifera appeared between the early Jurassic and the early Triassic periods. The oldest, still extant family of Ensifera, the Prophalangopsidae, appeared in the early Jurassic. The oldest extant family of Caelifera, the Eumastacidae, appeared in the middle Jurassic, followed by the Tetrigidae and the Tridactylidae at the beginning of the Cretaceous.

The suborder Ensifera (long-horned grasshoppers) is divided into 6 superfamilies and 21 families, with approximately 1,900 genera and 11,000 described species. The largest superfamily of this suborder, Tettigonioidea (katydids or bushcrickets), includes over 1,000 genera and over 7,000 known species; Grylloidea (crickets) includes over 500 genera and 3,500 described species. Other subfamilies of long-horned grasshoppers are the Stenopelmatoidea (Jerusalem and camel crickets), the Gryllotalpoidea (mole crickets), the Mogoplistoidea (scale crickets), and the Hagloidea (grigs or humpwinged crickets).

The Caelifera (short-horned grasshoppers) includes 8 superfamilies, 22 families, nearly 2,400 genera, and over 10,400 described species. The Acridoidea (true grasshoppers and locusts) is the largest superfamily of Caelifera, divided into over 1,600 genera and about 7,200 described species. Other superfamilies of this suborder are the Pyrgomorphoidea (lubber and bush grasshoppers), the Trigonopterygoidea, the Tanaoceroidea (desert grasshoppers), the Eumastacoidea (monkey grasshoppers), the Pneumoroidea (bladder grasshoppers), the Tetrigoidea (grouse or pygmy grasshoppers), and the Tridactyloidea (pygmy mole crickets and sandgropers).

Rhinoceros katydid (*Copiphora rhinoceros*) threat display. (Photo by Michael Fogeden. Bruce Coleman, Inc. Reproduced by permission.)

Physical characteristics

Most species of the Orthoptera are large- or medium-sized insects. Body lengths of less than 0.4 in (10 mm) are uncommon, while many exceed 2 in (50 mm) in length, with some having bodies over 3.9 in (100 mm) long and a wingspan of 7.9 in (200 mm) or more. The smallest Orthoptera are ant-associated crickets (*Myrmecophilus* and other genera), whose body length rarely exceeds 0.08 in (2 mm); the largest are katydids of the genera *Phyllophora* and *Macrolyristes*. The heaviest of all Orthoptera (and also the heaviest living insect) is the New Zealand giant weta (*Deinacrida heteracantha*), with a recorded body weight of 0.16 lb (71 g).

Orthopterans are hemimetabolous insects, with larvae resembling adult forms in their general appearance but lacking fully developed wings and reproductive organs. The overall body shape varies dramatically depending on the lifestyle of the species. Grass inhabitants, such as those of the genus *Lepacritis*, generally tend to have slender, stick- or bladelike body shapes, while arboreal species, such as *Steirodon careovirgulatum*, are often leaf-shaped. Several groups of desert grasshoppers are perfect mimics of pebbles and moss, and lichen mimicry is common among species inhabiting high-elevation tropical cloud forests, such as *Championica montana*.

The mouthparts of orthopterans are of the chewing/biting type. The head is hypognathous (mouthparts pointing down), rarely prognathous (mouthparts pointing forward); the antennae are usually long, threadlike, consisting of fewer than ten to several hundred articles. The pronotum, the part of the body immediately behind the head, is usually large, often shieldlike, and in extreme cases covers a large part (as in many katydids) or the entire body of the insect (as in pygmy grasshoppers). The front and middle legs are cursorial, or adapted for walking, yet in some cases the front pair of legs may be modified for digging (as in mole crickets, pygmy mole crickets, sandgropers) or both the front and middle pairs may be modified for grasping (as in predatory katydids). In some orthopterans (most katydids and crickets) the front legs have tibial auditory organs (the ear). The hind legs of most orthopterans are saltatorial, or modified for leaping, with large, muscular femora and long, slender tibiae. Some grasshoppers can perform repeated leaps of 8.5 ft (2.6 m) without any obvious signs of fatigue. This is possible primarily because of the presence of the protein resilin in their back legs. Resilin has superb elastic properties, with a 97% efficiency in returning stored energy. This allows for an explosive release of energy that catapults the insect, a task impossible with muscle power alone. Certain groups of orthopterans, especially those leading a subterranean life, lost their ability to jump, and their hind legs resemble typical cursorial legs.

The wings of orthopterans are either fully developed or reduced to various degrees. Wing polymorphism, or the occurrence of individuals with well-developed and reduced wings within the same species, is not uncommon. The forewings are somewhat thickened, forming leathery tegmina. In most katydids and crickets, parts of the tegmina are modified for stridulation. The hindwings, when present, are fanlike, folded under the first pair in the resting position. The hindwings are often longer than the tegmina and protrude behind their apices. The wing buds of larval stages are always positioned in such a way that the second pair of wings overlaps the first, whereas in adult individuals of micro- and brachypterous species, the first pair of wings always overlaps the second, despite their nymphal appearance. The base of the abdomen in grasshoppers has lateral auditory organs known as abdominal tympana. Most female orthopterans have a prominent ovipositor at the end of the abdomen, derived from the eight and ninth abdominal segments. Katydids and crickets usually have well-developed ovipositors that are sword-, sickle-, or needle-shaped, whereas female grasshoppers and their relatives usually lack a long, external ovipositor.

Body coloration of species of the Orthoptera varies greatly, usually being cryptic, thus resembling the species' immediate surroundings. Arboreal forms are mostly green, often exhibiting a remarkable similarity to leaves, both fresh and those in various states of decomposition. Grass-inhabiting species tend to be either brown or yellow, whereas grasshoppers found in sandy habitats have the coloration of the substrate. Aposematic, or warning coloration, is rare, most often found in toxic or distasteful grasshoppers of the families Pyrgomorphidae and Romaleidae.

Distribution

Orthopteran species occur in nearly all parts of the world and in almost all habitats where insects are found. They are absent only from the polar regions of the globe, the oceans, and extreme alpine zones. The highest species diversity is present in the tropical and subtropical areas, although dry temperate zones, such as southern Europe, the southwestern United States, or Western Australia show remarkably high species diversity of the Orthoptera. Many taxa of this order, including entire families or even superfamilies, show a high degree of endemicity. Two superfamilies of the Orthoptera are very restricted in their distribution: the Tanaoceroidea are found only in the southwestern deserts of North America, the Pneumoroidea occur only in arid zones of southern Africa. The continent of Australia has proportionately the highest number of endemic higher taxa, such as the family Cooloolidae (Cooloola monsters), several subfamilies (including Phasmodinae and Zaprochilinae), and many tribes (including Kakaduacridini and Praxibulini). The island of Madagascar also has a number of endemic taxa, such as the family Malgasiidae (Malagasy crickets), the tribe Aspidonotini (Malagasy helmet katydids), and a very large number of genera and species, whereas New Guinea is home to the subfamily Phyllophorinae (giant helmet katydids). The Caribbean island of

Hispaniola has most of the known species of the katydid tribe Polyancistrini. Grasshoppers of the families Lithidiidae, Pneumoridae, and Charilaidae are restricted to southern Africa, Ommexechidae are known only from South America, and Xyronotidae have been found only in Mexico.

On the other hand, some taxa exhibit very wide distribution patterns. Meadow katydids of the genus *Conocephalus* are found on all continents and most islands, from the subarctic circle to the equator. Even individual species can be cosmopolitan, as is the case of crickets *Gryllodes sigillatus* and *Acheta domesticus* (both species are associated with human domiciles), and some grasshoppers (including *Tetrix subulata*) have Holarctic distribution.

Habitat

Members of the order Orthoptera inhabit virtually all terrestrial habitats, from the rock crevices of the littoral zone of the oceans, subterranean burrows, and caves, to rainforest treetops and peaks of the alpine zones of mountain ranges.

There are few aquatic forms, such as *Paulinia acuminata*, among the Orthoptera, but many are associated with marshes and other semi-aquatic environments. Reed beds are home to numerous conehead katydids (including *Conocephalus*, *Ruspolia*, and *Pyrgocorypha*), while on muddy banks of rivers, both pygmy grasshoppers (Tetrigidae) and pygmy mole crickets (Tridactylidae) occur. The true mole crickets (Gryllotalpidae) and sandgropers (Cylindrachetidae) are perfectly designed for life underground, where they dig tunnels with their enlarged, shovel-like front legs, and feed on insect larvae and earthworms.

Meadows and savannas are populated by hundreds of species of orthopterans adapted to life among blades of grass. Many of these forms, especially some katydids (including *Megalotheca* and *Peringueyella*) and grasshoppers (including *Acanthoxia*, *Leptacris*, and *Acrida*) have body shapes that perfectly mimic the blades and stalks of these plants. Deserts of the world also have unusually rich faunas of orthopterans, with many taxa, such as *Comicus* and *Urnisilla*, showing adaptations to life in fine sand. Others, such as *Trachypetrella* and *Lathicerus*, prefer rocky deserts, perfectly blending in among pebbles. A number of orthopteran taxa can be found in caves across the globe. Most troglophilous species belong to camel and cave crickets (Rhaphidophoridae) or various families of true crickets (Gryllidae, Phalangopsidae, and Mogoplistidae), although one species of katydids, the Cedarberg katydid (*Cedarbergeniana imperfecta*), was recently discovered in a cave in South Africa. A few species of crickets, such as *Myrmecophilus*, are inquilines in ant colonies, and some, including *Tachycines* and *Acheta*, are associated with human habitats, such as greenhouses or basements of houses.

The greatest diversity of the Orthoptera, however, is found in tropical forests, both dry and humid. Thousands of species of katydids, crickets, and grasshoppers have already been described from forest habitats, ranging from crickets, such as *Luzara* and *Eneoptera*, in the forest-floor litter, to katydids, such as *Sphyrometopa* and *Euthypoda*, on low understory plants, to grasshoppers, such as *Ommatolampis* and *Eumastax*, in the forest canopy. Yet thousands more still await discovery, even as large portions of their habitat are destroyed at an accelerating rate.

Behavior

The circadian rhythms of activity among the Orthoptera vary greatly from group to group. Diurnal activity is prevalent in short-horned grasshoppers. During the day most grasshoppers and locusts feed and mate, reserving the night for activities that would be very dangerous during the day, such as molting or laying eggs. The North American grasshopper *Cibolacris parviceps* is one of the few members of the Caelifera that prefer to feed at night. The reverse is true for katydids and crickets, with most species choosing nocturnal activity. In temperate zones, a greater proportion of species of crickets and katydids are active during the day than in the tropics, where few species can be found stridulating or feeding before dusk. An interesting case of different behavioral patterns exhibited during different times of day is a wasp-mimicking katydid *Agacris insectivora* from Central America. These black and orange katydids are excellent mimics of large pompilid wasps of the genus *Pepsis*. During the day these harmless katydids move in a jerky, wasplike fashion, rapidly flickering their orange-tipped antennae. They are not afraid to be seen, in fact they prefer to forage in well-lit gaps in the understory of lowland rainforests. At night, however, when they cannot rely on their pretend warning coloration, their movements are slow and deliberate, similar to those of their cryptically colored relatives.

Most orthopteran species are solitary animals, although gregarious tendencies are common among many crickets (especially members of the family Phalangopsidae), and cave and camel crickets (Rhaphidophoridae). Some crickets (Brachytrupinae) even form small family groups. But the most spectacular example of gregarious behavior in the Orthoptera is that of locusts. Locusts are not members of any particular genus or subfamily of grasshoppers, but the name is applied to those species of grasshoppers that exhibit a clearly defined shift in their behavior, morphology, and physiology, from a solitary to a migratory phase. One example of such a species is the desert locust (*Schistocerca gregaria*) from arid regions of Africa and the southwest of Asia. During most of the year these insects lead a solitary life, but the spring rains trigger a remarkable transformation in their behavior. Females respond to the rains and the resulting abundance of plant food with a production of larger-than-usual eggs. As the number of insects in the population grows, so does the chance of physically running into another member of their own species. This casual contact triggers a change in the endocrine system of young locusts, which starts to produce hormones that turn these green, cryptically colored insects vividly yellow and black. They begin seeking each other's company and form dense clusters, feeding and marching together. The adults are also different: from sandy gray they turn bright yellow, their wings become longer, and their body larger and more streamlined. Males start producing pheromones that accelerate the development of other individuals, leading to synchronized maturation across the entire population. At the same time, fe-

males' pheromones attract other females, causing them to lay eggs close to each other in dense groups. Soon after their final molt, the fully winged adults start flying erratically. One group of flying adults stimulates others to take wing and move off in the direction of the wind. The size of a single swarm can be larger than any other single congregation of organisms on Earth. Desert locust swarms can range in size from 100,000 to 10 billion insects, making them, in terms of the number of individuals, greater than the entire human population. They can stretch from a mere 0.38 mi^2 (1 km^2) to about 385 mi^2 (1,000 km^2) and weigh more than 77,161 tons (70,000 metric tons). In 1794 a particularly large swarm that spread over 1,930 mi^2 (5,000 km^2) succumbed to the wind and drowned in the sea off the coast of South Africa. Within days, a 4 ft (1.2 m) deep wall of insect corpses covered 50 miles (80 km) of the shore.

Swarming behavior is not restricted to the suborder Caelifera. Within the Ensifera certain katydids can produce huge swarms. The North American Mormon cricket (*Anabrus simplex*; despite the common name it is a large, wingless katydid) forms large marching bands that can bring devastation to crops. Conehead katydids (*Ruspolia* spp.) occasionally form large flying swarms in Africa.

The single characteristic most frequently associated with grasshoppers and their relatives is their ability to produce sounds. Although less widespread than generally believed, this is nonetheless quite common in some groups of the Orthoptera. The role of sound production is similar in some respects to that of birdcalls: for the attraction of mates, the defense of territory, and to raise the alarm when seized by a predator (release calls). The calls of orthopterans are usually species specific and play a very important role in species recognition. The information in the call may be coded in the form of frequency modulation (the pitch of the call changes through time, a mechanism best known in birds) or time modulation (the pitch of the call remains the same throughout its duration, but its temporal pattern is unique to the species), or a combination of both modes.

The dominant mechanism of sound production in Orthoptera is stridulation, which involves rubbing one modified area of the body against another. Contrary to popular belief, no orthopterans produce sound by rubbing their hind legs against each other. Katydids (Tettigoniidae) and crickets (Grylloidea) produce sound by rubbing a modified vein (the stridulatory vein) of one tegmen (front wing) against a hardened edge of the second tegmen (the scraper). The stridulatory vein is equipped with a filelike row of teeth, the number of which varies from a few to a few hundreds. In most katydids, the stridulatory area is situated at the base of the tegmina, except in short-winged species, such as *Thyridoropthrum*, where it covers their entire surface. In crickets, virtually the entire surface of tegmina is modified for stridulation. As a rule, katydids have the stridulatory file situated on the left tegmen and the scraper on the right one; in crickets, the situation is reversed. A membranous area at the base of the tegmen, the mirror, amplifies the sound. In addition, some species of katydids, such as *Thoracistus* and *Polyancistrus*, use their enlarged, shield-like pronotum as an additional sound amplifier. Crickets, generally lacking the enlarged pronotum, use other methods of

A Jerusalem cricket (*Stenopelmatus fuscus*) with a rat skull. (Photo by D. R. Thomson & G. D. Dodge. Bruce Coleman, Inc. Reproduced by permission.)

sound amplification, such as singing from burrows, the shape and size of which is attuned to boost certain frequencies (*Gryllotalpa*), or using the surface of a leaf for the same purpose (*Oecanthus*). The ability to stridulate is restricted almost exclusively to males, although in many groups of katydids, such as Phaneropterinae, Pseudophyllinae, and Ephippigerinae, females respond to the male's calls by producing short calls themselves. Their sound apparatus is different from that of the males, and is usually quite simple, lacking the sophisticated mechanism for sound amplification. In addition to tegminal sound apparatus, a few groups of katydids have other mechanisms of stridulation. For example, all members of the Australasian subfamily Phyllophorinae lack the typical wing stridulation and produce sound by rubbing their hind coxae against modified thoracic sterna. Mandibular sound production occurs in some members of Mecopodinae.

Grasshoppers use the same principle of stridulation, but instead of rubbing their tegmina against each other, they produce sound by rubbing the inner surface of the hind femur against one of the veins of the tegmen. In the slant-faced grasshoppers (Gomphocerinae) the inner surface of the femur possesses a file of small knobs and the vein on the tegmen acts as the scraper. In band-winged grasshoppers (Oedipodi-

nae), the vein has a row of pegs and the femur plays the role of the scraper. In addition to these two principal mechanisms, some grasshoppers stridulate by rubbing their hind legs against the sides of the abdomen (Pamphagidae) or by kicking their legs feet against a modified area at the apex of the tegmen (*Stethophyma*). Australian false mole crickets (Cylindrachetidae) have a stridulatory file at the base of their maxillary palps, and some species of pygmy mole crickets (Tridactylidae) produce sound by rubbing a modified vein on the dorsal side of the tegmen against another vein at the base of the hindwing.

The sound frequencies produced by orthopterans during stridulation vary, from a few kHz in most crickets and grasshoppers, to well above 100 kHz in some katydids. Cricket calls are characterized by their tonal purity, with most energy of the call allocated within a narrow range of frequencies. Katydid calls vary from tonally pure (although often well above the human hearing range) to broad, noiselike signals. Grasshoppers produce mostly broad spectrum, noiselike calls. Unlike vertebrates, many orthopterans produce time-modulated rather than frequency-modulated signals. Crickets are a notable exception, and most species produce melodious, birdlike, frequency-modulated chirps.

A red-legged locust (*Melanoplus femurrubrum*) rests on stem of plant. (Photo by Susan Sam ©2001. Reproduced by permission.)

In addition to stridulation, some grasshoppers crepitate, or make a crackling sound, in flight. In this case the sound is produced by rapidly flexing the hind wings while in the air. This behavior is especially common among band-winged grasshoppers (Oedipodinae) and plays an important role in courtship and territorial displays.

A few members of normally acoustic orthopterans have lost their ability to produce airborne signals and instead have developed a number of substitute mechanisms of substrateborne communication. Males of the oak katydids (*Meconema*) lack the typical tegminal sound apparatus, and instead produce sound by drumming with their hind legs against tree bark. Similar drumming behavior, although still accompanied by the typical stridulation, is a component of the courtship behavior of the pitbull katydid (*Lirometopum*). Males of the cricket *Phaeophilacris spectrum* have lost their ability to stridulate, and instead signal by rapidly flicking their tegmina back and forth while holding them in a vertical position. The near-field motion is detected by the cerci of the female cerci, rather

than by her ears. Despite having a fully developed stridulatory apparatus, many neotropical members of the katydid subfamilies Pseudophyllinae and Conocephalinae spend little or no time stridulating, relying instead on substrateborne tremulations. In this case, a male stands rigidly on a leaf or stem of a plant and violently shakes his entire body. The low-frequency waves are then transmitted along the branches of the plant. The reason for this behavior is unclear, although a few explanations have been proposed. The most widely accepted of these is the avoidance of predation in the case of foliage-gleaning bats that are known to use insect sounds to locate their prey. Others include eluding satellite males (nonsinging males of the same species trying to intercept a female), avoiding of parasitoid flies, and helping females locate males on multibranched plants. Thanks to the rapid development of recording techniques in recent years, many groups of orthopterans previously believed to be silent appear to employ a number of techniques of substrate communication.

Feeding ecology and diet

The food preferences and foraging behavior of orthopterans are as diverse as their habitats. Virtually all Caelifera (short-horned grasshoppers) are herbivorous, with only a few observations indicating that under some conditions, such as overcrowding or dehydration, grasshoppers may attack each other, especially molting or injured individuals. However, in the great majority of cases grasshoppers feed on leaves and other parts of plants. Many savanna and grassland species are obligatory grass feeders, while arboreal forms feed on tree leaves or the lichens and mosses covering branches. Most grasshoppers are polyphagous, feeding on a variety of species of plants. However, a number of taxa, especially those in the Neotropical subfamilies Ommatolampinae and Rhytidochrotinae, are restricted to feeding on single or only few species of plants, frequently those with high levels of toxic secondary compounds, such as the Solanaceae. *Bootettix argentatus* is the only North American grasshopper known to be restricted to feeding on a single plant species, in this case the creosote bush (*Larrea tridentata*). South American grasshopper (*Paulinia acuminata*) specializes on the giant salvinia (*Salvinia molesta*), a serious aquatic weed species, and has been used, rather unsuccessfully, in biocontrol efforts to eradicate this plant in Africa, Southeast Asia, and Australia. Pygmy grasshoppers (Tetrigidae) are some of the few insects that feed on mosses and lichens.

Ensifera (katydids, crickets, and their relatives) range from herbivorous, to omnivorous, to strictly predaceous. Some katydids specialize on somewhat unique food sources, for example, members of the Australian genus *Zaprochilus* feed exclusively on pollen and nectar of flowers. Others, such as many species of the genera *Neoconocephalus* and *Ruspolia*, feed mostly on seeds of grasses, whereas many members of the subfamily Phaneropterinae specialize in eating broad leaves. One of a few katydids feeding on pine trees and other conifers is *Barbitistes constrictus* from Europe.

Most katydids and crickets, however, are opportunistic in their culinary practices and feed on a wide range of organic material. For example, the Central American rhinoceros katydid (*Copiphora rhinoceros*) is known to feed on flowers, fruits, hard seeds, caterpillars, other katydids, snails, frog eggs, and even small lizards. Strictly predaceous katydids employ both the "sit-and-wait" strategy (Saginae) or actively forage and hunt living insects (Listroscelidinae). Specialized predatory species usually show distinct modifications of their front and middle legs, which may be very long and equipped with sharp spines to facilitate grasping and holding their prey. Some raspy crickets (Gryllacrididae) also actively search for insect prey by rapidly running along branches and grasping any sitting insect they encounter. Crickets and cave crickets tend to be generalists in their culinary preferences, but also exhibit tendencies to feed on live prey; tree crickets (*Oecanthus*) are known to feed on aphids. Some mole crickets have a unique behavior among orthopterans (and insects in general) of gathering and storing germinating seeds in circular chambers below ground for later consumption.

Reproductive biology

The courtship and mating behaviors of orthopterans are some of the most complex and fascinating spectacles of the insect world. As well as sound production, many species employ visual, tactile, and olfactory signals in their mating strategies. Visual communication is especially well developed in the predominantly diurnal grasshoppers, where males often have bright, species-specific markings on different parts of their body, and display them in carefully choreographed sequences during courtship. Grasshoppers of the genus *Syrbula* are champions in this respect, and males of some species, in addition to calling, perform a dance consisting of 18 distinct movements. The visual signals employed by many diurnal grasshoppers include flight displays, in which males flash their colorful hind wings (this is sometimes accompanied by crepitation), flagging with distinctly colored hind legs, and displays involving brightly colored, and often enlarged antennae. Courtship in katydids and crickets relies less on visual signals and more on sound and chemical cues, which are more appropriate for these mostly nocturnal animals. In both groups, males sometimes produce two types of calls, a long-range advertisement call and a more quiet, courtship song, which is performed only in the presence of a female. Female in some species may reply using either airborne signals or tremulation.

Chemical communication in Orthoptera has been little studied, but there is evidence that at least some species employ it during courtship. Females of the New Zealand giant weta produce a musky substance used by males to locate females; male camel crickets of the genus *Ceuthophilus* have thoracic glands that may also play a role in courtship. The field crickets *Teleogryllus comodus* use pheromones covering the female antennae to initiate courtship. Some other crickets use airborne pheromones in locating members of the opposite sex.

Copulation in orthopterans involves the transfer of a spermatophore, or sperm sac, which in some groups is accompanied by the spermatophylax, a large packet of nutritious proteins. The size of the spermatophylax can approach 60% of the male's body mass, making it an extremely costly and significant contribution to egg production. This causes the males of many species to be quite choosy when selecting their mating partners, and under certain circumstances, the females may compete for males, a role reversal remarkably rare in the animal world. Some male orthopterans also allow females to feed on parts of their own bodies during copulation. Males of grigs (*Cyphoderris*) have their hindwings modified into thick, fleshy lobes, the sole purpose of which is to be eaten by the female during copulation. Males who have already mated once and lack these courtship "snacks" must resort to other methods for holding the female's attention. They instead use the "gin trap," a complex system of cuticular modifications whose role is to hold the female's abdomen firmly in place during copulation. Female tree crickets (*Oecanthus*) feed on the males' thoracic glands during copulation, and in some crickets of the subfamily Nemobiinae, the females feed on enlarged spines on males' hind tibia. Males of other orthopterans, lacking such tasty incentives, must rely on their strong grasp or modified cerci at the end of their abdomen to hold the female during copulation.

Oviposition takes place in a variety of substrates, such as soil, plant tissues, or rock crevices. In some cases, eggs are protected from desiccation by a foamy mass produced by the female. Larvae usually hatch within a few weeks or months, but sometimes the eggs undergo a yearlong, or longer, diapause. Few orthopterans display any kind of parental behavior, although some crickets (*Anurogryllus*) lay eggs in burrows guarded by the female. Female mole crickets (*Gryllotalpa*) not only lay eggs in special egg chambers underground, but also actively care for the eggs by licking and removing fungal spores from their surfaces. The hatchlings stay with their mother for a few weeks before dispersing.

Conservation status

As is the case with most invertebrate taxa, there is little information about individual species and population sizes of the Orthoptera on which to precisely assess their conservation status. As of 2002, the IUCN Red List included 74 species of the Orthoptera. Two of these species, the central valley grasshopper (*Conozoa hyalina*) and Antioch dunes shieldback (*Neduba extincta*), are listed as Extinct, and the Oahu deceptor bush cricket (*Leptogryllus deceptor*) is listed as Extinct in the Wild. Eight species are listed as Critically Endangered, eight as Endangered, and 50 as Vulnerable.

The single most critical threat to the survival of orthopteran species is habitat loss. In central Europe, most populations of the heath bush-cricket (*Gampsocleis glabra*) are extinct or severely reduced in size due to the loss of the original steppe habitats. Invasive species are also extremely serious factors leading to species decline and/or extinction. For example, in Hawaii, the introduced ants *Pheidole megacephala* have been shown to severely reduce population sizes of native *Laupala* crickets, and in New Zealand, native wetas (Deinacridinae) are decimated by introduced rats and other mammals.

Significance to humans

Locusts and grasshoppers have been part of human history from the very beginning of our agricultural tradition. They still pose a great risk for agriculture in many parts of the world, although they are less of a problem now than a few hundreds years ago, thanks mostly to a better understanding of their population dynamics and the application of various chemical and biological control measures. Swarms of desert locusts can range in size from 100,000 to 10 billion insects. The amount of food such a mass of insects requires is staggering: in one day a very large swarm can devour the equivalent of food consumed daily by about 20 million people. Such staggering numbers of plant-feeding insects cause similarly unfathomable devastation to crops, and specialized government agencies worldwide constantly monitor the points of origin and movements of locust swarms. Unfortunately, once the swarms are airborne, little can be done to stop the advancing force. Instead, most of the effort goes toward detecting the most likely areas of swarm formation and eradicating young larvae and eggs. Recently, a promising pathogenic fungus *Metarhizium anisopliae* var. *acridum* provided a viable alternative to harmful and costly pesticide treatments by selectively targeting locusts without any harm to other organisms. In addition to locusts, a few species of shield-backed katydids are agricultural pests, the best known being the Mormon cricket of the western United States.

A few cultures have realized the nutritional value of locusts, which in some cases can counterbalance the complete devastation of the crops. The Jewish Torah made an exception to the law forbidding eating any insects by stating "[...] The only flying insects with four walking legs that you may eat are those which have knees extending above their feet, [us-ing these longer legs] to hop on the ground." Some tribes in southern Africa eat locusts boiled or roasted, and grilled locusts are often consumed in Cambodia. Mole crickets (Gryllotalpidae) and some armored katydids (Hetrodinae) are also eaten in some parts of Africa.

Despite the devastation caused by some Orthoptera to agriculture, songs of katydids and crickets had a remarkable impact on the poetry and other arts of China and Japan. The Japanese have listened to and appreciated the calls of various Orthoptera, both those in the wild and those kept in cages as pets, for hundreds of years. This activity was popular with the Japanese court, which probably imported some of the customs associated with orthopterans from China, and with the common people. For the Chinese and Japanese, visiting places known for the abundance and high quality of their singing insects was a seasonal pleasures, such as viewing cherry blossoms and autumn leaf. Even now, selling caged singing crickets and katydids is a thriving business in China, and in Japan it is possible to buy a digital replica of a singing katydid.

There are no venomous species of Orthoptera, although a few species of bushhoppers (*Phymateus* spp.; Pyrgomorphidae) can be toxic if ingested, as they feed on milkweed and other species of Asclepiadaceae, plants that contain high levels of cardiac glycosides. In southern Africa, where bushhoppers are common, children sometimes may become seriously ill or even die after eating these candy-colored insects.

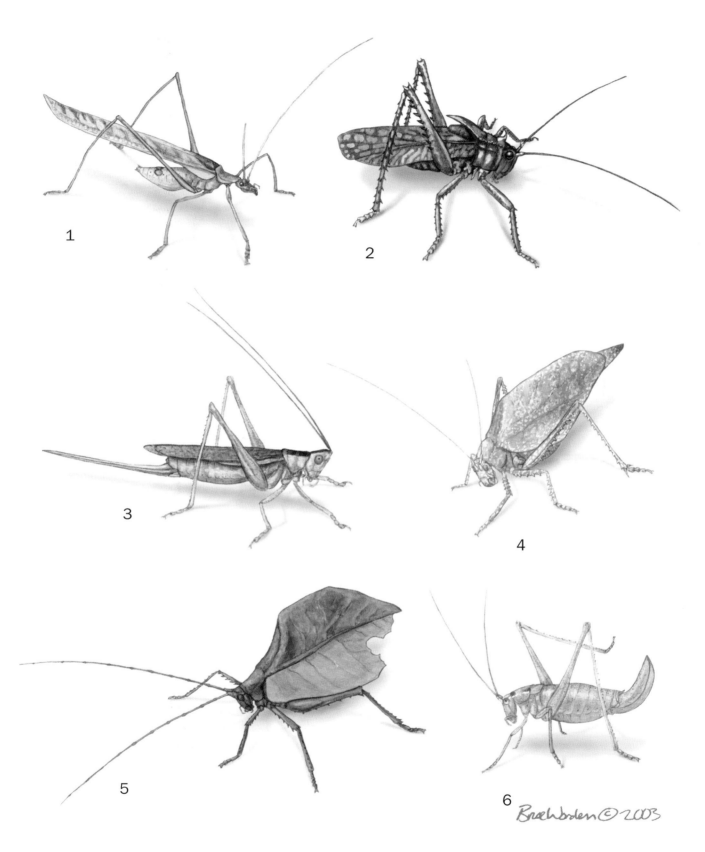

1. Balsam beast (*Anthophiloptera dryas*); 2. Hispaniola hooded katydid (*Polyancistrus serrulatus*); 3. Long-winged conehead (*Conocephalus discolor*); 4. Speckled rossophyllum (*Rossophyllum maculosum*); 5. Dead leaf mimetica (*Mimetica mortuifolia*); 6. Speckled bush-cricket (*Leptophyes punctatissima*). (Illustration by Bruce Worden)

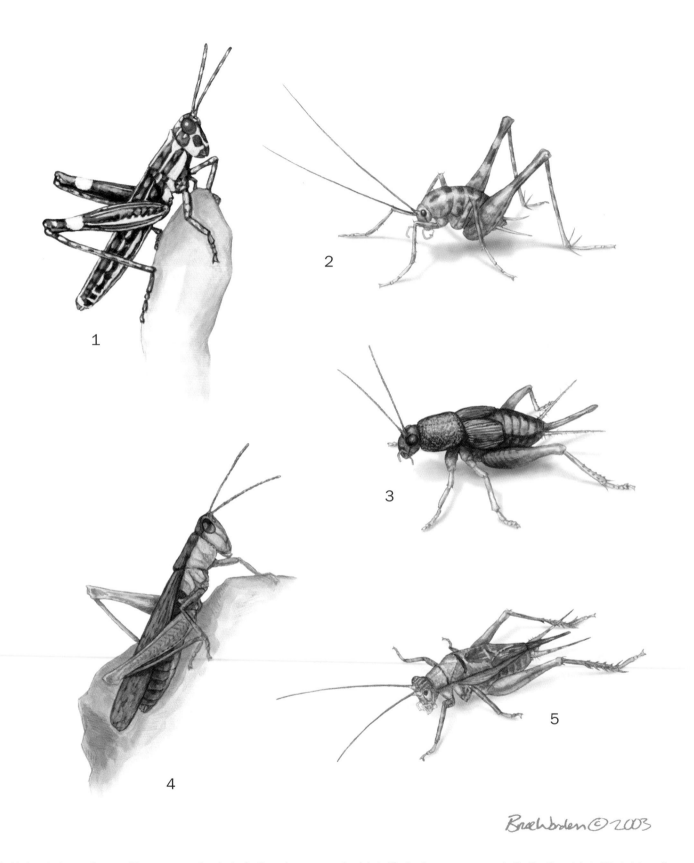

1. Variegated grasshopper (*Zonocerus variegatus*); 2. Greenhouse camel cricket (*Tachycines asynamorus*); 3. Beetle cricket (*Rhabdotogryllus caraboides*); 4. Field grasshopper (*Chorthippus brunneus*); 5. Suriname clicking cricket (*Eneoptera surinamensis*). (Illustration by Bruce Worden)

Grzimek's Animal Life Encyclopedia

Species accounts

Field grasshopper
Chorthippus brunneus

FAMILY
Acrididae

TAXONOMY
Chorthippus brunneus Thunberg, 1815, Sweden.

OTHER COMMON NAMES
None known.

PHYSICAL CHARACTERISTICS
Small, 0.5–1 in (14–25 mm); long wings. Coloration extremely varied, especially in females, from light brown to black to green to rose red.

DISTRIBUTION
Northern and central Europe.

HABITAT
Dry, sunny meadows, roadsides, and forest edges.

BEHAVIOR
Diurnal; males produce loud calls of hard "sst" sounds of about 0.2 sec duration.

FEEDING ECOLOGY AND DIET
Feeds primarily on grasses.

REPRODUCTIVE BIOLOGY
Eggs laid in soil, enclosed in foamy egg pods.

CONSERVATION STATUS
Not threatened. One of the most common grasshoppers in its distribution.

SIGNIFICANCE TO HUMANS
None known. ◆

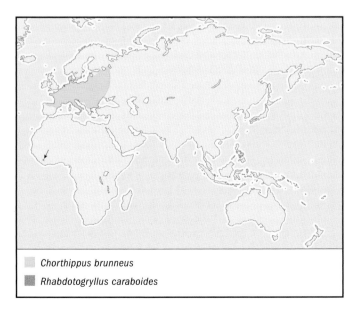

☐ *Chorthippus brunneus*
■ *Rhabdotogryllus caraboides*

Suriname clicking cricket
Eneoptera surinamensis

FAMILY
Eneopteridae

TAXONOMY
Eneoptera surinamensis De Geer, 1773, Suriname.

OTHER COMMON NAMES
None known.

PHYSICAL CHARACTERISTICS
Small, 1–1.4 in (25–35 mm); fully developed wings and large, protruding eyes. Female ovipositor long and needle shaped. Body brown.

DISTRIBUTION
South and Central America (exact boundaries unknown).

HABITAT
Feeds on herbaceous plants in understory of lowland tropical rainforests.

BEHAVIOR
Strictly nocturnal. Males produce very short, one-syllable calls unlike those of most other crickets.

FEEDING ECOLOGY AND DIET
Little known, observed feeding on decaying organic material.

☐ *Rossophyllum maculosum*
■ *Eneoptera surinamensis*

REPRODUCTIVE BIOLOGY
Nothing is known.

CONSERVATION STATUS
Not listed by the IUCN, but like most tropical insects can be threatened by loss of natural habitat.

SIGNIFICANCE TO HUMANS
None known. ◆

Variegated grasshopper
Zonocerus variegatus

FAMILY
Pyrgomorphidae

TAXONOMY
Zonocerus variegatus Linnaeus, 1758, Africa.

OTHER COMMON NAMES
None known.

PHYSICAL CHARACTERISTICS
Medium, 1.4–2.2 in (35–55 mm); wings greatly shortened, reaching only to about middle of abdomen, but long-winged forms also occur. Adult coloration aposematic, yellow-green, with yellow, orange, white, and black markings; nymphs black with bright yellow speckles.

DISTRIBUTION
Sub-Saharan Africa.

HABITAT
Savannas, pastures, and agricultural fields.

BEHAVIOR
Larvae exhibit strong gregarious behavior and may cluster in tens or even hundreds on a single plant. Nymphs and adults move rather slowly, and even fully winged individuals are reluctant to take flight, trusting in their own unpalatability.

FEEDING ECOLOGY AND DIET
Feeds on a variety of plants, including many Leguminosae, from which they sequester pyrrolizidine alkaloids, secondary compounds that make them unpalatable to many predators (a

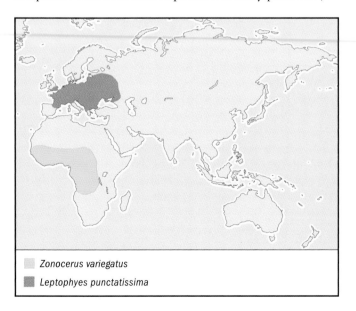

◻ *Zonocerus variegatus*
◼ *Leptophyes punctatissima*

fact the grasshoppers advertise with their warning coloration). Commonly eaten roasted by local people in southern Nigeria, suggesting level of toxic secondary compounds in their bodies varies depending on plant species fed upon.

REPRODUCTIVE BIOLOGY
Females lay eggs in soil, enclosed in foamy egg pods.

CONSERVATION STATUS
Not threatened.

SIGNIFICANCE TO HUMANS
Serious pest of cassava, maize, and other crops in sub-Saharan Africa. ◆

Beetle cricket
Rhabdotogryllus caraboides

FAMILY
Gryllidae

TAXONOMY
Rhabdotogryllus caraboides Chopard, 1954, Mt. Nimba area; Keoulenta, Guinea, Africa.

OTHER COMMON NAMES
None known.

PHYSICAL CHARACTERISTICS
Both sexes with shortened, thick tegmina covering half of abdomen, venation consists of many straight, parallel veins. Males do not have sound-producing modifications on wings and are presumably silent. Black, shiny, resembles small beetle.

DISTRIBUTION
Guinea (West Africa).

HABITAT
Litter of the lowland and midelevation tropical rainforest, sometimes observed on termite mounds.

BEHAVIOR
Almost nothing is known; may be associated with termites but nature of association is unknown.

FEEDING ECOLOGY AND DIET
Nothing is known.

REPRODUCTIVE BIOLOGY
Nothing is known.

CONSERVATION STATUS
Not listed by the IUCN, but likely threatened by habitat loss.

SIGNIFICANCE TO HUMANS
None known. ◆

Greenhouse camel cricket
Tachycines asynamorus

FAMILY
Rhaphidophoridae

TAXONOMY
Tachycines asynamorus Adelung, 1902, St. Petersburg botanical garden, Russia.

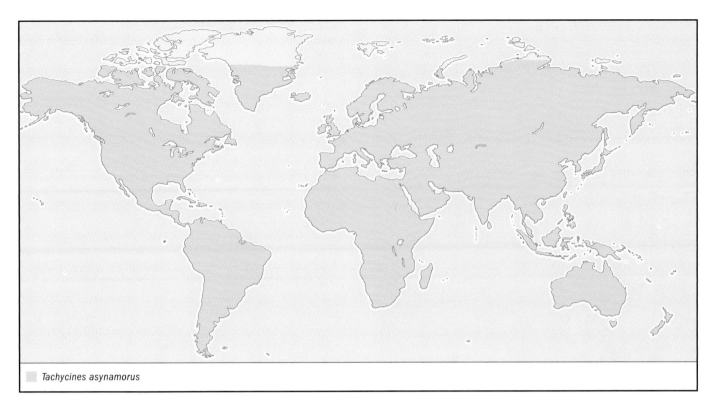

Tachycines asynamorus

OTHER COMMON NAMES
None known.

PHYSICAL CHARACTERISTICS
Small, 0.5–0.7 in (13–19 mm); completely wingless. Legs and all appendages very long and slender, giving the appearance of a long-legged spider. Extremely agile and can jump long distances. Body is yellow brown with dark mottling.

DISTRIBUTION
Originally from Far East (probably China), now cosmopolitan.

HABITAT
Wild populations probably inhabited caves, now found in greenhouses and warm, humid cellars and basements of houses.

BEHAVIOR
Exclusively nocturnal, spends the day hidden in crevices and under large objects; strongly gregarious.

FEEDING ECOLOGY AND DIET
Feeds on variety of organic matter, including other insects and plants.

REPRODUCTIVE BIOLOGY
Females lay eggs in soil; larvae hatch and join groups of older individuals.

CONSERVATION STATUS
Not threatened.

SIGNIFICANCE TO HUMANS
Can injure young plants in greenhouses. Disliked by humans because of its agility and spiderlike appearance. ◆

Balsam beast
Anthophiloptera dryas

FAMILY
Tettigoniidae

TAXONOMY
Anthophiloptera dryas Rentz and Clyne, 1983, Turramurra, Sydney, New South Wales, Australia.

OTHER COMMON NAMES
None known.

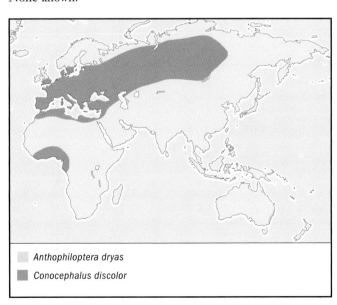

Anthophiloptera dryas
Conocephalus discolor

PHYSICAL CHARACTERISTICS
Large, 2–2.75 in (50–70 mm); long, pointed wings and prognathous head. Legs and antennae slender and very long. General coloration green or brown, leaflike.

DISTRIBUTION
New South Wales and Queensland (Australia).

HABITAT
Wooded suburbs and gardens of coastal southeastern Australia.

BEHAVIOR
Nocturnal; active primarily high in the treetops.

FEEDING ECOLOGY AND DIET
Feeds on flowers and variety of trees, but particularly fond of garden balsam (*Impatiens* sp.)

REPRODUCTIVE BIOLOGY
Eggs laid singly in the bark of trees, especially near the base.

CONSERVATION STATUS
Not listed by the IUCN. Considered "controlled specimens" by the Minister of the Environment and Heritage of Australia.

SIGNIFICANCE TO HUMANS
Occasionally damages garden flowers. ◆

Long-winged conehead
Conocephalus discolor

FAMILY
Tettigoniidae

TAXONOMY
Conocephalus discolor Thunberg, 1815, Sweden.

OTHER COMMON NAMES
None known.

PHYSICAL CHARACTERISTICS
Small, 0.5–0.7 in (12–17 mm), wings longer than body; hind wings protrude beyond apices of front wings when folded. Ovipositor straight, nearly as long as body. Light green with characteristic dark brown stripe on back.

DISTRIBUTION
Widespread in Europe and western Asia, also found in West Africa.

HABITAT
Meadows, marshes, reed beds, and near water.

BEHAVIOR
Very agile, diurnal, with good vision and thus difficult to approach. At slightest indication of danger, quickly moves to opposite side of the plant stem it is sitting on and clings to it, becoming virtually invisible. Males produce soft, continuous, buzzing call.

FEEDING ECOLOGY AND DIET
Feeds mostly on grasses and other plants, but also catches small insects such as caterpillars and aphids.

REPRODUCTIVE BIOLOGY
Eggs laid in grass or reed stems; females sometimes chew small holes in stems through which they insert the ovipositor. Larvae light green with characteristic black stripe on back.

CONSERVATION STATUS
Not threatened.

SIGNIFICANCE TO HUMANS
None known. ◆

Speckled bush-cricket
Leptophyes punctatissima

FAMILY
Tettigoniidae

TAXONOMY
Leptophyes punctatissima Bosc d'Antic, 1792, Paris, France.

OTHER COMMON NAMES
None known.

PHYSICAL CHARACTERISTICS
Small, 0.4–0.7 in (10–17 mm); reduced, scalelike wings. Legs long and slender; antennae several times longer than body. Females have broad, sickle-shaped ovipositor, very finely toothed at the tip. Light green.

DISTRIBUTION
Widespread in Europe, from southern Scandinavia in the north to the southern peninsulas in the Mediterranean. Recently introduced into the United States.

HABITAT
Sunny meadows, gardens, and orchards.

BEHAVIOR
Active at dusk and at night; males produce calls consisting of series of soft, short syllables, females respond to male calls with short clicks.

FEEDING ECOLOGY AND DIET
Feeds on leaves and flowers of a variety of plants, including clover, dandelion, roses, snowberry, and many others.

REPRODUCTIVE BIOLOGY
Eggs laid in tree bark and fruits of plants such as snowberry.

CONSERVATION STATUS
Not threatened.

SIGNIFICANCE TO HUMANS
Occasionally damages orchard trees. ◆

Dead leaf mimetica
Mimetica mortuifolia

FAMILY
Tettigoniidae

TAXONOMY
Mimetica mortuifolia Pictet, 1888, Guatemala.

OTHER COMMON NAMES
None known.

Polyancistrus serrulatus

Mimetica mortuifolia

PHYSICAL CHARACTERISTICS
Superb mimic of live and dead leaves. Tegmina resemble leaves to a degree capable of deceiving a botanist, complete with leaflike venation and fake "herbivory"; hind wings strongly reduced and hidden under tegmina. Ovipositor of female is strongly curved, with thickened, strongly serrated tip. Coloration varies greatly; individuals with green and brown, or even half-green and half-brown wings, may occur within same population.

DISTRIBUTION
Costa Rica and Panama.

HABITAT
Occurs in lowland and midelevation tropical rainforests, from the understory to highest levels of the forest canopy.

BEHAVIOR
Extremely cryptic, impossible to locate during the day, which it spends completely motionless, in a position that breaks the symmetry of its outline. At night feeds on leaves of plants, and males produce short, buzzing calls.

FEEDING ECOLOGY AND DIET
Little known, has been seen feeding on leaves of various trees.

REPRODUCTIVE BIOLOGY
Strongly curved ovipositor of females is shaped to penetrate tissues of plant stems where eggs are laid. To lay eggs, female bends abdomen down and forward until ovipositor faces forward between her front legs. She then uses legs to guide ovipositor and insert it into plant tissue. Eggs are laid individually, left partially protruding from the plant to allow for easy exchange of oxygen for developing embryo.

CONSERVATION STATUS
Not listed by the IUCN. Locally common, but threatened by habitat loss.

SIGNIFICANCE TO HUMANS
None known. ◆

Hispaniola hooded katydid
Polyancistrus serrulatus

FAMILY
Tettigoniidae

TAXONOMY
Polyancistrus serrulatus Beauvois, 1805, Santo Domingo, Dominican Republic, Hispaniola.

OTHER COMMON NAMES
None known.

PHYSICAL CHARACTERISTICS
Medium, 1.4–2.5 in (35–65 mm); enlarged, hoodlike pronotum in males and females. Ovipositor of females is long, sword shaped. Body coloration generally brown but green forms also occur.

DISTRIBUTION
Dominican Republic (Hispaniola).

HABITAT
Trees and tall bushes in tropical forests.

BEHAVIOR
Strictly nocturnal, spends the day in rolled-up leaves or under loose strips of tree bark. At night males produce long and very loud buzzing calls, females also capable of producing loud calls.

FEEDING ECOLOGY AND DIET
Feeds on leaves, fruits, and flowers of a wide variety of plants.

REPRODUCTIVE BIOLOGY
Females probably lay eggs in soil.

CONSERVATION STATUS
Not listed by the IUCN, but like most tropical insects can be threatened by loss of natural habitat.

SIGNIFICANCE TO HUMANS
None known ◆

Speckled rossophyllum
Rossophyllum maculosum

FAMILY
Tettigoniidae

TAXONOMY
Rossophyllum maculosum Bowen-Jones, 2000, Sirena Research, Corcorvado National Park, Costa Rica.

OTHER COMMON NAMES
None known.

PHYSICAL CHARACTERISTICS
Long, 2.5–3 in (65–75 mm); long, threadlike antennae. Ovipositor in females strongly reduced, adapted to depositing eggs on surface of leaves or stems of forest-canopy plants. Cryptically colored, resembles lichen-covered leaves.

DISTRIBUTION
Costa Rica.

HABITAT
Occurs only in canopy of lowland tropical rainforests of Costa Rica.

BEHAVIOR
Nothing is known.

FEEDING ECOLOGY AND DIET
Nothing is known.

REPRODUCTIVE BIOLOGY
Nothing is known.

CONSERVATION STATUS
Not listed by the IUCN, but like most tropical insects can be threatened by loss of natural habitat.

SIGNIFICANCE TO HUMANS
None known. ◆

Resources

Books
Field, L., ed. *The Biology of Wetas, King Crickets and Their Allies.* Oxford: CAB International, 2001.

Gangwere, S. K., et al., eds. *The Bionomics of Grasshoppers, Katydids and Their Kin.* Oxford: CAB International, 1997.

Gwynne, D. T. *Katydids and Bush-Crickets: Reproductive Behavior and Evolution of the Tettigoniidae.* Ithaca: Cornell University Press, 2001.

Kevan, D. K. McE. "Orthoptera." In *Synopsis and Classification of Living Organisms*, Vol. 2, edited by S. P. Parker. New York: McGraw-Hill, 1982.

Otte, D. *The North American Grasshoppers*, Vol. I. Cambridge, MA: Harvard University Press, 1981.

———. *The North American Grasshoppers*, Vol. II. Cambridge: Harvard University Press, 1984.

———. *The Crickets of Hawaii: Origin, Systematics & Evolution.* Philadelphia: The Orthopterists' Society, 1994. Rentz, D. C. F. *Grasshopper Country: The Abundant Orthopteroid Insect Fauna of Australia.* Sydney: University of New South Wales Press, 1986.

Uvarov, B. P. *Grasshoppers and Locusts: A Handbook of General Acridology*, Vol. 1. Cambridge, UK: Cambridge University Press, 1966.

Other
Orthoptera Species File Online [May 12, 2003]. <http://osf2.orthoptra.org>.

The Orthopterists' Society [May 12, 2003]. <http://www.orthoptera.org>.

Piotr Naskrecki, PhD

Mantophasmatodea

(Heel-walkers or gladiators)

Class Insecta

Order Mantophasmatodea

Number of families 3

Photo: A mating pair of *Karoophasma biedouwensis*. The smaller male assumes sigmoid mating posture, curling his abdomen around and under that of the female (on the side of the female). The bulbous phallus with its white lobes is seen between the cerci and ovipositor of the female. The characteristic terminal bend of the antennae are seen on the female. (Photo by Mike Picker. Reproduced by permission.)

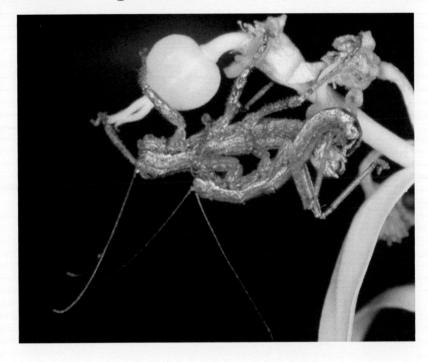

Evolution and systematics

The order Mantophasmatodea (heel-walkers, gladiators) was discovered in 2001 and formally described in 2002 based on two specimens that had been stored for decades in the natural history museums of Berlin and Lund (Sweden). With its 12 extant species, assigned to nine genera and three families, the Mantophasmatodea is the smallest insect order. The closely related genera *Namaquaphasma* (one species), *Karoophasma* (two species), *Lobophasma* (one species), *Hemilobophasma* (one species), and *Austrophasma* (three species) comprised in the family Austrophasmatidae, differ in details of the male and female genitalia. The genera *Mantophasma* (one species) and *Sclerophasma* (one species), which together constitute the family Mantophasmatidae, as well as *Tanzaniophasma* (one species) all display striking percularities in the copulatory organs. *Praedatophasma* (one species), whose systematic placement is unresolved, is unique in its spiny thorax ("gladiator").

The only known fossil genus is *Raptophasma* (two species) from Baltic amber, which is about 45 million years old. These creatures resemble extant Mantophasmatodea, but the thornless legs constitute a conspicuous difference. Using other insect orders as a measure, the origins of Mantophasmatodea probably go back to much earlier times.

Anatomical details of the abdominal spiracles and the ovipositor of the female, for example, show that despite lacking wings, the Mantophasmatodea belong to the Pterygota, the group comprising all winged insects as well as many that have lost their wings secondarily (such as Mantophasmatodea). It is unclear which other insect order is the closest relative of the Mantophasmatodea; the ice-crawlers (Grylloblattodea) and the stick insects (Phasmida) are the most likely candidates.

Physical characteristics

The consistently wingless Mantophasmatodea have a very uniform body shape, which superficially resembles that of certain grasshoppers or stick insects. The body length (without the antennae) ranges from 0.35 to 0.94 in (9–24 mm), males usually being somewhat smaller than females. The basic color of the body is brown, gray, green, or yellow; different tints of these colors, a whitish component of varied extent, and—in some species—black dots form a pattern of mottles and/or longitudinal stripes. Coloration varies between and also within species. Nymphs resemble the adults in appearance.

The antennae are long and have many segments. The compound eyes are well developed, albeit of varied size, but ocelli are lacking. The chewing mouthparts are generalized and directed downward. The femora are distinctly thickened in the forelegs and somewhat thickened in the midlegs, but they are very slender in the hind legs. The tibiae of the forelegs and midlegs bear on their inner surfaces (opposing the femora) two rows of short thorns, which render the legs suitable for grasping other insects. The tarsi comprise five tarsomeres, but the three basal ones are fused (with grooves indicating the borders). A large adhesive lobe (arolium) originates from between the claws.

The abdomen consists of 10 well-developed segments and a reduced eleventh segment. In the females, segments eight and nine bear three pairs of processes (valves) that together

1. *Lobophasma redelinghuisensis*; 2. *Praedatophasma maraisi*; 3. *Karoophasma biedouwensis*. (Illustration by Bruce Worden)

form a short ovipositor. The second and third valves are largely fused, and the latter are very hard and conspicuously claw-shaped. The genital opening lies behind the eighth sternite, which forms a subgenital plate. In the male, the ninth sternite is elongated to form a subgenital plate, which covers the retracted, largely membranous male genitalia ventrally. For copulation the genitalia become everted, thus displaying two or three small sclerites and two small sclerotized hooks. The one-segmented cerci are short in the female but fairly long and curved in the male, which uses them as accessory copulatory structures. The tenth abdominal tergum, which bears the muscles moving the cerci, is much wider in the male than in the female. While the middle part of the abdomen is widest in females (because it harbors the ovaries), in males the hindmost part usually is widest due to the expanded tenth tergum.

Derived features distinguishing Mantophasmatodea from other insects are the location of the anterior tentorial pits (anterior invagination points of the endoskeleton of the head) far above the mandibular articulation, a peculiarly shaped differentiation of the antennae (e.g., with differently shaped basal and distal segments) and a rounded projection on the middle of the male's subgenital plate.

Distribution

Extant Mantophasmatodea are known only from Africa: Namibia (two species), westernmost South Africa (eight species), and Tanzania (one species). *Raptophasma* from Baltic amber shows, however, that in the early Tertiary the group also occurred in Europe. The species recorded from South Africa seem to occupy fairly restricted, mutually largely exclusive geographic areas. This high degree of endemism renders the Mantophasmatodea a very interesting group for studies of the biogeographical history of southern Africa.

Habitat

Mantophasmatodea inhabit three of the major biomes found in western South Africa and Namibia: nama karoo, succulent karoo, and fynbos. The former location is characterized by (poor) summer rainfall and the two latter by winter rainfall (poor in succulent karoo), but all are moderately or extremely dry throughout the rest of the year and almost bare of trees. Broad differences between temperatures during the day and at night (with frost in winter) are normal. During and shortly after the humid season, the species-rich

Mantophasmatodea

Reproductive biology

Courtship is unknown. For copulation the male jumps onto the female and bends his abdomen aside and downward to bring his copulatory organs into contact with the hind part of the female's abdomen. The act of copulation, with the male sitting on the female's back, can last up to three days. The female then produces several sausage-shaped egg pods, which contain about 10–20 long, oval eggs covered by a hard envelope made from sand and gland secretions. The ovipositor serves for shaping the egg pod and for proper placement of the eggs into it. The nymphs hatch at the beginning of the rainfall period, molt several times, reach maturity near the end of the humid season, and die during the early dry period. Consequently, life cycles are diametrically different depending on whether precipitation falls predominantly in summer or in winter.

Conservation status

As of 2002, no species of Mantophasmatodea were listed on the IUCN Red List. Mantophasmatodea are not uncommon in Namibia and western South Africa. Nothing is known, however, about how common the various species are. The apparently small distribution areas of some species might make them sensitive to extinction. Nonetheless, the finding of a rich population immediately alongside a road indicates that not all species suffer from this aspect of human civilization. Only a single specimen has been recorded from Tanzania, and it is unknown how common Mantophasmatodea are in this part of Africa.

Significance to humans

None known.

flora and fauna bloom. Localities harboring Mantophasmatodea show a wide range of vegetation density. Some species frequently are found in tufts of grass (Poaceae) or Cape reed (Restionaceae, which superficially resemble grasses); they occur either at the bottom between the culms or on top of the culms, where they are well camouflaged by their mottled and striped coloration. The habitat of the Tanzanian *Tanzaniophasma subsolana* is unknown.

Behavior

Mantophasmatodea essentially live singly, but a male and a female often can be found together in the same tuft of grass, and within suitable habitats an area of several hundred square yards (meters) may harbor a sound population while no heel-walkers are found in the surrounding areas. There appears to be both nocturnal and diurnal activity. Movements typically are quite slow, but they can be rapid when prey is caught or a male mounts a female for copulation. When walking, Mantophasmatodea generally have their basal tarsomeres on the ground, while the fifth tarsomeres and the terminal claws and arolia are held in an elevated position (hence the name heel-walkers). The males use the projection of their subgenital plates to knock on the ground, which may serve for communication.

Feeding ecology and diet

Mantophasmatodea eat other insects of various kinds, up to their own size. They grasp and hold them by means of their powerful, spiny forelegs, additionally using the midlegs for large prey. Each mandible has a sharp edge for cutting the victim to pieces; all parts, except the legs and wings are devoured. Cannibalism also occurs.

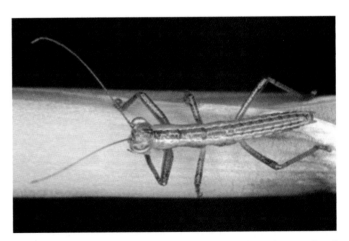

A green female nymph of *Austrophasma redelinghuisensis*, associated with evergreen fynbos vegetation of Western Cape Province, South Africa. Like most of the other species of Mantophasmatodea, there are various color morphs in each population; shown here is a brown morph. (Photo by Mike Picker. Reproduced by permission.)

Resources

Periodicals

Adis, Joachim, Oliver Zompro, Esther Moombolah-Goagoses, and Eugène Marais. "Gladiators: A New Order of Insect." *Scientific American* 287, no. 5 (November 2002): 60–65.

Arillo, Antonio, Vicente M. Ortuño, and André Nel. "Description of an Enigmatic Insect from Baltic Amber." *Bulletin de la Société Entomologique de France* 102 (1997): 11–14.

Klass, Klaus-Dieter. "Mantophasmatodea: A New Insect Order?" Response to technical comment by Erich Tilgner. *Science* 297 (August 2, 2002): 731.

Klass, Klaus-Dieter, Mike D. Picker, Jakob Damgaard, Simon van Noort, and Koji Tojo. "The Taxonomy, Genitalic Morphology, and Phylogenetic Relationships of Southern African Mantophasmatodea (Insecta)." *Entomologische Abhandlungen* 61 (in press).

Klass, Klaus-Dieter, Oliver Zompro, Niels Peder Kristensen, and Joachim Adis. "Mantophasmatodea: A New Insect Order with Extant Members in the Afrotropics." *Science* 296, no. 5572 (May 24, 2002): 1456–1459.

Picker, Mike D., Jonathan F. Colville, and Simon van Noort. "Mantophasmatodea Now in South Africa." *Science* 297, no. 5586 (August 30, 2002): 1475.

Tilgner, Erich. "Mantophasmatodea: A New Insect Order?" Technical comment. *Science* 297, no. 5582 (August 2, 2002): 731.

Zompro, Oliver. "The Phasmatodea and *Raptophasma* n. gen., Orthoptera incertae sedis, in Baltic Amber (Insecta: Orthoptera)." *Mitteilungen des Geologisch-Paläontologischen Instituts der Universität Hamburg* 85 (2001): 229–261.

Zompro, Oliver, Joachim Adis, and Wolfgang Weitschat. "A Review of the Order Mantophasmatodea (Insecta)." *Zoologischer Anzeiger* 241, no. 3 (2002): 269–279.

Other

Iziko Museums of Cape Town. "Biodiversity Explorer, Order: Mantophasmatodea (mantos)" [March 12, 2003]. <http://www.museums.org.za/bio/insects/mantophasmatodea>.

Zompro, Oliver, and Joachim Adis. "Mantophasmatodea: A New Order of Insects" [March 12, 2003]. <http://www.sungaya.de/oz/gladiator>.

Klaus-Dieter Klass, PhD

Phasmida
(Stick and leaf insects)

Class Insecta
Order Phasmida
Number of families 8

Photo: A leaf insect (*Phyllium bioculatum*) in Japan. (Photo by ©Ron Austing/Photo Researchers, Inc. Reproduced by permission.)

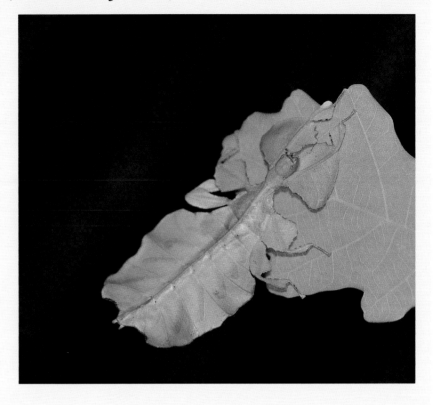

Evolution and systematics

The oldest fossil specimens of Phasmida date to the Triassic period—as long ago as 225 million years. Relatively few fossil species have been found, and they include doubtful records. Occasionally a puzzle to entomologists, the Phasmida (whose name derives from a Greek word meaning "apparition") comprise stick and leaf insects, generally accepted as orthopteroid insects. Other alternatives have been proposed, however. There are about 3,000 species of phasmids, although in this understudied order this number probably includes about 30% as yet unidentified synonyms (repeated descriptions). Numerous species still await formal description.

Extant species usually are divided into eight families, though some researchers cite just two, based on a reluctance to accept Bradley and Galil's 1977 rearrangement of the order. There are three suborders, the Anareolatae, Areolatae, and Timematodea. The Areolatae has five families, thought to be distinguished from the Anareolatae by the presence of a sunken area on the underside of the mid-tibia and hind tibia, but this is not always a reliable feature. The families include the Phylliidae (4 genera, 36 known species), Aschiphasmatidae (18 genera, c. 100 species), Bacillidae (48 genera, c. 230 species), Heteronemiidae (1 genus, 2 species), and Pseudophasmatidae (51 genera, c. 300 species). Two large families, the Diapheromeridae (180 genera, c. 1,600 species) and the Phasmatidae (100 genera, c. 700 species), represent the Anareolatae. The Timematodea has only one family, the Timematidae (1 genus, 21 species). These small stick insects are not typical phasmids, having the ability to jump, unlike almost all other species in the order. It is questionable whether they are indeed phasmids, and phylogenetic research is not conclusive. Studies relating to phylogeny are scarce and limited in scope. The eggs of each phasmid are distinctive and are important in classification of these insects.

Physical characteristics

Stick insects range in length from *Timema cristinae* at 0.46 in (11.6 mm) to *Phobaeticus kirbyi* at 12.9 in (328 mm), or 21.5 in (546 mm) with legs outstretched. Numerous phasmid "giants" easily rank as the world's longest insects. The largest leaf insect is *Phyllium (Pulchriphyllium) giganteum* from Malaysia, at 4.4 in (113 mm), while the smallest is *Nanophyllium pygmaeum*, at only 1.1 in (28 mm).

The elongate stick insects typically resemble twigs, and leaf insects (Phylliidae) look like broadened, flat leaves, providing one of the best camouflages in the animal world. Phasmids are smooth, scarcely or heavily granulated, and sometimes with extensive spines and tubercles. The legs are similar to one another. Cerci usually are short. Females often are larger than the typically very thin males. Wings are present in many species, but they may be shortened or even absent. While they

A Malaysian species, *Tagesoidea nigrofascia*, flashes its brilliant yellow wings in the face of a predator. (Illustration by Emily Damstra)

frequently are green or brown to better match vegetation, phasmids sometimes are brightly colored or boldly striped. Species with colorful hind wings rank among the most spectacular of all insects.

Most stick insects belong to the Diapheromeridae and Phasmatidae. The Diapheromeridae are a mixture of winged and wingless species, while the Phasmatidae include the world's longest insects as well as more bulky, often winged, insects. The Asian Aschiphasmatidae are mainly winged, although the forewings are usually only a stalk-like structure. The Bacillidae includes spectacular, broad-bodied species. The Neotropical Pseudophasmatidae have many beautiful winged species to rival Asian winged representatives of the Diapheromeridae. The Heteronemiidae has recently been reduced to only a single wingless genus. The leaf insects (Phylliidae) are mainly from Southeast Asia, although some have successfully spread elsewhere.

Distribution

Phasmids are mostly tropical and subtropical throughout the world, although some hardy species persist in temperate areas. There even are three species of stick insects well established in the United Kingdom. Pet keepers occasionally deliberately discard stocks in the wild, and there also are accidental releases. Tropical species do not normally survive long in the United Kingdom, but escapees can become established in warmer climates.

Habitat

Found in a variety of habitats, phasmids can be abundant in wet and dry forests and in grasslands. In some countries they are common in gardens. Although they sometimes are found resting on or near their food plants in the daytime, they often are well hidden under leaves on the forest floor or in crevices. Tropical forests teem with these insects at night, when they move from their hiding places under cover of darkness. Some species frequent treetops and hence are seen rarely.

Behavior

Mostly nocturnal and remaining motionless in the daytime, phasmids blend in with the background; hence procrypsis (concealment from predators) is the primary defense. Some species even have the ability to change color to match their surroundings better, perhaps becoming a darker shade toward nighttime. Two-tone species with darker undersides are not uncommon. Many species feign death when disturbed, falling to the ground and remaining motionless; they may willingly shed a limb ("autotomy") in an effort to escape. The nymphs of some species (*Extatosoma tiaratum*) are thought to mimic ants. Many winged species flash open brightly colored wings or rustle their wings to startle predators; others fly away and "vanish" suddenly within vegetation. Some species, such as the *Eurycnema* from Australasia, use spines on their legs to strike out in attack. Certain stick insects, such as *Anisomorpha*

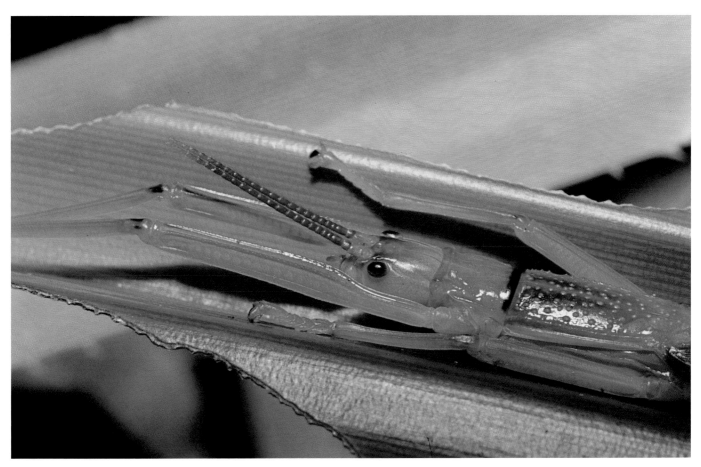

The peppermint stick insect (*Megacrania batesii*) is very rare and is found only on the pandanus plant. (Photo by ©B. G. Thomson/Photo Researchers, Inc. Reproduced by permission.)

buprestoides from the southeastern United States, exhibit warning colors on their bodies, perhaps stripes, and eject foul-smelling fluid from glands on the upper part of the thorax or from their mouthparts.

Feeding ecology and diet

Phasmids feed on leaves, taking large, circular bites out of the edges. A few species also eat flowers or bark. Some species have very few host plants, whereas many others accept the leaves of numerous different plants. It therefore is not surprising that pet keepers in different parts of the world often successfully rear phasmids in captivity on *Eucalyptus*, *Psidium*, *Rubus*, and *Quercus* species, regardless of their natural host plants.

Reproductive biology

Although they usually reproduce bisexually, many phasmids are able to reproduce parthenogenetically if males are absent, producing all-female offspring and thus enabling the species to survive. Only females are known in some species. Where males occur, mating typically involves the transfer of a spermatophore (sperm sac) from the male to the female. Hy-

bridization has been reported in *Bacillus* species from Europe. Between 100 and 2,000 eggs per species are either dropped or flicked to the ground; glued to surfaces, such as leaves, singly or in batches; or pierced into leaves. The eggs of many phasmids have a caplike structure, or operculum, on top, which assists ants in transferring the eggs to their underground nests. This benefits phasmids, since the ants eat only the caps, leaving the egg capsules to hatch later rather than being eaten on the forest floor. Nymphs hatch after a month to more than a year, depending on species, and frequently look rather like a miniature version of the adult. They can regrow lost limbs at the next nymphal molt. Nymphs typically molt six to seven times. Adults often live for several months and up to three years in a few species. In some genera, such as *Timema* in western North America, adult males mount females and remain there throughout their life span, in an effort to prevent the females from mating with rival males.

Conservation status

Many phasmids are known only from the originally described specimen(s), and their status is not known. The pet trade relies mainly on insects reared from captive stock. Showy species sometimes are imported in great numbers from

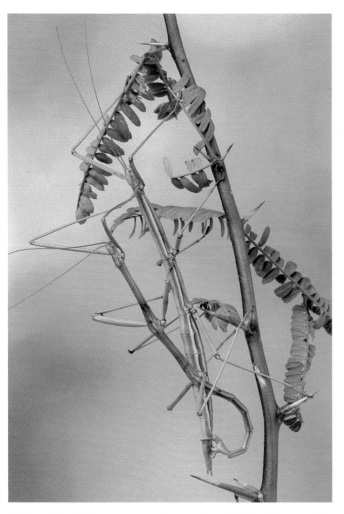

Walkingsticks (*Diapheromera arizonensis*) mating in Arizona, USA. (Photo by Bob Jensen. Bruce Coleman, Inc. Reproduced by permission.)

the wild, however, as with *Heteropteryx dilatata* from the Cameron Highlands of Malaysia. Trade declined by the early twenty-first century, although they still are collected widely for use in the framing industry. The spectacular Lord Howe Island (Australia) stick insect, *Dryococelus australis* (listed as Endangered by the IUCN), was thought to have become extinct following the introduction of rats to the island in 1918. The rediscovery of this species on a rugged and barren volcanic spire known as Balls Pyramid in 2001 caused great excitement. A captive-breeding project is under way by Australian authorities. The import of stick insects is regulated strictly in many countries, including the United States, as the insects may become pests.

Significance to humans

There are few reports of phasmids being eaten by humans. An old report states that natives of Goodenough Island, New Guinea, used the boldly spined hind legs of a *Eurycantha* species as fishhooks. The spectacular appearance of phasmids has led to their commercial use for framing (like butterflies, they are sold mainly to tourists as home decorations), in films, and on T-shirts, postcards, and toys. Phasmids are showy and relatively easy to look after, making them very popular insects in the pet trade. Nearly all species are harmless, but some need to be handled carefully, since they have shown aggressive defensive behavior. In extreme cases a few species squirt defensive sprays that have been known to cause temporary blindness in humans. Certain species are regarded as pests, with occasional population explosions resulting in severe defoliation of plants.

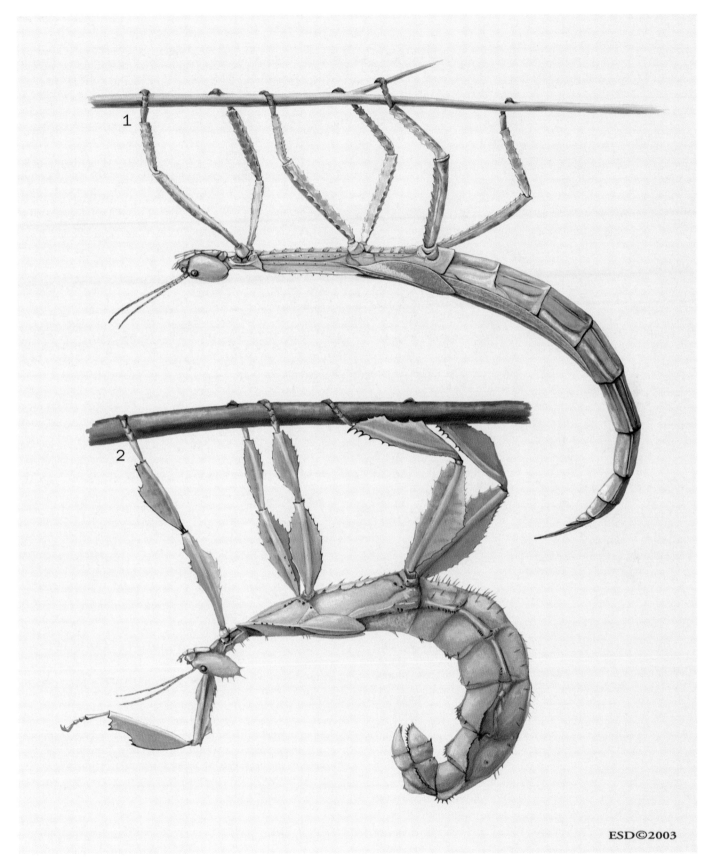

1. Female Darwin stick insect (*Eurychema osiris*); 2. Female Macleay's spectre (*Extatosoma tiaratum*). (Illustration by Emily Damstra)

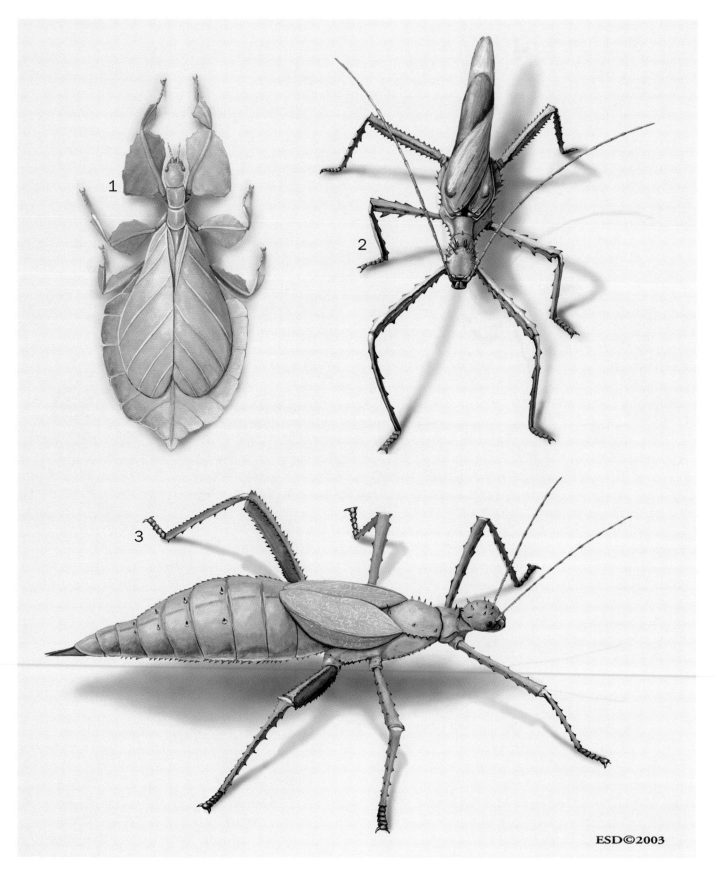

1. Javan leaf insect (*Phyllium [Pulchriphyllium] bioculatum*); 2. Male jungle nymph (*Heteropteryx dilatata*); 3. Female jungle nymph (*H. dilatata*). (Illustration by Emily Damstra)

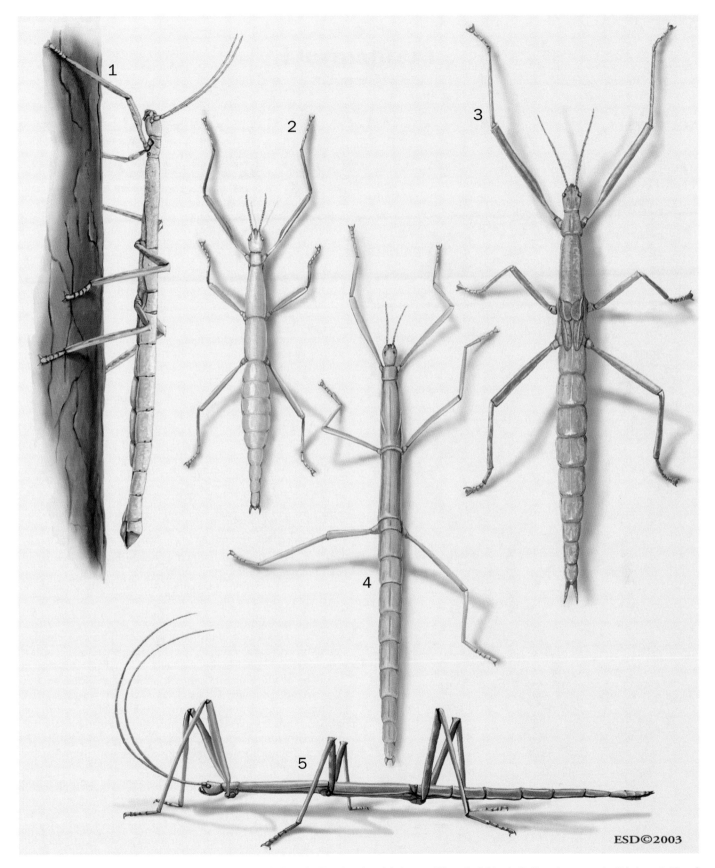

1. Female Indian stick insect (*Carausius morosus*); 2. Female Thunberg's stick insect (*Macynia labiata*); 3. Female coconut stick insect (*Graeffea crouanii*); 4. Female Mediterranean stick insect (*Bacillus rossius*); 5. Female common American walking stick (*Diaphecomera femorata*). (Illustration by Emily Damstra)

Species accounts

Mediterranean stick insect
Bacillus rossius

FAMILY
Bacillidae

TAXONOMY
Bacillus rossius Rossi, 1788, Pisa (Tuscany), Italy.

OTHER COMMON NAMES
English: Corsican stick insect.

PHYSICAL CHARACTERISTICS
Rather plain, wingless, elongate species with short antennae. The male is 2.0–3.1 in (52–79 mm) long; the female is 2.5–4.1 in (64–105 mm) long. Females are green or brown, and the much thinner males are brown.

DISTRIBUTION
Widespread throughout the warmer parts of Europe and the Mediterranean countries.

HABITAT
Found where bushes grow that support their food plants but prefers coastal areas.

BEHAVIOR
Nymphs and adults sometimes sway in the breeze to imitate moving twigs.

FEEDING ECOLOGY AND DIET
Widespread on bramble *Rubus fruticosus* and many species in the rose family (Rosaceae).

REPRODUCTIVE BIOLOGY
Adults are most common around August and September, although they can be seen all year in some areas. Each female lays about 1,000 eggs, which are dropped to the ground. The resulting nymphs hatch in three to eight months (some over winter), and nymphs mature in two to four months. Adults live several months.

CONSERVATION STATUS
Not threatened.

SIGNIFICANCE TO HUMANS
None known. ◆

Jungle nymph
Heteropteryx dilatata

FAMILY
Bacillidae

TAXONOMY
Phasma dilatatum Parkinson, 1798, Asia.

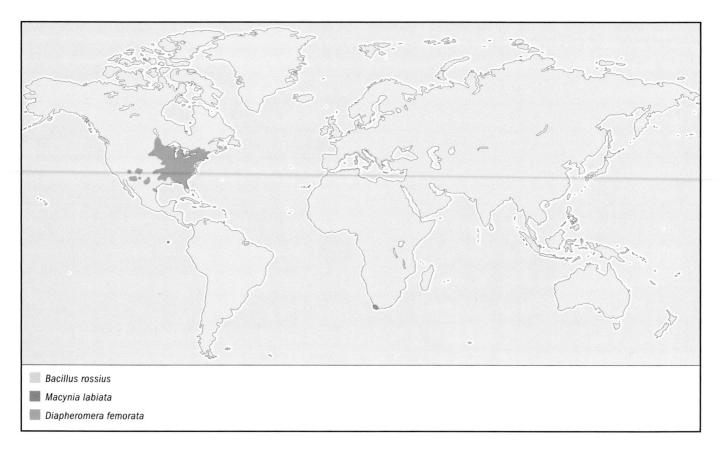

    Bacillus rossius

    Macynia labiata

    Diapheromera femorata

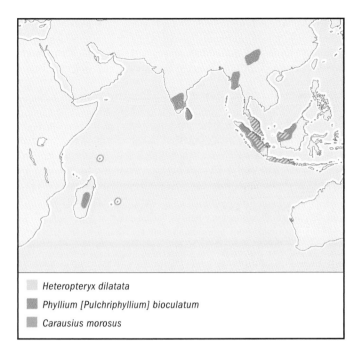

Heteropteryx dilatata

Phyllium [Pulchriphyllium] bioculatum

Carausius morosus

OTHER COMMON NAMES
None known.

PHYSICAL CHARACTERISTICS
Apple green females with large green forewings; hind wings are hidden beneath the forewings. Males are mottled brown with large wings. Males are 3.1–3.5 in (80–90 mm) long, and females are 5.4–6.3 in (140–160 mm) long. Females are very broadened and heavy, up to 2.3 oz (65 g). Both sexes with spiny legs.

DISTRIBUTION
Java, Malaysia, Sarawak, Singapore, Sumatra, Thailand.

HABITAT
Lives in rainforest.

BEHAVIOR
Females are camouflaged with a darker green underside. If disturbed, both sexes react, such as by arching the body forward and striking out with the spiny hind legs. A hissing sound is produced by rubbing the forewings and hind wings together. Biting is a last resort. Gynandromorphs (part male, part female) occasionally are reared or found in the wild.

FEEDING ECOLOGY AND DIET
Eats leaves of several bushes and trees, including *Eugenia*, *Grewia*, *Psidium*, *Rubus*, and *Uncaria*. Also feeds widely in captivity.

REPRODUCTIVE BIOLOGY
The life cycle is long. Females bury eggs in soil, and they take eight to 18 months to hatch.

CONSERVATION STATUS
Not threatened at present but heavily collected for use in the framing industry or for the pet trade.

SIGNIFICANCE TO HUMANS
Droppings of these phasmids, dried and mixed with herbs, are said to cure numerous ailments, such as asthma. Chinese families

therefore often rear them on guava leaves, from which the medicinal properties are thought to derive. This is a popular species for live displays in zoos and butterfly houses worldwide. ◆

Thunberg's stick insect
Macynia labiata

FAMILY
Bacillidae

TAXONOMY
Mantis labiata Thunberg, 1784. Type locality not recorded but probably Cape Town, South Africa.

OTHER COMMON NAMES
English: Green stick insect

PHYSICAL CHARACTERISTICS
The female is small and plump. The male is typically sticklike, with long cerci (claspers). Both sexes have short antennae. Male grow to 1.6–2 in (42–52 mm) and females to 2.1–2.2 in (54–56 mm). Head and pronotum are yellowish with green bands. Females usually are green and sometimes pink.

DISTRIBUTION
Cape Province, particularly around Cape Town, South Africa

HABITAT
Lives on natural fynbos vegetation.

BEHAVIOR
A master of camouflage. If disturbed, males sometimes walk away quickly.

FEEDING ECOLOGY AND DIET
Heathers and several plants associated with fynbos.

REPRODUCTIVE BIOLOGY
Females mate frequently, with different males. Adults are most common between September and January, although they can be found throughout the year. Adults live a few months, with females laying eggs that hatch after several months.

CONSERVATION STATUS
Not threatened.

SIGNIFICANCE TO HUMANS
None known. ◆

Indian stick insect
Carausius morosus

FAMILY
Diapheromeridae

TAXONOMY
Dixippus morosus Sinéty, 1901, Palni Hills, Tamil Nadu, southern India.

OTHER COMMON NAMES
English: Laboratory stick insect.

PHYSICAL CHARACTERISTICS
Plain, wingless species with medium-size antennae. Males grow to 1.9–2.4 in (48–61 mm) and females to 2.8–3.3 in (70–84 mm).

Females are various shades of dull green or brown, with some small tubercles (knobs); they are brightened up by a vivid red inner base of the foreleg. The much thinner males are brown.

DISTRIBUTION
Shembagonor and Trichinopoly in Madura Province, southern India. An established alien in Madagascar; the Cape Town suburbs, South Africa; and California, United States, they also are seen occasionally in Europe, including the United Kingdom, usually as a result of discarded cultures. In temperate climates they generally die out within a few years.

HABITAT
Lives on unnamed bushes and trees in India. Elsewhere found in gardens and natural vegetation on ivy and many other plants.

BEHAVIOR
A classic "stick" insect, which plays dead and can remain motionless for hours. Mainly parthenogenetic, males rarely are seen in captivity. When they do occur, they are believed to be genetic females with male characteristics and are sterile. Gynandromorphs (with part male and part female characteristics) also are reared occasionally. Males were present in the original culture but soon died out.

FEEDING ECOLOGY AND DIET
In captivity it accepts a vast range of plants but often is kept on privet *Ligustrum* species, often used as garden hedges in northern Europe.

REPRODUCTIVE BIOLOGY
Females drop several hundred eggs to the ground. The life cycle usually is completed in 12–16 months.

CONSERVATION STATUS
Not threatened.

SIGNIFICANCE TO HUMANS
The most written-about phasmid, frequently used in experiments as well as one of the most widely cultured insects. This hardy, prolific species is a wonderful introduction to the insect world for schoolchildren. ◆

Common American walkingstick
Diapheromera femorata

FAMILY
Diapheromeridae

TAXONOMY
Spectrum femoratum Say, 1824, Niagara Falls, United States.

OTHER COMMON NAMES
English: Walkingstick.

PHYSICAL CHARACTERISTICS
Long and slender, with long antennae. Males reach 2.2–3.3 in (55–84 mm) in length and females 2.8–4.0 in (70–101 mm). The hind femur has an apical spine. Midlegs of the male are banded, and the cerci are large and curved inward. Adult females are green, gray, or brown, and males are brownish with stripes. Both sexes are glossy.

DISTRIBUTION
Canada (Manitoba to Quebec) and distributed widely in the United States, south to Arizona and Florida.

HABITAT
Forests.

BEHAVIOR
Relies on its camouflage among vegetation.

FEEDING ECOLOGY AND DIET
Prefers oaks (*Quercus* species), though nymphs may feed on the lower vegetation of various plants.

REPRODUCTIVE BIOLOGY
Adults are most common in August and September, dropping 100–150 eggs to the ground. These overwinter and hatch in June or July of the next year.

CONSERVATION STATUS
Not threatened.

SIGNIFICANCE TO HUMANS
Sometimes regarded as a pest, this species can cause severe defoliation of host trees. ◆

Darwin stick insect
Eurycnema osiris

FAMILY
Phasmatidae

TAXONOMY
Phasma (Diura) osiris Gray, 1834, Melville Island, Australia.

OTHER COMMON NAMES
None known.

PHYSICAL CHARACTERISTICS
Large, spectacular, green-winged species. Males reach 4.5–5.3 in (115–134 mm), and females grow to 6.9–8.7 in (170–221 mm). The female's thorax has a bold pink central stripe. Legs are spiny.

DISTRIBUTION
Found in northern Australia (Northern Territory, Queensland, and Western Australia).

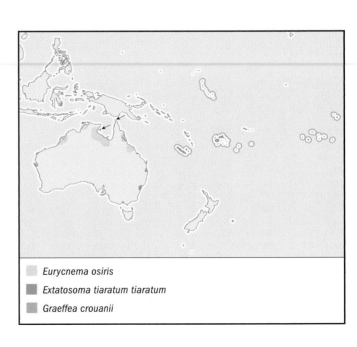

☐ *Eurycnema osiris*

▨ *Extatosoma tiaratum tiaratum*

▨ *Graeffea crouanii*

HABITAT
Lives in suitable bush, often high up in trees.

BEHAVIOR
Typically well camouflaged in vegetation, but if disturbed they resort to a startle display, showing the bright red underside of the wings. The spiny hind legs are spread and strike out.

FEEDING ECOLOGY AND DIET
Prefers *Acacia* and *Eucalyptus* species.

REPRODUCTIVE BIOLOGY
Adults are most common May to August but have been reported throughout the year. They live for several months, laying a few hundred eggs, which take several months to hatch.

CONSERVATION STATUS
Not threatened.

SIGNIFICANCE TO HUMANS
None known. ◆

Macleay's spectre
Extatosoma tiaratum tiaratum

FAMILY
Phasmatidae

TAXONOMY
Phasma tiaratum tiaratum Macleay, 1826, probably Parramatta, New South Wales, Australia.

OTHER COMMON NAMES
English: Giant prickly stick insect.

PHYSICAL CHARACTERISTICS
Large and spectacular winged species, though wings are rudimentary in the plump, heavy females (0.7–1.1 oz, or 20–30 g). Males reach 3.2–4.5 in (81–115 mm); females reach 3.9–6.3 in (100–160 mm). A leaf mimic. Legs and bodies of both sexes have leaflike expansions. Females are very spiny and brown or sometimes green. The subspecies *Extatosoma tiaratum bufonium* is a lichen mimic.

DISTRIBUTION
Parts of New South Wales and southeast and north Queensland, Australia.

HABITAT
Found in suitable bush or gardens. *Extatosoma tiaratum bufonium* usually is found at higher altitudes.

BEHAVIOR
In response to a perceived threat, the spiny hind legs are spread and strike out. Forelegs are waved, and sometimes the body sways from side to side. Females curl their abdomens, resembling scorpions. Defense glands release a chemical.

FEEDING ECOLOGY AND DIET
Eucalyptus species are included among their natural food plants.

REPRODUCTIVE BIOLOGY
It is possible that females attract males by flashes of ultraviolet light. Usually reproducing bisexually, females sometimes can reproduce parthenogenetically. Adults live for several months. Females lay several hundred eggs, which are flicked to the

ground. The eggs take five to eight months to hatch, and nymphs mature in three to six months.

CONSERVATION STATUS
Not threatened.

SIGNIFICANCE TO HUMANS
Popular in zoos worldwide. In parts of Papua New Guinea, its relative *Extatosoma popa popa* sometimes is cooked and eaten by natives. ◆

Coconut stick insect
Graeffea crouanii

FAMILY
Phasmatidae

TAXONOMY
Bacillus crouanii Le Guillou, 1841, Samoa Island.

OTHER COMMON NAMES
None known.

PHYSICAL CHARACTERISTICS
Broad green or pinkish-brown species, with bold pink, shortened wings. Males grow to 2.6–2.8 in (65–70 mm) and females to 4.1–4.6 in (105–116 mm).

DISTRIBUTION
South Pacific Islands.

HABITAT
Found wherever coconut palms grow.

BEHAVIOR
Well camouflaged on or near coconut palms, where they typically lay flat underneath leaflets. When disturbed, they may display their bright pink wings.

FEEDING ECOLOGY AND DIET
Feeds mainly on coconut palms (*Cocos nucifera*), devouring the green tissue of leaflets and sometimes the youngest fronds.

REPRODUCTIVE BIOLOGY
Females lay a few hundred eggs, which hatch in about three to four months.

CONSERVATION STATUS
Not threatened.

SIGNIFICANCE TO HUMANS
When population explosions of this species occur, they can result in defoliation of coconut palms, leading to serious losses to planters and village communities. ◆

Javan leaf insect
Phyllium (Pulchriphyllium) bioculatum

FAMILY
Phylliidae

TAXONOMY
Phyllium bioculatum Gray, 1832. Type locality not recorded but probably Southeast Asia or the Seychelles.

OTHER COMMON NAMES
English: Gray's leaf insect.

PHYSICAL CHARACTERISTICS
Rather flattened, with a broad body and legs. Color varies but often is green with or without mottling. Some variability in body shape, causing confusion among taxonomists. Females have very short antennae; those of the male are longer. Males grow to 1.8–12.7 in (46–68 mm) and females to 2.6–3.7 in (67–94 mm).

DISTRIBUTION
Widespread in Southeast Asia, in Borneo, China, India, Java, Malaysia, Singapore, and Sumatra. Also found in Madagascar, Mauritius, and the Seychelles.

HABITAT
Lives in rainforest.

BEHAVIOR
Relies on its excellent camouflage.

FEEDING ECOLOGY AND DIET
Food plants include *Nephelium lappaceum* and *Psidium guajava*. In captivity some accept *Quercus* and *Rubus* species.

REPRODUCTIVE BIOLOGY
The strangely shaped, large, seedlike eggs are dropped to the ground. Nymphs take several months to mature. Males can fly well and often are short-lived.

CONSERVATION STATUS
Not threatened.

SIGNIFICANCE TO HUMANS
None known. ◆

Resources

Books

Bragg, Philip E. *Phasmids of Borneo*. Kota Kinabalu, Borneo: Natural History Publications, 2001.

Brock, Paul D. *The Amazing World of Stick and Leaf Insects*. Orpington, U.K.: Amateur Entomologists' Society, 1999.

———. *Stick and Leaf Insects of Peninsular Malaysia and Singapore*. Kuala Lumpur: Malaysian Nature Society, 1999.

———. *A Complete Guide to Breeding Stick and Leaf Insects*. Havant, U.K.: T.F.H. Kingdom Books, 2000.

Grösser, Detlef. *Wandelnde Blätter*. (Book on Leaf Insects). Frankfurt, Germany: Edition Chimaira, 2001.

Otte, Daniel, and Paul Brock. *Phasmida Species File: A Catalog of the Stick and Leaf Insects of the World*. Philadelphia: Orthopterists' Society, 2003.

Periodicals

Bedford, Geoffrey O. "Biology and Ecology of the Phasmatodea." *Annual Review of Entomology* 23 (1978): 125–149.

Bradley, James Chester, and Bella S. Galil. "The Taxonomic Arrangement of the Phasmatodea with Keys to the Subfamilies and Tribes." *Proceedings of the Entomological Society of Washington* 79, no. 2 (1977): 176–208.

Brock, Paul D. "Studies on the Stick-Insect Genus *Eurycnema* Audinet-Serville (Phasmida: Phasmatidae) with Particular Reference to Australian Species." *Journal of Orthoptera Research* 7 (December 1998): 61–70.

———. "Studies of the Australian Stick-Insect Genus *Extatosoma* (Phasmida: Phasmatidae: Tropoderinae: Extatosomatini)." *Journal of Orthoptera Research* 10, part 2 (2002): 303–313.

Clark, J. T. "A Key to the Eggs of Stick and Leaf Insects (Phasmida)." *Systematic Entomology* 4, no. 4 (1979): 325–331.

Zompro, O. "A Generic Revision of the Insect Order Phasmatodea: The New World Genera of the Stick-Insect Family Diapheromeridae. Diapheromeridae = Heteronemiidae: Hemeronemiinae sensu Bradley & Galil (1977)." *Review Suisse de Zoologie* 108, no. 1 (2001): 189–255.

Organizations

The Phasmid Study Group. 40 Thorndike Road, Slough, Berkshire SL2 1SR United Kingdom. Phone: 01753 579447.

Paul D. Brock

Embioptera
(Webspinners)

Class Insecta

Order Embioptera

Number of families 8

Photo: A webspinner (*Clothoda urichi*) in its web. These insects avoid daylight and live in silken tunnels in associations of about 20 individuals. The two sexes have different mouthparts, and as a result, usually subsist on different foods. (Photo by ©George Bernard/Science Photo Library/Photo Researchers, Inc. Reproduced by permission)

Evolution and systematics

Fossil embiids date from the Oligocene, with controversial records from the Lower Permian of the Urals in Russia. The exact phylogenetic position of webspinners is uncertain, but they are considered orthopteroid insects, sharing some characters with earwigs (Dermaptera), stoneflies (Plecoptera), stick insects (Phasmatodea), zorapterans (Zoraptera), and termites (Isoptera). The order Embioptera (or Embiidina) includes eight living families with only 300 described species, although it is estimated that the true number of species is around 2,000.

Physical characteristics

Webspinners are small-to-medium-size insects, with lengths ranging from 0.06 to 0.78 in (1.5 to 20 mm). They have narrow, elongated bodies, which are usually brown or black, an anteriorly directed head with long filiform antennae, and short legs. Both larvae and adults can be easily recognized by their greatly swollen foretarsi. This enlarged foretarsi houses about 100 silk glands, which are used to spin silk galleries, inside which webspinners spent almost their entire life.

The ten-segmented embiid abdomen ends in two-articulated cerci, which are surrounded with sensory hairs and provide tactile orientation against the walls of the galleries, especially when webspinners run backward. Males have large compound eyes and asymmetric external genitalia, and lack wings in some species. When present, the wings have only a few veins, are flexible, and can be folded forward over the body at any point, allowing for reverse

movement in the narrow galleries. Some veins are transformed into blood sinuses, which provide temporary stiffness to the wings when used for flight. Females always lack wings, have smaller compound eyes, and are usually larger than males. Larvae look like the adult female, and only male larvae that will develop into winged adult males have external wing buds. Eggs are elongated, with a rimmed, circular operculum.

Distribution

The bulk of the embiid diversity is found in the tropics worldwide, although some species extend into the southern United States and temperate regions of Europe. Embiids are usually absent or poorly represented on islands. Some species have spread through commerce and as a result are widely distributed.

Habitat

Embiids spin their silk galleries on exposed bark or rock surfaces in humid areas; hidden under bark flakes, stones, or leaf litter; and in crevices or cracks in the soil, rocks, or termite mounds. Their galleries can also be found in hanging moss in mountain rainforests.

Behavior

With the exception of occasional dispersal, all embiid activity takes place inside the silk galleries. Webspinners are gregarious, with one or more adult females and their broods sharing a branched system of galleries, which they all spin and extend continuously. The galleries are just slightly wider than the body of the embiids, allowing their sensory body hairs to be in constant contact with the walls. Some species add vegetable detritus and frass to the galleries, providing additional cover and camouflage. Galleries are used as protected routes to food sources, to avoid desiccation, and as escape routes from predators. When an embiid is threatened, it rapidly retreats backward inside the labyrinth of silken tubes.

Feeding ecology and diet

Larvae and adult female embiids feed on vegetable matter, from live moss and lichens to dead leaves and bark. Adult males do not feed.

Reproductive biology

Males use their modified mandibles to hold the female's head during copulation. After mating, males soon die, and fe-

males lay a layer of eggs on a silken surface, which is used as a starting point for a new colony or an extension of an old one. Females show parental care, guarding their eggs and young larvae. In some species, females coat the eggs with feces, masticated vegetable matter, or leaf fragments; those in other species move the eggs about in the galleries. There are some known cases of parthenogenetic females, in which the young develop from unfertilized eggs. Development is through gradual metamorphosis.

Conservation status

No webspinner is listed by the IUCN.

Significance to humans

Webspinners are scarcely noticed by humans because of their secluded habits in galleries. They are of no economic importance, as they feed on dead vegetable matter in uncultivated areas.

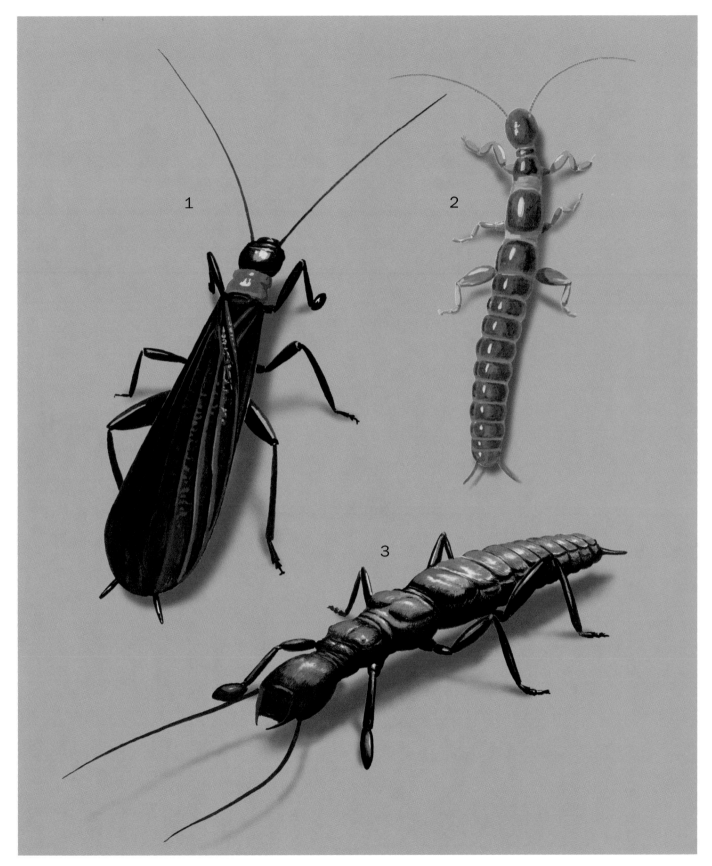

1. Adult male *Antipaluria urichi*; 2. Adult female Saunders embiid (*Oligotoma saundersii*); 3. Adult female *Australembia rileyi*. (Illustration by John Megahan)

Species accounts

No common name
Australembia rileyi

FAMILY
Australembiidae

TAXONOMY
Metoligotoma rileyi Davis, 1940, Queensland, Australia.

OTHER COMMON NAMES
None known.

PHYSICAL CHARACTERISTICS
Both sexes apterous; males with large maxillary palp, elongate submentum, and one-segmented left cercus.

DISTRIBUTION
Restricted to savanna zones of North Queensland, Australia.

HABITAT
Under leaf litter.

BEHAVIOR
Nothing is known.

FEEDING ECOLOGY AND DIET
Feeds on vegetable detritus.

REPRODUCTIVE BIOLOGY
Nothing is known.

CONSERVATION STATUS
Not threatened.

SIGNIFICANCE TO HUMANS
None known. ◆

No common name
Antipaluria urichi

FAMILY
Clothodidae

TAXONOMY
Clothoda urichi Saussure, 1896, Trinidad.

OTHER COMMON NAMES
None known.

PHYSICAL CHARACTERISTICS
Exceptionally large body, at least 0.6 in (16 mm) long. Hind basitarsi with two conspicuous ventral papillae. Strong wing venation. Male caudal tergal processes short and of almost equal size.

DISTRIBUTION
Trinidad.

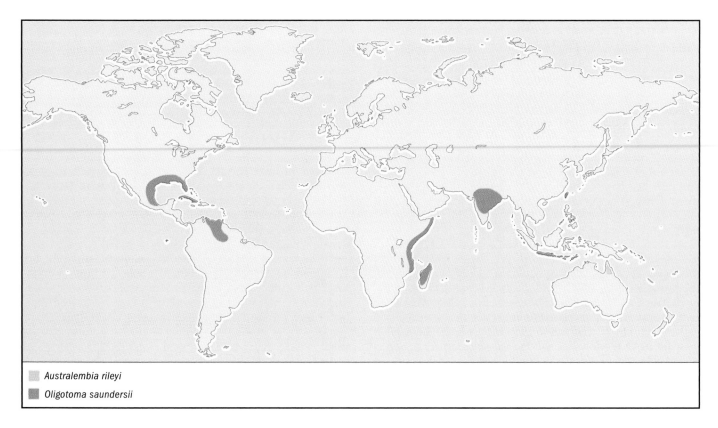

Australembia rileyi
Oligotoma saundersii

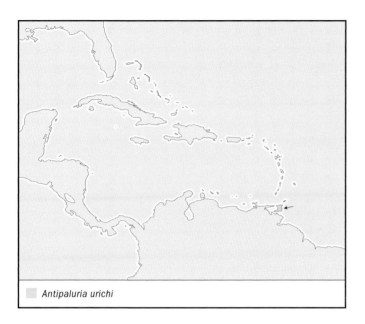

Antipaluria urichi

HABITAT
Conspicuous colonies on trunks of mountain rainforest trees.

BEHAVIOR
Silk galleries are bright and visible against the substrate. They have retreat areas with thickened reinforced silk coverings, foraging zones at edges of silk, and silken tunnels built up as embiids travel back and forth between foraging sites and retreats, where they remain during the day. Three styles of spinning behavior: construction of silk scaffold around body and over back; mending of holes by repeatedly attaching silk to existing silk around the hole and stretching new silk across opening; and reinforcement of silk covering by facing it and stepping up and down while releasing many strands of silk.

FEEDING ECOLOGY AND DIET
Feeds on lichens and algae growing on bark.

REPRODUCTIVE BIOLOGY
Forms colonies with maternal care of offspring.

CONSERVATION STATUS
Not threatened.

SIGNIFICANCE TO HUMANS
None known. ◆

Saunders embiid
Oligotoma saundersii

FAMILY
Oligotomidae

TAXONOMY
Embia saundersii Westwood, 1837, Australia and Brazil.

OTHER COMMON NAMES
None known.

PHYSICAL CHARACTERISTICS
Adult males have dentate mandibles and unbranched anterior medial wing vein; reddish brown body.

DISTRIBUTION
Native to north central India; widespread by means of commerce throughout warm and tropical regions, even to North America and Southeast Asia.

HABITAT
Common on trunks of rainforest trees and royal palms (*Roystonea* spp.) in landscaped areas.

BEHAVIOR
Tends to live in colonies of mothers and offspring. Area of silk reflects number of gallery occupants, although many silk coverings can be empty, suggesting embiids failed to establish successful colony, have been killed, or dispersed. Gallery surfaces almost completely camouflaged with fecal pellets or finely masticated chips of wood and bark from substrate spun into surface of silk. Males frequently fly to lights.

FEEDING ECOLOGY AND DIET
Feeds on lichens and algae on bark of trunk of host tree.

REPRODUCTIVE BIOLOGY
Exhibits maternal care.

CONSERVATION STATUS
Not threatened.

SIGNIFICANCE TO HUMANS
None known. ◆

Resources

Books

Ross, E. S. "Embiidina (Embioptera, Webspinners)" In *Encyclopedia of Insects*, edited by V. H. Resh and R. T. Cardé. San Diego: Academic Press/Elsevier Science, 2003.

———. "Embioptera: Embiidina (Embiids, Web-Spinners, Foot-Spinners)." In *The Insects of Australia: A Textbook for Students and Research Workers*. Vol. 1, 2nd edition, edited by CSIRO. Carlton, Australia: Melbourne University Press, 1991.

———. "Order Embiidina (Embioptera)." In *Immature Insects*. Vol. 1, edited by F. W. Stehr. Dubuque, IA: Kendall/Hunt Publishing, 1987.

———. "Web Spinners of Panama (Embiidina)." In *Insects of Panama and Mesoamerica*, edited by D. Quintero and A. Aiello. New York: Oxford University Press, 1992.

Periodicals

Edgerly, J. S. "Maternal Behavior of a Webspinner (Order Embiidina)." *Ecological Entomology* 12 (1987): 1–11.

Ross, E. S. "Embia. Contributions to the Biosystematics of the Insect Order Embiidina." Part 1, "Origin, Relationships and Integumental Anatomy of the Insect Order Embiidina."

Occasional Papers of the California Academy of Sciences 149 (2000): 1–53.

———. "Embia. Contributions to the Biosystematics of the Insect Order Embiidina." Part 2, "A Review of the Biology of Embiidina." *Occasional Papers of the California Academy of Sciences* 149 (2000): 1–36.

Valentine, B. D. "Grooming Behavior in Embioptera and Zoraptera (Insecta)." *Ohio Journal of Science* 86 (1986): 150–152.

Natalia von Ellenrieder, PhD

Zoraptera
(Zorapterans)

Class Insecta
Order Zoraptera
Number of families 1

Photo: *Zorotypus hubbardi* are found under bark or in rotten wood. This specimen was found in Gainesville, Florida, USA. (Photo by ©David R. Maddison, 2003. Reproduced by permission.)

Evolution and systematics

Zoraptera is a small order of exopterygote neopteran insects, for which the only general common name is zorapterans, although "angel insects" occasionally has been used. The affinities of Zoraptera have long been debated, and the order has been regarded as distinct but difficult to place phylogenetically. The major point of contention is whether they are primitive Paraneoptera (and thereby allied with bugs, lice, and their relatives), are related to cockroaches (Blattodea), or perhaps are descendants of some pre-dictyopteran insect stock. Other suggested alliances, based in part on the presence of distinct cerci, are as a sister group of the termites (Isoptera) or the webspinners (Embioptera). Although it has not yet been studied for many species, wing structure allies them most closely with cockroaches and their allies, with recent authors believing them to be primitive Blattoneoptera that separated early from the main lineage. There are no known fossils, other than for one species (*Zorotypus palaeus*) from Dominican amber (Lower Miocene/Upper Eocene).

All Zoraptera have been placed in the single family Zorotypidae. Suggestions that the New World taxa may constitute a separate family, as yet unnamed, need further verification. Until recently, all the 30 described species were allocated to the single genus *Zorotypus*. In 1993 six additional genera (*Brazilozoros*, *Centrozoros*, *Floridazoros*, *Latinozoros*, *Meridozoros*, and *Usazoros*) were erected, based on wing venation. All of these genera contain one species and are from the New World. All Old World species are placed in *Zorotypus*. *Zorotypus* is now a "holding genus" for the many species for which venation is unknown. Species identification often is difficult, with the females of some species almost impossible to place without associated males.

Physical characteristics

Zoraptera are small, elongate insects, ranging from 0.08 to 0.16 in (2–4 mm) in length. All species are similar in appearance and require dissection and microscopic examination for accurate identification. Adults usually are brown but range from pale brown to almost black and typically are not strongly patterned. The nymphs are pale creamy brown. Two distinct body forms occur in both males and females: wingless individuals (by far the more common and frequently encountered form) and winged individuals, which are generally darker and hold the wings flat over the abdomen. Winged individuals can shed their wings like termites and are recognized by the small wing "stubs" that remain.

The hypognathous head, usually narrowed anteriorly, has chewing mouthparts and a pair of antennae with nine segments. Males of many species have a cephalic gland (fontanelle) on the dorsal surface of the head. Mandibles have several "teeth"; maxillary palpi have five segments, and labial palpi have three segments. The well-defined prothorax is more or less square, and the pterothorax is more developed in winged than in wingless forms. The wings are broad and elongate, with reduced venation, and the hind wing is considerably shorter than the forewing and has less venation. Wingless forms lack both compound and simple eyes (ocelli); winged forms have well-developed compound eyes and three simple eyes. The legs are unremarkable other than for the strongly broadened hind femora, which can bear a row of thickened spines. Their number and arrangement are of taxonomic relevance; the tarsi have two segments with two claws. The oval abdomen is somewhat flattened dorsoventrally, and the short, unsegmented cerci (posterior sense organs) usually have a long apical bristle. Female genital structures are simple, with a well-sclerotized internal spermatheca and a long spermathecal duct. Male terminal structures are more complex, often with distinctive genital details. The pattern of thickened bristles on the posterior apex of the abdomen is also distinctive; the most universal male genital feature is a hooklike projection used to link with the female during copulation.

Distribution

The order is distributed widely but predominantly is tropical and subtropical. Zoraptera occur on all continents but have not yet been found on mainland Australia. Their greatest diversity appears to be in the neotropics, with a few species known from North America, Southeast Asia, Africa, or the Pacific islands. *Zorotypus hubbardi* is widespread in the southeastern United States and ranges northward to beyond forty-one degrees north latitude.

Habitat

Zoraptera are predominantly subcortical and are thus found under the bark of dead and fallen timber. They appear to be most common in rotting wood and logs with loose bark but with the cambium relatively intact. Their presence may help identify a particular phase in wood decomposition. They sometimes occur with termites. Others are found in ground litter, and *Z. hubbardi* has been reported from piles of decaying sawdust on old milling sites in North America. Winged individuals occasionally are attracted to light or taken in flight interception traps.

Behavior

Zoraptera are gregarious and occur in colonies probably founded by single females. Although winged females have been seen to mate soon after they emerge, no such colony establishment has been noted in the wild. Social structure may be well defined, with formation of dominance hierarchies among males in *Z. gurneyi*. These hierarchies are determined in part by size and age, with older males becoming more dominant. Aggressive encounters between males of this species in-

volve contact avoidance, head butting, grappling, chasing, and kicking. Grooming behavior, including mutual grooming, is well developed.

Feeding ecology and diet

Zoraptera are primarily fungus feeders, eating both hyphae and spores. Examination of the gut contents of particular species also has revealed arthropod fragments and pieces of wood. Two New World species, *Z. barberi* and *Z. gurneyi*, have been observed ingesting small nematodes. Captive zorapterans feed on yeast and crushed rat chow and may engage in cannibalism.

Reproductive biology

Zoraptera usually reproduce sexually. Males of some species are much rarer than the females. The Panamanian *Z. gurneyi* normally reproduces by parthenogenesis, although mating occurs when males and females meet. Males of *Z. gurneyi* are larger than females and may fight to gain access to females. The dominance hierarchy of this species may be pronounced and may represent a form of female defense polygyny. Females may mate every few days, either with the same male or with a variety of partners. In contrast, the smaller *Z. barberi* does not form hierarchies. Courtship involves nuptial feeding, by presentation of a drop of liquid secreted from the opening of the cephalic gland as a courtship gift. In mating of this species, the male is upside down, and copulation lasts for about one minute, occurring after periods of antennation. A single mating pair may copulate several times in succession. There appears to be no well-defined seasonal development in the tropics; eggs take several weeks to hatch. About four or five larval stages (instars) occur; the earlier stages have eight-segmented antennae. Adults may live for about three months.

Conservation status

Most species appear to be rare, with some known from single individuals and localities. Regional or local endemism is common for most species. No formal evaluation of conservation status has been advanced for any member of the order, and no species is listed by the IUCN or in any individual country legislation as having conservation significance. Occasionally, more local concerns (such as for *Zorotypus swezeyi* in Hawaii) have been expressed because of loss of habitat. No estimates of population size or trends in abundance are available for any species of the order.

Significance to humans

None known.

Species accounts

Hubbard's zorapteran
Zorotypus hubbardi

FAMILY
Zorotypidae

TAXONOMY
Zorotypus hubbardi Caudell, 1918, Florida, United States.

OTHER COMMON NAMES
English: Hubbard's angel insect.

PHYSICAL CHARACTERISTICS
No superficial, macroscopic, specific characters; recognizable by distribution. Body length 0.10–0.11 in (2.6–2.9 mm). Color medium to dark brown.

DISTRIBUTION
Southeastern United States. The most widely distributed zorapteran in North America.

HABITAT
Under bark in moist logs and in sawdust piles on old mill sites.

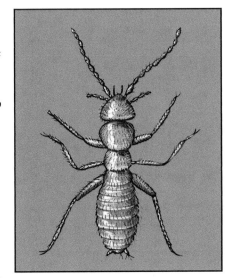
Zorotypus hubbardi

BEHAVIOR
Occurs in colonies that may persist for several years.

FEEDING ECOLOGY AND DIET
Feeds on fungal spores and hyphae; ingestion of other foods, such as arthropod fragments, may be fortuitous and may represent a more general scavenging habit.

Zorotypus hubbardi

REPRODUCTIVE BIOLOGY
Bisexual or facultatively parthenogenetic. Oviparous, with four or five nymphal instars.

CONSERVATION STATUS
Not threatened.

SIGNIFICANCE TO HUMANS
None known. ◆

Resources

Books

Choe, Jae C. "Zoraptera of Panama with a Review of the Morphology, Systematics and Biology of the Order." In *Insects of Panama and Mesoamerica: Selected Studies*, edited by D. Quintero and A. Aiello. Oxford: Oxford University Press, 1992.

Periodicals

Choe, Jae C. "Courtship, Feeding and Repeated Mating in *Zorotypus barberi* (Insecta: Zoraptera)." *Animal Behaviour* 49, no. 6 (1995): 1511–1520.

Gurney, A. B. "A Synopsis of the Order Zoraptera, with Notes on the Biology of *Zorotypus hubbardi* Caudell." *Proceedings of the Entomological Society of Washington* 40 (1938): 57–87.

Hubbard, M. D. "A Catalog of the Order Zoraptera (Insecta)." *Insecta Mundi* 4 (1990): 49–66.

Kukalová-Peck, Jarmila, and Stewart B. Peck. "Zoraptera Wing Structures: Evidence for New Genera and Relationships with the Blattoid Orders (Insecta: Blattoneoptera)." *Systematic Entomology* 18, no. 4 (1993): 333–350.

New, T. R. "Notes on Neotropical Zoraptera, with Descriptions of Two New Species." *Systematic Entomology* 3 (1978): 361–370.

Timothy R. New

Psocoptera
(Book lice)

Class Insecta

Order Psocoptera

Number of families 36 or 37

Photo: A giant psocopteran, *Poecilopsorus iridescens* (family Psocidae), from the tropical rainforests of Peru. (Photo by Rosser W. Garrison. Reproduced by permission.)

Evolution and systematics

Phylogenetically, the order Psocoptera (psocids, barklice, and booklice) is closely related to the Phthiraptera (lice), Thysanoptera (thrips), and Hemiptera (bugs, cicadas, aphids, etc.). These four orders compose a monophyletic group, the Paraneoptera (hemipteroid insects), and psocids retain the most primitive features in the group. Within the Paraneoptera, psocids are most closely related to lice, with which they compose a monophyletic taxon, Psocodea. Furthermore, the family Liposcelididae (booklice) of the Psocoptera is regarded as the sister group of the Phthiraptera. Therefore, the order Psocoptera is probably a paraphyletic taxon.

The Psocoptera has been divided into three suborders, Trogiomorpha, Troctomorpha, and Psocomorpha. The Trogiomorpha consists of two infraorders and five families. The Troctomorpha, in which the Liposcelididae is classified, consists of two infraorders and eight families. The Psocomorpha, the largest suborder in the Psocoptera, consists of four or six infraorders and 23 or 24 families. Monophyly of suborders other than the Troctomorpha is well supported by morphology and DNA data.

The oldest fossil psocid is recorded from the lower Permian deposits in Kansas (about 290 million years ago). Fossil psocopterans in the Permian to Cretaceous periods represent extinct families. In contrast, fossil psocopterans in the Oligocene to Holocene periods are all assigned to extant families.

Physical characteristics

Psocids range from small (about 0.04 in/1 mm) to medium (about 0.4 in/10 mm) in size. Most of them are brownish or whitish with blackish brown markings, but some species of

Caeciliusidae, Amphipsocidae, and Stenopsocidae have very colorful bodies and wings with markings. Morphologically, psocids are characterized by a well-developed postclypeus, long antennae, pick-like laciniae, a reduced prothorax, a well-developed pterothorax, and membranous wings (when present) held rooflike over the abdomen. The biting-type mouthparts with pick-like lacinia in psocids are considered to represent the intermediate condition between initial biting mouthparts in orthopteroids and the piercing and sucking type mouthparts in thrips and bugs, and this characteristic strongly supports the monophyly of the Paraneoptera. Psocids and lice share the specialized hypopharynx to uptake water vapor from the air, which supports monophyly of the Psocodea (psocids and lice). Most psocids are macropterous (fully winged), but brachypterous (short winged) or apterous (without wings) forms are also known.

Distribution

Psocids have been recorded from all zoogeographical regions, from tropical to subarctic zones.

Habitat

Psocids are found on a wide range of terrestrial habitat, such as on dead or living leaves, on stone or bark surfaces, and in leaf litter. Cave dwellers (such as *Prionoglaris stygia*) and wood borers (*Psilopsocus mimulus*) are also known. Some psocids are frequently discovered from food storage or other domestic environments.

Behavior

Most psocids are solitary, but the aggregation of larvae (e.g., *Sigmatoneura*, *Psococerastis*, and *Metylophorus* of Psocidae)

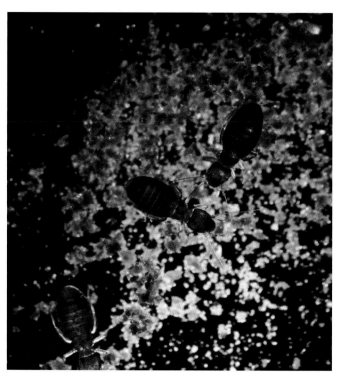

Book lice among grains of flour. (Photo by Kim Taylor. Bruce Coleman, Inc. Reproduced by permission.)

or colony-forming psocids (including subsociality of *Archipsocus*) are also known. Some psocids weave silk nests of various shapes and sizes, and from one to numerous individuals may live under it.

Sound production in psocids is widely known. The sound is generally considered to be a courtship song, but this behavior is very poorly understood.

Feeding ecology and diet

Psocids feed chiefly on lichen, fungi, or organic debris such as flour and scurf (skin flakes).

Reproductive biology

Most psocids are bisexual, but parthenogenesis (thelytoky) is also known in 12 families of all three suborders. Courtship involves various behavioral patterns. Eggs are laid singly or in groups, and may be bare, covered with silk webbing, or encrusted with fecal material. A few psocids are viviparous (such as *Archipsocopsis* and *Phallocaecilius* species). Larvae hatch from eggs using a specialized egg-burster. The usual number of instars is six, but this is sometimes reduced to five, four, or rarely three. Psocids have no pupal stage.

Conservation status

No species of psocids are included on the IUCN Red List. Most leaf- and bark-dwelling psocids are common and seem to endure human influence. However, some psocids are clearly sensitive to human impact. For example, many psocids are endemic to a single small island; some cave-dwelling psocids are known only from a single cave; all species of the genus *Psilopsocus* are known to be rare, possibly because of their specialized wood-boring behavior.

Significance to humans

Most psocids live in the wild and are thus harmless to humans. Some domestic psocids, such as species of *Liposcelis*, are common household insects. Under warm, humid conditions, they can reproduce very rapidly, becoming serious pantry pests. Psocids also occur in stored food, and at high densities can contaminate foodstuffs. Domiciliary psocids are known to cause allergic responses in sensitized people.

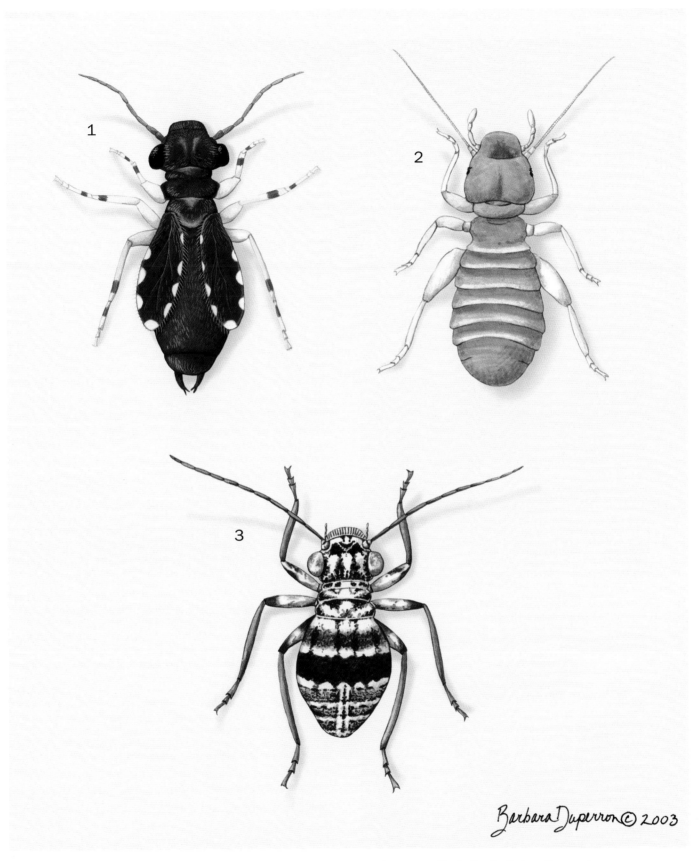

1. *Psoquilla marginepunctata*; 2. *Liposcelis bostrychophila*; 3. *Mesopsocus unipunctatus*. (Illustration by Barbara Duperron)

Species accounts

No common name
Psoquilla marginepunctata

FAMILY
Psoquillidae

TAXONOMY
Psoquilla marginepunctata Hagen, 1865, Germany.

OTHER COMMON NAMES
None known.

PHYSICAL CHARACTERISTICS
Small (about 0.04 in/1 mm long). Body and wings black in basal color, with white distinctive markings. Wing polymorphism is known.

DISTRIBUTION
Primarily wet tropics. Introduced by humans to some countries in the temperate zone, including Europe, the United States, and Japan.

HABITAT
Primarily under bark or bird nests; also found in greenhouse and domestic environments.

BEHAVIOR
Crowding provides increased numbers of macropterous forms and females.

FEEDING ECOLOGY AND DIET
Feeds on fungi and organic debris.

REPRODUCTIVE BIOLOGY
Bisexual. Egg laid singly and covered by debris. Six larval instars.

CONSERVATION STATUS
Not listed by the IUCN.

SIGNIFICANCE TO HUMANS
None known. ◆

No common name
Liposcelis bostrychophila

FAMILY
Liposcelididae

TAXONOMY
Liposcelis bostrychophila Badonnel, 1931, Mozambique.

OTHER COMMON NAMES
None known.

PHYSICAL CHARACTERISTICS
Small, about 0.04 in (1 mm) long; flattened dorsoventrally, pale brown in color, apterous.

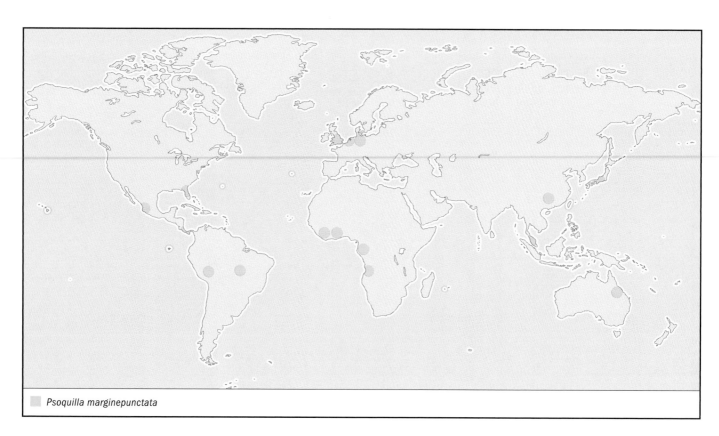

◻ *Psoquilla marginepunctata*

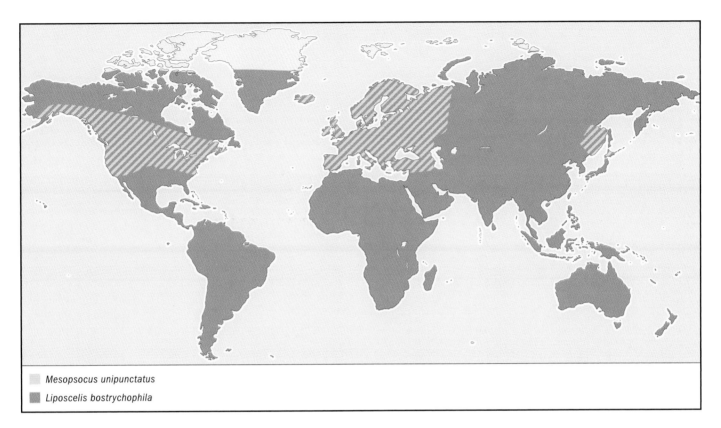

Mesopsocus unipunctatus
Liposcelis bostrychophila

DISTRIBUTION
Worldwide.

HABITAT
Common in stored food found in domestic environments; also found in the wild in bird nests, etc.

BEHAVIOR
Variations in color, size, egg production, and tolerance to pyrethroid insecticides are known among different populations.

FEEDING ECOLOGY AND DIET
Feeds on fungi and organic debris.

REPRODUCTIVE BIOLOGY
Parthenogenesis (thelytoky) is caused by *Wolbachia* infection. Eggs laid singly or in small batches and covered with powdery dusts. Four larval instars.

CONSERVATION STATUS
Not listed by the IUCN.

SIGNIFICANCE TO HUMANS
Primarily nuisance and stored food pest. ◆

No common name
Mesopsocus unipunctatus

FAMILY
Mesopsocidae

TAXONOMY
Hemerobius unipunctatus Müller, 1764, Europe.

OTHER COMMON NAMES
None known.

PHYSICAL CHARACTERISTICS
Relatively large psocid, about 0.2 in (5 mm) long. Sexual dimorphism present: male macropterous, female micropterous. Dimorphic in body color, light and melanic morphs.

DISTRIBUTION
Holarctic, north to Alaska.

HABITAT
Found in tree bark.

BEHAVIOR
Industrial melanism is known.

FEEDING ECOLOGY AND DIET
Feeds on lichens.

REPRODUCTIVE BIOLOGY
Bisexual. Eggs laid in groups, covered with incrustation and silk webs. Six larval instars.

CONSERVATION STATUS
Not listed by the IUCN.

SIGNIFICANCE TO HUMANS
None known. ◆

Resources

Books

Lienhard, C. "Psocopteres Euro-Mediterraneens." In *Faune de France* No. 83. Paris: Fédération Française des Sociétés de Sciences Naturelles, 1998.

Lienhard, C., and C. N. Smithers. "Psocoptera: World Catalogue and Bibliography." In *Instrumenta Biodiversitatis*, Vol. 5. Geneva: Museum d'histoire naturelle, 2002.

Mockford, E. L. "North American Psocoptera." In *Flora & Fauna Handbook*, No. 10. Gainesville, FL: Sandhill Crane Press, 1993.

Periodicals

New, T. R. "Biology of the Psocoptera." *Oriental Insects* 21 (1987): 1–109.

Smithers, C. N. "Keys to the Families and Genera of Psocoptera (Arthropoda: Insecta)." *Technical Reports of the Australian Museum* no. 2 (1990): 1–82.

Yoshizawa, K. "Phylogeny and Higher Classification of Suborder Psocomorpha (Insecta: Psocodea: "Psocoptera")." *Zoological Journal of the Linnean Society* 136 (2002): 371–400.

Other

"Psocid as Pests" [March 31, 2003]. <http://www.kcl.ac.uk/ip/bryanturner/other/index-psocids.html>.

"PsocoNet" [March 31, 2003]. <http://insect3.agr.hokudai.ac.jp/psoco-web/psoc.html>.

Kazunori Yoshizawa, PhD

Phthiraptera
(Chewing and sucking lice)

Class Insecta

Order Phthiraptera

Number of families 24

Photo: Scanning electron micrograph of a female pubic or crab louse (*Pthirus pubis*). The short crablike body form and oversized claws are an adaptation to the coarse, thick hairs present in the pubic and perianal region of its human hosts. (Image by Vince Smith. Reproduced by permission.)

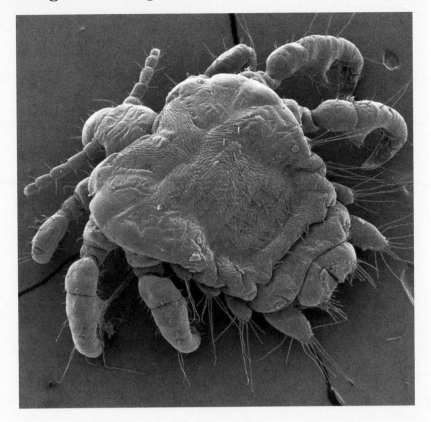

Evolution and systematics

It generally is accepted that lice (Phthiraptera) are derived from the insect order Psocoptera (the so-called book lice or bark lice), and speculative estimates place their origin between the Late Carboniferous and the end of the Cretaceous, 66–320 million years ago. With the exception of a louse egg found in Baltic amber and a louselike insect from the Lower Cretaceous of Transbaikalia in Russia, there are no fossils that might provide direct information on the evolution of lice. The host distribution of lice is, however, analogous in some ways to a fossil record.

The order Phthiraptera comprises four suborders, three of which (the Amblycera, Ischnocera, and Rhynchophthirina) are known as chewing or biting lice and the fourth (the Anoplura) as sucking lice. All species of Anoplura and Rhynchophthirina are restricted to mammals, whereas species of Ischnocera and Amblycera are known from both mammals and birds. Sucking lice and the so-called elephant and warthog lice (suborder Rhynchophthirina) are one another's closest relatives and share more derived morphological features. These have a common ancestor with the Ischnocera and together form a sister taxon to the most basal louse suborder—the Amblycera.

By 1999, 4,384 valid species of chewing lice had been recorded from 3,910 different hosts (3,508 bird and 402 mammal species). The sucking lice are a much smaller group and, as of the year 2002, 543 valid species were described from 812 different species of mammals. These figures are a small fraction of the true species diversity among lice, and many new species await formal description or discovery.

Physical characteristics

Lice are wingless, typically flat-bodied insects with three to five short, segmented antennae, highly modified mouthparts, and six relatively short legs modified for clinging to their host's feathers or pelage. Their size ranges from just 0.01 in (0.3 mm) long for the nymphs of some *Hoplopleura* (Anoplura) species on rodents, to 0.4 in (11 mm) for the giant adults of *Laemobothrion* (Amblycera) from birds of prey. Within many host groups there is a strong correlation between the size of the host and the size of their lice. Sexual dimorphism also influences louse size and is evident in many species, with females typically 10–20% larger than males.

The body is dorsoventrally flattened, with a horizontally positioned head. These are adaptations for lying flat against

Scanning electron micrograph of a lemur louse (*Trichophilopterus babakotophilus*) attached to a single hair shaft from a Madagascan lemur (*Propithecus verreauxi*). (Image by Vince Smith. Reproduced by permission.)

a hair shaft or between feather barbules, and they reduce the chance of the louse becoming dislodged during grooming or preening by the host. Coloration varies from pale white through shades of yellow and brown to black. Patterns are evident on some species, and cryptic coloration sometimes is employed, allowing the louse to match the coloration of its host's plumage or pelage.

Lice have mandibles that have been variously modified in each of the suborders. Asymmetric opposing mandibles are present in Amblycera and Ischnocera. In Amblycera, the mandibles articulate horizontally from the head, whereas in Ischnocera, the mandibles articulate vertically. These mandibles are involved in feeding and play a vital secondary role in anchoring the louse to the host. Within the Rhynchophthirina, the mandibles are much reduced, occurring at the end of a long rostrum and articulating outward rather than opposing each other. In most Anoplura, the mandibles have been completely lost. However, some species have tiny mandibular vestiges present internally within the anterior section of the head. These vestiges are just one of the many characters that highlight the common ancestry of chewing and sucking lice.

Distribution

As permanent, obligate (host-specific) ectoparasites, lice have distributions that essentially mirror those of their hosts, with very few exceptions. As such, they are found worldwide and are present on every continent and in virtually every habitat occupied by birds and mammals. All orders and most families of birds have records of host-specific lice; of the few groups that do not, it is likely either that their lice are extinct

or that the hosts have been searched insufficiently. Similarly, all major groups of mammals have lice, with the exception of species belonging to the orders Chiroptera (bats), Cetacea (whales and dolphins), Microbiotheria (Chilean colocolos), Monotremata (echidna and platypus), Notoryctemorphia (marsupial moles), Pholidota (pangolins), and Sirenia (dugongs and manatees).

Within the range of a host species, louse distribution often is patchy, and not every individual harbors all the lice previously recorded from the host. A true geographic distribution within the range of a host also has been noted for some species of lice. The size of a louse population varies enormously on different individuals, sometimes seasonally. Sick animals and, in particular, birds with damaged bills or feet may have abnormally large numbers, owing to their inability to groom or preen effectively.

Habitat

All lice complete their entire life cycle from egg to adult on the body of the host. The constant temperature and relative humidity that this environment affords lice may account for their success on mammals and birds. This seemingly uniform environment is, in fact, a series of interconnected microhabitats, and different species of lice are morphologically and behaviorally adapted to exploiting these niches on their host. This allows several species of lice to coexist on the same host species. These microhabitats are most evident on birds and are partitioned by the different feather types present on the wings, back, head, and rump. The differential ability of birds and mammals to preen or groom various parts of their bodies also exerts a major selection pressure on louse morphological characteristics and influences the microhabitat occupied by most lice. Extreme examples are the species of *Piagetiella* (Amblycera), which live inside the throat pouches of pelicans and cormorants, feeding on blood and serum within the pouch, but return to the head feathers to lay their eggs. Some species of *Actornithophilus* (Amblycera) have adapted to living inside the quills of wing feathers, thus completely escaping the effects of preening by their shorebird hosts.

Most species of lice are highly host specific, restricted to a single host species or a handful of closely related hosts. In several cases, host specificity extends to the host subspecies, and for this reason it often is possible to judge the identity of the host from the assemblage of lice present on its body. There are some notable exceptions to this trend, and a few louse species are recorded from hosts spanning several bird or mammal orders. There are also many anomalies in host-louse associations that can be explained only by accepting that there has been some interchange of lice between major host groups. For example, *Trichophilopterus babakotophilus* from Madagascan lemurs is a species of ischnoceran louse that belongs to a group otherwise restricted to birds. Similarly, *Heterodoxus spiniger* (Amblycera) on the domestic dog and other carnivores is a secondary infestation derived from an Australian marsupial. Thus, the axiom of strict host specificity that once was thought to be the rule of host-louse associations is a generalization, for which there are exceptions.

Behavior

Observations on the behavior of lice are limited to generalizations obtained from a few louse species. Amblyceran lice typically are more mobile than the other louse suborders, and some species are known to make short forays away from the host. For the majority of species, however, remaining attached to the host is critical for survival, and lice have a variety of behavioral and morphological adaptations to ensure that they never become parted from their host, except during dispersal. Wing lice of birds escape preening by inserting between feather barbs of the wing feathers, and there is a strong correlation between the size of the interbarb space and the size of the lice for different host and louse species. Similarly, "fluff lice" that occupy the fine feathers close to the abdomen, escape the preening activity of their hosts by burrowing down into the downy basal regions of these feathers. Lice grip feather barbs or hair shafts with the aid of their tightly locking mandibles. Even when dead, the dried exoskeletons of lice can remain fixed firmly to the host's body and frequently are recovered from museum collections of mammal and bird skins, long after the death and preservation of the host.

Successful transmission is perhaps the greatest challenge faced by any parasite, and lice are no exception. Direct physical contact between host individuals remains the principal route of dispersal for lice within a host species. Shared nest holes and nest material, predator-prey interactions, and mixed species use of dust baths all provide opportunities for dispersal to a new host species. Arguably the most unusual means of dispersal between hosts involves hitchhiking on the abdomen of hippoboscid flies, a phenomenon known as "phoresy." Records of phoresy are relatively common for louse species belonging to the ischnoceran louse genera *Degeeriella* and *Brueelia* but are rare for other taxa, and this is unlikely to represent a major means of dispersal for most lice.

Feeding ecology and diet

Specializations in the diet of lice underpin their major taxonomic divisions. The Anoplura, or sucking lice, are the only group that feeds exclusively on mammalian blood, drawn up from small vessels located close to the surface of the host's skin. Small buccal teeth are used to pierce the skin's surface, and a bundle of sharp stylets are extruded from the haustellar sheath of the louse into the host. This flexible bundle can be driven in different directions until a suitable blood vessel is located. Once the tip of the bundle enters a vessel, feeding can commence. The other three louse suborders possess a pair of distinct mandibles.

Rhynchophthirina (elephant lice and their relatives) have mandibles located at the end of a long rostrum. They each have been rotated 180 degrees so that they articulate outward, rather than oppose each other. Thus, Rhynchophthirina cannot "bite" or "chew" in the traditional sense of the word, but instead they use the sharply serrated mandibles to rasp at the skin, allowing a pool of blood to form that is sucked up through an opening at the end of the rostrum. These are true pool feeders (telmophages), unlike the vessel-feeding Anoplura (solenophages).

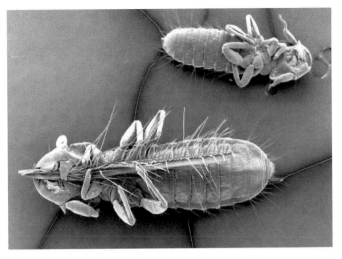

Scanning electron micrograph of two species of rhea louse (*Archolipeurus nandu*, bottom, and *Struthiolipeurus stresemanni*, top). The lower specimen is gripping a feather barbule between its legs and mandibles, illustrating a typical feeding and attachment behavior. (Image by Vince Smith. Reproduced by permission.)

All species of avian Ischnocera are believed to feed exclusively on feather barbules, the remains of which usually can be seen in crops of specimens that have fed recently. These are sheared from feathers by toothed mandibles. Mammalian Ischnocera (family Trichodectidae) feed on skin debris and hair (although this is disputed by some researchers), and at least one species is known to take blood meals. Amblycera have more generalist feeding habits, and in addition to feathers or hair, are known to feed on flakes of dead skin, blood, and skin secretions. Most chewing lice that do not partly blood-feed have an efficient water-vapor uptake system that extracts water from the atmosphere. For this reason, these species are particularly sensitive to the ambient humidity.

Reproductive biology

Lice are difficult to study under natural conditions, and for this reason most information on the bionomics of lice (i.e., their relations with other organisms and the environment) is obtained from in vitro studies, with the attendant disadvantages of applying laboratory results to natural situations. Separate male and female sexes are known for most louse species, but a few species reproduce parthogenetically. Lice eggs are attached to feather barbs; to hairs; or, in the case of the human body louse, to projecting clothing fibers with a drop of glandular cement. This surrounds the substrate and the base of the egg. Eggs, also known as nits, usually are attached close to the host's skin and remain firmly fastened as the feather barb or hair grows outward. Typically whitish in color, the eggs require four to 10 days of incubation before hatching, depending upon species and ambient temperature. When the nymph is ready to hatch, air is drawn in though the mouth and accumulates behind the nymph. When sufficient pressure is reached, the caplike operculum on the noncemented end of the egg is forced open, and the first-stage nymph crawls out. Three nymphal stages follow, each lasting three to 12

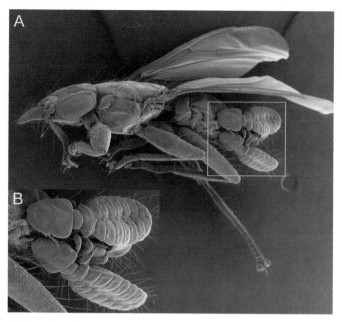

A. Scanning micrograph of two lice (*Brueelia* sp.) riding on the abdomen of a hipposcid fly. B. Inset shows close up of the attachment site on the fly. This phoretic association may help distribute lice between hosts. (Image by Vince Smith. Reproduced by permission.)

days. In some cases the first-stage nymphs, while lacking functional genitalia, may be miniature versions of the adult; in others, the nymphs successively become more adultlike through each instar (developmental stage). Adult lice live for about one month, and a female human body louse may produce 50–150 eggs during her lifetime.

Conservation status

It is seldom appreciated that the extinction of a mammal or bird species also results in the extinction of all associated host-specific parasites. The extreme host specificity of lice, along with other ectoparasite groups, such as feather mites and fleas, leaves them particularly vulnerable to co-extinction.

At least eight species of lice were known to be extinct by 1990, and this is almost certainly a gross underestimate of true loss of louse species diversity within the past century. Perhaps as the result of ignorance or the negative human perception of parasites, just one louse species (*Haematopinus oliveri*—an anopluran whose host species is the pygmy hog, *Sus salvanius*) is listed on the IUCN Red List of threatened species as of 2002, yet this same list defines 185 bird and 184 mammal species as Extinct in the Wild or Critically Endangered. If we assume that these species have an average rate of host-specific louse infestation, there are at least 50 species of lice that face a significant and immediate threat of extinction within the next 10 years.

Significance to humans

The human body louse (*Pediculus humanus* "*humanus*"; quote marks are used to indicate that this is not considered a valid taxonomic species) is the principal vector for *Rickettsia prowazekii*, which causes louse-borne typhus; *Bartonella quintana*, which causes trench fever; and *Borrelia recurrentis*, which causes epidemic or louse-borne relapsing fever. Epidemic and endemic infections can occur in conditions that foster the prevalence of lice, such as among homeless populations, in refugee camps, and during times of war or natural disasters. Humans can be infested with the human head louse (*Pediculus humanus* "*capitis*"), which is confined to the scalp and is common among schoolchildren worldwide, or the pubic louse (*Pthirus pubis*), which normally is transmitted through sexual contact.

Lice are important pests of domesticated mammals and poultry, and while modern insecticides have proved highly effective at controlling louse infestations, concerns over the safety of these chemicals, insecticide resistance, and the difficulty of treating large numbers of animals on a regular basis mean that lice will continue to be a major problem for livestock farmers. In 1994, for example, lice cost the Australian sheep industry an estimated $100 million U.S. dollars through lost production and control costs. Similar losses are likely in other countries where sheep, cattle, or poultry are farmed intensively.

1. Slender pigeon louse (*Columbicola columbae*); 2. Elephant louse (*Haematomyzus elephantis*); 3. Wandering seabird louse (*Ancistrona vagelli*); 4. Human head/body louse (*Pediculus humanus*). (Illustration by Jacqueline Mahannah)

Species accounts

Elephant louse
Haematomyzus elephantis

FAMILY
Haematomyzidae

TAXONOMY
Haematomyzus elephantis Piaget, 1869. Type host: *Loxodonta africana*.

OTHER COMMON NAMES
None known.

PHYSICAL CHARACTERISTICS
Distinct triangular head with the pre-antennal region elongated into a long rostrum bearing a pair of outward-facing mandibles. Short, broad thorax without a sternal plate and long, slender legs with a single serrated claw.

DISTRIBUTION
Restricted to the Indian and African elephant but not recorded throughout the host's range and less common on African elephants.

HABITAT
Common on the more hairy regions, especially in folds of soft skin of juvenile elephants. They are most common on the ears, groin, or axilla (armpits) or at the base of the tail.

BEHAVIOR
Nothing is known.

FEEDING ECOLOGY AND DIET
Sharp, outward-facing mandibles rasp at the surface of the skin, causing blood to flow. The louse sucks the blood up through a median notch at the end of the rostrum.

REPRODUCTIVE BIOLOGY
Nothing is known.

CONSERVATION STATUS
Not listed by the IUCN but should be considered threatened in those areas where the host population is considered endangered.

SIGNIFICANCE TO HUMANS
None known. ◆

Wandering seabird louse
Ancistrona vagelli

FAMILY
Menoponidae

TAXONOMY
Pediculus vagelli Fabricius, 1787. Type host: *Fulmarus g. glacialis*.

OTHER COMMON NAMES
None known.

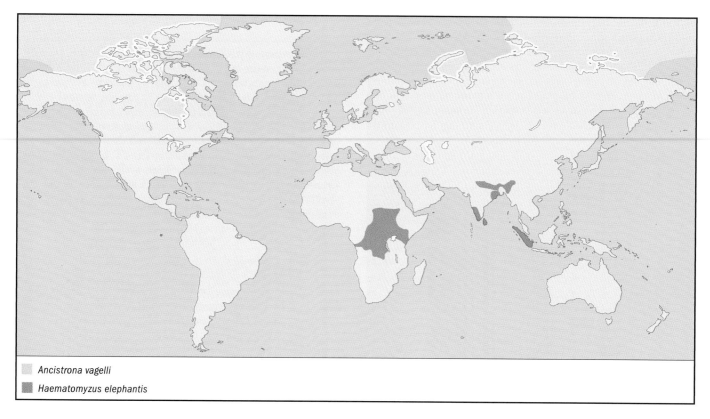

Ancistrona vagelli
Haematomyzus elephantis

PHYSICAL CHARACTERISTICS
Distinguished by its large size (0.1–0.2 in, or 3–6 mm), large triangular or rectangular postnotum, characteristic gular processes, and the absence of setal brushes on the venter of the third femur and abdominal sternites.

DISTRIBUTION
Recorded from more than 45 species of seabird in the families Procellaridae and Hydrobatidae. These include species of fulmar, petrel, prion, and shearwater, with a combined distribution covering virtually every patch of seawater in the world.

HABITAT
Restricted to the host's plumage, particularly regions surrounding the head and neck of the bird that are preened infrequently or are difficult to preen.

BEHAVIOR
Particularly vagile; hence its name. Capable of moving rapidly over the skin between feathers. This behavior may explain why this species has one of the widest host distributions of any louse in the suborder Amblycera.

FEEDING ECOLOGY AND DIET
Mainly soft feathers, particularly those close to the skin. As with most Amblycera, however, this probably is supplemented with traces of host blood, serum, and skin debris.

REPRODUCTIVE BIOLOGY
Nothing specific is known. As with most avian lice, the female is considerably larger than the male and always is found in greater numbers on the host. Eggs are cemented onto feather barbules and hatch within five to 10 days.

CONSERVATION STATUS
Not threatened. However, some populations of the species are restricted to rare or endangered hosts, such as the magenta petrel (*Pterodroma magentae*) and as such should be considered vulnerable.

SIGNIFICANCE TO HUMANS
None known. ◆

Human head/body louse
Pediculus humanus

FAMILY
Pediculidae

TAXONOMY
Pediculus humanus Linnaeus, 1758. Type host: *Homo sapiens*.

OTHER COMMON NAMES
English: Cootie

PHYSICAL CHARACTERISTICS
Head with distinctive dark eyes. Abdomen elongate and lacking distinct tubercles. The head louse variant is typically 20% smaller than the body louse form.

DISTRIBUTION
Worldwide. Lives as an ectoparasite on humans but also is recorded from gibbons and New World monkeys.

HABITAT
Two morphological variants exist that were once thought to be separate species or subspecies. The form commonly referred to as "head lice" is restricted to the human scalp, whereas the form referred to as "body lice" is restricted to clothing and the human torso.

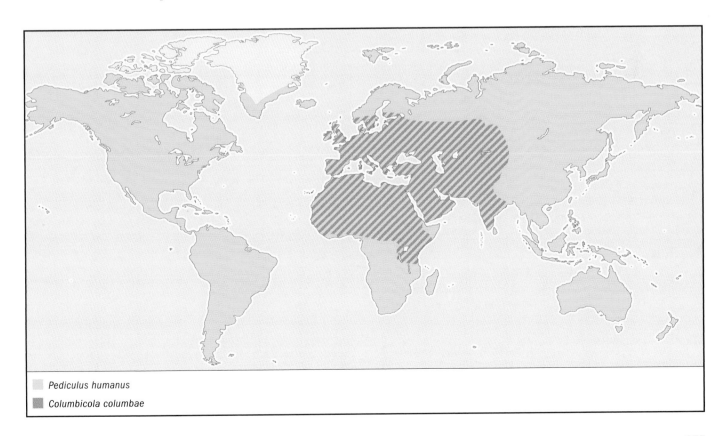

Pediculus humanus
Columbicola columbae

BEHAVIOR

Human head and body lice are ecological variants of the same species, capable of adapting to the different ecological conditions on the human scalp, torso, or clothing. They are intermixed genetically, and their behavioral ecology has been studied extensively, as they are the vector for several important human diseases.

FEEDING ECOLOGY AND DIET

Both forms feed on blood, but their ecology differs between body morphs. Head lice feed at regular intervals every few hours, whereas body lice feed only once or twice per day during periods of host inactivity.

REPRODUCTIVE BIOLOGY

The head louse variant attaches eggs (nits) to the base of hair shafts, whereas body lice glue their eggs to projecting fibers of clothing on the host. Eggs typically hatch in five to seven days, and the nymphs reach maturity after another 10–12 days.

CONSERVATION STATUS

Not threatened. However, populations restricted to isolated human tribes and nonhuman primates should be considered vulnerable.

SIGNIFICANCE TO HUMANS

As the principal vector of epidemic typhus (*Rickettsia prowazekii*), the body louse variant was responsible for hundreds of millions of deaths up until the early 1900s. Since World War II, large outbreaks of typhus have occurred mainly in Africa, with reported cases coming predominantly from Burundi, Ethiopia, and Rwanda. The head louse variant is common in the Western World, with infection rates exceeding 20% reported from selected primary schools in Australia, the United Kingdom, and the United States. This form is not known to transmit louse-borne diseases in natural circumstances. ◆

Slender pigeon louse
Columbicola columbae

FAMILY
Philopteridae

TAXONOMY
Pediculus columbae Linnaeus, 1758. Type host: *Columba livia domestica*.

OTHER COMMON NAMES
None known.

PHYSICAL CHARACTERISTICS
A long, slender louse, readily distinguished from other ischnoceran genera by the presence of two bladelike dorsal setae on the anterior margin of the head.

DISTRIBUTION
This louse is restricted to four species of pigeon (*Columba eversmanni*, *C. guinea*, *C. livia*, and *C. oenas*), including the widely distributed common rock dove. The latter has been introduced throughout the world and has a cosmopolitan distribution, living commensally with humans in temperate and tropical areas. Louse distribution is assumed to mirror the distribution of the host. The distribution map accompanying this text shows only the host's native distribution.

HABITAT
Restricted to the wing feathers of their hosts, either on the undersurface of the wing coverts or at the base of secondary feathers.

BEHAVIOR
Its slender shape allows the louse to live between the feather barbs. The edge of the barb is grasped with the mandibles and legs, protecting it from the preening activities of the host.

FEEDING ECOLOGY AND DIET
Feeds on the downy part of the feathers and may migrate from the wings to feed on the fluffy basal portions of body feathers.

REPRODUCTIVE BIOLOGY
Females deposit their eggs on the underside of the wing feathers, next to the pigeon's body. Eggs are attached to a feather in the space between feather barbs and hatch between three and five days at 98.6°F (37°C).

CONSERVATION STATUS
Not threatened. However, populations present on *Columba eversmanni* (the pale-backed pigeon) should be considered vulnerable.

SIGNIFICANCE TO HUMANS
Used as a "model organism" by biologists to address questions on the evolution and ecology of host-parasite interactions. ◆

Resources

Books

Hopkins, G. H. E., and T. Clay. *A Checklist of the Genera and Species of Mallophaga*. London: British Museum of Natural History, 1952.

Kim, K. C., H. D. Pratt, and C. J. Stojanovich. *The Sucking Lice of North America*. University Park: Pennsylvania State University Press, 1986.

Ledger, J. A. *The Arthropod Parasites of Vertebrates in Africa South of the Sahara* Vol. 4, *Phthiraptera (Insecta)*. Johannesburg: South African Institute for Medical Research, 1980.

Palma, R. L., and S. C. Barker. "Phthiraptera." In *Psocoptera, Phthiraptera, Thysanoptera*, edited by A. Wells. Zoological Catalogue of Australia, vol. 26. Melbourne, Australia: CSIRO Publishing; 1996.

Price, Roger D., Ronald A. Hellenthal, Ricardo L. Palma, Kevin P. Johnson, and Dale H. Clayton. *The Chewing Lice: World Checklist and Biological Overview*. Illinois Natural History Survey Special Publication no. 24. Champaign-Urbana: Illinois Natural History Survey, 2003.

Resources

Periodicals

Clay, T. "Some Problems in the Evolution of a Group of Ectoparasites." *Evolution* 3 (1949): 279–299.

———. "The Amblycera (Phthiraptera: Insecta)." *Bulletin of the British Museum (Natural History) Entomology* 25 (1970): 73–98.

Durden, L. A, and G. G. Musser. "The Sucking Lice (Insecta, Anoplura) of the World: A Taxonomic Checklist with Records of Mammalian Hosts and Geographical Distributions." *Bulletin of the American Museum of Natural History* 218 (1994): 1–90.

Price, M. A., and O. H. Graham. "Chewing and Sucking Lice as Parasites of Mammals and Birds." *USDA Agricultural Research Service Technical Bulletin* 1849 (1997): 1–309.

Smith, V. S. "Avian Louse Phylogeny (Phthiraptera: Ischnocera): A Cladistic Study Based Morphology." *Zoological Journal of the Linnean Society* 132 (2001): 81–144.

Other

"Tree of Life—Phthiraptera" [January 14, 2003]. <http://tolweb.org/tree/eukaryotes/animals/arthropoda/hexapoda/phthiraptera/phthiraptera.html>.

"National Pediculosis Association" [January 14, 2003]. <http://www.headlice.org>.

"Phthiraptera Central." November 8, 2002 [January 14, 2003]. <http://www.phthiraptera.org>.

"Phthiraptera Research" [January 14, 2003]. <http://darwin.zoology.gla.ac.uk/~vsmith>.

Vincent S. Smith, PhD

Hemiptera
(*True bugs, cicadas, leafhoppers, aphids, mealy bugs, and scale insects*)

Class Insecta
Order Hemiptera
Number of families More than 140

Photo: Whitefly larvae *Aleurothrixus antidesmae* (family Alerodidae) from Hawaii, USA. (Photo by Rosser W. Garrison. Reproduced by permission.)

Evolution and systematics

Most of the diversification of the Hemiptera started in the late Paleozoic (Upper Permian), and the major lineages diverged early in the Mesozoic (Triassic). Regrettably, some key fossils come from beds of uncertain age. Well-preserved, complete fossils are scarce—for example *Karabasia evansi* (family Karabasiidae, Upper Jurassic?); *Architettix compacta* (family Cicadoprosbolidae, Cretacic); *Incertametra santanensis* (family Hydrometridae, Cretacic). For the most part only isolated structures have been reported, and many of them have not been assigned with certainty to groupings below the family. Certain very-well-preserved specimens are known from amber, such as *Metrocephala anderseni* (family Hydrometridae) and *Succineogerris larssoni* (family Gerridae) from Baltic amber (Eocene), and *Brachymetroides atra* (family Gerridae) and *Halovelia electrodominica* (family Veliidae) from Dominican amber (Oligocene-Miocene). Several schemes of phylogenetic relationships among Hemiptera have been proposed, based on varying criteria. In the last decades of the twentieth century, they succeeded at short intervals, with remarkable discrepancies among them, and the proposed schemes are in steady flux.

The partitioning into two suborders, Homoptera and Heteroptera, or into three, segregating the Coleorrhyncha, does not express the inferred (and most accepted) paths of evolution. The relationships between Heteroptera and Homoptera and between suborders and infraorders of both major groupings have been discussed in many papers by several hemipterists; interesting results were obtained by the end of the 1980s, based on important paleontological, chemical, morphological, cytogenetical, and behavioral facts of living Hemiptera. The Heteroptera plus the Coleorrhyncha represent the most advanced grouping, derived from an extinct homopterous stock of Scytinopteroidea. Updated and adjusted results were edited by Schaefer.

The Hemiptera are divided into four undergroupings: the aphids and scale insects (and perhaps the whiteflies), included in Sternorrhyncha; the cicadas, leafhoppers, treehoppers, plant-hoppers (and perhaps the whiteflies), included in Auchenorrhyncha; the conenoses, water bugs, stink bugs, and others, included in Heteroptera, or true bugs; and the moss bugs, included in Coleorrhyncha. The Sternorrhyncha and the Auchenorrhyncha together are called Homoptera; the systematic position of the whiteflies is still dubious. The Coleorrhyncha and Heteroptera make up the suborder Prosorrhyncha. The Sternorrhyncha are clearly monophyletic (derived from a single ancestor). The Auchenorrhyncha do not appear to be monophyletic; they comprise two infraorders, the more primitive Fulgoromorpha and the more advanced Cicadomorpha. The monophyletic Heteroptera, or true bugs, are grouped into eight infraorders: Enicocephalomorpha, Dipsocoromorpha, Gerromorpha, Nepomorpha, Leptopodomorpha, Cimicomorpha, Pentatomomorpha, and Aradomorpha. The Coleorrhyncha, with a single family, restricted to the Southern Hemisphere, evolved parallel to both Homoptera and Heteroptera. The exact number of families of Hemiptera is largely a matter of opinion; by 2003 approximately 140

A thornlike treehopper (*Umbonia crassicornis*) in Florida, USA. (Photo by James H. Carmichael. Bruce Coleman, Inc. Reproduced by permission.)

extant families were accepted, some 60 in Homoptera, about 80 in Heteroptera, and one in Coleorrhyncha.

Physical characteristics

The body shapes are extremely diverse, ranging from plump, short, and cylindrical—such as the terrestrial scutellerid bugs and the aquatic pygmy backswimmers—to very slender—such as the semiaquatic water measurers, and some members of the assassin bug family—or even extremely flat—such as the aradid bugs. The sizes range from 0.03 in (0.8 mm: litter-dwelling bugs and some plant lice) to about 4.3 in (110 mm: giant water bug); of course, larvae in every case are much smaller. The head bears a beak, bent backward against the venter and varying from almost inconspicuous, as in the family Corixidae, to very long, reaching the rear end. It is a gutter-shaped, articulated labium holding the distal part of the very long and slender stylets. The antennae may be very short and concealed under the head border in aquatic bugs or longer and exposed in almost all other Hemiptera species. Although most adult hemipterans bear two pairs of wings, some are always wingless, like the females of scale insects. In Heteroptera the forewings are called hemelytra (derived from the classical Greek *hemi*, half, and *elytron*, case or

etui), because the basal half is mostly stiff and the distal one is membranous. In some Auchenorrhyncha they are called tegmina (classical Greek for carapace, a somewhat flexible but fairly stiff structure), since they are entirely stiff. Hind wings are membranous and translucent or whitish; some Homoptera have all four wings of the same texture (aphids and cicadas, among others). The legs have a wide adaptive spectrum: for walking, running, jumping, swimming, or skating on water; for grasping prey; or for digging. Only seldom are legs missing. Larvae generally are similar to small adults but lack wings and genital appendages; the larvae of some Homoptera, however, are quite different from adults. In the whiteflies, adults are slender, with long legs and ample membranous wings, and the larvae are broad and flat, scale-like, and lack legs.

Many hemipterans are dull-colored, for instance, most aquatic species; others display bright and contrasting colors, sometimes with a showy metallic shine. The color may be uniform, but stripes, dots, or extensive contrasting and brightly stained areas frequently appear. Larvae can be similar to adults in color or differ greatly; some evenly stained species may have spotted larvae, among them, many stinkbugs. Underground-dwelling cicada larvae are colorless or pale yellowish. The body surface sometimes is obscured by a whitish, compact, powdery, woolly, or filamentous wax cover, as in the mealy bugs and whiteflies.

Numerous hemipterans, such as the members of the stinkbug family, have scent glands, which are used against predators, as an alarm signal, or as a call to aggregation. These glands open ventrally or laterally in the hind thorax in adults, and dorsally on the abdomen in larvae. The strong smell justifies the name "stink bugs" for these offensive stinking species. Many families display alary polymorphism—individuals may have complete wings, reduced wings, or no wings. Fliers and nonfliers may coexist in a population, and proportions seem to be linked to the habitat. Among bugs living in the water, reduction affects only the hind wings; the forewings are complete, holding air for buoyancy. Several bugs living under water have adaptations for taking in oxygen at the water surface, such as siphons or dry, hairy areas. Some larvae take oxygen from the water itself.

Adult harlequin stink bug (*Murgantia histrionica*) on a flower, from southern California, USA. (Photo by Rosser W. Garrison. Reproduced by permission.)

Distribution

The Hemiptera are distributed broadly, both on land and in freshwater; moreover, the only insects dwelling on the open ocean are all bugs. On all continents families are more diverse in the tropics. In high boreal latitudes and high altitudes the ranges of some families extend close to permanent ice. Even remote oceanic islands such as Tristan da Cunha in the Atlantic and the Easter Islands in the Pacific are populated.

Habitat

Hemiptera may be either terrestrial or aquatic. They occur in almost all habitats, including deserts and at high altitudes. Those that have aquatic tendencies occur in every freshwater, brackish water, and saline habitat, including the open sea. Most hemipterans are terrestrial and dwell on plants (including roots), on the ground, in soil litter, or as external parasites on vertebrates. Many are linked to running or standing freshwater, living on the surface (semiaquatic bugs) or in the water (aquatic). Some live in water-filled tree holes or in epiphytic plants. Few are marine—only five species live on the surface of the open ocean. Certain species dwell in natural caves or those excavated by crabs. Others live in nests of social insects (ants and termites), or of birds. Still others live in spider webs.

Behavior

Most species are diurnal and dwell on the ground or on plants, searching for food or prey, for a mate, or for a suitable egg-laying site; every part of a plant, including the roots, may serve these purposes. Aquatic bugs thrive in or on water, frequently among aquatic plants, and almost all are predaceous. Stones, twigs, and other substrates may serve as perches or shelters, especially in swift-running brooks, or as egg-laying sites. In the water they swim or crawl on the bottom or on supports (aquatic bugs); on the surface they walk or skate (semiaquatic bugs). Females and advanced larvae of most scale insects do not move from the site upon which they set

down (only the first larval instar has active legs and moves around); the tiny males are winged and fly in search of ripe females. No hemipteran is truly social, as are ants and termites and some bees and wasps, but some live in dense aggregations, sometimes only transitorily.

Visual displays are achieved by expanded legs, wings, or antennae, which sometimes also are brightly or contrastingly stained or very hairy. Sounds are produced by scraping together two sclerotized parts of the body or by vibrating the tymbals, as in the cicadas; specialized hairs receive the sound waves. Some bugs drum with their legs on the substrate, and others (e.g., certain assassin bugs) scrape their beaks against their own chests, the vibration being transmitted via the legs to the substrate. Males of many water striders produce ripples on the surface with their legs, which are detected at a distance. Scent substances, the sex pheromones, usually are present too. All these signals are highly specific for attracting sexes to each other; some individuals can identify them and decide to flee, avoiding competition with individuals that had previously arrived and may have already established their territories.

Many hemipterans disperse by flight, especially water-dwelling species. Dispersal capacity may be retained for life or soon lost. Rain-pool species can fly almost their whole life span, and they successively colonize habitats avoided by stability-loving species, thus lessening the risk of competition because they are able to support environmental instability. No hemipterans have spectacular migratory swarms, and they do not fly great distances, as do certain butterflies, but some observations suggest that giant water bugs can fly many hundreds of kilometers at a time. In the dry and warm season, plant lice start shorter migrations by the thousands, producing a sort of fog. Territoriality was studied in only a very few Hemiptera species, and the results are not understood clearly. Signals, both mechanical and chemical, play an important role in marking territorial boundaries, as seems to be the case among certain water striders and large coreids.

A seventeen-year cicada emerges from its pupa. (Photo by Dr. E. R. Degginger, FPSA. Bruce Coleman, Inc. Reproduced by permission.)

Feeding ecology and diet

All the Homoptera are sap feeders. Heteroptera are primarily predaceous, and in the course of evolution several groups have evolved to exploit plants; some of these groups evolved and returned to the predatory condition or shifted to parasitism on higher animals. All take only liquids in the form of plant juices, chiefly sap, or animal juices, chiefly predigested tissues or blood; only a few water bugs are able to add small particles to their diet, such as algae. Plant juices are taken from leaves, stems, buds, flowers, fruits, or roots. Some species suck the contents of fungal threads under rotting bark; others suck from moss cells. Predators suck from almost every arthropod; some prefer snails; and others may attack small fish, frogs, and tadpoles. Certain bugs feed on dead or half-dead arthropods, mainly insects. Those that suck blood take only warm blood from birds or mammals, mostly bats. Cannibalism may occur, primarily in gregarious species. Some Hemiptera do not feed as adults, for example, the males of scale insects. Food is obtained in the places they live, but blood-sucking bugs rest in shelters and leave their shelters to bite warm-blooded hosts.

Water striders' main food consists of aerial insects blown onto the water surface; they localize these prey by the ripples they produce with their helpless movements. Oceanic bugs feed on dead, floating jellyfish, and on planktonic microcrustaceans and fish larvae trapped in the surface film. Larvae of most Hemiptera thrive and feed in the same manner as adults, but some stink bugs take sap at the first larval instar and later become predaceous like the adults. Adult cicadas take sap from twigs, but larvae dwell underground and take it from roots. The first-instar larvae of some species eat no food. Some Sternorrhyncha produce anomalous growths in leaves, stems, or roots, called "galls"; larvae and newly molted adults feed and develop therein, without seriously interfering with the plant's physiology. Some Homoptera that take dilute resources, such as plant sap, eliminate much of the water immediately, concentrating the nutritive substances. They excrete some sugar with the liquid feces, which, if abundant, can be used as food by mammals and humans, as the biblical "manna." Often a black mold that resembles soot develops on that substance and affects plant growth as it restricts the amount of light that reaches leaves.

The saliva may form a small cone on the plant surface, helping hold the slender mouthparts in position. Other saliva components break the cell walls down, to release the contents. Some bugs feed on dry seeds, piercing them to inject a digestive saliva and sucking the resulting fluid. Still others inject a plant hormone mimic, which mobilizes nitrogen-rich substances to the wound.

Reproductive biology

Courtship and mating take place mostly on perches among terrestrial species, generally on plants but sometimes on the ground or in shelters. Courtship, if any, is frequently brief and consists of chirps, scent emissions, or displays with the legs or antennae or a combination of these types of behavior. Courtship activities are undertaken by the males, but in some species females collaborate; in most boatmen the loud male chirps are answered weakly by females, allowing for mutual localization. Some bugs living in the water alternately take and expel air, thus rising and sinking at the surface and forming coarse waves, which orient the partner. Water bugs mate above or below the surface; submerged and floating plants and logs may serve as perches.

At mating the male mounts the female, but sometimes he shifts to an end-to-end position. The male also may lie at an angle across the female or beside her. Coupling may be brief or may last for hours. Insemination is internal, the male transferring sperm with specialized intromittent organs (a rather complex aedeagus). So-called traumatic insemination is known in a couple of families: males slash the abdominal wall of females with their swordlike claspers, discharging sperm in the general cavity. The number of scars shows how often a female has mated. Immobile, wingless female scale insects are mated by the winged males, which may have a very long copulating organ to reach beneath the shield or the wax cover.

Parthenogenesis (reproduction without mediation of males) is extremely rare in Heteroptera; it is mentioned in only a couple of unrelated species. It is frequent, however, in certain Sternorrhyncha. Most plant lice, for example, alternate yearly between one bisexual reproductive phase and few to many parthenogenetic phases, the females giving off living larvae, which in turn give off living larvae. At the end of the season, a bisexual, frequently winged phase reappears. Some pest plant lice are permanently parthenogenetic in warm areas. Some scale insects may be bisexual or parthenogenetic, depending on environmental or geographic factors.

Eggs are often barrel-shaped, more or less elongated, and sometimes weakly curved; they may bear a long stalk. The shell is translucent or whitish, but frequently the stained vitellum makes them brightly colored. The surface is smooth or rugose or is ornamented with spines, tubercles, crests, and so on, giving them a bizarre appearance. Eggs laid under the water's surface have a thick and complex shell, formed by extremely fine spun work, which traps the tiniest air bubbles, providing oxygen for the embryo. Most eggshells have a weak strip, which breaks at hatching; it may form a distal ring, defining a cap, which drops away. In some species the strips form a distal rosette: the shell "bursts," and the larva emerges through the opening. Some banana-shaped eggs are embedded in aquatic plants, and the strip is longitudinal, allowing for easy emersion (emergence from the egg). Many first-instar larvae bear a sharp "egg burster" on the head, which tears the eggshell.

Eggs may be laid singly or in groups in suitable sites for the emerging young to find food soon. Most plant-feeding bugs glue eggs with a special secretion; aquatic and aerial bugs embed them into tissues. Predators and blood-feeders glue them to a firm substrate or let them drop to the floor of the host's resting site. Many scale insects hide the egg batch below themselves or below the rigid dorsal shield or below a wax cover. Among certain giant water bugs the female lays the eggs on the back of the male. Bat parasites do not lay eggs; instead, they retain and nourish the young in the genital tract.

No hemipteran species spins nests or egg cases. Only a few species care for eggs or young or both. Males of some giant waterbugs protect the eggs glued to their backs and clean them using their legs, which also ensures an oxygen supply for the embryos. Males of other giant water bugs stay near the egg clusters laid on twigs emerging out of the water, guard them, and readily threaten potential egg predators. Egg guarding has been observed in isolated cases among several terrestrial families; the female or male covers the egg batches until they hatch. This behavior perhaps is more frequent than usually is assumed. The best protective action probably is to select a proper site for laying the eggs, lowering the risks at hatching.

Many hemipterans living in temperate zones have a single yearly reproductive cycle, but some have two or more generations per year. Exceptions are the cicadas, which have an extended larval development lasting several years. Embryos or adults, but rarely larvae, overwinter; some species aggregate en masse to pass this period. In the tropics generations frequently overlap, and insects at all stages of development are found there year-round. If marked climatic seasonality occurs (humid vs. dry and warm vs. cold), estivation or hibernation may take place. The unfavorable season sometimes is passed

Larval stink bug (family Pentatomidae) resting on a leaf, from Mexico. (Photo by Rosser W. Garrison. Reproduced by permission.)

in a dormant, hormonally driven diapause; the embryo, the larva of every instar, or the adult may engage in this diapause, which may last up to half a year or even more (nine months in one Alaskan shore bug).

Conservation status

Among insects, hemipterans are not a frequently mentioned group of conservation concern. By 2002 the IUCN Red List had cited only five species of Homoptera (two Extinct and three Near Threatened) and no species of Heteroptera. There are some regional listings for Europe, North America, New Zealand, and other countries, which also include very few Hemiptera. More extensive samplings might indicate that more species are of conservation concern.

Among wild species, decline follows environmental damage. For aquatic species, water pollution is a widely occurring alteration. For water striders and the like, any surfactant present in the water, even in low amounts, is lethal. Extensive control of pests with insecticides, which are never specific, put at risk every insect, whether a pest or not. Extensive forest clearing eliminates many scarce trees and shrubs, perhaps the only food resource of some bugs, and also water-filled epiphytic plants where some bugs live. Forest clearing also increases the light intensity at the ground level, affecting shade-adapted plants and insects. Only in protecting ecosystems and ensuring their sustainable use will the Hemiptera be protected.

Significance to humans

Humankind profits from bugs by eating them, by using them against pests, and as models in art. Bugs are also used as a source of amusement and entertainment; for example, cicadas are often tied together with thread to make decorations or jewelry, and children have races with insects. On the other hand, humankind suffers from bug bites, from illnesses transmitted by certain species, and from agricultural pests. Some Mexicans eat the egg masses of the boatmen species, called "ahuautle." Shrub twigs are drowned in ponds after large amounts of eggs have been glued to them; eggs are later gathered and either fried or dried in the sun. In India the adults of one species of giant waterbug are cooked in syrup, which is considered an expensive delicacy. Roasted, egg-filled female cicadas are eaten in some Asian countries. In several places, stink bugs are dried and powdered for use as a condiment. People have engaged in other forms of recreational and ceremonial insect eating.

Cicadas have received considerable attention, as it is said that their chirps predict warm or stormy weather. They have been highly regarded as symbols of resurrection. In the Far East jade carvings in the shape of cicadas once were put in the mouth of dead princes and other important people. The Chinese keep cicadas in cages, like singing birds, and also fly kites made in their shape. Kissing bugs, especially the domestic species that transmit Chagas' disease, are reputed to bring luck and happiness, which has made them popular and has contributed to the dissemination of the illness and constraining sanitary controls. A few predaceous bugs are used as biological control agents against crop pests and are reared to be released in the field. Some species of stink bugs, chiefly a plant feeder family, are efficient at attacking pest caterpillars on soybean plants; certain species ingest and transmit caterpillar disease viruses (e.g., polyhedrosis virus).

Carmine is a very valuable dyestuff produced by an American cactus-dwelling scale insect; the plant and the insect were introduced to other continents by the eighteenth century, producing a noxious spread of the weed that rendered vast tracts of land unusable. Shellac is produced by another tropical scale insect that dwells on fig trees. Several Hemiptera occurring on different wild plants shifted to crop plants once they were introduced into new areas and then turned into pests. Some are genus specific or family specific; others are generalists. The ensemble is extensive and includes many of the plant-feeding bug families. The pest condition is achieved through the generous food supply offered by cultivation, encouraging insect numbers to rise quickly. Families with many serious pest species are Aphididae, Coccidae, Diaspididae, Delphacidae, Pentatomidae, Miridae, and Alydidae, among others. Underground-dwelling larvae of some cicadas may cause locally severe losses in sugarcane fields. Often the damage caused to crops is not so much due to the removed sap but to the transmission of microbial diseases, mainly plant viruses and fungi. Species of the family Delphacidae cause losses in corn and other cereals by transmission of viruses; some Pentatomidae transmit disease-producing fungi. One American species of *Phylloxera* may produce severe damage in the production of grapes by attacking the roots and leaves of the vines; they have caused extensive losses in European countries where the species was accidentally introduced.

Some water bugs, and members of the assassin bug family sting very quickly with their mouthparts when they are taken by hand, injecting a poisonous saliva and causing intense pain and sometimes swelling. Some exclusively blood-sucking bugs attack humans and may transfer parasitic microbes, causing diseases; indeed, Chagas' disease, widespread in tropical and temperate South America, is due to a flagellate, *Trypanosoma cruzi*, which also is a common parasite of many wild and domestic mammals.

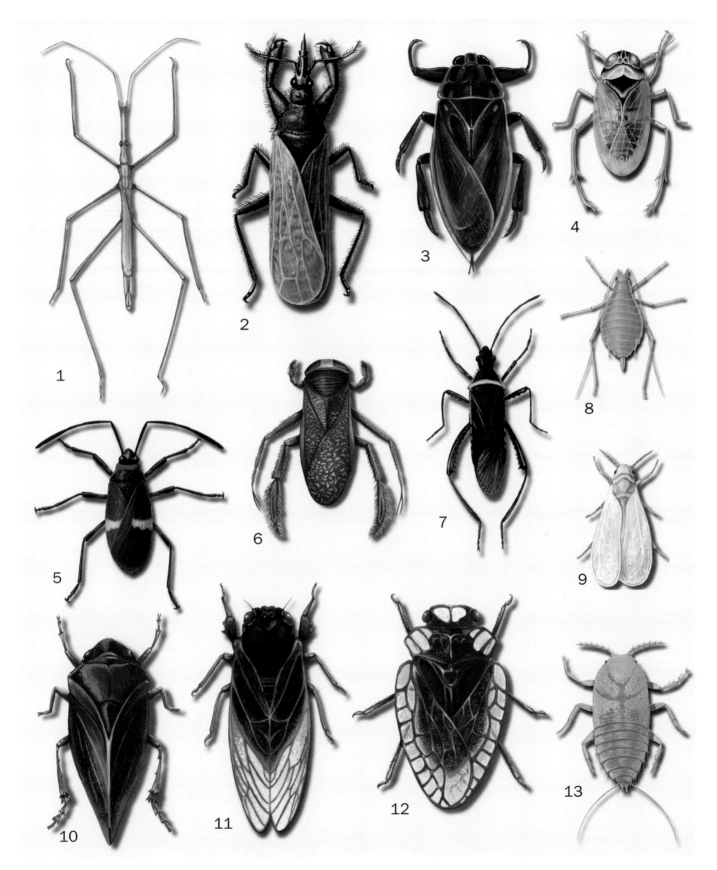

1. Water measurer (*Hydrometra argentina*) 2. Unique-headed bug (*Gamostolus subantarcticus*); 3. Giant water bug (*Lethocerus maximus*); 4. Delphacid treehopper (*Delphacodes kuscheli*); 5. Staining bug (*Dysdercus albofasciatus*); 6. Water boatman (*Sigara platensis*); 7. Tomato bug (*Phthia picta*); 8. Pea aphid (*Acyrthosiphon pisum*); 9. Greenhouse whitefly (*Trialeurodes vaporariorum*); 10. Spittle bug (*Cephisus siccifolius*); 11. Seventeen-year cicada (*Magicicada septendecim*); 12. Moss bug (*Peloridium hammoniorum*); 13. Rhodesgrass mealybug (*Antonina graminis*). (Illustration by Katie Nealis)

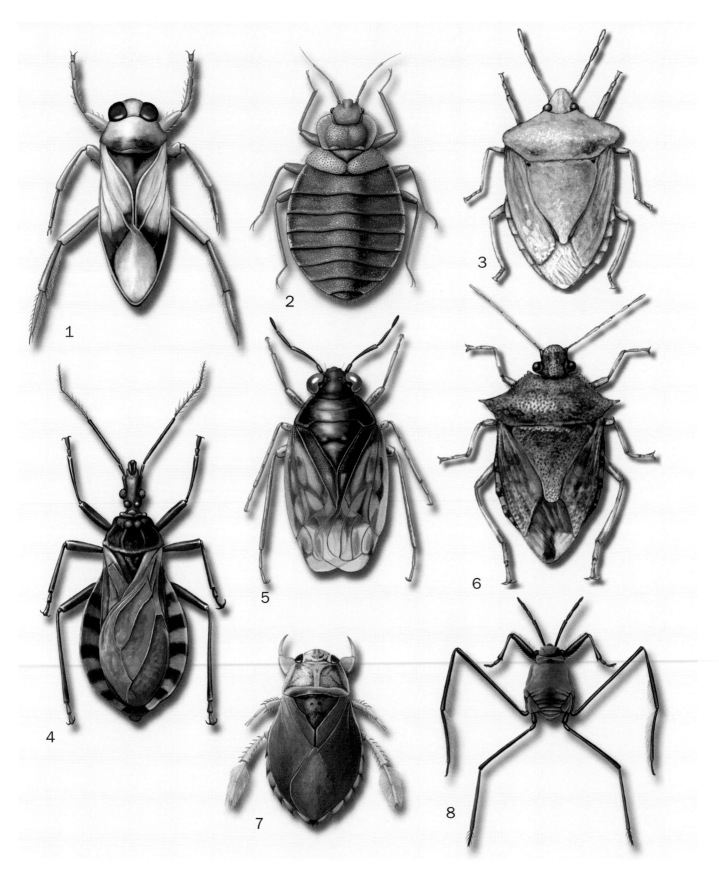

1. Backswimmer (*Notonecta sellata*); 2. Bed bug (*Cimex lectularius*); 3. Southern green stink bug (*Nezara viridula*); 4. Kissing bug (*Triatoma infestans*); 5. Shore bug (*Saldula coxalis*); 6. Spiny soldier bug (*Podisus maculiventris*); 7. Creeping water bug (*Ilyocoris cimicoides*); 8. Sea skater (*Halobates micans*). (Illustration by Emily Damstra)

Species accounts

Greenhouse whitefly
Trialeurodes vaporariorum

FAMILY
Aleyrodidae

TAXONOMY
Aleurodes vaporariorum Waterhouse, 1856, England.

OTHER COMMON NAMES
English: Glasshouse whitefly; Spanish: Mosca blanca de invernaderos.

PHYSICAL CHARACTERISTICS
Small, about 0.06 in (1.5 mm) long. Body and wings are powdered with white wax, which masks the yellowish to pale brown surface. Sexes are similar, both winged. Larvae scalelike and yellowish.

DISTRIBUTION
Cosmopolitan and intertropical; almost exclusively found in greenhouses in temperate zones.

HABITAT
Leaves (mostly on the underside) and twigs of a great variety of plants, including many cultivated ones.

BEHAVIOR
First-instar larvae walk for a couple of hours and then fix the beak at the underside of a leaf and remain there through four molts. The last instar serves as a puparium, inside which the winged adult develops. Gregarious, mostly at the underside of leaves; adults fly quickly if disturbed.

FEEDING ECOLOGY AND DIET
Once the ambulatory first-instar larva finds a suitable place on a leaf, it buries its beak and starts feeding. Adults also suck sap but move around.

REPRODUCTIVE BIOLOGY
Mating and egg laying take place on the plants. The yellow eggs are glued to the surface in curved rows; they turn black before hatching. Reproduction occurs year-round.

CONSERVATION STATUS
Not threatened.

SIGNIFICANCE TO HUMANS
A serious pest in greenhouses. Large numbers reduce plant vigor. Sooty mold develops on the honeydew from adults and larvae, reducing marketability; some cultures must be abandoned. ◆

Pea aphid
Acyrthosiphon pisum

FAMILY
Aphididae

TAXONOMY
Aphis pisum Harris, 1776. Type locality not specified.

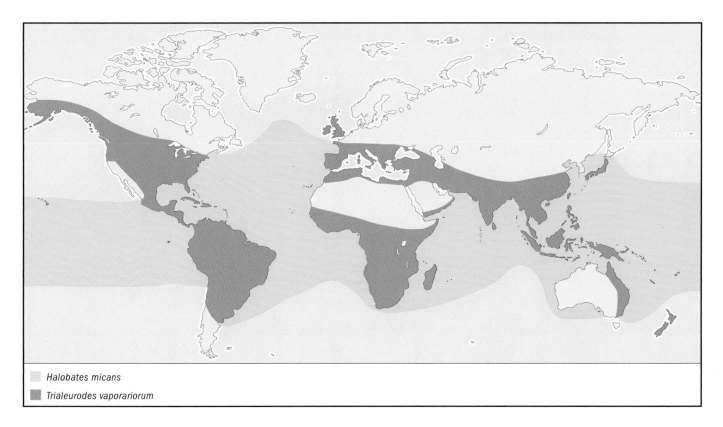

Halobates micans
Trialeurodes vaporariorum

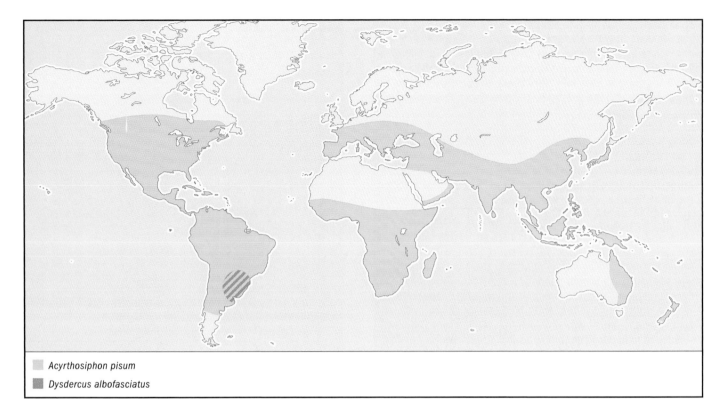

☐ *Acyrthosiphon pisum*

■ *Dysdercus albofasciatus*

OTHER COMMON NAMES
English: Green alfalfa aphid; French: Puceron vert, puceron du pois; German: Grüne Erbsenblattlaus; Spanish: Pulgón verde de alfalfa; Italian: Afidone verdastro del pisello; Portuguese: Piolho verde da ervilla.

PHYSICAL CHARACTERISTICS
Small, 0.08–0.16 in (2–4 mm) long. Wingless and winged individuals are light green, with red eyes. Antennae are slender and at least as long as the body. Legs long and slender. Wings, if present, are translucent. Abdominal siphons are long and slender. Larvae like small wingless adults.

DISTRIBUTION
Intertropical and temperate zones.

HABITAT
Canopy of leguminous weeds and shrubs. Prefer areas with alfalfa crops.

BEHAVIOR
In winter they hide as wingless parthenogenetic females on leguminous shrubs. In spring, winged migratory individuals appear. Adults and young form dense patches.

FEEDING ECOLOGY AND DIET
Adults and larvae take sap from leaves, stems, and flowers of alfalfa, pea, bean, clover, and other crops.

REPRODUCTIVE BIOLOGY
The species is permanently parthenogenetic in most temperate and subtropical zones, although in some areas one bisexual and many parthenogenetic generations alternate every year; females produce young, which in turn give off more young after a week.

CONSERVATION STATUS
Not threatened. Control is achieved chiefly by parasitoid wasps.

SIGNIFICANCE TO HUMANS
It is adequately controlled in most countries, but it has been a serious pest of alfalfa crops. ◆

Giant water bug
Lethocerus maximus

FAMILY
Belostomatidae

TAXONOMY
Lethocerus maximus De Carlo, 1938, Santa Cruz, Bolivia.

OTHER COMMON NAMES
English: Giant electric light bug, giant toebiter, giant fishkiller; German: Riesenwasserwanze; Spanish: Chinche de agua gigante, cucaracha de agua gigante; Portuguese: Barata d'água gigante.

PHYSICAL CHARACTERISTICS
The largest bug: adults can exceed 4.3 in (110 mm). Color an almost uniform grayish-brown; individuals may be darker or paler. Forelegs are robust, adapted for grasping prey; the middle and hind legs are flattened, adapted to swimming. Ventral parts covered with very short hairs that retain air for breathing, which is taken in by two extensible posterior appendices. Sexes alike. Larvae are like adults but smaller and wingless.

DISTRIBUTION
Tropical South America from the West Indies to northern Argentina.

HABITAT
Margins of pools and lakes, especially dam lakes, in the water among plants.

BEHAVIOR
Prey, sometimes larger than the insects themselves, are grasped with the front legs, a poison is injected, and the digested tissues are sucked up. For taking in air, the giant water bug swims backward until the breathing appendages reach the surface. Courtship and egg laying have not been described. Males guard the eggs until they hatch. Dispersion flights occur at night, during which the bugs frequently are attracted to artificial lights.

FEEDING ECOLOGY AND DIET
A strong hunter of aquatic insects, fish, frogs, and tadpoles. Prey is killed by digestive saliva, and the liquefied tissues then are sucked up.

REPRODUCTIVE BIOLOGY
Mating takes place in the water among plants; clusters of several dozens eggs are laid around a twig above the surface. Larvae disperse at hatching.

CONSERVATION STATUS
Not threatened.

SIGNIFICANCE TO HUMANS
Predation on young fish may seriously constrain production in fish culture. Bites are very painful, but infrequent. ◆

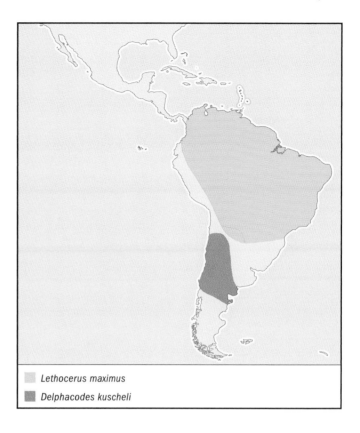

Lethocerus maximus
Delphacodes kuscheli

Spittle bug
Cephisus siccifolius

FAMILY
Cercopidae

TAXONOMY
Aphrophora siccifolia Walker, 1851, "West Africa" (probably erroneous).

OTHER COMMON NAMES
English: Frog-hopper, froth bug, spittle-insect, cuckoo's spittle; Spanish: Cotorrita de la lluvia, cotorrita de las tipas.

PHYSICAL CHARACTERISTICS
Elongated and anteriorly rounded, the hind portion acutely stretched. About 0.4–0.5 in (10–12 mm) long. Dull brown, with indefinite darker areas and minute whitish dots. Larvae are plumper, soft-bodied, and whitish with dark markings.

DISTRIBUTION
Mexico through Argentina and Uruguay.

HABITAT
Canopy of several trees, mainly Leguminosae, in dense and open forests. Adults are solitary and larvae gregarious, living in the abundant froth they produce. Common in parks and avenues of cities.

BEHAVIOR
Adults fly frequently, mate, and search for egg-laying sites. As most homopterans do, spittle bug larvae quickly eliminate most of the water contained in the sap they consume. The water is excreted as watery feces, to which they add a thickening secretion and blow air bubbles into. The air bubbles produce a froth that builds up to form a large mass, which eventually begins to condense very slowly and produces drops that fall from

the canopy of trees to the ground. This process has given the bug its Spanish common name "cotorrita de la lluvia."

FEEDING ECOLOGY AND DIET
They feed on sap from twigs of trees; the amount removed is the highest known among the Homoptera.

REPRODUCTIVE BIOLOGY
Mating occurs on twigs, in which the eggs are embedded in summer. Larvae hatch the next spring and develop quickly, reaching adulthood in a few weeks.

CONSERVATION STATUS
Not threatened.

SIGNIFICANCE TO HUMANS
Abundant dropping of condensed froth sometimes causes trouble in cities. ◆

Seventeen-year cicada
Magicicada septendecim

FAMILY
Cicadidae

TAXONOMY
Cicada septendecim Linné, 1758 Boreal America.

OTHER COMMON NAMES
English: Periodical cicada, locust (this name should be restricted to grasshoppers); French: Cigale de dix-sept ans.

Cephisus siccifolius

Magicicada septendecim

Peloridium hammoniorum

PHYSICAL CHARACTERISTICS
Plump, 1.37–1.57 in (35–40 mm) long. Dull dark brown to
black and shiny, with reddish eyes and legs. Wings transparent,
with well-marked, reddish veins. Venter has broad orange
stripes. Antennae very short and hairlike. Legs short, adapted
for walking; when at rest wings extend far beyond the rear of
the abdomen. Larvae almost colorless, living underground.

DISTRIBUTION
United States east of the Great Plains.

HABITAT
Canopy of deciduous trees, in temperate forests and rainforests.

BEHAVIOR
Adults are diurnal. Males chirp loudly, attracting other males
as well as females to the chorus. Larvae remain at the same
site on a root for a long period of time and move stepwise to
thicker roots as they grow and molt. At maturity they dig a
tunnel to emerge. At sunset they emerge and molt to the
adult form on stems. Most of every brood molts at the same
time, thus appearing simultaneously in great numbers every
17 years at each location (but not always in the same year in
distant locales).

FEEDING ECOLOGY AND DIET
Plants are pierced to suck liquid; adults feed from twigs and
larvae from roots.

REPRODUCTIVE BIOLOGY
The males' chirping attracts females, and mating occurs on
stems in the end-to-end position. Eggs are embedded in plant
tissues in the spring, up to twenty in each nest. Young larvae
drop from the trees and burrow underground, searching for a
suitable rootlet. They need 17 years to reach maturity.

CONSERVATION STATUS
Not threatened, but destruction of the forests can affect populations.

SIGNIFICANCE TO HUMANS
If abundant on plantations, larvae are harmful to roots. End
twigs replenished with eggs fail to grow the next season, especially in young trees. Choruses of cicadas may be annoyingly
loud. The rigid periodical abundance is astonishing. ◆

Bed bug
Cimex lectularius

FAMILY
Cimicidae

TAXONOMY
Cimex lectularius Linné, 1758, "Habitat in domibus exoticus,
sed ante epocham salutarem in Europa, at in Anglia vix ante
1670 visus teste Southall" [It lives in exotic countries' housings,
but before the healthful epoch in Europe, as it was seen in
England somewhat before 1670, according to Southall].

OTHER COMMON NAMES
English: Red coat, mahogany flat, wall louse; French: Punaise
des lits; German: Bettwanze; Spanish: Chinche de cama; Portuguese: Percevejo de cama.

PHYSICAL CHARACTERISTICS
Oval to round, very flat when not fed, globose and longer
when fully fed. 0.16–0.19 (4–5 mm) long. Rusty brown to dull
red. Rather short beak. Four-jointed antennae. Fore thorax
roundly expanded around the rear of the head. Legs short.
Flightless. Larvae look like small adults.

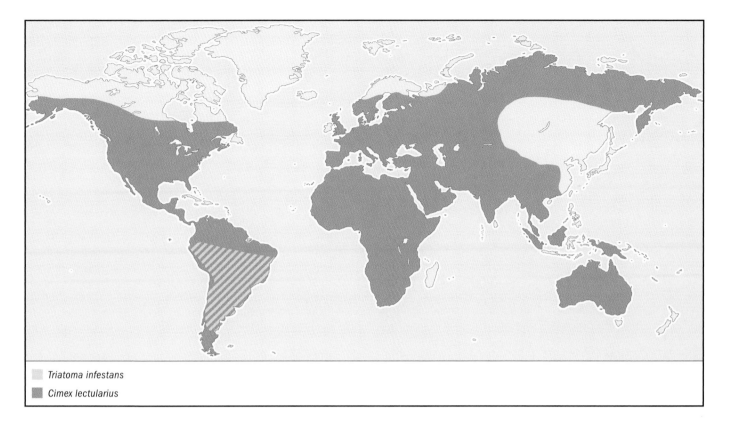

Triatoma infestans
Cimex lectularius

DISTRIBUTION
Almost cosmopolitan, including cold areas, but rare or absent in large parts of Asia because of hybridization with another species.

HABITAT
Human-made housings, mostly in bedrooms, crevices, bed frames, and mattresses or under wallpaper.

BEHAVIOR
They creep from their shelters and walk among bedding and clothing to the host, biting to take the blood of resting humans. They avoid humid surfaces.

FEEDING ECOLOGY AND DIET
Adults and larvae hide in the daytime and emerge at night, searching for human blood; they also may feed on poultry, dogs, and bats. They can take a blood meal four to five times as heavy as themselves. Starving individuals survive a long time, more than a year in cold climates, but do not reproduce.

REPRODUCTIVE BIOLOGY
Mating occurs in the shelters; the male mounts the female, puncturing her abdomen with his large clasper and discharging sperm into her cavity. Eggs are glued to any support at or near the bed. Young larvae start feeding soon.

CONSERVATION STATUS
Not threatened.

SIGNIFICANCE TO HUMANS
Bed bugs were known in ancient Egypt and classical Greece. Rapid expansion of populations can be expected in crowded conditions. Bed bugs can transmit some microbial parasites. Biting itself is painless, but the saliva produces an uncomfort-

able itching. Some people used (and perhaps still use) bed bugs as a medicine. ◆

Tomato bug
Phthia picta

FAMILY
Coreidae

TAXONOMY
Cimex pictus Drury, 1770, Meridional America.

OTHER COMMON NAMES
English: Potato bug; Spanish: Chinche del tomate, chinche de la papa; Portuguese: Chupador do tomate.

PHYSICAL CHARACTERISTICS
Elongate, reaching some 0.59 in (15 mm) in length. Its abdomen is nearly triangular in shape. Dark grayish brown to almost black, frequently with a transverse yellowish to orange cross stripe. Long and slender antennae and legs. Sexes alike. Larvae wingless, brightly colored, with silky, whitish areas.

DISTRIBUTION
Most of tropical and temperate America from southern United States (California through Florida) through Uruguay and central Argentina in lowlands.

HABITAT
The canopy of bushes and dense grasses; frequently seen on tomato and other solanaceous crops.

BEHAVIOR
Adults are mostly solitary, moving about on plants and flying readily if disturbed. Larvae are gregarious, forming dense

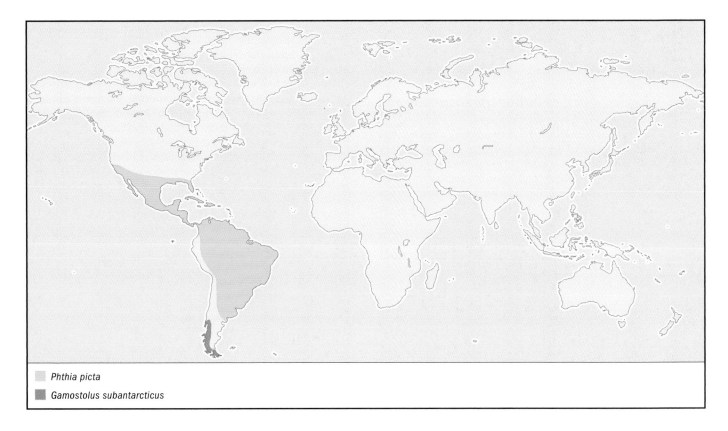

Phthia picta

Gamostolus subantarcticus

groups, which remain for extended periods of time at the same place.

FEEDING ECOLOGY AND DIET
Adults and larvae take sap from shrubs and bushes, mainly from the unripe fruits; they prefer solanaceous species, but they are found on squash too.

REPRODUCTIVE BIOLOGY
Mating occurs on the plants, in an end-to-end position. Some 20–40 pale yellowish eggs are glued to leaves, in straight rows, and become ochre in color. After hatching, larvae remain aggregated almost until the last instar. Three to five generations per year.

CONSERVATION STATUS
Not threatened.

SIGNIFICANCE TO HUMANS
A minor pest, mainly of tomato crops; little effort is needed to control populations. ◆

Water boatman
Sigara platensis

FAMILY
Corixidae

TAXONOMY
Sigara platensis Bachmann, 1962, Buenos Aires, Argentina.

OTHER COMMON NAMES
None known.

PHYSICAL CHARACTERISTICS
Elongate, some 0.24–0.28 in (6–7 mm) long. Dull brown, composed of very fine dark and pale vermiculate lines. Beak very short, not prominent. Forelegs spoon-shaped and adapted for sweeping small particles; middle legs long and very slender, adapted for anchoring to the bottom; hind legs extended to the sides like the oars of a boat, flattened, and with long hairs, adapted for swimming. Larvae similar to adults but much smaller and wingless.

DISTRIBUTION
Lowlands of southern South America, from central Bolivia, Paraguay, and southern Brazil to northern Patagonia, east of the Andes.

HABITAT
Shallow ponds and pools, especially rain pools, in water with few plants. A pioneer species.

BEHAVIOR
Most of their time is spent at the bottom, anchored to the substrate. Every few minutes they loosen from the bottom, reaching the surface by buoyancy; take air very quickly; and return to the bottom by swimming. They leave habitats that are drying up, flying in search of another one. They are attracted en masse to artificial lights.

FEEDING ECOLOGY AND DIET
Small prey are swept up with the forelegs and pierced by the buccal stylets; digestive juices are injected, and once the contents are digested they are swallowed. Very small items can be ingested whole. Algae and detritus may be taken too.

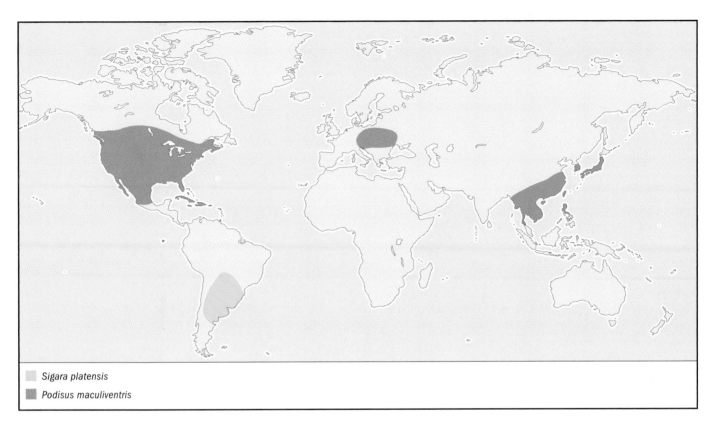

Sigara platensis
Podisus maculiventris

REPRODUCTIVE BIOLOGY
Courtship begins with loud chirping by the male, answered
gently by the female, which orients the insects to each other.
The male mounts the back of the female, bends his abdomen
laterally, and mates with her from the right side. Eggs are
glued to any support. Males soon die; adult proportions are
usually biased toward females.

CONSERVATION STATUS
Not threatened.

SIGNIFICANCE TO HUMANS
The appearance of masses of insects around artificial lights at
night may be troublesome. ◆

Delphacid treehopper
Delphacodes kuscheli

FAMILY
Delphacidae

TAXONOMY
Delphacodes kuscheli Fennah, 1955, Juan Fernández Islands,
Chile.

OTHER COMMON NAMES
Spanish: Cotorrita del maíz.

PHYSICAL CHARACTERISTICS
Elongate, some 0.31–0.39 in (8–10 mm) long. Brownish to al-
most black, with a narrow yellowish stripe on the head.
Forewings each have a black spot. Beak reaches the midlegs.
Legs adapted for walking. Hind legs also adapted for jumping,

as the ends of the tibiae contain conspicuously flattened, dentic-
ulate spurs. Short-winged specimens are more common than
winged ones. Sexes alike. Larvae resemble adults but are
smaller and wingless.

DISTRIBUTION
Central Chile and central to northwestern Argentina, mainly in
corn crop areas, up to an elevation of 6,562 ft (2,000 m).

HABITAT
On stems and blades of grasses.

BEHAVIOR
Long-winged adults fly easily from grass to grass; all forms
(both flying specimens and those that do not fly) and larvae
walk along stems and blades and jump quickly if disturbed.
Eggs are embedded in grass stems.

FEEDING ECOLOGY AND DIET
Larvae and adults suck sap from grasses, frequently of cereals.

REPRODUCTIVE BIOLOGY
Mating takes place on grasses. Slots are cut into blades of grass
and two eggs are embedded in each slot; if the blades are too
tough to be cut, the bugs move onto softer grass.

CONSERVATION STATUS
Not threatened.

SIGNIFICANCE TO HUMANS
An important cereal crop pest, causing heavy losses of yield,
mainly by transmission of pathogenic viruses such as *MRCud*;
this endemic disease is called "Mal de Río Cuarto del maíz"
(Río Cuarto maize disease). ◆

Unique-headed bug
Gamostolus subantarcticus

FAMILY
Enicocephalidae

TAXONOMY
Henicocephalus subantarcticus Berg, 1883, Isla de los Estados, Tierra del Fuego, Argentina.

OTHER COMMON NAMES
English: Four-winged fly.

PHYSICAL CHARACTERISTICS
Small, elongate, 0.24–0.31 in (6–8 mm) long. Dull yellowish to light brown in color. Prothorax is constricted twice, forming three lobes. Forelegs are thicker, adapted for grasping. Wings have somewhat reduced veins; both sexes are winged.

DISTRIBUTION
From southern tip of South America north to Osorno, Chile.

HABITAT
Hidden under stones in humid, cold-temperate forests.

BEHAVIOR
Nothing is known.

FEEDING ECOLOGY AND DIET
Generalized predator on small prey.

REPRODUCTIVE BIOLOGY
Little is known. Both sexes form dense swarms. Mating has not been observed. Eggs probably are laid in shelters.

CONSEZTION STATUS
Not threatened.

SIGNIFICANCE TO HUMANS
None known. ◆

Sea skater
Halobates micans

FAMILY
Gerridae

TAXONOMY
Halobates micans Eschscholtz, 1822, southern Pacific and Atlantic Oceans.

OTHER COMMON NAMES
English: Sea water-strider.

PHYSICAL CHARACTERISTICS
Plump, with a very short abdomen. About 0.16–0.18 in (4.0–4.6 mm) long. Body is dull brown to black with a very fine silvery coating; there is a pair of yellow marks on the head. Antennae are long. Forelegs short; middle and hind legs very long and slender. Wings absent.

DISTRIBUTION
Intertropical zones of all oceans, ranging to temperate latitudes.

HABITAT
The open ocean surface.

BEHAVIOR
They skate swiftly on the water surface, with the aid of a fringe of hair on the middle legs. Behavioral adaptations to the open ocean life may include a means of communication between individuals, to keep the population in one place.

FEEDING ECOLOGY AND DIET
Adults and larvae feed on planktonic microcrustaceans and fish larvae trapped in the surface film, and on floating jellyfish.

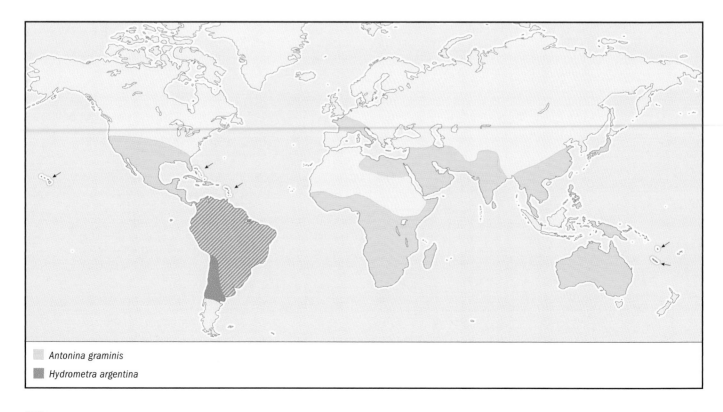

Antonina graminis
Hydrometra argentina

REPRODUCTIVE BIOLOGY

Mating and egg laying occur at the surface; the laying site may be any floating object, such as bird feathers, seaweed, timber, cork, coal, shells of cuttlefish, tar lumps, or even the feathers of living seabirds; eggs often are found attached in great numbers. There are five larval instars.

CONSERVATION STATUS

Not threatened.

SIGNIFICANCE TO HUMANS

None known. ◆

Water measurer
Hydrometra argentina

FAMILY

Hydrometridae

TAXONOMY

Hydrometra argentina Berg, 1879, Buenos Aires, Argentina.

OTHER COMMON NAMES

English: Marsh treader, walking water-stick.

PHYSICAL CHARACTERISTICS

Slender and cylindrical with a very long head, the eyes placed midway along its length. About 0.39 in (10 mm) long. Uniformly pale brown. Antennae and legs are very slender. Long-winged, short-winged, and wingless individuals coexist. Sexes are alike. Larvae resemble adults but are much smaller and wingless.

DISTRIBUTION

Lowlands of most of South America on both sides of the Andes but not in Patagonia.

HABITAT

Surface of ponds and dead waters of small brooks, without plants or with a few plants.

BEHAVIOR

They walk slowly on the water surface in the daytime, searching for prey or for a mate. Long-winged individuals disperse by flight and are attracted to artificial lights.

FEEDING ECOLOGY AND DIET

Adults and larvae feed on small insects, aquatic or not—mainly those blown onto the water surface.

REPRODUCTIVE BIOLOGY

Mating takes place on the water surface. Large, elongated, ornamented eggs are glued singly to emerging sticks.

CONSERVATION STATUS

Not threatened.

SIGNIFICANCE TO HUMANS

Perhaps controls populations of mosquito larvae. ◆

Creeping water bug
Ilyocoris cimicoides

FAMILY

Naucoridae

TAXONOMY

Nepa cimicoides Linné, 1758, Europe.

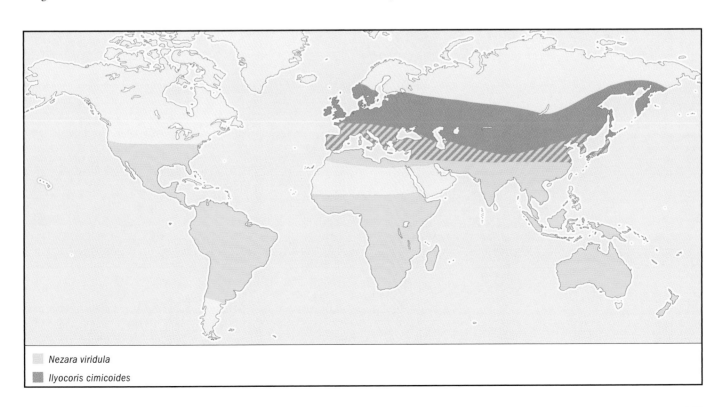

Nezara viridula

Ilyocoris cimicoides

OTHER COMMON NAMES
English: Saucer water bug, water bee; German: Schwimmwanze.

PHYSICAL CHARACTERISTICS
Rounded, oval, and beetle-like, reaching 0.59 in (15 mm) in length. Dull green, with somewhat darker forewings. Beak is short and conical. Forelegs very robust, adapted for grasping prey; middle and hind legs have long swimming hairs. Venter is covered by air-retaining hairs. Sexes are alike. Larvae resemble adults but are much smaller and wingless.

DISTRIBUTION
Southern Palearctic region, from southern Britain and the northern Iberian Peninsula eastward through China.

HABITAT
Freshwater ponds among submerged plants.

BEHAVIOR
They swim swiftly around. Males chirp to attract females. They lie in ambush, awaiting and quickly attacking prey.

FEEDING ECOLOGY AND DIET
Adults and larvae prey on insects and snails. Prey is killed with a poison injected with the beak and is sucked empty.

REPRODUCTIVE BIOLOGY
Mating occurs on the bottom or on supports in the water. The male mounts the back of the female, grasps her with the forelegs and middle legs with his axis at a slight angle to hers, and copulates at the left side of her abdomen. Eggs are embedded in rows into stems or leaves.

CONSERVATION STATUS
Not threatened.

SIGNIFICANCE TO HUMANS
They readily bite if carelessly handled; bites are extremely painful. They may be a nuisance to fish fry in fishery ponds. ◆

Backswimmer
Notonecta sellata

FAMILY
Notonectidae

TAXONOMY
Notonecta sellata Fieber, 1851, Buenos Aires, Argentina.

OTHER COMMON NAMES
None known.

PHYSICAL CHARACTERISTICS
Spindle shaped and somewhat rounded in front; some 0.31–0.35 (8–9 mm) in length. Typically marked with dark blue and white, but colorless specimens occur. Beak is short and conical. Front and middle legs are rather short; hind legs are very long, extended to the sides like the oars of a boat and with long hairs adapted for swimming. Larvae are similar to adults but smaller and wingless.

DISTRIBUTION
Lowlands of southern South America from central Bolivia, Paraguay, and southern Brazil to northern Patagonia east of the Andes.

HABITAT
Shallow ponds and pools, especially rain pools, in water with few plants. Prefers turbid water.

BEHAVIOR
They swim slowly, always on their backs near the surface, searching for prey, but they escape very quickly if disturbed. They fly away from pools that are drying up. Often attracted to electric lights.

Notonecta sellata

Saldula coxalis

FEEDING ECOLOGY AND DIET
A generalist predator on small insects and crustaceans in the upper water layers. Prey is seized and held by the front and middle legs.

REPRODUCTIVE BIOLOGY
Mating takes place near the surface of the water, males mounting the (downturned) backs of females. Eggs are glued singly to submerged supports, such as twigs and even algal filaments. There are five larval instars.

CONSERVATION STATUS
Not threatened.

SIGNIFICANCE TO HUMANS
Some damage to fish fry may be expected in fish ponds. Perhaps controls populations of mosquito larvae. ◆

Moss bug
Peloridium hammoniorum

FAMILY
Peloridiidae

TAXONOMY
Peloridium hammoniorum Breddin, 1897, Navarino Island (Tierra del Fuego), Chile.

OTHER COMMON NAMES
None known.

PHYSICAL CHARACTERISTICS
Small and flat, 0.16–0.19 in (4–5 mm) long. Dull colored, cryptic on mosses. Head and prothorax are expanded laterally. Forewings have coarse reticulation. Short-winged specimens are more common than winged ones.

DISTRIBUTION
From the southern tip of South America north to Chubut Province (Argentina) and Aysén (Chile).

HABITAT
Humid forests on the moss *Polytrichum strictum.*

BEHAVIOR
Almost nothing is known. Found quiescent or climbing among mosses.

FEEDING ECOLOGY AND DIET
Feed on liquids they suck from mosses.

REPRODUCTIVE BIOLOGY
Mating has not been studied. The smooth-shelled eggs are laid on mosses. Larvae cohabit with adults.

CONSERVATION STATUS
Not threatened.

SIGNIFICANCE TO HUMANS
None known. ◆

Southern green stink bug
Nezara viridula

FAMILY
Pentatomidae

TAXONOMY
Cimex viridulus Linné, 1758, "Indiis" (may mean India, Southeastern Asia, or the East Indies).

OTHER COMMON NAMES
English: Southern green bug, green vegetable bug, tomato bug, bean bug; French: Punaise verte; Spanish: Chinche verde; Portuguese: Percevejo verde.

PHYSICAL CHARACTERISTICS
Broad and medium sized, 0.47–0.51 in (12–13 mm) long. Green, with dark red eyes. Other forms may be green with a yellowish collar, pink, golden, or liver brown. Antennae have five joints. All the legs are about the same shape and size, and adapted for walking. Basal part of the forewing is stiff; the distal one is translucent. Sexes are alike. Larvae wingless and brightly stained with red, yellow, white, or black.

DISTRIBUTION
Temperate and tropical zones of both hemispheres, but not in very cold and desert areas. Worldwide spread started some 250 years ago from an uncertain origin, perhaps eastern Africa or the Far East.

HABITAT
Canopy of many field and ornamental crops and weeds.

BEHAVIOR
Active mainly in the daytime, climbing to the canopy early in the morning; egg laying is mostly a nocturnal activity. Newly born larvae remain densely grouped; adults are solitary, flying readily if disturbed.

FEEDING ECOLOGY AND DIET
Sap is sucked from leaves, twigs, buds, growing shoots, flowers, and fruits; succulent parts are preferred. The list of food plants includes more than 80 species from more than 30 families; the species is a generalist. First-instar larvae do not feed.

REPRODUCTIVE BIOLOGY
Mating occurs on plants, with the partners in an end-to-end position. Egg batches with 30–130 (average 70–75) drum-shaped, yellowish eggs in several rows are glued to the undersides of leaves; they turn pink and then red-orange. Females lay one or two egg batches. Up to five generations per year in warm climates but fewer in temperate areas.

CONSERVATION STATUS
Not threatened.

SIGNIFICANCE TO HUMANS
Severe damage is done in several cultures, with yield loss and damage to quality; viruses and disease-producing fungi are transmitted. Much effort is invested in controlling these pests. Adults are controlled mainly by species of parasitic flies; eggs are efficiently controlled (95%) by tiny wasps. ◆

Spiny soldier bug
Podisus maculiventris

FAMILY
Pentatomidae

TAXONOMY
Pentatoma maculiventris Say, 1831, Louisiana, United States.

OTHER COMMON NAMES
English: Spiny predator stink bug.

PHYSICAL CHARACTERISTICS
Elongate, about 0.35–0.55 in (9–14 mm) long. Dull yellow to tan with small black punctures and a dark spot at the rear; sometimes marked with purplish-red. Fore thorax with a prominent spine on each "shoulder." Antennae are five-jointed. All the legs are about the same shape and size, and adapted for walking. Basal part of the forewing is stiff; the distal one is translucent. Sexes are alike. Larvae are broader, wingless, and dotted with red, cream, and black.

DISTRIBUTION
Temperate North America and the West Indies; introduced for pest control to European and Far East countries.

HABITAT
Canopy of trees and shrubs, several field crops, fruit orchards.

BEHAVIOR
Diurnal, surveying on foot the canopy of plants and searching for prey and egg-laying sites. Prey is located by sight, and tracks are followed with the antennae and beak. Adults fly quickly when disturbed. Young larvae are gregarious, dispersing progressively as they grow and molt; they quickly walk from plant to plant. A secretion produced by males attracts all instars acting as a cue to the presence of prey. Hibernation in litter, soil, or bark or under stones.

FEEDING ECOLOGY AND DIET
Adults and larvae prey mainly on caterpillars and beetle larvae but also on eggs, sucking them out; this is a fairly generalist species. Densely hairy caterpillars are avoided, but size of prey is unimportant. If prey items are bigger than the insects themselves, they slowly approach the prey, extend their beak forward, and pierce it from the side without touching it with their legs. Attacks sometimes are accomplished collectively, and other insects may come to share an already attacked prey item. If food is scarce, cannibalism occurs. Some water or juices are taken; if prey are scanty, they can survive a long time.

REPRODUCTIVE BIOLOGY
Mating occurs in the end-to-end position on plants and lasts many hours. Some 15–30 cream to black, barrel-shaped eggs are glued onto leaves in loose oval patches. They reproduce throughout their lifetime, which may last several months. Two or three generations per year are produced in temperate zones, but continuous reproduction takes place in warmer ones.

CONSERVATION STATUS
Not threatened.

SIGNIFICANCE TO HUMANS
Useful controllers of pests, chiefly the Colorado potato beetle and caterpillars. Reared to be released in the field. Their powerful pheromone is produced commercially to attract specimens to orchards. A light and uncomfortable irritation may occur on human skin, but true bites have not been reported. ◆

Rhodesgrass mealybug
Antonina graminis

FAMILY
Pseudococcidae

TAXONOMY
Sphaerococcus graminis Maskell, 1897, Hong Kong.

OTHER COMMON NAMES
English: Grass-crown mealybug, felted grass coccid, Rhodesgrass scale.

PHYSICAL CHARACTERISTICS
Ellipsoid to nearly spherical and about 0.11–0.16 in (3–4 mm) long. Purplish brown and covered dorsally by a felted, waxy, brittle, whitish to yellowish coating. Openings at the anterior and posterior ends expose the body, with a tubular, waxy white filament protruding from the anal end. First larval instar is ambulatory; the remaining instars and adults are legless.

DISTRIBUTION
Of Asiatic origin but now almost cosmopolitan, chiefly in warmer parts of temperate zones.

HABITAT
Crown, base, leaf sheaths, and stolons of forage grasses and lawns.

BEHAVIOR
The first instar larva walks and changes resting sites; thereafter, they are legless and remain in place for life, soon starting to secrete their felted, waxy coating.

FEEDING ECOLOGY AND DIET
Adults and larvae take sap from several grasses, chiefly cereals. Newly born larvae walk around and set themselves down in a leaf sheath, engaging in active feeding until death.

REPRODUCTIVE BIOLOGY
Parthenogenetic, without males. The female produces living larvae; only newly born larvae walk. There are about five generations per year.

CONSERVATION STATUS
Not threatened.

SIGNIFICANCE TO HUMANS
A minor pest of field crops. Yield losses can be expected in cereal cultures, as well as withering of lawns. The sugary feces favor sooty mold development. ◆

Staining bug
Dysdercus albofasciatus

FAMILY
Pyrrhocoridae

TAXONOMY
Dysdercus albofasciatus Berg, 1878, Corpus, Misiones Province, Argentina.

OTHER COMMON NAMES
English: Cotton bug, cotton stainer; Spanish: Chinche negra (general), chinche guacha (general), chinche tintórea (Uruguay), chinchorro (Ecuador), churumbo (Peru); Portuguese: Barbeiro, percevejo manchador.

PHYSICAL CHARACTERISTICS
Nearly oval and elongate. Back is black, with a transverse whitish-yellowish stripe; venter is bright red. Antennae and legs are long and slender. Larvae smaller, wingless, bright red.

DISTRIBUTION
Tropical South America, following the distribution of their host plants.

HABITAT
Canopy and flowering and fruiting twigs of malvaceous shrubs.

BEHAVIOR
Adults are mostly solitary; larvae often are gregarious. Adults fly readily if disturbed.

FEEDING ECOLOGY AND DIET
Both adults and larvae prefer maturing or mature seeds; they inject saliva into them and suck the resulting liquid. When no fruits are available, they go to flowers, buds, or growing twigs.

REPRODUCTIVE BIOLOGY
They mate end to end. Eggs are laid on twigs or leaves. Larvae aggregate at the site and disperse progressively as they grow and molt.

CONSERVATION STATUS
Not threatened.

SIGNIFICANCE TO HUMANS
An uncommon secondary pest of cotton crops. ◆

Kissing bug
Triatoma infestans

FAMILY
Reduviidae

TAXONOMY
Reduvius infestans Klug, 1834, South America.

OTHER COMMON NAMES
English: Bloodsucking conenose; Spanish: Vinchuca (Bolivia, Paraguay, Chile, Argentina, and Uruguay), Chinchorro (Ecuador), Churumbo (Peru), Chinche negra, Chinche gaucha; Portuguese: Barbeiro.

PHYSICAL CHARACTERISTICS
Flat when unfed and globose when fully fed. Grows to 0.83–1.18 in (21–30 mm) long. Dark brown to almost black, with yellow square spots along the abdominal margins; legs somewhat paler and wings dark brown. Head long and cylindrical. Compound eyes are large and set nearer the base of the head; also a pair of prominent simple eyes. Antennae and legs are long and slender. Sexes similar. Larvae look like adults but are smaller, dull grayish brown, and wingless.

DISTRIBUTION
Most of tropical and temperate South America south of the equator, from northern Brazil and Ecuador through Patagonia (southern Argentina) up to an elevation of 13,123 ft (4,000 m); their distribution is linked to the movement of humans with baggage.

HABITAT
Human dwellings and poultry yards. Forest populations apparently are rare. Even if straw roofs or crevices are smoked—a process that repels most insects—the insects are still able to sustain their colony.

BEHAVIOR
Adults chirp by scraping the end of the beak against a ventral, transversely furrowed sulcus at the breast, the vibration being transmitted via the legs to the substrate. They orient themselves to their prey by heat and emission of carbon dioxide and to their shelters by the smell of their own excreta. Sleeping humans often are reached by flying. Adults may live two years.

FEEDING ECOLOGY AND DIET
Nocturnal, feeding on hosts as they sleep. They may feed on domestic animals, such as dogs and poultry. One blood supply is needed at each larval instar to reach adulthood.

REPRODUCTIVE BIOLOGY
Mating occurs in shelters. The whitish eggs drop to the ground and change to pink. There is one generation per year.

CONSERVATION STATUS
Not threatened.

SIGNIFICANCE TO HUMANS
An efficient transmitter of Chagas' disease. People believe that having kissing bugs in the home will bring good luck and happiness, a belief that interferes with sanitary control measures. ◆

Shore bug
Saldula coxalis

FAMILY
Saldidae

TAXONOMY
Acanthia coxalis Stål, 1873, Cuba.

OTHER COMMON NAMES
None known.

PHYSICAL CHARACTERISTICS
Small, reaching 0.14–0.16 in (3.5–4 mm) in length; broad and flattened. Color varies widely from light to dark brown, mottled; lateral margins of pronotum are pale. Dorsum densely covered with short hairs. Legs are short, robust, adapted for walking and jumping. Wings complete or partially reduced (proportions of morphs vary). Forewings have well-marked veins and translucent membranes. Larvae are like adults but much smaller and wingless.

DISTRIBUTION
Texas (United States) through Chile and Argentina and the West Indies, principally along coasts; lives at elevations up to 11,480 ft (3,500 m) in Peru.

HABITAT
Salty, brackish, and freshwater swamps and muddy beaches of seas, lakes, and rivers; at floodtide they climb on grasses. Hardly seen on the mud and seldom collected.

BEHAVIOR
They walk, run, or jump quickly on the soil and fly readily if disturbed. Much time is spent grooming the legs and antennae.

FEEDING ECOLOGY AND DIET
They feed on dead insects on the soil or drive the beak into the surface mud and suck out small invertebrates such as insect larvae and small earthworms.

REPRODUCTIVE BIOLOGY
At mating the male positions himself side by side with the female, grasping her ovipositor with the claspers. Eggs are laid on grass blades; they can withstand drowning. Reproduction occurs year-round.

CONSERVATION STATUS
Not threatened.

SIGNIFICANCE TO HUMANS
None known. ◆

Resources

Books

Andersen, Nils M. *The Semiaquatic Bugs (Hemiptera, Gerromorpha): Phylogeny, Adaptations, Biogeography, and Classification.* Entomonograph 13. Klampenborg, Denmark: Scandinavian Science Press, 1982.

Schaefer, Carl W. "Prosorrhyncha (Heteroptera and Coleorrhyncha)." In *Encyclopedia of Insects*, edited by V. H. Resh and R. T. Cardé. Amsterdam: Academic Press, 2003.

———, ed. *Studies on Hemipteran Phylogeny.* Thomas Say Publications in Entomology. Lanham, MD: Entomological Society of America, 1996.

———. "True Bugs and Their Relatives." In *Encyclopedia of Biodiversity*, vol. 5. San Diego: Academic Press, 2000.

Schaefer, Carl W., and A. R. Panizzi, eds. *Heteroptera of Economic Importance.* Boca Raton, FL: CRC Press, 2000.

Schuh, R. T., and J. A. Slater. *True Bugs of the World (Hemiptera: Heteroptera): Classification and Natural History.* Ithaca, NY: Cornell University Press, 1995.

Periodicals

Evans, J. W. "A Review of Present Knowledge of the Family Peloridiidae and New Genera and Species from New Zealand and New Caledonia (Hemiptera: Insecta)." *Records of the Australian Museum* 34, no. 5 (1981): 381–406.

Wheeler, W. C., R. T. Schuh, and R. Bang. "Cladistic Relationships among Higher Groups of Heteroptera: Congruence between Morphological and Molecular Data Sets." Entomologia Scandinavica 24 (1993): 121–137.

Other

"Fulgoromorpha Lists on the Web." November 30, 1999 [May 12, 2003]. <http://flow.snv.jussieu.fr/>.

"Halobates—Oceanic Insects." June 26, 2002 [May 12, 2003]. <http://www.zmuc.dk/EntoWeb/Halobates/HALOBAT1.HTM>.

"The International Heropterist's Society." January 3, 2002 [May 12, 2003]. <http://entomology.si.edu/ihs/home.lasso>.

"Periodical Cicada Page." December 16, 2002 [May 12, 2003]. <http://insects.ummz.lsa.umich.edu/fauna/Michigan_Cicadas/Periodical/Index.html>.

"Scale Net." March 19, 2003 [May 12, 2003]. <http://www.sel.barc.usda.gov/scalenet/scalenet.htm>.

"Whitefly Taxonomic and Ecological Website." November 5, 2002 [May 12, 2003]. "http://www.fsca-dpi.org/Homoptera_Hemiptera/Whitefly/whitefly_catalog.htm>.

Axel O. Bachmann, Doctor en Ciencias Biológicas
Silvia A. Mazzucconi, Doctor en Ciencias Biológicas

Thysanoptera
(Thrips)

Class Insecta
Order Thysanoptera
Number of families 9

Photo: Giant thrips of the Thysanoptera order, in Java, Indonesia. (Photo by ©Simon D. Pollard/Photo Researchers, Inc. Reproduced by permission.)

Evolution and systematics

Although recorded as fossils from the lower Permian, most fossil thrips prior to those in Cretaceous Lebanese amber are equivocal. Currently the group is considered part of the Paraneoptera, together with the Psocodea and Hemiptera. Two suborders are recognized: the Tubulifera with one family (Phlaeothripidae) and the Terebrantia with eight families (Merothripidae, Melanthripidae, Aeolothripidae, Adiheterothripidae, Fauriellidae, Heterothripidae, Thripidae, and Uzelothripidae). Species in the Merothripidae and Melanthripidae retain more ancestral features than other thrips, and the Uzelothripidae is represented by a single highly aberrant species. Currently, about 5,500 species are recognized, but there are many undescribed species in tropical areas.

Physical characteristics

Adult and larval thrips are unique among insects in retaining in the head only the left mandible, the right one being resorbed by the embryo. The maxillary stylets form a suctorial feeding tube, but this has only one channel for both food and saliva, unlike hemipterans. The wings are slender and fringed with long cilia, hence the ordinal name meaning "fringed-wings," but similar narrow wings with long, fringing cilia occur in unrelated small insects, and many adult thrips are wingless. The German name for the group, Blassenfusse, refers to the adhesive tarsal pads found in adults, and the common name, thrips, is Greek for "woodworm," refer-

ring to the fact that many species live on dead branches. Adults are usually flattened dorsoventrally, ranging in length from 0.02–0.6 in (0.5–15 mm), and although commonly black, many species are yellow to white, and others exhibit typical aposematic colors black, red, and white.

Distribution

Thrips are essentially tropical. For example, the combined area of Costa Rica and Panama is roughly equal in size to Britain, but despite limited study, more than 300 species are described from these two tropical countries, whereas 100 years of active study in Britain found less than 150 native thrips species. About 700 species are described from North America, 1,500 from South America, and 500 from Australia, but these figures probably represent scarcely 50% of the real fauna. The thrips fauna of tropical Africa is virtually unknown.

Habitat

About 40% of thrips species live on dead branches or in leaf litter, whereas about 30% live on green leaves. Most of the remaining 30% live in flowers, many being specific to grass florets. A few species live in mosses. Larvae and adults occupy the same habitat, but larvae commonly fall to the ground to pupate. Diversity is greatest in tropical forests, but populations are largest in open habitats, and vast numbers of individuals occur sometimes in alpine meadows.

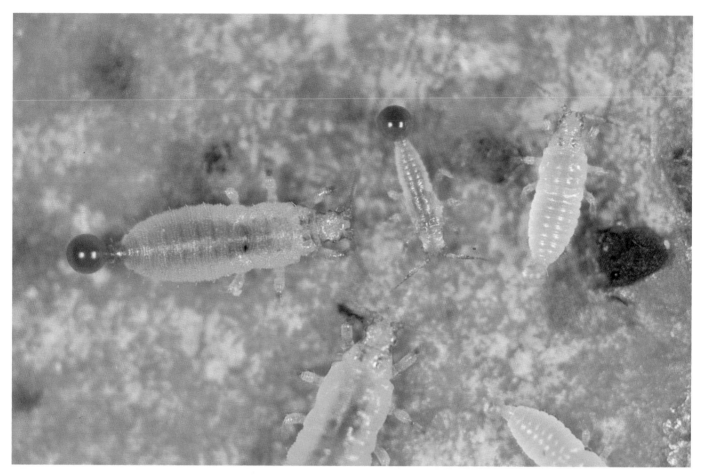

A group of black thrips (*Echinothrips americanus*) nymphs, some with secretion droplets. (Photo by ©Holt Studios Int./Photo Researchers, Inc. Reproduced by permission.)

Behavior

Fighting between males is probably widespread amongst thrips species, particularly in Phlaeothripidae, and the body size of males often varies greatly within species. In some fungus feeding species a male may fight to defend his mate, or alternatively to defend a clutch of eggs to which several females contribute after mating with him. Male lekking has been observed in two unrelated species, of which one, the Australian pest species Kelly's citrus thrips (Thripidae), has females visiting a male aggregation briefly to copulate. Thigmotaxis, the habit of crawling into confined spaces, is widespread amongst thrips species.

Feeding ecology and diet

At least 2,000 described species feed on fungi (most on hyphae), but with about 700 taking whole fungal spores into their gut. Such species live on freshly dead leaves and branches, as well as in leaf litter. In the tropics, many species feed on the leaves of trees, some inducing galls, but in temperate regions thrips are usually associated with flowers, feeding on pollen and other floral tissues. Some species are predatory on other small arthropods, and a few pest species are so adaptable that by their feeding they may control populations of pest mites on crop plants as well as damage such crops directly.

Reproductive biology

Sex determinism in thrips is haplodiploid, that is, males have half the number of chromosomes of females and develop from unfertilized eggs. Thrips metamorphosis is unique, with two larval instars followed by two (or even three) nonfeeding pupal instars. In fungus-feeding Phlaeothripidae, males are commonly larger than females, but in gall thrips and most Terebrantia, males are smaller than females. Many flower-living species have only a single generation in a year, but most species breed whenever suitable conditions exist. Pest thrips often breed more or less continuously, with a new generation developing every three weeks. In arid parts of Australia, a considerable number of phlaeothripine species construct a domicile in which to breed by gluing or sewing together pairs of leaves.

Conservation status

Thrips faunal diversity is dependent on conserving the diversity of the native flora. Thus large areas of Australia or

North America that are intensively disturbed commonly have few or no native thrips species. No Thysanoptera are included on the IUCN Red List.

Significance to humans

Thrips are commonly considered pest insects, although fewer than 10% of known species have been recorded as causing crop damage. When in high population numbers, some species may sometimes bite humans by probing the skin with their mouthparts, and adult thigmotactic behavior can result in thrips triggering smoke detectors when entering these for shelter during massed flights in late summer. Pest species are usually highly adaptable insects that can feed on a wide range of plants under varying conditions, whereas most thrips are relatively host and habitat specific. Ten species of Thripidae, including the western flower thrips, are known to infect plants with virus diseases known as tospoviruses, and worldwide such thrips are among the most serious of insect pests.

1. Crowned thrips (*Ecacanthothrips tibialis*); 2. Western flower thrips (*Frankliniella occidentalis*). (Illustration by John Megahan)

Species accounts

Western flower thrips
Frankliniella occidentalis

FAMILY
Thripidae

TAXONOMY
Euthrips occidentalis Pergande, 1895, California, United States.

OTHER COMMON NAMES
None known.

PHYSICAL CHARACTERISTICS
About 0.04 in (1 mm) long; color yellow to dark brown.

DISTRIBUTION
Throughout temperate parts of the world, but originally western United States.

HABITAT
Flowers and leaves of many plants.

BEHAVIOR
Individuals commonly fly for hundreds of yards when their host plant is disturbed, but long-distance transport is due to the horticultural trade in plants. Males sometimes compete for territories on a leaf, but only when the population density is low.

FEEDING ECOLOGY AND DIET
Feeds on pollen, also flower and young leaf tissues; sometimes predatory on mites.

REPRODUCTIVE BIOLOGY
Usually bisexual, but males develop from eggs that have not been fertilized.

CONSERVATION STATUS
Not threatened; a serious pest that requires control throughout much of the world.

SIGNIFICANCE TO HUMANS
This is one of the most important horticultural pests in the world, causing serious damage to flower crops, tomatoes, capsicums and cucumbers, as well as stone fruits and table grapes. As well as direct feeding damage, the thrips infects many plants with viruses. ◆

Ectoparasitic thrips
Aulacothrips dictyotus

FAMILY
Heterothripidae

TAXONOMY
Aulacothrips dictyotus Hood, 1952, Brazil.

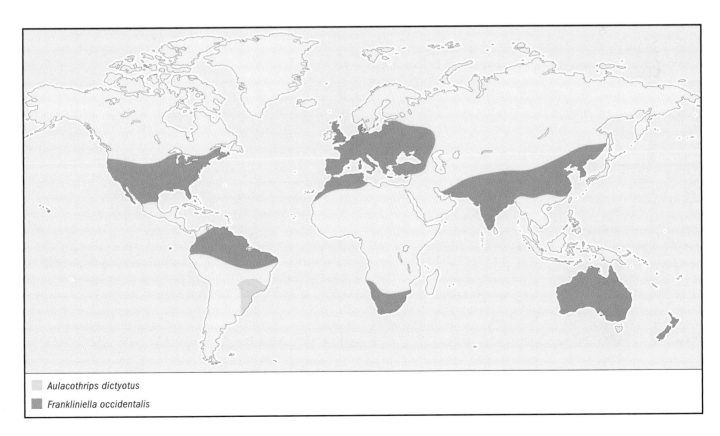

☐ *Aulacothrips dictyotus*
■ *Frankliniella occidentalis*

OTHER COMMON NAMES
None known.

PHYSICAL CHARACTERISTICS
About 0.06 in (1.5 mm) long; dark brown, with third and fourth antennal segments larger than remaining seven segments and both bearing a greatly elongate, convoluted sensory area.

DISTRIBUTION
Southern Brazil.

HABITAT
Under the wings on the upper surface of the abdomen of a plant-feeding treehopper, *Aetalion*, (Aetalionidae) unlike the flower-feeding habit of all other Heterothripidae.

BEHAVIOR
The host insect lives in colonies, thus making it easier for the larval thrips to transfer between individuals when these molt. The bugs become disturbed when the adult thrips walk over them.

FEEDING ECOLOGY AND DIET
All life stages, from the youngest larvae to pupae and adults, live under the wings of the host insect, presumably sucking its blood.

REPRODUCTIVE BIOLOGY
Possibly parthenogenetic; the male is not known. Eggs have not been observed, but newly emerged first instar larvae occur on the abdomen of their host, together with later instar larvae and pupae. Each pupa is enclosed in a transparent cocoon fixed to the surface of the host's abdomen.

CONSERVATION STATUS
Not listed by the IUCN, although currently reported from only two localities.

SIGNIFICANCE TO HUMANS
None known. ◆

Crowned thrips
Ecacanthothrips tibialis

FAMILY
Phlaeothripidae

TAXONOMY
Idolothrips tibialis Ashmead, 1905, Philippines.

OTHER COMMON NAMES
None known.

PHYSICAL CHARACTERISTICS
About 0.08–0.16 in (2–4 mm) long; third antennal segment bears crown of at least 10 large sensoria, (most of the 3,500 members of the family have only one to three small sensoria on this segment). Front legs or large males are greatly enlarged and bear a stout tooth on the tarsus; smallest males and females have slender front legs. The last abdominal segment is tubular (as in all Phlaeothripidae); forewings do not have veins bearing setae.

DISTRIBUTION
Old World tropics, from Africa to northern Australia.

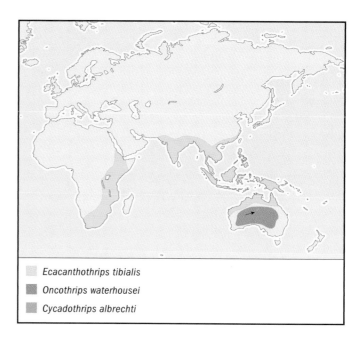

- Ecacanthothrips tibialis
- Oncothrips waterhousei
- Cycadothrips albrechti

HABITAT
On dead twigs and branches.

BEHAVIOR
Males exhibit great variation in size and are known to be subsocial or even truly social. Males fight other males to defend either a single female or a single egg mass to which several females will progressively contribute. Smallest males may attempt to sneak mate while their larger siblings are fighting. Small males have a nutritional advantage in that they require less food to achieve maturity; thus there is a balance between feeding and mating advantages.

FEEDING ECOLOGY AND DIET
Adults and larvae feed on fungal hyphae growing on freshly dead wood.

REPRODUCTIVE BIOLOGY
Eggs are laid on the surface of dead twigs among the fungus on which adults and larvae feed.

CONSERVATION STATUS
Not listed by the IUCN.

SIGNIFICANCE TO HUMANS
None known. ◆

Australian acacia gall thrips
Oncothrips waterhousei

FAMILY
Phlaeothripidae

TAXONOMY
Oncothrips waterhousei Mound and Crespi, 1995, Australia.

OTHER COMMON NAMES
None known.

PHYSICAL CHARACTERISTICS
Winged and wingless morphs of both sexes are about 0.06 in (1.5 mm) long. Wingless adults mainly yellow, with thorax and forelegs enlarged, antennae short. Winged adults are dark brown with slender forelegs and thorax, long slender antennae.

DISTRIBUTION
Widespread across the arid zone of Australia.

HABITAT
Gall-inducing on the phyllodes (leaves) of several Acacia species.

BEHAVIOR
Wingless adults of both sexes function as soldiers to defend their gall from invading kleptoparasitic thrips and also moth larvae.

FEEDING ECOLOGY AND DIET
When a female feeds on a young phyllode, this is induced to form a small pouch that rapidly encloses the insect.

REPRODUCTIVE BIOLOGY
Female lays eggs within the hollow gall soon after becoming enclosed. These hatch, and larvae of this first small generation develop into wingless soldiers. A second generation is then produced in which the adults are fully winged and disperse to found new galls.

CONSERVATION STATUS
Not listed by the IUCN. Presumably threatened by progressive destruction by farmers of host plants to feed livestock, particularly during years of drought.

SIGNIFICANCE TO HUMANS
None known. ◆

Australian cycad thrips
Cycadothrips albrechti

FAMILY
Aeolothripidae

TAXONOMY
Cycadothrips albrechti Mound and Terry, 2001, Australia.

OTHER COMMON NAMES
None known.

PHYSICAL CHARACTERISTICS
About 0.08 in (2 mm) long, color varies from golden yellow to light brown. Forewings are broad, as in other members of the family, but third antennal segment bears two unusually inflated sensoria. Large males have pair of stout, thornlike setae near tip of abdomen; these setae scarcely developed in small males.

DISTRIBUTION
Central Australia, near Alice Springs.

HABITAT
Breeds in the male cones of the cycad *Macrozamia macdonnellii*.

BEHAVIOR
Breeds in a particularly hot and arid area, has only been seen to fly late in the afternoon when the humidity rises. At that time, both sexes fly in swarms from the male cones on which they have fed and produced larvae, and each adult carries about 20 pollen grains on its body. Concurrently, a powerful odor is given off by any mature female cone, and this attracts the flying thrips, and induces them to crawl inside with their pollen load.

FEEDING ECOLOGY AND DIET
Adults and larvae suck the contents from the pollen grains of *M. macdonnellii*; populations have been estimated at 50,000 adults in a single cone.

REPRODUCTIVE BIOLOGY
Variation in size of males and the powerful setae near the tip of their abdomen suggests some sort of male competitive behavior in mating, but this has not been observed. Larvae develop only in male cones, and subsequently fall to the ground to pupate around the base of the cycad.

CONSERVATION STATUS
Not listed by the IUCN. The only known pollinator of this species of cycad, a plant that grows only in a restricted area of central Australia. In their mutual adaptation, the future existence of both is threatened by human activity.

SIGNIFICANCE TO HUMANS
None known. ◆

Resources

Books

Crespi, B. J., and L. A. Mound. "Ecology and Evolution of Social Behaviour Among Australian Gall Thrips and Their Allies." In *Evolution of Social Behaviour in Insects and Arachnids*, edited by J. Choe and B. J. Crespi. Cambridge, UK: Cambridge University Press, 1997.

Lewis, T., ed. *Thrips as Crop Pests*. Wallingford, UK: CAB International, 1997.

Mound, L. A. "Thysanoptera." In *Zoological Catalogue of Australia*. Vol. 26, *Psocoptera, Phthiraptera, Thysanoptera*, edited by A. Wells. Melbourne: CSIRO Publishing, 1996.

Periodicals

Izzo, T. J., S. M. J. Pinent, and L. A. Mound. *"Aulacothrips dictyotus* (Heterothripidae), the First Ectoparasitic Thrips (Thysanoptera)." *Florida Entomologist* 85 (2002): 281–283.

Mound, L. A., and R. Marullo. "The Thrips of Central and South America: An Introduction." *Memoirs on Entomology, International* 6 (1996): 1–488.

Mound, L. A., and I. Terry. "Pollination of the Central Australian Cycad, *Macrozamia macdonnellii*, by a New Species of Basal Clade Thrips (Thysanoptera)." *International Journal of Plant Sciences* 162 (2001): 147–154.

Other

Moritz, G., D. C. Morris, and Mound, L. A. "ThripsID: Pest Thrips of the World." *CD-ROM. Melbourne: CSIRO Publishing, 2001.*

Laurence A. Mound, DSc

Megaloptera
(Dobsonflies, fishflies, and alderflies)

Class Insecta
Order Megaloptera
Number of families 2

Photo: A fishfly (*Neohermes* sp.) on a flower in Washington, USA. (Photo by Steve Solum. Bruce Coleman, Inc. Reproduced by permission.)

Evolution and systematics

The oldest known fossils of Megaloptera date from the Permian. The Megaloptera are often considered to be the most primitive group of endopterygote (with internal development of wings as imaginal discs in the larvae) insects. The order Raphidioptera constitutes their sister group, and together with Neuroptera they form the monophyletic group of the Neuropterida. Megaloptera includes only two families, Corydalidae and Sialidae (alderflies). Corydalidae, in turn, comprises two subfamilies, Corydalinae (dobsonflies and hellgrammites) and Chauliodinae (fishflies), which are considered families by some authors.

Physical characteristics

Adult alderflies range from 0.4 to 0.6 in (10–15 mm), and dobsonflies range from 1.6 to 2.4 in (40–60 mm); larvae reach a maximum length of 1 in (25 mm) and 1.2–2.6 in (30–65 mm), respectively. Adults are soft-bodied insects, black, brown, or yellowish orange to dark green in color. They possess filiform, moniliform, or pectinate antennae and large compound eyes, and they either have (Corydalidae) or lack (Sialidae) ocelli. They have two pairs of membranous wings

A fishfly (*Chauliodes* sp.) on leaves in Maryland, USA. (Photo by Ron Goor. Bruce Coleman, Inc. Reproduced by permission.)

An alder fly (*Sialis lutaria*) with a batch of eggs. (Photo by J. Markham. Bruce Coleman, Inc. Reproduced by permission.)

with a complex nervelike pattern of veins, although longitudinal veins are not branched near the wing margin. The hind wings are broader at the base than the front wings, and this enlarged anal area is kept folded like a fan at rest. Larvae are elongated, flattened, and prognathous, with chewing mouthparts. They have seven or eight lateral pairs of abdominal gills. In sialids the abdomen ends in a median unsegmented filament, whereas in corydalids it terminates in a pair of anal prolegs. Pupae have free appendages that are not fastened to the body and are not enclosed in a cocoon.

Distribution

They occur in the New World, South Africa, Madagascar, and Asian and Australian regions and are most diverse in temperate regions, with fewer species in the tropics.

Habitat

Larvae are aquatic, inhabiting both lentic (still water) and lotic (moving water) environments, including streams, spring seeps, rivers, lakes, ponds, swamps, and even temporarily dry streambeds. They burrow into soft substrate or in crevices or hide under stones or bark. Eggs, pupae, and adults are ter-

restrial. Eggs can be found in masses on leaves, branches, rocks, and bridges overhanging the aquatic habitat. Pupae are found in the shoreline soil and litter adjacent to the larval habitat, and adults are found in the same general area where larvae live.

Behavior

Adults seldom are seen in large numbers because they are short-lived and secretive. Sialids are active during the warm midday hours, engage infrequently in brief flights, and can be found resting near their larval habitats with the wings held rooflike over the abdomen. Most corydalids have nocturnal habits and are attracted to light, although there are some diurnal species. Their flight is slow and irregular, but some may fly considerable distances. At rest they keep the wings flat over the abdomen.

Feeding ecology and diet

Larvae are entirely predaceous, feeding nonselectively on a wide variety of small aquatic invertebrates, such as insect larvae, crustaceans, mollusks, and annelids. Adults apparently do not feed.

Reproductive biology

Communication between sexes before mating is known to occur in some sialids and corydalids. *Corydalus* species show sexual dimorphism; males have very long mandibles with which they duel with each other and prod the female. Copulation takes place in vegetation near the water. Females deposit masses of one to five layers of 200 to 3,000 eggs on objects overhanging the water, in locations protected from insolation during the hottest part of the day. After opening the egg with an egg burster, the larvae drop to the water after hatching, where they undergo 10 to 12 molts. Fully grown larvae leave the water and pupate in an unlined chamber in the soil or litter. Adults usually emerge from late spring to midsummer. Life cycles range from one to five years.

Conservation status

About 300 species of megalopterans have been described, none of which is listed by the IUCN. Some species with re- stricted distributions in small rivers and streams probably are sensitive to deforestation, pollution, and eutrophication (involving depletion of oxygen that is normally taken from the water by aquatic larvae) and could be potential candidates for conservation programs.

Significance to humans

Larvae are important predators in aquatic food chains. Some species are economically important as trout food, and their larvae are used as fish bait (i.e., *Archichauliodes diversus* from New Zealand). In the Japanese tradition, dried larvae of some dobsonflies are believed to be a remedy for emotional problems in children.

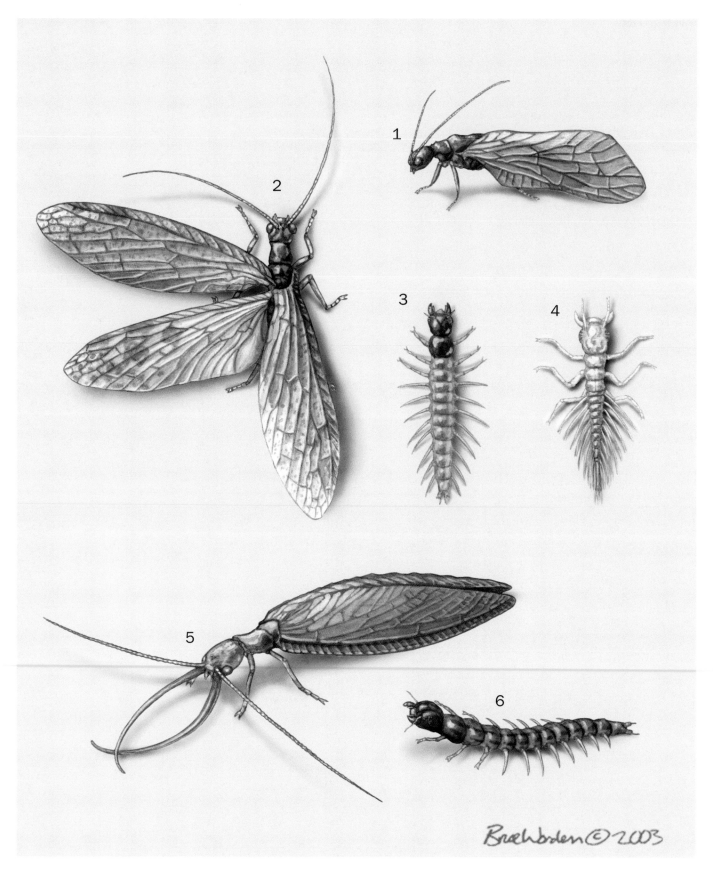

1. Alderfly (*Sialis lutaria*) adult; 2. Dobsonfly (*Archichauliodes diversus*) adult; 3. Black creeper (*A. diversus* larva); 4. *S. lutaria* larva; 5. Eastern dobsonfly (*Corydalus cornutus* adult); 6. *C. cornutus* larva. (Illustration by Bruce Worden)

Species accounts

Eastern dobsonfly
Corydalus cornutus

FAMILY
Corydalidae (Corydalinae)

TAXONOMY
Hemerobius cornutus Linneus, 1758, Pennsylvania.

OTHER COMMON NAMES
English: Hellgrammite, toebiter, bass bait (larva); French: Grande mouche Dobson (adult).

PHYSICAL CHARACTERISTICS
The adult is 2 in (50 mm) long, with a wingspan up to 5 in (125 mm), and the larva is 2.6 in (65 mm) long. The head is almost circular and the prothorax square and slightly narrower than the head. The wings are translucent gray with dark veins and cells with white spots. The mandibles of the male are as long as half of the body length, curved and tapering to the tips and held crossing each other. The mandibles of the female are shorter.

DISTRIBUTION
Occur east of the Rocky Mountains in the United States and Canada.

HABITAT
Larvae live in fast-flowing water.

BEHAVIOR
Adults are nocturnal and secretive and are seldom seen during the daytime, when they hide under leaves in the canopy of trees. Larvae have been seen swimming forward or backward in a snakelike fashion, but they usually crawl.

FEEDING ECOLOGY AND DIET
Adults do not feed; larvae consume other insects and small invertebrates.

REPRODUCTIVE BIOLOGY
Mating behavior is stereotypical; males flutter their wings, display genital appendages, and fight over females. Rounded masses containing 100–1,000 or more eggs are laid on rocks, branches, and objects close to the water. Each mass is coated with a whitish secretion. Larvae drop into the water or crawl to reach feeding grounds. After two or three years, they crawl out of the water and prepare pupal chambers under stones or logs, where they overwinter. Adults emerge in early summer.

CONSERVATION STATUS
Not threatened.

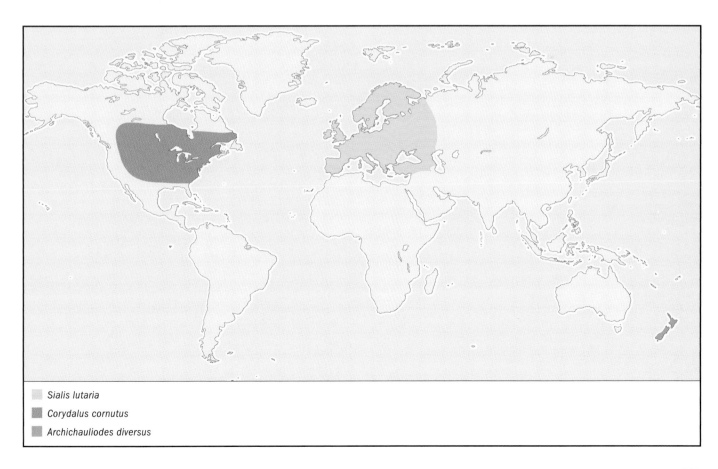

Sialis lutaria
Corydalus cornutus
Archichauliodes diversus

SIGNIFICANCE TO HUMANS
Fishermen use dobsonfly larvae, called hellgrammites, as bait for trout, largemouth bass, catfish, and other fishes. Hellgrammites also help control populations of pest aquatic insects, such as the Asian tiger mosquito. ◆

Black creeper (larva), Dobsonfly (adult)
Archichauliodes diversus

FAMILY
Corydalidae (Chauliodinae)

TAXONOMY
Chauliodes diversus Walker, 1853, New Zealand.

OTHER COMMON NAMES
English: Toebiter, black fellow (larva).

PHYSICAL CHARACTERISTICS
Larvae have a long, thin, gray body up to 2 in (50 mm) long and a subtriangular jet-black head. Adults are about 2–3 in (50–75 mm) long with large, clear wings with a 2–3.2 in (50–80 mm) wingspan.

DISTRIBUTION
New Zealand.

HABITAT
Stony streams and rivers. Older larvae are found beneath dry stones close to the water's edge.

BEHAVIOR
Larvae cling to the stream bottom in all but the strongest flows and become dislodged only with flooding. Adults fly in a slow, clumsy manner and only for short distances at low heights over streams.

FEEDING ECOLOGY AND DIET
Larvae feed on mayfly larvae; adults do not feed.

REPRODUCTIVE BIOLOGY
Larvae take from 18 months to three years to reach full size. Adults emerge in summer and live for only six to 10 days. Eggs that are laid late in the season undergo obligate diapause (suspension of development, which starts again once climatic conditions become more favorable).

CONSERVATION STATUS
Not threatened.

SIGNIFICANCE TO HUMANS
Larvae often are used by anglers as trout bait. ◆

Alderfly
Sialis lutaria

FAMILY
Sialidae

TAXONOMY
Hemerobius lutaria Linnaeus, 1758, Europe.

OTHER COMMON NAMES
French: Mouche du Saule; German: Gemeine Wasserflorfliege; Danish: Dovenflue.

PHYSICAL CHARACTERISTICS
Adults are 0.52–0.72 in (13–18 mm) in length and blackish-brown in color.

DISTRIBUTION
Occurs in Europe and into Russia.

HABITAT
Larvae inhabit the depths of still waters and muddy backwaters of rivers.

BEHAVIOR
During spring and early summer adults are found on plants near the water; they fly only when it is sunny and warm.

FEEDING ECOLOGY AND DIET
Larvae feed on worms, insect larvae, and other small freshwater animals. Adults occasionally take nectar from flowers with easily accessible nectaries.

REPRODUCTIVE BIOLOGY
After mating, the female deposits dark gray eggs on leaves of littoral vegetation and then cleans the newly laid eggs. Larvae crawl into the water and scurry to the bottom, where they tunnel in the silt. After two winters the larva leaves the water and pupates for two weeks. The greatest numbers of adults are seen on the wing from May through mid-June.

CONSERVATION STATUS
Not threatened.

SIGNIFICANCE TO HUMANS
They are bioindicators in water quality assessment. ◆

Resources

Books

Brigham, W. U. "Megaloptera." In *Aquatic Insects and Oligochaetes of North and South Carolina*, edited by A. R. Brigham, W. U. Brigham, and A. Gnilka. Mahomet, IL: Midwest Aquatic Enterprises, 1982.

Chandler, H. P. "Megaloptera." In *Aquatic Insects of California*, edited by R. L. Usinger. Berkeley: University of California Press, 1956.

Contreras-Ramos, A. *Systematics of the Dobsonfly Genus Corydalus Latreille (Megaloptera: Corydalidae).* Thomas Say

Monographs. Lanham, MD: Entomological Society of America, 1998.

Evans, E. D., and H. H. Neunzig. "Megaloptera and Aquatic Neuroptera." In *An Introduction to Aquatic Insects of North America*, edited by R. W. Merritt and K. W. Cummins. 3rd. edition. Dubuque, IA: Kendall/Hunt Publishing Company, 1996.

Henry, C. S., N. D. Penny, and P. A. Adams. "The Neuropteroid Orders of Central America (Neuroptera and Megaloptera)." In *Insects of Panama and Mesoamerica*, edited

Resources

by D. Quintero and A. Aiello. Oxford: Oxford University Press, 1992.

Penny, N. D. "Neuroptera." In *Aquatic Biota of Tropical South America*. Part 1, *Arthropoda*, edited by S. B. Hurlbert, G. Rodríguez, and N. Dias dos Santos. San Diego: San Diego State University, 1981.

———. "Neuroptera." In *Aquatic Biota of Mexico, Central America and the West Indies*, edited by S. B. Hurlbert and A. Villalobos-Figueroa. San Diego: San Diego State University, 1982.

Periodicals

Contreras-Ramos, A. "Mating Behavior of *Platyneuromus* (Megaloptera: Corydalidae), with Life History Notes on

Dobsonflies from Mexico and Costa Rica." *Entomological News* 110 (1999): 125–135.

Davis, K. C. "Sialididae of North and South America." *New York State Museum Bulletin* 68 (1903): 442–486, 499.

Stewart, K. W., G. P. Friday, and R. E. Rhame. "Food Habits of Hellgrammite Larvae, *Corydalus cornutus* (Megaloptera: Corydalidae), in the Brazos River, Texas." *Annals of the Entomological Society of America* 66 (1973): 959–963.

Natalia von Ellenrieder, PhD

Raphidioptera
(Snakeflies)

Class Insecta
Order Raphidioptera
Number of families 2

Photo: A snakefly (*Agulla* sp.) on vegetation in California, USA. (Photo by ©Gilbert Grant/Photo Researchers, Inc. Reproduced by permission.)

Evolution and systematics

An enormous abundance and diversity of raphidiopteran fossils occur in Jurassic and Cretaceous deposits, within the Mesozoic era. The extraterrestrial impact that occurred 65 million years ago probably led to the extinction of most snakeflies. Raphidioptera from the Tertiary belong to the two extant (non-extinct) families. They represent living fossils. Raphidioptera are currently considered the sister group of Megaloptera and Neuroptera, and all three orders constitute the superorder Neuropterida at the base of the Holometabola—the group of orders containing species that undergo complete metamorphosis. The order Raphidioptera comprises two homogeneous families: Raphidiidae, with 185 species, and Inocelliidae, with 21 described species. The estimated number of extant species may be around 250.

Physical characteristics

Raphidioptera have narrow bodies with an elongate pronotum, movable head, and two pairs of subequal wings whose forewings range from 0.20 to 0.79 in (5 to 20 mm). The head is prognathous, flattened dorsoventrally, with chewing mouthparts, large compound eyes, and in the Raphidiidae, with three dorsal ocelli. The wing venation is net-like, and the hyaline wing membrane contrasts with a bright yellow, brown, or bicolored pterostigma. Females have a long ovipositor, and male genitals may be spectacularly shaped. Larvae are terrestrial and elongate, with a flat prognathous head, chewing mouthparts, and 4 to 7 stemmata. The head and prothorax are strongly sclerotized (hardened), while the meso-

and metathorax and abdomen are soft-bodied. Preimaginal pupae are extremely mobile and, except for their small wing pads, are similar to adults.

Distribution

Extant Raphidioptera occur throughout the Holarctic region but have not been found in the northern and eastern parts of North America. Their distribution comprises almost all arboreal parts of the Palaearctic region, including fringes of the Oriental region, with the southernmost records (at higher altitudes) in Morocco, northern Algeria, northern Tunisia, Israel, Syria, northern Iraq, northern Iran, northern Pakistan, northern India, Bhutan, Myanmar, northern Thailand, and Taiwan. In the Nearctic region their distribution is restricted to the southwestern and southern parts of North America; the southernmost records are from the Mexican-Guatemalan border. Most species have small distribution areas and are often restricted to a single mountain range.

Habitat

Snakeflies are restricted to bushy arboreal woodland habitats and to latitudes and/or altitudes with a pronounced winter period, and from sea level to more than 9,840 ft (3,000 m). Larvae live under the bark of trees or shrubs (many Raphidiidae, all Inocelliidae) or in the top layer of soil (Raphidiidae). Larvae of a few species sometimes also live in rock crevices.

A snakefly (*Raphidia notata*) perched on a dried leaf. (Photo by ©Stephen Dalton/Photo Researchers, Inc. Reproduced by permission.)

Behavior

Snakeflies are solitary. Premating communication via antennae and movements of abdomen and wings has been observed in several species. Adults are active only during daytime, where they spend much of their time grooming themselves, combing their foretibiae against parts of the ventrally bent head and then pulling their antennae between their tibiae. As they retract their legs, the tibiae and tarsi are pulled through the mouthparts. Cleaning of the mouthparts starts with moving one mandible aside and proceeds with moving the palps and all other parts against each other. Snakeflies are poor flyers with an extremely low vagility and thus have little capacity for dispersal.

Feeding ecology and diet

Larvae of all snakeflies and adults of Raphidiidae are predaceous, feeding on softskinned arthropods. Adults of both families have been observed feeding on pollen.

Reproductive biology

Two positions of copulation have been found: a "wrecking position" (in Raphidiidae), in which the male hangs head first from the female, being carried by her; and a "tandem position" (in Inocelliidae), in which the male crawls under the female, attaching his head in fixed connection to the fifth abdominal sternite of the female. Copulation lasts a few minutes to 1.5 hours in Raphidiidae and up to three hours in Inocelliidae. Oval cigarlike eggs are laid singly or in a batch by the long ovipositor into crevices of bark or under litter. The egg stage lasts from a few days up to three weeks. The larval period lasts at least one year and at most two or three years, although under experimental conditions the period can stretch up to six years. The number of larval instars is generally 10 or 11 but may reach 15 or more. Pupation occurs in spring, and the pupal stage lasts from a few days up to three weeks. In some species pupation takes place in summer or autumn, and the pupal stage lasts several months, up to 10. The extreme mobility of the pupa is a remarkable primitive feature of snakeflies.

Hibernation may occur in the last larval stage, the penultimate larval stage, or the pupa stage, but never in the egg, prepupa, or adult stages. Snakeflies require a low-temperature period (around 32°F [0°C]) to induce pupation or hatching of the imago. Larvae are often parasitized by various species of the families Ichneumonidae (in particular of the genus *Nemeritis*) and Braconidae.

Conservation status

Almost all species are restricted to limited areas of refugial character, and a high number of species are endemic to certain mountain ranges or islands. Only three species occur throughout northern Asia to northern and central Europe, while a few species in North America, with distribution centers in the Southwest, have succeeded in reaching southern Canada. Although no species are listed by the IUCN, several species with small distributions are seriously endangered by habitat destruction.

Significance to humans

Woodcut illustrations of snakeflies appeared in the seventeenth century, and first species descriptions were published in the eighteenth century. Several species are beneficial to humans as predators of aphids and other soft-bodied pest arthropods, but trials to introduce them into Australia for biological control failed.

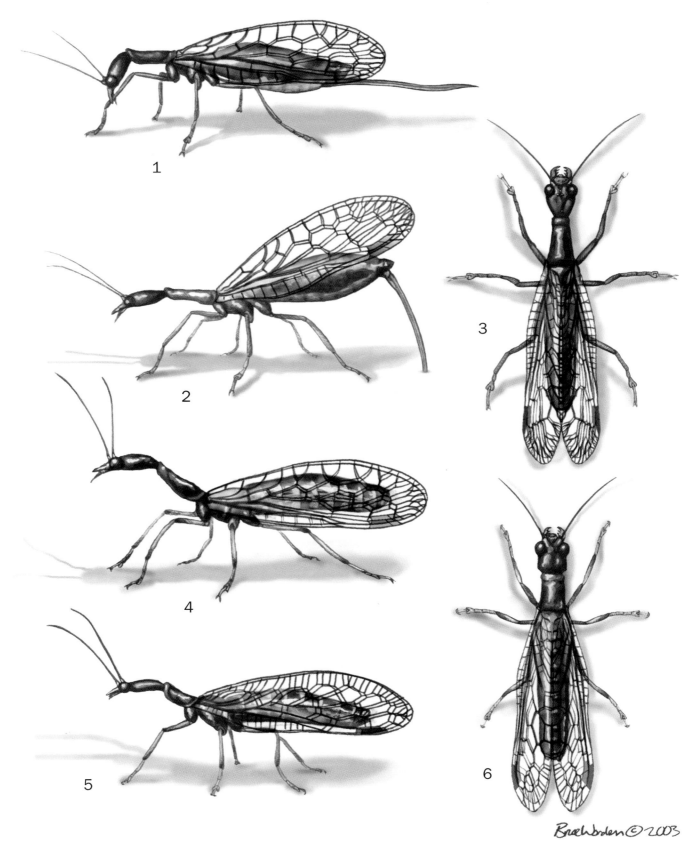

1. Female Schremmer's snakefly (*Alena [Aztekoraphidia] schremmeri*); 2. Female yellow-footed snakefly (*Dichrostigma flavipes*); 3. Female common European snakefly (*Phaeostigma [Phaeostigma] notata*); 4. Male wart-headed Uzbekian snakefly (*Mongoloraphidia [Usbekoraphidia] josifovi*); 5. Male Schummel's inocelliid snakefly (*Inocellia crassicornis*); 6. Male Brauer's inocelliid snakefly (*Parainocellia braueri*). (Illustration by Bruce Worden)

Species accounts

Schummel's inocelliid snakefly
Inocellia crassicornis

FAMILY
Inocelliidae

TAXONOMY
Raphidia crassicornis Schummel, 1832, near Scheitnich (Silesia), Poland.

OTHER COMMON NAMES
German: Schummels Inocelliide.

PHYSICAL CHARACTERISTICS
Small to medium-sized species with forewings of 0.31–0.43 in (8–11 mm) in males and 0.41–0.59 in (10.5–15 mm) in females. Pterostigma is dark brown. The "rectangular" head and the lack of ocelli are typical for the family.

DISTRIBUTION
Central and northern Europe to eastern Asia.

HABITAT
Coniferous, particularly pine, forests from low elevations up to 3,280 ft (1,000 m).

BEHAVIOR
Little is known, though females are clumsy flyers.

FEEDING ECOLOGY AND DIET
Larvae are found under the bark of conifers, particularly pines, feeding on soft-bodied arthropods. Adult diet is unknown but may consist of pollen.

REPRODUCTIVE BIOLOGY
Life cycle is usually two to three years.

CONSERVATION STATUS
Not listed by the IUCN. Rare but not threatened.

SIGNIFICANCE TO HUMANS
None known. ◆

Brauer's inocelliid snakefly
Parainocellia braueri

FAMILY
Inocelliidae

TAXONOMY
Inocellia braueri Albarda, 1891, "Southern Europe."

OTHER COMMON NAMES
German: Brauers Inocelliide.

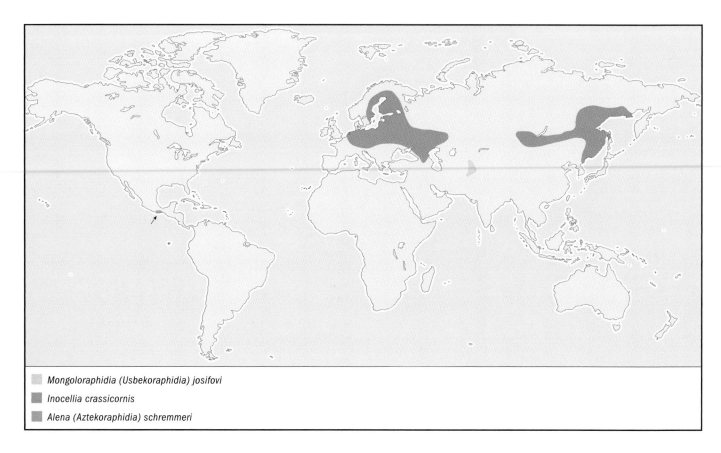

☐ *Mongoloraphidia (Usbekoraphidia) josifovi*
■ *Inocellia crassicornis*
■ *Alena (Aztekoraphidia) schremmeri*

Phaeostigma (Phaeostigma) notata

Parainocellia braueri

PHYSICAL CHARACTERISTICS
Small to medium-sized species with forewings of 0.27–0.34 in
(7–8.7 mm) in males and 0.35–0.47 in (8.8–12 mm) in females.
Characterized by a yellowish pattern on its otherwise black
head.

DISTRIBUTION
Central and southeastern Europe.

HABITAT
Open deciduous forests in warm localities, from below 656 ft
(200 m) in central Europe to more than 3,280 ft (1,000 m) in
the southern-most parts of its range.

BEHAVIOR
Larvae are sometimes found in high densities. Adults are poor
flyers.

FEEDING ECOLOGY AND DIET
Larvae are found under the bark of various deciduous trees
(pears, oaks, maples, etc.) feeding on soft-bodied arthropods.
Adult diet is unknown but may consist of pollen.

REPRODUCTIVE BIOLOGY
The life cycle is at least two and generally three to four years.

CONSERVATION STATUS
Not listed by the IUCN, but rare and possibly threatened.

SIGNIFICANCE TO HUMANS
None known. ◆

Schremmer's snakefly
Alena (Aztekoraphidia) schremmeri

FAMILY
Raphidiidae

TAXONOMY
Alena (Aztekoraphidia) schremmeri Aspöck, Aspöck, and Rausch,
1994, Sierra de Miahuatlán, Oaxaca, México.

OTHER COMMON NAMES
German: Schremmers Kamelhalsfliege.

PHYSICAL CHARACTERISTICS
Slender snakefly with forewings of 0.27–0.31 in (7–8 mm) in
males and 0.31–0.37 in (8–9.5 mm) in females and a
pterostigma that is mostly yellow but brown at its borders.

DISTRIBUTION
Known only from the mountain range of its type locality
(Sierra de Miahuatlán, Oaxaca, México). This species has the
southernmost-known distribution of any extant Raphidiid
snakefly.

HABITAT
Inhabits pine forests from 6,235–6,890 ft (1,900–2,100 m).

BEHAVIOR
Little is known aside from feeding ecology and reproductive
biology.

FEEDING ECOLOGY AND DIET
Larvae live under bark and adults on trees, both feeding on
soft-bodied arthropods.

REPRODUCTIVE BIOLOGY
The life cycle lasts two years, with pupation and appearance of
adults in spring or summer.

CONSERVATION STATUS
Not listed by the IUCN. Although adults are encountered
rarely, the species likely is not threatened.

SIGNIFICANCE TO HUMANS
Beneficial as a predator of pine-tree pests. ◆

Yellow-footed snakefly
Dichrostigma flavipes

FAMILY
Raphidiidae

TAXONOMY
Raphidia flavipes Stein, 1863, Greece.

OTHER COMMON NAMES
German: Gelbfuss-Kamelhalsfliege.

PHYSICAL CHARACTERISTICS
A slim species with forewings of 0.31–0.52 in (8–13.3 mm) in
males and 0.39–0.57 in (10–14.5 mm) in females. Characterized
by a bicolored (proximally brown, distally yellow) pterostigma
and (with the exception of the black coxae) predominantly yellow legs.

DISTRIBUTION
Central, eastern, and southeastern Europe.

HABITAT
Prefers warmer habitats, particularly open forests with scattered trees and shrubs.

Dichrostigma flavipes

BEHAVIOR
Occasionally reaches swarm-like population densities on whitethorns.

FEEDING ECOLOGY AND DIET
Larvae are found under layers of soil around trees and bushes, while adults are found mainly on shrubs; both feed on soft-bodied arthropods.

REPRODUCTIVE BIOLOGY
The life cycle lasts two years.

CONSERVATION STATUS
Not listed by the IUCN. Common in many parts of Europe.

SIGNIFICANCE TO HUMANS
This was the first snakefly found to have soil-dwelling larvae. ◆

Wart-headed Uzbekian snakefly
Mongoloraphidia (Usbekoraphidia) josifovi

FAMILY
Raphidiidae

TAXONOMY
Raphidia (Bureschiella) josifovi Popov, 1974, Kondara-Kwack, Gissarskij chrebet, Tajikistan.

OTHER COMMON NAMES
German: Usbekische Warzenkopf-Kamelhalsfliege.

PHYSICAL CHARACTERISTICS
This stout medium-sized species is characterized by the wart-like sculpture of its head.

DISTRIBUTION
Tajikistan and Uzbekistan.

HABITAT
Juniperus stands in mountainous areas from 4,595–6,890 ft (1,400–2,100 m) in elevation.

BEHAVIOR
Little is known beside feeding ecology and reproductive biology.

FEEDING ECOLOGY AND DIET
Larvae are found predominantly under the bark of *Juniperus* and adults on twigs, both feeding on soft-bodied arthropods.

REPRODUCTIVE BIOLOGY
Observed copulation follows the "wrecking position," in which the male hangs head first from the female, who carries him. The life cycle is at least two years.

CONSERVATION STATUS
Not listed by the IUCN. Dependent on conservation of *Juniperus*.

SIGNIFICANCE TO HUMANS
None known. ◆

Common European snakefly
Phaeostigma (Phaeostigma) notata

FAMILY
Raphidiidae

TAXONOMY
Raphidia notata Fabricius, 1781, Coombe Hurst, Croydon, Surrey, England.

OTHER COMMON NAMES
German: Gemeine europäische Kamelhalsfliege.

PHYSICAL CHARACTERISTICS
Robust snakefly with forewings of 0.35–0.51 in (9–13 mm) in males and 0.39–0.57 in (10–14.5 mm) in females and a dark brown pterostigma.

DISTRIBUTION
From Great Britain to northeastern Spain, including central and northern Europe as well as parts of eastern Europe.

HABITAT
Inhabits a wide range of coniferous and deciduous forests from low elevations to almost 6,560 ft (2,000 m).

BEHAVIOR
The grooming behavior comprises a complex ritual of cleaning the head, antennae, mouthparts, and legs.

FEEDING ECOLOGY AND DIET
Larvae live under bark and adults on trees, and both feed on soft-bodied arthropods.

REPRODUCTIVE BIOLOGY
The life cycle lasts either two or three years.

CONSERVATION STATUS
Not listed by the IUCN. Rare but not threatened.

SIGNIFICANCE TO HUMANS
Beneficial to humans as a predator of pests. ◆

Resources

Books

Aspöck, Horst, Ulrike Aspöck, and Hubert Rausch. *Die Raphidiopteren der Erde.* 2 vols. Krefeld, Germany: Goecke und Evers, 1991.

Tauber, Catherine A. "Order Raphidioptera." In *Immature Insects,* edited by Frederick W. Stehr. 2 vols. Dubuque, IA: Kendull/Hunt Publishing Company, 1987–1991.

Periodicals

Acker, Thomas S. "Courtship and Mating Behavior in *Agulla* species (Neuroptera: Raphidiidae)." *Annals of the Entomological Society of America* 59 (1966): 1–6.

Aspöck, Horst. "The Biology of Raphidioptera: A Review of Present Knowledge." *Acta Zoologica Academiae Scientiarum Hungaricae* 48 (2002): 35–50.

———. "Distribution and Biogeography of the Order Raphidioptera: Updated Facts and a New Hypothesis." *Zoologica Fennica* 209 (1998): 33–44.

Kovarik, Peter W., Horace R. Burke, and Charles W. Agnew. "Development and Behavior of a Snakefly, *Raphidia bicolor* Albarda (Neuroptera: Raphidiidae)." *Southwestern Entomologist* 16 (1991): 353–364.

Organizations

IAN: International Association of Neuropterology. Web site: <http://www.neuroptera.com>

Other

Oswald, John D. "Bibliography of the Neuropterida." 2003 [May 8, 2003]. <http://entowww.tamu.edu/research/neuropterida/neur_bibliography/bibhome.html>.

Ulrike Aspöck, PhD
Horst Aspöck, PhD
Hubert Rausch

Neuroptera
(Lacewings)

Class Insecta
Order Neuroptera
Number of families 17

Photo: A mantisfly (family Manitispidae) near the Bayer River of Papua New Guinea. (Photo by George D. Dodge & Dale R. Thompson. Bruce Coleman, Inc. Reproduced by permission.)

Evolution and systematics

Neuroptera represents one of the oldest and most archaic lineages of endopterygote (= holometabolous, or undergoing complete metamorphosis) insects. The oldest known fossils of Neuroptera suggest that the group had its origin in the early Permian period, with significant family level diversification in the late Triassic and early Jurassic. At the end of the Jurassic, however, all extant lineages had appeared, and at least three lineages (14 families) had become extinct. Truly spectacular lacewings that we know only from fossils include *Kalligramma haeckelli* and *Lithogramma oculatum* (Kalligrammatidae, Upper Jurassic), two species with distinctive eyespots in their large, brightly colored wings. Numerous families of lacewings are known only from fossils, including Kalligrammatidae, Nymphitidae, Permithonidae, Mesopolystoechotidae, Solenoptilidae, Allopteridae, and Osmylitidae. Moreover, many described fossil lacewings can be placed in extant families, including *Permithonopsis obscura* (Polystoechotidae), *Embaneura vachrameevi* (Psychopsidae), *Plesiorobius* (Berothidae), and *Euporismites balli* (Osmylidae).

Neuroptera comprises 17 extant families containing more than 6,000 species worldwide divided into three superfamilies. Myrmeleontidae (more than 2,000 spp.) and Chrysopi-

dae (more than 1,200 spp.) are the most species-rich families, followed by Hemerobiidae (about 550 spp.) and Ascalaphidae (some 400 spp.). The superfamily Nevrorthiformia, with the single family Nevrorthidae, represents the most basal group; members of this family are sporadically found in Japan, Taiwan, Australia, and Europe. The Myrmeleotiformia contains five families (Myrmeleontidae, Ascalaphidae, Nemopteridae, Psychopsidae, and Nymphidae). It is a well-defined group of generally large lacewings with soil-dwelling or arboreal larvae. The Hemerobiiformia, made up of 11 families (Hemerobiidae, Chrysopidae, Sisyridae, Berothidae, Mantispidae, Rhachiberothidae, Ithonidae, Polystoechotidae, Dilaridae, Coniopterygidae, and Osmylidae), is a morphologically diverse assemblage of lacewings, many of which have unique and highly specialized life cycles. Ithonidae are robust, moth-like lacewings with fossorial, scarab-like larvae associated with roots of trees and bushes (e.g., creosote). This family and the sister family Polystoechotidae sometimes are considered the most basal clade of lacewings. Another clade, or group of closely related families, is the Dilaridae clade. This group comprises Dilaridae, Rhachiberothidae, Mantispidae, and Berothidae and is united by particular larval head characteristics.

Antlion larvae (*Myrmeleon* spp.) make traps. (Illustration by Barbara Duperron)

Physical characteristics

Lacewing is a common name that describes the lace-like venation pattern of the relatively large, delicate wings of most adult Neuroptera. The adult body shape is relatively uniform across the order. The head is well defined, typically without setae or bristles (although dense tufts of long setae are found in many Ascalaphidae). The eyes are large, well defined, and spherical in shape; some ascalaphids have a horizontal sulcus (line) dividing the eyes into upper and lower regions. Well-defined ocelli (simple eyes) are present on the vertex only in the family Osmylidae. The mouthparts are chewing and directed either anteriorly or ventrally; in many Nemopteridae the mouthparts are extremely elongated. Antennae typically are very elongate and moniliform (simple) and rarely flabellate (Dilaridae); sometimes they have apical clubs (Ascalaphidae) or are thickened (Myrmeleontidae).

The thorax is divided into three segments, the posterior, wing-bearing segments being much larger than the anterior segment. Legs usually are slender and elongate; forelegs sometimes are raptorial (Rhachiberothidae and Mantispidae). The wings almost always are large and broad and rarely are reduced. Some species are brachypterous or apterous. The wing shape is elongate and oblong to elliptical or ovate; the margin either is round or falcate (hooked or curved). Venation is highly reticulate in most groups, although it is reduced to only a few veins in some (Coniopterygidae). The abdomen usually is long, and the genitalia on terminal segments either are reduced and concealed (Myrmeleontidae) or elaborate and exposed (Osmylidae and Nymphidae).

Larval body shape typically is campodeiform or, rarely, scarabaeiform. The head is dorsoventrally flattened, with mouthparts projecting anteriorly. The mouthparts are mod-

ified uniquely, such that the buccal (mouth) cavity is closed and sucking tubes are formed laterally by the interlocking of the mandibles and maxillae. Jaws are elongate, simple, or toothed and used to impale prey and suck out its contents. In Sisyridae and Osmylidae the jaws are extremely long and slender. In Myrmeleontiformia the jaws often are held open at more than 180 degrees and snapped closed to trap prey. Eyes comprise a group of five, six, or seven stemmata (rudimentary eyes), but eyesight is poor.

Antennae either are short or elongate. The thorax usually is short and broad, but in some Nemopteridae, the prothorax (anterior segment) may be lengthened into a neck. Legs are long in the active arboreal larvae (Chrysopidae), short in fossorial larvae (Myrmeleontidae and Ascalaphidae), or rudimentary in egg sac predators (Mantispidae). The abdomen may be elongate or ovate. The thorax and abdomen often have fleshy lateral projections (scoli) (Nymphidae) or long and recurved (Chrysopidae) or ornately shaped (Ascalaphidae) setae used to hold items of debris on the dorsum. Another unique characteristic of neuropteran larvae is that the midgut is discontinuous with the hindgut; solid waste is not passed until the adult emerges from the pupal case with a fully formed digestive system. Larvae spin pupal cases with silk produced from modified Malphigian tubules. The pupa is exarate and decticous and emerges from the pupal case to molt into a fully winged adult.

Distribution

Neuropterans are distributed throughout tropical and temperate regions, with the greatest species richness and diversity in the tropics. Several families (Myrmeleontidae, Chrysopidae, Hemerobiidae, Coniopterygidae, Mantispidae, and Ascalaphidae) are distributed widely, although particular subfamilies, genera, and species within each family are much more restricted. From fossil evidence it is clear that the former distribution of some groups was more extensive than today. Nemopteridae is a group of lacewings distributed throughout Africa, the Palaearctic region, Australia, and South America but absent in North America (although a fossil nemopterid has been found in Colorado). Psychopsidae are restricted to Africa, Asia, and Australia, while Nymphidae are found only in the Indonesian Archipelago, Papua New Guinea, and Australia. Some groups have highly disjunct, apparently relict distributions, also evidence of more extensive past distributions. Nevrorthidae, comprising 10 species with aquatic larvae, are found in Australia, Japan, Taiwan, and the Mediterranean region. Ithonidae (including Rapismatidae) are diverse in Australia but also are found in mountainous regions of Central America and Asia and in the southwestern United States.

Habitat

Lacewings are found in a wide variety of habitat types, from arid desert plains to montane rainforests. While adults typically are found on vegetation, larvae are more specific in their habitat requirements and often are associated with a particular substrate or prey type. Families such as

Myrmeleontidae and Nemopteridae, with larvae that live in sandy soils, are adapted for existence in deserts and dry savannas and are particularly diverse and numerous in these regions around the world. Larvae of some species of antlion (Myrmeleontidae: Acanthoclisinae) can swim through loose sand in search of prey. Some lacewings, such as Chrysopidae and Hemerobiidae, are strictly arboreal as larvae on trees, shrubs, and grasses. Larvae of Psychopsidae are recorded inhabiting deep crevices in *Eucalyptus* spp. trees in dense forests in Australia. Larvae and adults of Coniopterygidae are found on foliage of trees and bushes, including some specific to particular vegetation types (e.g., *Aleuropteryx juniperi*). Larvae of both Nevrorthidae and Sisyridae are predators in freshwater streams, while larvae of some Osmylidae (e.g., *Osmylus fulvicephalus* in Europe and unidentified Kempyninae in Australia) live in the littoral zone of such streams under rocks and among leaf litter. Larvae of moth lacewings (Ithonidae) are fossorial, living among the roots of trees and shrubs. One example is *Oliarces clara*, the only North America ithonid, which is associated with the roots of the creasote bush (*Larrea tridentata*).

Antlions (family Myrneleonidae) live in the savanna of southern Africa. (Photo by Michael Fogden. Bruce Coleman, Inc. Reproduced by permission.)

Behavior

Several neuropteran families possess various anatomical characteristics that apparently are involved in chemical communication between sexes during courtship. Males of Nevrorthidae and some Myrmeleontidae possess eversible pleurocavae on the abdomen, whereas males of Nemopteridae have a bulla on the wing margin or wing base; both are used to disperse chemical pheromones. Males of some Mantispidae possess an Eltringham organ on the abdomen, which also is used in the dispersal of mating chemicals. There is complicated communication among Chrysopidae involving abdominal vibrations (*Chrysoperla*) or wing "rapping" (*Mallada* spp.), resulting in complex "calls" that are communicated via the substrate. Various defensive behavioral mechanisms are employed by different lacewings to evade or deter predators.

While most adult neuropterans remain inactive during the day, relying on camouflage to escape detection, some engage in various forms of behavior to make the deception more complete. The families that have setae holding items of debris on the dorsum use this "trash packet" as camouflage and as a shield against predators. Beaded lacewings (Berothidae) commonly begin gently swaying when a potential predator is detected, apparently to simulate a twig being moved by a breeze. When disturbed, certain lacewings feign death (some Hemerobiidae and Chrysopidae), whereas others emit an offensive odor (*Nymphes* spp. of Nymphidae and *Plesiochrysa* spp. of Chrysopidae). Some Mantispidae (*Euclimacia* spp. and *Climaciella* spp.) are effective mimics of paper wasps (Vespidae) in color and shape and also adopt postures and movements resembling the paper wasp when disturbed. Some first instar larvae of Mantispidae have been shown to follow a series of obligatory behavioral cues when searching for a suitable host spider egg sac. Larvae of *Climaciella brunnea* must board an adult spider before they enter the egg sac. If placed on a spider and not allowed to board, they simply climb to the highest point of the spider and assume a questing posture to look for another passing spider they can board. Upon hatching, larvae of Ascalaphidae and Nymphidae (*Nymphes*) group together for a period of time with jaws outstretched in an apparent defensive posture.

Feeding ecology and diet

Lacewings typically are generalist predators as larvae and adults, although there are exceptions; in several families, the larva has become highly specialized in its feeding ecology and diet. Many lacewing adults may be generalist omnivores, feeding opportunistically on soft-bodied insects, pollen, and honeydew. Adults of Nemopteridae and some Chrysopidae are obligate feeders on pollen and nectar from flowers, with many nemopterids having greatly elongated mouthparts modified for insertion into flowers with long corollas (where separate or fused flower petals come together and often form a long tube). Predatory adult Mantispidae and Rhachiberothidae have raptorial forelegs for seizing and holding prey, and many adult antlions (Myrmeleontidae: Acanthoclisinae) and owlflies (Ascalaphidae) have elongated claws and long, stiff bristles and spines on their legs for capturing prey in flight.

In most families the larvae are either sedentary "sit-and-wait" predators, waiting for hapless prey to walk into their open jaws (Ascalaphidae and Nymphidae), or active foragers, scouring the tactile landscape in search of prey items (Chrysopidae, Hemerobiidae, and Coniopterygidae). Larvae of some antlions (Myrmeleontidae: *Myrmeleon* spp.) construct conical pits in the sand and wait at the bottom with only their large jaws exposed. When a suitable prey falls into the pit, the antlion larva tosses sand upward with its head to dislodge the prey, so that it falls into its open jaws and then is dragged beneath the surface of the sand to be consumed. Larvae of Ithonidae and, presumably, Polystoechotidae are the only non-carnivorous lacewings, with short blunt jaws not suitable

A green lacewing (*Chrysopa*) resting on seeds of black bindweed in Europe. (Photo by Kim Taylor. Bruce Coleman, Inc. Reproduced by permission.)

for impaling prey as other lacewings do. On the contrary, Ithonid larvae feed on root exudates of trees and bushes.

Among the families with aquatic larvae, Nevrorthidae are generalist predators in fast-flowing streams, while spongilla flies (Sisyridae) are obligate specialist predators of freshwater sponges and bryozoans. Sisyrid larvae have highly modified jaws that are extremely long (often longer than the body) and narrow and are used to pierce individual cells of their sponge prey delicately. Semiaquatic larvae of Osmylidae use their long jaws to probe the wet soil and mud in search of soft-bodied prey, for example, larvae of Chironomidae (Diptera). Larvae of some families (Dilaridae, Osmylidae, and Psychopsidae) live under bark and in deep bark crevices, where they are generalist predators on a variety of arthropods living there. Beaded lacewing (Berothidae) larvae live as specialized predators on subterranean termites. Fragmentary evidence suggests that the larvae either secrete an allomone or inject a neurotoxin to immobilize their termite prey, on which they then can feed safely. The first and third instar larvae actively feed on termites, but the second instar is an inactive resting stage.

Larvae of Mantispidae are obligate, specialized predators on either social hymenoptera nests (Symphrasinae) or spider egg sacs (Mantispinae). In larvae that prey on spider eggs sacs, the active first instar triungulin larva seeks out a suitable host spider, which it boards and ultimately enters the egg sac. Once it is in the egg sac, the second and third instars take on a physiogastric form (i.e., hypermetamorphic development, or dramatic change in overall morphological features between instars). The larvae also may arrest the development of the spider eggs by chemical means to prevent them from hatching. There appears to be some level of host specificity.

Reproductive biology

Mating either is brief or takes place over an extended period of time, usually through solitary encounters between the sexes. Mating swarms have been recorded for *Ithone* in Australia and *Oliarces* in the United States. Eggs are laid either solitarily or in batches on substrate, in crevices, or on silken stalks (certain Nymphidae, Mantispidae, and Chrysopidae). Ascalaphidae often deposit infertile eggs (repagula) between their fertile eggs and the probable path of potential egg predators. Some species of Mantispidae lay hundreds of eggs, presumably because of the high mortality rate faced by the minute first instar larvae as they seek out suitable spider egg sacs. There is little or no parental care after oviposition.

Conservation status

Owing to the often high degree of regional endemicity and low degree of vagility, many neuropterans are particularly vulnerable to extinction from pollution and habitat alteration by human activities. Other species are large and brightly colored (*Libelloides* spp. and *Nemoptera* spp.), making localized populations susceptible to being overly collected by amateur collectors. At present no species of lacewing is listed by the IUCN worldwide. In various countries around the world, there are national, state, or local regulatory lists of protected species or populations of neuropterans.

In the United States several species of neuropterans are considered endangered, with legislated protection in some states (e.g., California and Hawaii). These species include *Nothochrysa californica* (Chrysopidae), *Oliarces clara* (Ithonidae),

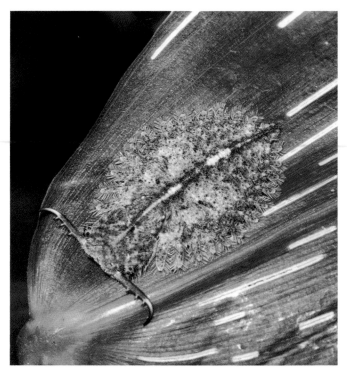

Owlfly (family Ascalaphidae) larva in the Costa Rican rainforest. (Photo by Michael Fogden. Bruce Coleman, Inc. Reproduced by permission.)

Distoleon perjerus (Myrmeleontidae), *Pseudopsectra usingeri* (Hemerobiidae), and three species of *Micromus* (Hemerobiidae). In South Africa species of lacewing considered critically endangered due to habitat destruction include *Pamexis bifasciatus, P. contamminatus, Exaetoleon obtabilis* (all Myrmeleontidae); *Sicyoptera dilatata, S. cuspidata, Halterina pulchella;* and *H. purcelli* (all Nemopteridae).

Significance to humans

As beneficial generalist predators, lacewings from at least three families (all with arboreal larvae) have been used in biological control of arthropod pests in agriculture. Coniopterygidae have long been recognized to have considerable potential for biological control, particularly of spider mites (Tetranychidae) in greenhouses and orchards. Two families, Chrysopidae and Hemerobiidae, are used on a commercial scale to control arthropod pests in numerous field and greenhouse crop situations. Chrysopids from various genera (*Mallada, Chrysoperla,* and *Chrysopa*) are reared in large numbers in commercial insectaries for inundative release among various crops for successful control of many arthropod pests. Hemerobiids also are reared for inundative release but are used less commonly in commercial situations.

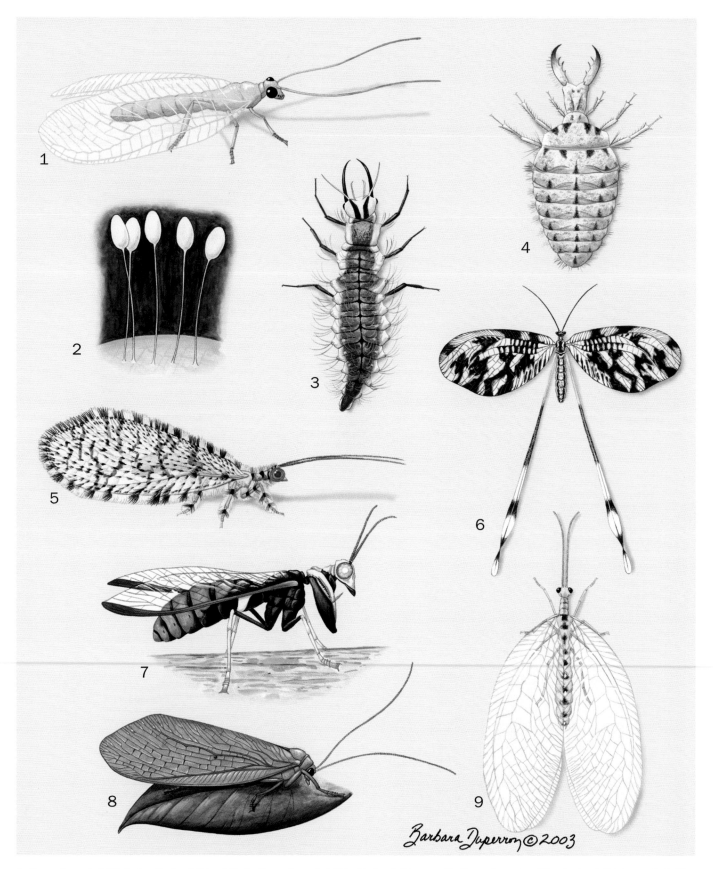

1. Green lacewing (*Mallada albofascialis*) adult; 2. Green lacewing (*M. albofascialis*) eggs; 3. Green lacewing (*M. albofascialis*) larva; 4. Antlion (*Myrmeleon formicarius*) larva; 5. Beaded lacewing (*Spermophorella maculatissima*); 6. Spoonwing lacewing (*Nemoptera sinuata*); 7. Mantid lacewing (*Euclimacia torquata*); 8. Moth lacewing (*Megalithone tillyardi*); 9. Norfolius (*Norfolius howensis*). (Illustration by Barbara Duperron)

Species accounts

Beaded lacewing
Spermophorella maculatissima

FAMILY
Berothidae

TAXONOMY
Spermophorella maculatissima Tillyard, 1916, Brisbane, Queensland, Australia.

OTHER COMMON NAMES
None known.

PHYSICAL CHARACTERISTICS
Relatively small lacewings. Short, narrow body, with wings held vertically over the abdomen. Wings have speckled black, brown, and white pattern on wing veins to aid in camouflage. Body and wing veins are covered with long setae. Newly hatched larvae are elongate with short jaws. Later instars unknown.

DISTRIBUTION
Queensland, Australia.

HABITAT
Arid regions, particularly in open sclerophyll forests.

BEHAVIOR
Adults remain motionless during the day, with the antennae held out in front of the head. They sway their bodies when potential predators are near.

FEEDING ECOLOGY AND DIET
Adults probably are generalist feeders. The larvae of all known berothids are obligate predators on subterranean termites. Larvae presumably use an allomone to subdue their termite prey. First and third instars are active feeders, whereas the second instar is a sedentary, resting stage.

REPRODUCTIVE BIOLOGY
Adult females lay solitary eggs on long, silken stalks. It is thought that the first instar triungulin larvae seek out and enter a suitable termite colony, where they can remain undetected by the termites. The larvae undergo hypermetamorphic development.

CONSERVATION STATUS
Not listed by the IUCN.

SIGNIFICANCE TO HUMANS
None known. ◆

Green lacewing
Mallada albofascialis

FAMILY
Chrysopidae

TAXONOMY
Mallada albofascialis Winterton, 1995, Brisbane, Queensland, Australia.

OTHER COMMON NAMES
None known.

PHYSICAL CHARACTERISTICS
Relatively small lacewings. The body is elongate and delicate, with broad wings that have characteristic open, "chrysopid-type" venation. The body is light to dark green in color, with red patches on the prothorax and head. The face also has a dis-

Euclimacia torquata
Myrmeleon formicarius
Spermophorella maculatissima

Megalithone tillyardi
Mallada albofascialis

tinctive white area above the mouth. The larva has an elongate body and carries a trash packet within special curved hairs on its back, used for camouflage.

DISTRIBUTION
Northern Territory and coastal Queensland, Australia.

HABITAT
Forested areas.

BEHAVIOR
Nothing is known.

FEEDING ECOLOGY AND DIET
Adults feed on honeydew and flower nectar. The larva is an arboreal generalist predator feeding on a variety of soft-bodied arthropods, especially mealybugs (Hemiptera: Margarodidae).

REPRODUCTIVE BIOLOGY
Adult females lay eggs on long silken stalks in patches of 10–15 eggs.

CONSERVATION STATUS
Not listed by the IUCN.

SIGNIFICANCE TO HUMANS
None known. ◆

Moth lacewing
Megalithone tillyardi

FAMILY
Ithonidae

TAXONOMY
Megalithone tillyardi Riek, 1974, Cunningham's Gap, Queensland, Australia.

OTHER COMMON NAMES
None known.

PHYSICAL CHARACTERISTICS
The moth lacewing is a relatively large, robust insect with an appearance similar to that of a dull hepialid moth. The wings and body are dull brown, and the body is covered with numerous long hairs. The wings are folded over the body. The larva is fossorial and scarabaeiform in body shape.

DISTRIBUTION
Southeastern Queensland and northern New South Wales, Australia.

HABITAT
Higher elevations, often on sandy soils.

BEHAVIOR
Adults emerge in masses to form large mating aggregations or swarms composed of many more males than females.

FEEDING ECOLOGY AND DIET
It is not clear if adults feed, but the larvae eat root exudates of plants. Ithonids have been recorded erroneously as predators of scarab larvae.

REPRODUCTIVE BIOLOGY
The female has a genital plug upon emergence, which is apparently displaced during copulation.

CONSERVATION STATUS
Not listed by the IUCN. The conservation status of the moth lacewing is difficult to assess, because larvae are fossorial and adult swarms are infrequent. Habitat destruction appears to be the main threat to individual populations.

SIGNIFICANCE TO HUMANS
Swarms have been recorded hitting the metal roofs of houses and sounding like a hail storm. While they are rare, swarms are a nuisance to humans, because adults also enter houses and gather in dark places. Such plagues are known to last as long as three weeks. ◆

Mantid lacewing
Euclimacia torquata

FAMILY
Mantispidae

TAXONOMY
Euclimacia torquata Navás, 1914, Queensland, Australia.

OTHER COMMON NAMES
None known.

PHYSICAL CHARACTERISTICS
Medium-size lacewings. The body is robust, with relatively narrow wings that are darkly pigmented along the anterior portion. This species has strongly contrasting black, yellow, and orange coloration, enabling it to be an effective wasp mimic. The front legs are raptorial. The immature stages are unknown.

DISTRIBUTION
Queensland, Australia, and Papua New Guinea.

HABITAT
Forested areas.

BEHAVIOR
Adults fly and walk in a way similar to paper wasps, as a defense against predators.

FEEDING ECOLOGY AND DIET
Adults are generalist predators, using their raptorial forelegs to capture prey. Based on evidence from other mantid lacewings, the larva is presumably a specialized predator of spider egg sacs and probably also is highly host specific.

REPRODUCTIVE BIOLOGY
Nothing is known.

CONSERVATION STATUS
Not listed by the IUCN.

SIGNIFICANCE TO HUMANS
None known. ◆

Antlion
Myrmeleon formicarius

FAMILY
Myrmeleontidae

TAXONOMY
Myrmeleon formicarius Linnaeus, 1767, Europe.

OTHER COMMON NAMES
English: Doodlebug.

PHYSICAL CHARACTERISTICS
Large, very long lacewing. The head and thorax are short and stout, and the abdomen is very elongate. Body is brown with tan markings; antennae are thickened apically. Wings very elongate, narrow, and hyaline. Venation is mottled brown, black, and white. Larvae are robust and ovoid-shaped with large curved jaws. Body is adapted for burrowing backward through sandy soil.

DISTRIBUTION
Western Europe.

HABITAT
A wide variety of habitats, especially sandy desert or savanna regions.

BEHAVIOR
Adults are active at night and sit on foliage during the day. The elongated body and brown coloration of the adult serve as crypsis as they lie flat against a twig or branch. Larvae quickly burrow deep into the sand to avoid predation when the pit is disturbed by anything larger than a small prey item.

FEEDING ECOLOGY AND DIET
Adults are generalist predators, capturing prey on the wing. Larvae construct conical pits in fine, sandy soil by flicking sand out of the pit with rapid upward movements of the head. The larva waits at the bottom of the pit with only its large jaws exposed. When a suitable prey falls into the pit, the larva quickly seizes the prey by impaling it and injects paralyzing venom before dragging it below the sand surface.

REPRODUCTIVE BIOLOGY
The adult female lays egg in sandy soil. Larvae burrow through the soil and form conical pits, typically under overhangs or in caves, to avoid precipitation that may disturb the pit and drown the larva. When development is complete, the larva spins a spherical cocoon from silk produced from the anus, which is impregnated with sand particles.

CONSERVATION STATUS
Not threatened.

SIGNIFICANCE TO HUMANS
Antlion larvae, or "doodlebugs," have long been established in human folklore, particularly in children's chants or charms, typically from countries around the world that have been influenced by Europeans. Many charms referring to doodlebugs specifically cite the conical pits formed by the larva or their peculiar reverse-burrowing behavior. ◆

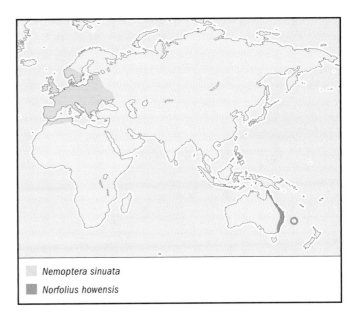

Nemoptera sinuata
Norfolius howensis

Spoonwing lacewing
Nemoptera sinuata

FAMILY
Nemopteridae

TAXONOMY
Nemoptera sinuata Olivier, 1811, eastern Mediterranean.

OTHER COMMON NAMES
None known.

PHYSICAL CHARACTERISTICS
Large to medium-size lacewings. The body is elongate and robust, with relatively broad, rounded forewings. The hind wings are highly modified, narrow, and petiolate basally and slightly dilated apically, so that the wing is somewhat spoon-shaped. Wings are strikingly marked with irregular yellow and black banding. The forewings are held above the body, while the hind wings project posteriorly. The larvae are broad and oval shaped, with short necks and short jaws.

DISTRIBUTION
Europe and the Mediterranean region.

HABITAT
Open grasslands and forests.

BEHAVIOR
Adults are active during the day, feeding at flowers.

FEEDING ECOLOGY AND DIET
Adults feed at flowers on pollen and nectar. Larvae lie buried in the sand, remaining inactive for long periods of time. When prey movement is detected, larvae approach slowly and attack with a single impaling of the prey with their jaws, during which time it is presumed that paralyzing venom is injected. Larvae are occasionally cannibalistic.

REPRODUCTIVE BIOLOGY
Females lay eggs in sand. Larval development is prolonged and probably univoltine in nature. Adults are active in late spring.

CONSERVATION STATUS
Not listed by the IUCN.

SIGNIFICANCE TO HUMANS
None known. ◆

Norfolius
Norfolius howensis

FAMILY
Nymphidae

TAXONOMY
Myiodactylus howensis Tillyard, 1917, Lord Howe Island, New South Wales, Australia.

OTHER COMMON NAMES
None known.

PHYSICAL CHARACTERISTICS
Medium-size to large lacewings, with large, broad wings. Body is elongate and yellow-green in color, with a series of brown spots along the dorsal surface of thorax and abdomen. Antennae are long and yellow. Wings are transparent, with densely reticulate venation and black spots located on the pterostigma. Larva is discoid, with a quadrangular head and large, scythe-like jaws.

DISTRIBUTION
Eastern coastal regions of mainland Australia and Lord Howe Island (Australia).

HABITAT
Dense forested areas.

BEHAVIOR
Nothing is known.

FEEDING ECOLOGY AND DIET
Adult and larva are generalist predators. Larva is a sedentary leaf litter dweller.

REPRODUCTIVE BIOLOGY
Nothing is known.

CONSERVATION STATUS
Not listed by the IUCN. Norfolius is not uncommon in densely forested (e.g., rainforest) areas along coastal eastern Australia. Habitat destruction appears to be the only real threat to this species.

SIGNIFICANCE TO HUMANS
None known. ◆

Resources

Books
Aspöck, H., U. Aspöck, and H. Hölzel. *Die Neuropteren Europas.* 2 vols. Krefeld, Germany: Goecke and Evers, 1980.

McEwen, P. K., T. R. New, and A. E. Whittington, eds. *Lacewings in the Crop Environment.* Cambridge, U.K., and New York: Cambridge University Press, 2001.

New, T. R. "Planipennia (Lacewings)." In *Handbuch der Zoologie: Eine Naturgeschichte der Stämme des Tierreiches.* Vol. 4, *Arthropoda: Insecta*, edited by Max Beier, Maximilian Fischer, Johann-Gerhard Helmcke, Dietrich Starch, and Heinz Wermuth. Berlin and New York: W. de Gruyter, 1989.

————. "Neuroptera (Lacewings)." In *The Insects of Australia*, edited by CSRIO. 2nd edition. Vol. 1. Carlton, Australia: Melbourne University Press, 1991.

Periodicals
Aspöck, U. "Male Genital Sclerites of Neuropterida: An Attempt at Homologisation (Insecta: Holometabola)." *Zoologischer Anzeiger* 241, no. 2 (2002): 161–171.

Aspöck, U., J. D. Plant, and H. L. Nemeschkal. "Cladistic Analysis of Neuroptera and Their Systematic Position within the Neuropterida (Insecta: Holometabola: Neuropterida: Neuroptera)." *Systematic Entomology* 26 (2001): 73–86.

Oswald, J. D. "Revision and Cladistic Analysis of the World Genera of the Family Hemerobiidae (Insecta: Neuroptera)." *Journal of the New York Entomological Society* 101 (1993): 143–299.

Oswald, J. D., and N. D. Penny. "Genus-Group Names of the Neuroptera, Megaloptera and Raphidioptera of the World." *Occasional Papers of the California Academy of Sciences* 147 (1991): 1–94.

Withycombe, C. L. "Some Aspects of the Biology and Morphology of the Neuroptera, With Special Reference to the Immature Stages and Their Possible Phylogenetic Significance." *Transactions of the Entomological Society of London* (1924) 303–411.

Other
Oswald, J. D. "NeuroWeb: The Neuropterists' Home Page." [3 Apr. 2003] <http://insects.tamu.edu/research/neuropterida/neuroweb.html>.

Shaun L. Winterton, PhD

•
Coleoptera
(Beetles and weevils)

Class Insecta
Order Coleoptera
Number of families 166

Photo: A weevil (*Eupholus bennetti*) in Papua New Guinea. (Photo by Bob Jensen. Bruce Coleman, Inc. Reproduced by permission.)

Evolution and systematics

The remains of ancient beetles are abundant in the fossil record, entombed in amber or as impressions in sedimentary rock. Fossils of beetle-like insects are known from the Lower Permian rocks of eastern Europe, dating back about 250 million years ago (mya). These ancient insects were placed in the extinct family Tshekardocoleidae in the order Protocoleoptera.

Protocoleopterans resembled the modern insect order of Megaloptera (alderflies, dobsonflies, and fishflies) and most likely were an assemblage of precursors to several modern insect orders. Many Protocoleoptera are flattened, suggesting that they occupied tight spaces, such as those under loose bark. The tshekardocoleids had elytra with distinct ribbing and sculpturing, resembling extant beetles of the family Cupedidae. Tshekardocoleids differ from modern beetles in having elytra that are less regularly sculptured, loose fitting, and extending beyond the abdomen.

All protocoleopterans disappeared at the beginning of the Triassic, some 240 mya, replaced by the modern Coleoptera. By the middle of the Triassic, all four of today's suborders of beetles (Archostemata, Myxophaga, Adephaga, and Polyphaga) were present. By the end of the Jurassic period, 210–145 mya, the evolutionary lineages of all of the modern Coleoptera were established.

Deposits of resin secreted by conifers began to appear during the Jurassic, suggesting that these ancient trees were attempting to defend themselves from the attacks of wood-boring insects similar to modern bark beetles (Curculionidae). At least 60 families of beetle are preserved in petrified sap deposits, or amber. The majority of these taxa are attributable to extant tribes and genera. Amber with fossil intrusions formed in tropical forests and other ancient habitats usually is poorly represented in the fossil record. Habitat types are inferred by comparing fossil beetles with the habits and distributions of extant species. Amber occurs throughout the world but is best known from deposits in the Baltic region, Dominican Republic, Mexico, China, Burma, Sicily, Canada, and Siberia. In the United States amber deposits are found in Alaska and New Jersey. Most fossil beetles from the Quaternary period, 1.6–0.5 mya, are identical to modern beetles. The remains of these beetles were not fossilized and instead were preserved among permanently frozen detritus, water-lain sediments, prehistoric dung middens, or asphalt seeps.

The relationship between beetles and plants is highly significant to their evolution. Already feeding on fungi, mosses, and algae for some 240 mya, beetles were poised to exploit the new range of vegetative structures offered by the appearance of angiosperms, or flowering plants, some 125 mya. The long and dynamic relationship between beetles and angiosperms has contributed to the amazing diversification of both groups. With approximately 350,000 species described thus far, beetles make up the largest known group of organisms.

Beetles are placed in the subclass Holometabola, also known as the Endopterygota. They are related closely to the

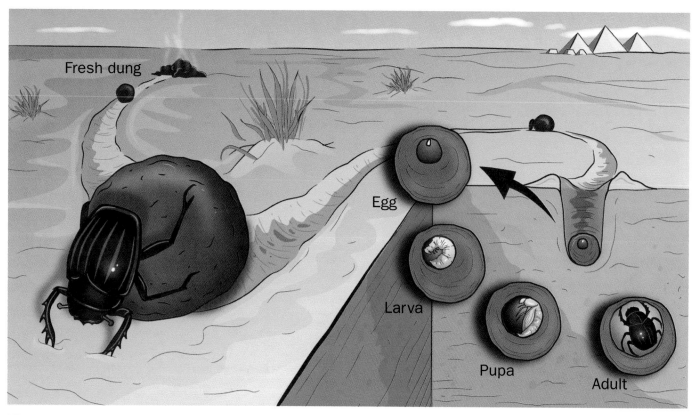

Life stages of a sacred scarab dung beetle. (Illustration by Katie Nealis)

neuropteroid orders (Neuroptera, Megaloptera, and Rhaphidioptera) and the twisted-wing parasites (Strepsiptera). Adults are distinguished from other holometabolous orders by the presence of chewing mouthparts, with a sclerite underneath known as the gula. The mesothorax and metathorax fuse to form the pterothorax. The pterothorax is connected broadly to the abdomen and typically is covered by the elytra. The hind wings have reduced venation and are folded or rolled under the elytra. The typical three-part body plan for adult beetles therefore consists of a head, prothorax, and elytra covering the pterothorax and abdomen. The antennae usually consist of 11 or fewer antennomeres. The genitalia are retracted completely within the abdomen.

The four living suborders of Coleoptera (Archostemata, Myxophaga, Adephaga, and Polyphaga) are divided into 16 superfamilies and 166 families. The suborders are distinguished primarily by the structure of the adult prothorax, hind wing, abdomen, and reproductive organs. The suborder Archostemata contains four families of beetles that are associated with wood and resemble some of the earliest beetle fossils. The Myxophaga includes four families with fewer than 100 species. Myxophagans are small to minute and live in the watery film on stream rocks, among stream and river debris, or in interstitial habitats among sand grains. The Adephaga is the second-largest suborder of beetles, consisting of nine families and more than 40,000 known species. Most are predatory, but a few feed on algae or seeds. The abdomen has pygidial glands that produce acrid defensive chemicals. The

suborder Polyphaga contains the majority of beetles, with about 300,000 described species placed in 149 families. Among the commonly encountered polyphagans are the rove beetles (superfamily Staphylinoidea), scarabs and stag beetles (Scarabaeoidea), metallic wood-boring beetles (Buprestoidea), and click beetles and fireflies (Elateroidea), as well as fungus beetles, grain beetles, ladybird beetles, darkling beetles, blister beetles, longhorn beetles, leaf beetles, and weevils (all Cucujoidea). Aquatic families have evolved independently in the Hydrophiloidea and Byrrhoidea.

The phylogenetic relationships among the beetle suborders are poorly known. Only recently have the morphological features of these groups been subjected to phylogentic analysis, and DNA sequencing information is only now being gathered. The two most widely discussed hypotheses regard the Polyphaga either as the sister group of Myxophaga or the sister group of all remaining beetles. Evidence for the Polyphaga plus Myxophaga hypothesis includes their shared reduction in the number of segments in the larval leg. The Adephaga have been proposed as sister group to Polyphaga plus Myxophaga based on the following shared characters: completely sclerotized elytra, reduced venation in the hind wings, and folded (as opposed to rolled) hind wings. Evidence for the Polyphaga as the sister group to all remaining beetles is grounded primarily on wing structure and the loss of the neck, or cervical sclerites, in the other three suborders. Recent phylogenetic analyses supports the Polyphaga plus Myxophaga hypothesis.

Physical characteristics

Beetles are very diverse in form and are elongate or spherical, cylindrical or flattened, slender or robust. The integument generally is tough and rigid, although in some families, such as the fireflies (Lampyridae), soldier beetles (Cantharidae), and net-winged beetles (Lycidae), it typically is soft and pliable. They range widely in length; some featherwing beetles (Anobiidae) measure less than 0.02 in (0.55 mm), while one of the world's largest species, *Titanus gigantea* (Cerambycidae), reaches 6.7 in (17 cm).

The beetle head is a hardened capsule bearing chewing mouthparts and is attached to the prothorax by a flexible, membranous neck. The clypeus is absent in weevils (Curculionidae) and other beetles with a rostrum, or elongated mouthparts. The mandibles usually are conspicuous and curved and sometimes toothed along the interior margin; they may be monstrously developed in some male stag beetles (Lucanidae) and longhorns (Cerambycidae). The mandibles are modified variously to cut, grind, and strain foodstuffs, but those of wrinkled bark beetles (Rhysodidae) are incapable of biting. The maxilla and labium may possess delicate fingerlike structures, or palpi, that manipulate food. The mouthparts may be directed downward or may be hypognathous, as in leaf beetles (Chrysomelidae) and weevils. The mouthparts of some net-winged beetles and weevils and their relatives are mounted at the very tip of an extended rostrum, an adaptation often associated with flower- or seed-feeding habits. In the predatory whirligigs and ground beetles, as well as in some longhorns, the mouthparts are prognathous, or directed forward.

The antennae are equipped with supersensitive receptors that detect food, locate egg-laying sites, identify vibrations, and assess temperature and humidity. Some ground beetles and rove beetles (Staphylinidae) possess specialized structures on their tibia and tarsi that are used to clean the antennae regularly to maintain their sensitivity. The antennae generally are shorter than the body but are much longer in some longhorns and brentid weevils (Brentidae). Males may have elaborate antennal structures that increase the surface area for special chemical receptors, a modification typical among species that use attractant odors, or pheromones, to locate mates.

Antennal segments lack internal musculature and are referred to as "antennomeres." The basic number of antennomeres for beetles is 11, but reductions to as few as seven are common. Each antenna consists of a basal scape that articulates with a smaller pedicel. The antennomeres are described variously as filiform (threadlike), moniliform (beadlike), serrate (saw-toothed), pectinate (comblike), flabellate (featherlike), clavate or capitate (distinctly clubbed), lamellate (terminal antennomeres flattened or platelike), or geniculate (elbowed). Antennal shapes sometimes are characteristic of families.

The shape of the compound eyes may be entire (rounded or oval), emarginate (kidney-shaped), or partially or completely divided. In whirligigs the eyes are completely divided, with the lenses of the exposed upper portion best suited for seeing in the air and those of the submerged portion of the eyes gathering images underwater. Compound eyes with few lenses are typical among flightless species, whereas those

A Colorado beetle (*Leptinotarsa decemlineata*) under the leaf of a potato plant. (Photo by J. C. Carton. Bruce Coleman, Inc. Reproduced by permission.)

species living in total darkness, such as cave and litter dwellers, may lack them altogether. Ocelli, or simple eyes, rarely are encountered in the Coleoptera but are present in most hide beetles (Dermestidae), a few rove beetles, some leiodids (Leiodidae), and derodontids (Derodontidae).

Horns are the most conspicuous and extraordinary modification of the beetle head and prothorax, jutting out from the upper surface of the prothorax or sweeping symmetrically back over the body. In the male Hercules beetle, *Dynastes hercules* (Scarabaeidae), the head and thoracic horns work in concert, like a vice. Horns generally are found in male beetles only and also may arise from the mandibles, legs, and elytra. Body size, nutrition, environmental factors, and heredity influence horn size and development.

The mesothorax supports a pair of elytra, whereas the more delicate and intricately folded flight wings are attached to the metathorax. Desert-dwelling darkling beetles and weevils frequently have elytra that are fused and lack hind wings. The legs are modified variously for burrowing, swimming, crawling, or running. The male harlequin beetle, *Acrocinus longimanus* (Cerambycidae); the long-armed chafer, *Euchirus longimanus* (Scarabaeidae); and several large weevils possess elongated front legs probably used for mating or defense. Several genera of darkling beetles living in the Kalahari Desert use their long, spindly legs as stilts to distance themselves along hot, blistering sands. The legs of aquatic beetles are flattened like oars, fringed with setae, and used like paddles to propel them through the water. The segments of the feet, or tarsi, lack musculature and are called tarsomeres. The tarsi, if present, are tipped with one or two claws. The tarsi of some leaf beetles are equipped with brushy, setose pads and oil

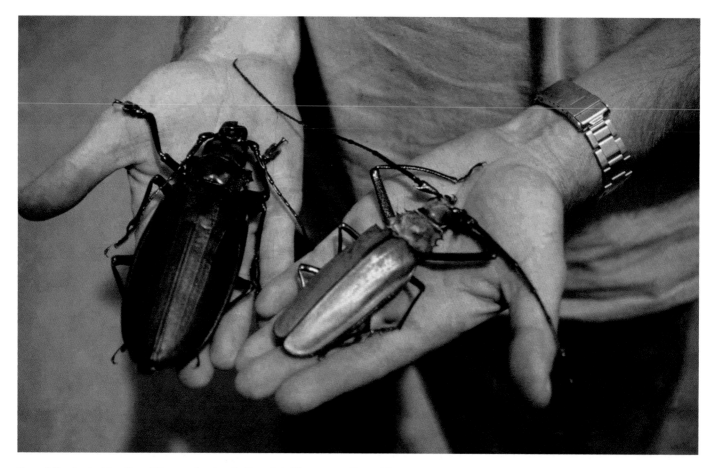

Two of the largest beetles, *Titanus gigantea* (left) and *Callipogon armillatus* (right), collected from the rain forests of French Guiana. (Photo by Rosser W. Garrison. Reproduced by permission.)

glands that adhere to plant surfaces. Certain male predaceous diving beetles have front tarsi similar to suction cups, to grasp the elytra of the female while copulating underwater.

The soft abdominal tergites usually are covered by the elytra. But in rove, clown (Histeridae), and many sap beetles (Nitidulidae), the elytra are short, leaving the tergites exposed. The tip of the abdomen is modified to facilitate egg laying and insemination. An elongate egg-laying tube, or ovipositor, is characteristic of many beetle families that develop in wood. Short, stout ovipositors are present in species adapted to lay their eggs directly on the surfaces of plants and rocks. The tip of the abdomen may be extremely flexible for use as a rudder (whirligigs) or to direct noxious defensive chemicals at attackers (ground beetles).

Beetle larvae usually are grublike or wormlike in appearance, but certain predatory forms may be flattened. They have distinct head capsules with chewing mouthparts modified for crushing, grinding, or tearing. Predatory species may have suctorial mouthparts for taking up the liquified tissues of their prey. The antennae vary in appearance, usually with two to four segments. The simple eyes vary in number from one to six on each side, but they may be absent altogether.

The three thoracic segments are very similar to one another, but the prothorax may have a more heavily sclerotized

dorsal plate. When present, the legs may have six (Adephaga), five (Polyphaga), or fewer segments. The abdomen usually is divided into 10 segments, but there may be nine or even eight segments; it typically is completely membranous in nature. The segments may bear ambulatory warts, or ampullae. Ampullae help the larvae gain purchase while moving through the substrate. Fleshy leglike structures, or prolegs, may appear on the first two abdominal segments, while the last segment sometimes has a fleshy lobe that forms an anal foot or pad. The ninth abdominal segment may possess a pair of fixed or articulated appendages, known as urogomphi.

The pupae of beetles are either adecticous, with legs tightly pressed against (appressed to) the body, or exarate, with the legs and wings held free from the body. The pupal abdomen may have functional muscles, allowing for some movement. "Gin-traps," or sclerotized teeth that form on the opposite surfaces of the abdominal segments of some beetles, snap together to protect the vulnerable intersegmental grooves that often are targets for small predators and parasitic mites.

The colors of adult beetles derive from the chemical (pigment) or physical (structural) properties of the outer layer of the cuticle. The vast majority of beetles are black owing to melanin deposition during the tanning, or hardening process, of the cuticle. Other pigments contribute to beetle colors, making ladybird beetles (Coccinellidae) reddish or yellowish or tortoise beetles (Chrysomelidae) green. Setae, scales, or microscopic sculpturing also influences color and pattern. Densely packed, recumbent setae may give beetles an overall velvety look, while fine, parallel rows of pits, or punctures, and ridges may produce a shimmering sheen. Black desert darkling beetles (Tenebrionidae) sometimes are covered with a waxy bloom secreted by an underlying layer of the cuticle (epidermis), which creates a distinct white or yellow pattern or an overall bluish-gray hue. These waxy coatings and paler colors reflect light and reduce the chances that the beetle will overheat in the hot sun.

Structural properties of the cuticle create the shiny, metallic colors of beetles. The iridescent qualities are the result of light interference from thin films in the cuticle. The colors vary with the incidence of light and among species. Because green pigments are especially rare in the Coleoptera, structural green is a useful alternative for achieving cryptic coloration among diurnal leaf beetles. Structural greens and blues are the result of concentrated reflected light, whereas the silvery and golden reflections of some neotropical jewel scarabs of the genus *Chrysina* are produced by broader, more scattered wavelengths.

Distribution

Beetles live in nearly every habitat on the earth, save for the oceans, polar ice caps, and some of the tallest mountain peaks. The vast majority of beetles are endemic to their habitat, species specifically adapted to soils, climates, and foods that restrict their distributions to a particular continent, mountain range, or valley. The distributions of plant-feeding and parasitic beetles are limited by (but seldom coincide perfectly with) those of their food plant or host. A few species are nearly cosmopolitan in distribution through human activity, either introduced purposely or accidentally transported from their native lands. Depending upon predators, parasites, and pathogens found in their new home, exotic species may become established or simply may vanish.

Habitat

Small and compact, beetles are well equipped for seeking shelter, searching for food, and reproducing in nearly every terrestrial and freshwater habitat, from coastal sand dunes to wind-swept rocky fields 10,000 ft (3,050 m) above sea level. They are equally well adapted to humid tropical forests, cold mountain streams, and parched deserts, where they forage high in the canopy, hunt in roiling water, or scavenge in deep, dark caverns. Others have managed to colonize nearby continental or distant oceanic islands.

Beetles have evolved numerous morphological and behavioral adaptations that enable them to survive and reproduce in widely different habitats. Desert darkling beetles and weevils have thickened exoskeletons with fused elytra and waxy coatings to prevent desiccation. By closing their spiracles, beetles can minimize water vapor loss through respiration. Other desert species may rely on a thick coating of insulating setae or may remain buried where temperatures are significantly

lower during the heat of the day. Some species, such as aquatic beetles, are flat and streamlined, enabling them to dive quickly into loose sand.

The elytra protect beetles from abrasion, desiccation, parasites, and predators. In addition, the space below, or the subelytral cavity, is an important adaptation used by both desert and aquatic species. In desert species this air space insulates the body from sudden changes in temperature, and predaceous diving beetles use it to store oxygen beneath the water. Some aquatic beetles are plastron breathers, trapping a layer of air elsewhere on their body. Water scavenger beetles (Hydrophilidae) capture a layer of air on the undersurface of the abdomen, whereas long-toed water beetles (Dryopidae) and riffle beetles (Elmidae) envelop their entire bodies with the aid of a thick layer of water-repellant setae.

Beetles capable of flight increase their chances of avoiding predation, locating mates, and finding food and also boost their ability to colonize new habitats. Weevils, ground beetles, and rove beetles living at high elevations in exposed and windy conditions often lose their wings. Flight is of little use in this type of harsh, restricted habitat.

Many cave-dwelling beetles have extremely thin and permeable cuticles that readily absorb moisture from the humid atmosphere. These species quickly die from desiccation if they are placed in outside air. In the absence of light, cave beetles must rely on organic materials transported into the cave by bats in the form of guano and carcasses. Water also carries small quantities of organic material into the deepest recesses of caves. Dry caves that lack water-carrying, nutrient-rich loads generally are devoid of beetles.

The beetle faunas of continental islands strongly resemble those of nearby continental fauna and harbor relatively few endemic species. Remote, volcanic oceanic islands tend to support the greater number of endemic beetles. Beetles may colonize islands by means of active transport, such as flying. Passive transport, such as rafting on logs, is primarily responsible for dispersing wood-boring species throughout the islands of the Caribbean. Larger beetles attracted to the lights may stow away on seafaring vessels and are carried considerable distances to remote islands, while wind currents generated by fierce storms sweep up smaller species.

Countless microhabitats also support beetles. Several families live in the nests of birds, reptiles, mammals, and social insects. Interstitial spaces, the nooks and crannies between sand grains at the seashore, adjacent dunes, and along watercourses, provide ample food and shelter to support numerous species. Leaf-mining species complete their entire life cycle between the upper and lower epidermal layers of plant leaves, whose tissues provide both food and shelter.

Behavior

Beetles communicate primarily to locate mates, using mechanical, visual, and chemical means. Several families of beetles use sound to locate one another. Bess beetles (Passalidae), longhorns, and bark beetles stridulate by rubbing parts of their bodies together. Male death watch beetles (Anobiidae) bang their heads against the walls of their wooden galleries to lure females, whereas the South African tok-tokkies, *Psammodes* (Tenebrionidae), drum their abdomens against soil and rocks to attract mates.

Bioluminescence in fireflies (Lampyridae) is the best-known form of visual communication. To attract and locate mates, each species has its own pattern and method of presentation. Males typically fly at night, emitting a species-specific flashing pattern. The number of flashes and the rate and duration of the signal are critical to mate recognition, as is the delay in and length of the female's response. The male repeats his signal at regular intervals until he picks up an answer from a receptive female. Upon locating a female, he continues flashing and flies toward the female's signal.

While beetles infesting stored products are literally wallowing about in mate-rich habitats, others must depend on chemical communication to find distant mates or those hidden nearby among tangles of vegetation. Females usually emit pheromones, chemical messengers that work outside the body, to elicit sexual excitement in males. Receptive males often are capable of detecting pheromones with just a few molecules in the air from considerable distances. Although it is the females that typically produce pheromones to attract males, some males produce pheromones that affect egg production in the female or attract other males. Large numbers of males and females may gather in these mating swarms, or leks, increasing the likelihood that each will find a mate. Some beetles are mutually drawn to odiferous food sources, such as carrion, dung, or sap flows, whereas others gather around prominent features in the landscape, such as open patches of ground, rocky outcrops, or lone sign posts. Horned males stake out an attractive feeding spot, such as a sapping wound on a tree trunk, to await the arrival of a sexually receptive female. They use their armament to defend the site against other insects or rival males.

Elaborate courtship behavior is uncommon among the Coleoptera. Male ground, tiger, and rove beetles may grasp the female's prothorax with their mandibles before mating, and some male blister beetles tug on the female's antennae before copulation. In most species the male simply mounts the female from behind and may remain in contact with her afterward, to block the advances of other males.

Beetles are capable of defending themselves with an array of structural, behavioral, and chemical defenses. Large tropical scarabs, stag beetles, and longhorns avoid predators simply through their size and frightening appearance, not to mention their sharp horns and powerful mandibles. Flattened species retreat into tight spaces, where they are out of reach of predators.

Mimicry occurs in the form of shape, color, or behavioral patterns that resemble those of stinging wasps, bees, or ants. Fast-moving, scarlet red and black–checkered beetles (Cleridae) resemble female velvet ants, which actually are wingless stinging wasps. Slender and quick, some rove beetles and longhorns also mimic wasps, while stout, hairy, flower-visiting scarabs of the European genera *Trichius* and the North American *Trichiotinus* seem to imitate bees. Another North Amer-

ican beetle, *Ulochaetes leoninus* (Cerambycidae) not only looks like a bumblebee, it buzzes like one too and even attempts to "sting" when captured.

The most common type of mimicry in beetles involves an unpalatable or pugnacious model and one or more palatable or docile mimics, a form of Batesian mimicry. Experienced predators quickly associate the color or behavioral patterns of the model with undesirable prey. Well-known Batesian mimicry complexes are found in the families Cantharidae, Lampyridae, Lycidae, Meloidae, Tenebrionidae, and Cerambycidae. In Müllerian mimicry complexes, all species share common aposematic patterns and are unpalatable. Net-wing beetles probably are the best-known Müllerian mimics in the Coleoptera, since all of them are probably unpalatable. Distinct color patterns also may be disruptive in nature, helping many tropical weevils and metallic jewel scarabs disappear by making them look less beetle-like to predators. By contrast, blotchy, somber colored patchworks of browns, blacks, and grays render many longhorns and weevils virtually invisible on a background of bark.

The chemical arsenals of beetles are synthesized in complex glands or obtained directly from foodstuffs. Dispensed by anal sprays or leaking leg joints, these defensive compounds function as repellents, insecticides, or fungicides. In bombardier beetles (Carabidae), hydroquinones, hydrogen peroxide, peroxidases, and catalases are stored separately. When the beetles are alarmed, these chemicals are combined in another chamber, resulting in a violent chemical reaction that makes an audible "pop," producing a boiling, acrid stream directed with some accuracy by a flexible abdominal turret. Ladybird, blister, and soldier beetles store defensive chemicals within their blood and release them through their leg joints when attacked, a process known as reflex bleeding. Beetles that feed on toxic plants, such as milkweeds, may incorporate the plant's chemical defenses as their own, shunting noxious compounds from the digestive tract into their body tissues.

Most beetles have fairly regular life cycles that are influenced heavily by or coincide with changes in temperature and precipitation. In temperate regions the onset of spring and summer, with their requisite combination of increased solar radiation and rain, triggers the emergence of adults from their pupal chambers. In tropical regions, with relatively constant temperatures, seasonal rains trigger emergence. One or more generations are produced per year, although some species may require two or more years to complete their life cycles.

Feeding ecology and diet

Beetles feed on a variety of fungi, plants, and animals, both living and dead. A dead tree branch presents a wealth of feeding niches, providing wood borers with stems of varying sizes and different layers of tissue, from the bark inward. Metallic wood-boring beetles (Buprestidae) may arrive within hours of a fresh cut, while others are drawn to burning wood through the use of special sensory structures that detect infrared radiation. As the branch ages, successive infestations of beetles reduce the wood to dust and are essential for nutrient recycling. Despite their dependence on wood as a food source, beetles are incapable of digesting cellulose and must rely on bacteria, yeasts, and fungi found in their digestive tracts. The eggs of wood-feeding beetles are inoculated with these gut symbionts as they pass through a residue lining the ovipositor. The larva's first meal is it own egg shell, laden with these microorganisms.

All parts of living plants are fodder for both larval and adult beetles. Leaf beetles may defoliate a plant completely or skeletonize the majority of its leaves, causing serious damage to garden plants and agricultural crops. Leaf miners leave in their wake a meandering and ever widening feeding track that traces their development from egg to adult. Floricolous, or flower-visiting, beetles may have tubular mouthparts for sucking up nectar. The mouthparts of pollen feeders sometimes resemble brushes, to facilitate the consumption of fine pollen grains. Beetles can hardly be considered pollinators in the same class as bees, but they can play a significant role in pollen transfer in some plant groups seldom visited by more traditional pollinators.

Carrion and burying beetles (Silphidae), hide beetles (Dermestidae), and others scavenge dead animal tissue, while keratin-feeding skin beetles (Trogidae) consume feathers, fur, horns, and hooves. Industrious dung beetles (Geotrupidae and Scarabaeidae) bury vast amounts of organic waste produced by vertebrates (especially large herbivores) for use as food by their larvae. They employ their membranous mandibles to strain out remnants of undigested food, bacteria, yeasts, and molds as food for themselves. Not all dung beetles feed on dung, preferring instead carrion, fungi, fruit, millipedes, and the slime tracks of snails.

Predatory species seldom hunt long distance, relying instead on habitat preferences to bring them into contact with their prey. The larvae of Lampyridae, however, are specialists, tracking snails by following their slime trails. Checkered (Cleridae) and bark-gnawing (Trogossitidae) beetles are reported to pursue plumes of pheromones to locate their prey, bark beetles (Curculionidae). The ant-loving scarab, *Cremastocheilus*, also locates its host nests by tracking ant pheromones. Whirligig beetles identify prey by using waves generated by struggling insects, which are transmitted across the surface of the water.

Ground and tiger beetles capture their prey on the run, biting them with their pronounced mandibles and tearing them into small chunks. They attack a broad range of beetles, other insects, and invertebrates, although some prey only on snails. Some rove and clown beetles (Histeridae) hunt for food among the labyrinth of detritus and decaying organic matter, while others live in ant or termite colonies or in the fur of small mammals. Carnivorous larvae of beetles are primarily liquid feeders and must first digest their prey "extra-orally," with digestive enzymes in their saliva. Extraoral digestion also occurs among certain adult ground and rove beetles as well as in some other families.

Reproductive biology

Most beetles engage in sexual reproduction. Parthenogenesis, development from an unfertilized egg, occurs rarely in

Mating wattle pigs (*Leptopius* sp.), a type of weevil, in Alice Springs, Northern Territory, Australia. (Photo by David C. Renitz. Bruce Coleman, Inc. Reproduced by permission.)

some families. Males of parthenogenetic species are rare or unknown. In sexually reproducing species, females generally need to mate only once, although several males may inseminate them. Females store sperm in an internal reservoir, or spermatheca. Fertilization takes place as the eggs are laid. Not all beetle eggs hatch outside the mother's body. In ovoviviparous species (some ground, rove, leaf, and darkling beetles), eggs are retained within the reproductive tract of the female until they hatch; later the larvae are "born."

The eggs usually are laid singly or in batches. Ground dwellers simply drop them on the ground, scatter them in rich organic soil, or place them in or near piles of dung or carrion. The eggs of herbivores and leaf miners are placed at the base of their food plant, glued to stems and leaves, or inserted in the crevices of bark. Leaf miners actually tear the tissue of the leaf surface to deposit an egg inside. Some female longhorns chew a channel around a branch, killing it by girdling to create a food source for their wood-boring larvae.

Upon hatching, the first-stage, or instar, larva begins its life with a single purpose: to eat. Larvae scavenge carrion and dung, attack roots, mine plant tissues, or tunnel through wood, taking weeks or years to mature. Three or more larval instars are required before the pupal stage. In temperate climates the pupa often is best equipped to survive harsh winter conditions when it is tucked carefully away in soil or humus or within the tissues of plants. Paedogenesis, the retention of immature features, occurs in micromalthids (Micromalthidae), glowworms (Phengodidae), and certain net-winged beetles and fireflies. In these families adult females are distinguished from the larvae by the presence of compound eyes and reproductive organs.

True social behavior is unknown among the Coleoptera. Subsocial behavior, exhibited by limited interaction between parents and young, is known, however, in at least 10 families. Some ground beetles construct soil depressions in which to lay their eggs and guard and clean them until they hatch. In several species of tortoise beetles (Chrysomelidae), the mother guards the eggs and remains with her larvae through pupation. Adult and larval bess beetles (Passalidae) bore in rotten logs, live in dense colonies, and stridulate continuously. This form of communication may serve to keep groups of adults and larvae together, since larvae seem to fare better ingesting wood that has been chewed, predigested, or converted into frass by the adults.

Female rove beetles of the European genus *Bledius* maintain and defend their intertidal brood tunnels, providing the larvae with algae for food. Bark and ambrosia beetles (Curculionidae) cultivate a mutualistic fungus for food for themselves and their larvae, much like some ants and termites. Adults carry fungal spores in the mycangia, specialized pits in their heads.

Among the earth-boring beetles, dung scarabs, and burying beetles, males and females cooperate in digging nests for their eggs and provisioning them with dung or carrion. Competition for this nutrient-rich, yet ephemeral resource is intense. They quickly sequester the "spoils" in underground chambers, or nests, to exclude flies, ants, mites, and other competitors and to maintain optimum moisture levels for successful brood development. Dung scarabs have evolved several strategies for nest building. Endocoprids live and breed directly in the dung pad, while paracoprids dig brood chambers underneath or immediately adjacent to the pad and provision them with dung. Telocoprids quickly carve out pieces of dung and shape them into balls that can be easily rolled away for burial later, some distance away from the source.

Burying beetles of the genus *Nicrophorus* exhibit the most advanced types of parental care known in the Coleoptera. Working in sexual pairs, adults locate and bury a carcass in an underground chamber. The carcass is meticulously prepared as food for both adults and larvae. All feathers or hair are removed, and the body is kneaded into a ball. Fungicides in the saliva of the adults retard decomposition. A conical depression is created on the upper surface of the carcass, to serve as a receptacle into which the adults regurgitate droplets of tissue as food for the newly hatched larvae. The female calls the larvae to the pool of tissue by rubbing a ridge on the elytra against the corresponding abdominal segment. The adults remain in the chamber with their brood until they pupate.

Conservation status

Habitat loss due to fire, urbanization, acid rain, electric lights, overgrazing, agricultural expansion, water impoundment, pollution, deforestation, soil erosion, persistent adverse weather, use of off-road recreational vehicles, exotic species, and logging—not collecting—are the greatest threats to beetle populations. The threat of habitat loss is exacerbated further by a steady decline in the number of trained taxonomists who provide critical data for beetle protection and habitat management.

The 2002 IUCN Red List contains 72 species of beetles, primarily from the families Dytiscidae, Carabidae, Lucanidae, Scarabaeidae, and Curculionidae. Listed species are categorized as Lower Risk (three), Vulnerable (27), Endangered (15), Critically Endangered (10), or Extinct (17). Most of the listed extant species have very restricted ranges within sensitive habitats, such as caves or sand dunes. All 14 species of the flightless South African genus *Colophon* (Lucanidae) are listed by the IUCN and in Appendix I of CITES primarily because of the high prices they command on the market.

The Endangered Species Act of the United States lists 16 species of American beetles, four as threatened and 12 as endangered. Of these species, the American burying beetle, *Nicrophorus americanus* (Silphidae), is probably one of the best documented, with a recovery plan in place for captive breeding and release. Only one species of beetle, *Phalacrognathus muelleri* (Lucanidae), is protected federally by the Australian Wildlife Protection Act, but in Western Australia collection of the entire family of jewel beetles (Buprestidae) is restricted. New Zealand's Department of Conservation recognizes 24 species of beetles as endangered.

The countries of the former Soviet Union, Finland, and Sweden also have produced Red Data books that include beetles. State and provincial governments throughout the world have enacted ordinances that prohibit the collection, trading, and export of species protected elsewhere by other conventions. Two European organizations actively promote the conservation of beetles on the basis of their ecological roles. The Water Beetle Specialist Group, part of the Species Survival Commission of the IUCN, recognizes the importance of aquatic beetles as bioindicators in wetland management in Europe and Southeast Asia. The Saproxylic Invertebrates Project focuses on selected groups of invertebrates, including beetles, dependent upon standing or fallen trees or wood-inhabiting fungi.

Significance to humans

The best-known example of a beetle in ancient mythology is the sacred scarab, *Scarabaeus sacer*. The Egyptian sun god Ra was symbolized as a great scarab, and images of scarabs appeared in much funerary art and hieroglyphs. Carved scarabs bore religious inscriptions from the Book of the Dead and were placed in tombs to ensure the immortality of the soul. Heart scarabs were stones placed with the dead that bore inscriptions admonishing the heart not to bear witness against its body on the day of judgment.

Beetles have been depicted in vase paintings, porcelain statuary, precious stones, glass paintings, sculptures, jewelry, coins, and illustrated manuscripts. Fireflies have long held a special fascination for the Chinese and Japanese and appear often in their art. One of the most notable examples of beetles in art is the German Renaissance artist Albrecht Dürer's 1505 watercolor of the European stag beetle, *Lucanus cervus*. The French artist E. A. Seguy's art deco insect portfolios, created in the 1920's, include several striking examples of Coleoptera. The durable bodies of the beetles themselves also are used in arts and crafts, especially the elytra, horns, mandibles, and legs. South American indigenous artisans use the elytra of the giant *Euchroma gigantea* (Buprestidae) for necklaces, head ornaments, and other decorative pieces. Today, in parts of Mexico and Central America, a zopherid beetle popularly known as the ma'kech, *Zopherus chilensis* (Zopheridae), is decorated with brightly colored glass beads, fixed to a short chain tether, and pinned to clothing as a reminder of an ancient legend.

Beetles are an important source of protein and fat for peoples around the world. Throughout the islands of the South Pacific, grubs of the palm weevil, *Rhynchophorus palmarum*, and rhinoceros scarabs, *Oryctes rhinoceros*, are roasted and relished. Larval and pupal predaceous diving and jewel beetles are eaten in Southeast Asia. The Chinese collect large aquatic beetles; remove the elytra, wings, legs, and head; and fry them in oil or soak them in brine. The Aborigines of Australia eat longhorn larvae, removing the large, nut-flavored grubs from rotten logs and roasting them like marshmallows over a fire. Even in the United States the mystique of "eating the worm"

from a bottle of Mexican mescal has resulted in the novelty of a tequila-flavored lollipop stuffed with a larva of the common mealworm, *Tenebrio molitor* (Tenebrionidae).

A small number of beetles have become important economic pests, as the result of their feeding and egg-laying activities on stored products, pastures, crops, and timber. Beetles feeding on legumes, tomatoes, potatoes, melons, gourds, and grains are some of humanity's greatest competitors for food. In temperate forests throughout the world, beetles generally attack trees that already are under stress from lack of water and nutrition. Bark beetles mine branches, twigs, cones, or roots, and the most destructive species attack the trunks of living trees.

Predatory beetles, especially ground beetles and ladybirds (Coccinellidae), are used to control insect pests around the world, and many herbivores have been recruited as biological control agents for plant pest projects. In the 1970s the Australians began a program to import exotic dung scarabs and predatory clown beetles (Histeridae) as biological agents to clean up dung pads and eat the maggots of the pestiferous flies breeding in them. The leaf, metallic wood-boring beetles and weevils are utilized to control the spread of noxious weeds throughout the world, as they feed on the leaves, bore into twigs and stems, or destroy the seeds.

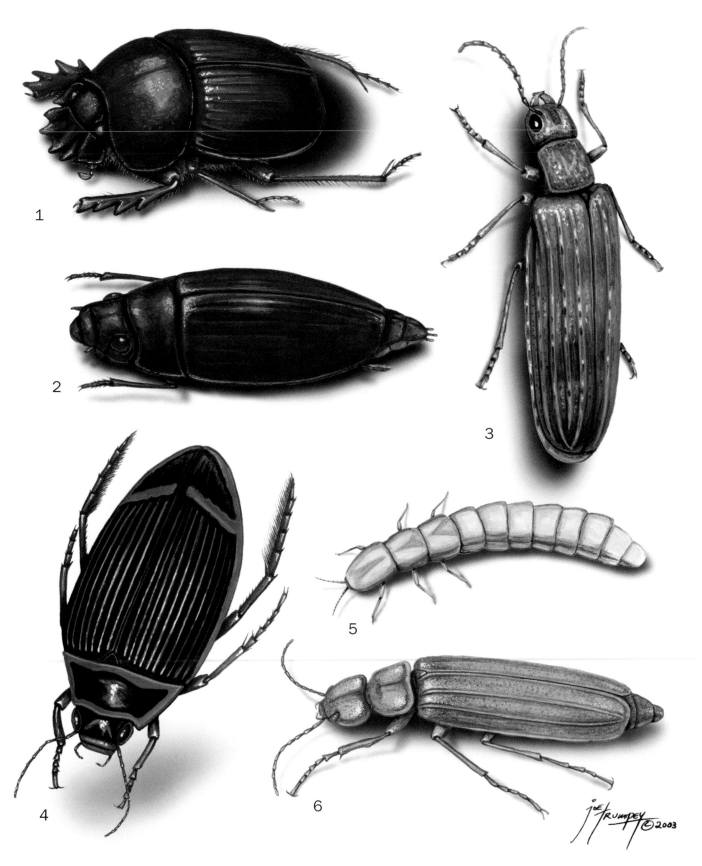

1. Sacred scarab (*Scarabaeus sacer*); 2. Whirligig beetle (*Dineutus discolor*); 3. Cupedid beetle (*Priacma serrata*); 4. Great water beetle (*Dystiscus marginalis*); 5. Pink glowworm (*Microphotus angustus*); 6. Spanish fly (*Lytta vesicatoria*). (Illustration by Joseph E. Trumpey)

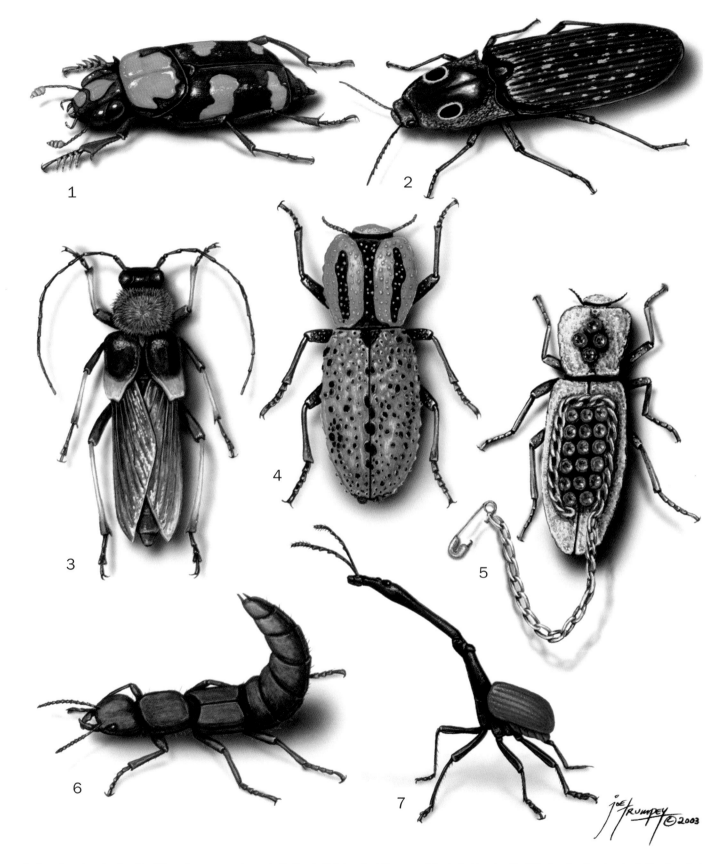

1. American burying beetle (*Nicrophorus americanus*); 2. Eyed click beetle (*Alaus oculatus*); 3. Lion beetle (*Ulochaetes leoninus*); 4. Non-jeweled ma'kech (*Zopherus. chilensis*); 5. Jeweled ma'kech (*Z. chilensis*); 6. Devil's coach-horse (*Ocypus olens*); 7. Giraffe-necked weevil (*Trachelophorus giraffa*). (Illustration by Joseph E. Trumpey)

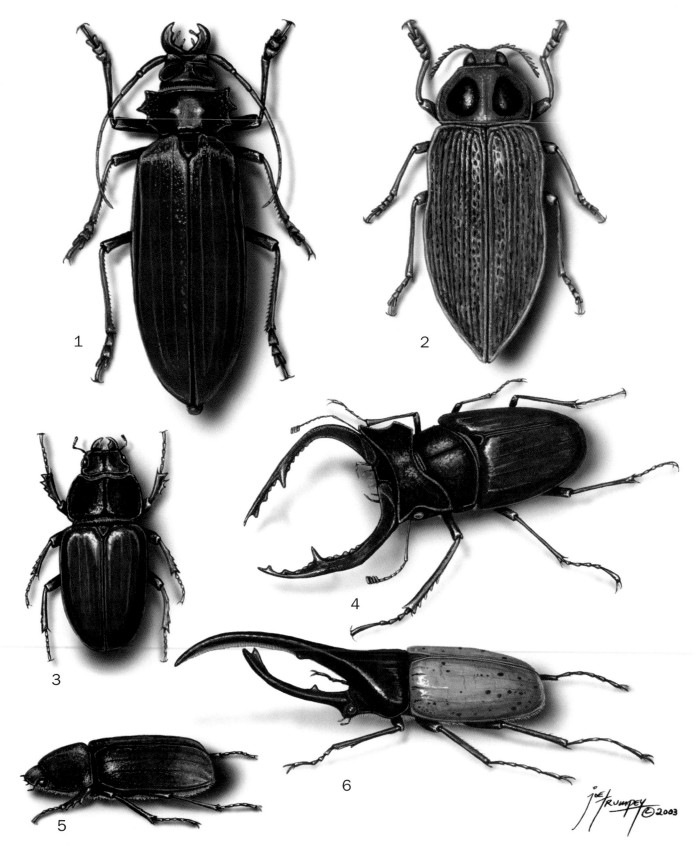

1. Titanic longhorn beetle (*Titanus gigantea*); 2. Giant metallic ceiba borer (*Euchroma gigantea*); 3. Female European stag beetle (*Lucanus cervus*); 4. Male European stag beetle (*L. cervus*); 5. Female Hercules beetle (*Dynastes hercules*) 6. Male Hercules beetle (*D. hercules*). (Illustration by Joseph E. Trumpey)

Species accounts

Giraffe-necked weevil
Trachelophorus giraffa

FAMILY
Attelabidae

TAXONOMY
Aploderus (Trachelophorus) giraffa Jekel, 1860, "Madagascar."

OTHER COMMON NAMES
English: Red-and-black giraffe beetle; French: Scarabée girafe; Dutch: Giraf nek kever; German: Giraffenrüssler.

PHYSICAL CHARACTERISTICS
Males up to 0.98 in (2.5 cm). Black with red elytra; only males have long "neck."

DISTRIBUTION
Madagascar.

HABITAT
Forests.

BEHAVIOR
Sits on leaves in open areas and along roadsides; rolls leaves.

FEEDING ECOLOGY AND DIET
Feeds on the leaves of *Dichaetanthera cordifolia*, a small tree in the family Melastomataceae.

REPRODUCTIVE BIOLOGY
Females lay their eggs on leaves; the leaves are rolled up into a tube to protect and nourish the larvae.

CONSERVATION STATUS
Not threatened.

Scarabaeus sacer
Lytta vesicatoria
Trachelophorus giraffa

SIGNIFICANCE TO HUMANS
None known. ◆

Giant metallic ceiba borer
Euchroma gigantea

FAMILY
Buprestidae

TAXONOMY
Buprestis gigantea Linnaeus, 1758, "America."

OTHER COMMON NAMES
Spanish: Eucroma, catzo; Portuguese: Mae do sol, ôlho do sol.

PHYSICAL CHARACTERISTICS
Grows to 2–2.8 in (5–7 cm). Wrinkled elytra, shining green infused with red. Back of prothorax has two large black spots. Freshly emerged specimens covered with yellow bloom.

Alaus oculatus
Ulochaetes leoninus
Euchroma gigantea

Priacma serrata

Zopherus chilensis

Dineutus discolor

DISTRIBUTION
Mexico to Argentina and the Antilles.

HABITAT
Tropical forests.

BEHAVIOR
Common on trunks of living or dead bombacaceous trees.

FEEDING ECOLOGY AND DIET
Larvae bore through trunks of dead bombacaceous trees.

REPRODUCTIVE BIOLOGY
Nothing is known.

CONSERVATION STATUS
Not threatened.

SIGNIFICANCE TO HUMANS
Elytra are made into jewelry and ornaments by peoples in Central and South America; adults are eaten by Tzeltal-Mayan Indians in Chiapas, Mexico. ◆

Titanic longhorn beetle
Titanus gigantea

FAMILY
Cerambycidae

TAXONOMY
Titanus gigantea Linnaeus, 1758, "Cayania."

OTHER COMMON NAMES
None known.

PHYSICAL CHARACTERISTICS
Grows up to 6.7 in (17 cm). Dark brown to black with faint, powerful mandibles and longitudinal ridges on the elytra.

DISTRIBUTION
Colombia, Ecuador, Peru, French Guiana, and northern Brazil.

HABITAT
Tropical forests.

BEHAVIOR
Adults are attracted to lights at night.

FEEDING ECOLOGY AND DIET
Larvae probably feed in rotten wood.

REPRODUCTIVE BIOLOGY
Nothing is known.

CONSERVATION STATUS
Not threatened.

SIGNIFICANCE TO HUMANS
One of the world's largest beetles, specimens are highly prized and sell for hundreds of dollars. ◆

Lion beetle
Ulochaetes leoninus

FAMILY
Cerambycidae

TAXONOMY
Ulochaetes leoninus LeConte, 1854, "Prairie Pass, Oregon Territory."

OTHER COMMON NAMES
None known.

PHYSICAL CHARACTERISTICS
Grows to 0.67–0.98 in (17–25 mm); hairy with black and yellow markings and short elytra.

DISTRIBUTION
Pacific coast, from British Columbia to southern California.

HABITAT
Pine forests.

BEHAVIOR
Look, sound, and behave like bumble bees.

FEEDING ECOLOGY AND DIET
Larvae bore into sapwood of conifers.

REPRODUCTIVE BIOLOGY
Eggs are laid at the base of standing dead trees and stumps.

CONSERVATION STATUS
Not threatened.

SIGNIFICANCE TO HUMANS
Interesting example of physical and behavioral mimicry. ◆

Cupedid beetle
Priacma serrata

FAMILY
Cupedidae

TAXONOMY
Cupes serrata LeConte, 1861, "East of Fort Colville," Oregon.

OTHER COMMON NAMES
English: cupesid beetle.

PHYSICAL CHARACTERISTICS
Reaches 0.4–0.9 in (11–22 mm). Irregularly marked with grayish-black scales; elytra with square pits.

DISTRIBUTION
Western North America.

HABITAT
Under tree bark in coniferous forests.

BEHAVIOR
During daylight hours, males sometimes are attracted to sheets freshly laundered with bleach.

FEEDING ECOLOGY AND DIET
Probably feed on fungus in rotten wood.

REPRODUCTIVE BIOLOGY
Nothing is known.

CONSERVATION STATUS
Not threatened.

SIGNIFICANCE TO HUMANS
A primitive family of beetles similar in appearance to the extinct Tshekardoleidae. ◆

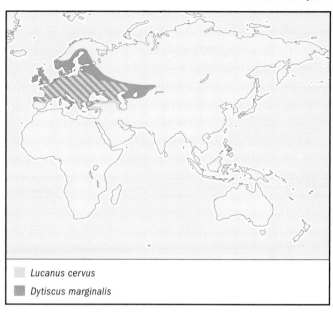

Lucanus cervus

Dytiscus marginalis

Great water beetle
Dytiscus marginalis

FAMILY
Dytiscidae

TAXONOMY
Dytiscus marginalis Linné, 1758, "Europae."

OTHER COMMON NAMES
English: Great water diving beetle.

PHYSICAL CHARACTERISTICS
Grows up to 1.4 in (35 mm). Prothorax and elytra have pale borders; male's elytra is smooth and female's usually grooved.

DISTRIBUTION
Europe.

HABITAT
Lakes and ponds with muddy bottoms.

BEHAVIOR
Obtains oxygen by breaking the water surface with the tip of the abdomen and trapping a bubble under the elytra.

FEEDING ECOLOGY AND DIET
Preys on aquatic insects, mollusks, crustaceans, and even tadpoles and small fish.

REPRODUCTIVE BIOLOGY
Eggs are deposited singly on stems of aquatic plants. Larvae molt three times in 35–40 days; pupation takes place in damp ground next to water. One generation per year

CONSERVATION STATUS
Not threatened.

SIGNIFICANCE TO HUMANS
One of the largest and most intensely studied European water beetles. ◆

Eyed click beetle
Alaus oculatus

FAMILY
Elateridae

TAXONOMY
Elater oculatus Linné, 1758, "America septentrionalis."

OTHER COMMON NAMES
None known.

PHYSICAL CHARACTERISTICS
Grows to 0.98–1.96 in (25–50 mm). Pronotum with two black spots encircled with white scales.

DISTRIBUTION
Eastern North America.

HABITAT
Woodlands.

BEHAVIOR
Adults found on hardwoods or under bark.

FEEDING ECOLOGY AND DIET
Larvae live in rotten hardwood. Adults and larvae feed on the larvae of wood-boring beetles.

REPRODUCTIVE BIOLOGY
Nothing is known.

CONSERVATION STATUS
Not threatened.

SIGNIFICANCE TO HUMANS
None known. ◆

Whirligig beetle
Dineutus discolor

FAMILY
Gyrinidae

TAXONOMY
Dineutus discolor Aube, 1838, "États-Unis d'Amerique."

OTHER COMMON NAMES
English: Gyrinids, apple bugs.

PHYSICAL CHARACTERISTICS
Reaches 0.4–0.5 in (11–13 mm). Black and shining, with pale underside.

DISTRIBUTION
Eastern North America and Mexico.

HABITAT
Surface of slow-moving ponds and streams.

BEHAVIOR
Lives singly or in groups.

FEEDING ECOLOGY AND DIET
Preys on insects trapped on the water surface.

REPRODUCTIVE BIOLOGY
Eggs are laid on submerged plants. Aquatic larvae prey on small invertebrates. Pupate in moist soil near water. Adults overwinter in debris at the edge of water.

CONSERVATION STATUS
Not threatened.

SIGNIFICANCE TO HUMANS
None known. ◆

Pink glowworm
Microphotus angustus

FAMILY
Lampyridae

TAXONOMY
Microphotus angustus LeConte, 1874, "Mariposa, California."

OTHER COMMON NAMES
None known.

PHYSICAL CHARACTERISTICS
Males reach 0.2 in (6 mm) and females 0.4–0.5 in (11–13 mm). Pink females are larviform; males look like fireflies with large eyes.

▢ *Dynastes hercules*
▪ *Microphotus angustus*

DISTRIBUTION
California and Oregon.

HABITAT
Forests and moist canyons in chaparral.

BEHAVIOR
Females hang with their heads pointed upward on rocks and stumps and "call" with continuous light. Males seldom are seen and flash weakly only when disturbed.

FEEDING ECOLOGY AND DIET
Unknown; larvae probably prey on small invertebrates.

REPRODUCTIVE BIOLOGY
Nothing is known.

CONSERVATION STATUS
Not threatened.

SIGNIFICANCE TO HUMANS
None known. ◆

European stag beetle
Lucanus cervus

FAMILY
Lucanidae

TAXONOMY
Scarabaeus cervus Linnaeus, 1758, "Europae."

OTHER COMMON NAMES
German: Donnerkafer, Hausbrenner, Feueranzunder, Köhler, Feuerschröter.

PHYSICAL CHARACTERISTICS
Males reach 1.4–2.95 in (35–75 mm) and females 1.2–1.8 in (30–45 mm). Dark brownish–black beetle. The male has a broad head and antler-like mandibles; female is smaller and more stout, with relatively small mandibles.

DISTRIBUTION
Central, southern, and western Europe; Asia Minor; Syria.

HABITAT
Old oak forests.

BEHAVIOR
Males use mandibles against rival males over females.

FEEDING ECOLOGY AND DIET
Adults feed on sap; larvae eat rotting wood.

REPRODUCTIVE BIOLOGY
Eggs are laid in old wood, larva take three to five years to mature. Adult matures in autumn but overwinters in pupal case.

CONSERVATION STATUS
Collection of this species is forbidden in several European countries.

SIGNIFICANCE TO HUMANS
Historically a symbol of evil and bad luck. ◆

Spanish fly
Lytta vesicatoria

FAMILY
Meloidae

TAXONOMY
Meloe vesicatoria Linné, 1758. Type locality not specified.

OTHER COMMON NAMES
None known.

PHYSICAL CHARACTERISTICS
Reaches 0.5–0.9 in (12–22 mm); slender, soft-bodied metallic golden-green beetle.

DISTRIBUTION
Throughout southern Europe and eastward to Central Asia and Siberia.

HABITAT
Scrublands and woods.

BEHAVIOR
When disturbed, meloids release the blistering agent cantharidin from their leg joints.

FEEDING ECOLOGY AND DIET
Adults feed on leaves of ash, lilac, amur privet, and white willow trees; larvae are parasitic on the brood of ground nesting bees.

REPRODUCTIVE BIOLOGY
Develop by hypermetamorphosis, a type of complete metamorphosis in which first larval instar (the triungulin) is very active,

while remaining instars are more sedentary and grublike. Eggs are laid near the entrance of host bee's nest; triungulins crawl into nest on their own.

CONSERVATION STATUS
Not threatened.

SIGNIFICANCE TO HUMANS
Bodies are filled with the toxin cantharidin; elytra were once pulverized and marketed as an aphrodisiac as well as a cure for various ailments. ◆

Hercules beetle
Dynastes hercules

FAMILY
Scarabaeidae

TAXONOMY
Scarabaeus hercules Linné, 1758, "America."

OTHER COMMON NAMES
French: Scieurs de long; Spanish: Tijeras.

PHYSICAL CHARACTERISTICS
Males reach 5.9–6.7 in (15–17 cm), up to half of which corresponds to the thoracic horn.

DISTRIBUTION
Mexico, Central and northern South America, Guadeloupe, and the Dominican Republic.

HABITAT
Humid tropical forests.

BEHAVIOR
Adults are attracted to oozing sap and sweet fruits; larvae develop in rotten logs.

FEEDING ECOLOGY AND DIET
Nocturnal and frequently attracted to lights.

REPRODUCTIVE BIOLOGY
Males defend feeding sights that will attract females; horns of males are used to grapple with other males over females.

CONSERVATION STATUS
Not threatened.

SIGNIFICANCE TO HUMANS
Ingesting the horn is thought by some people to increase their sexual potency. ◆

Sacred scarab
Scarabaeus sacer

FAMILY
Scarabaeidae

TAXONOMY
Scarabaeus sacer Linné, 1758, "Aegypto."

OTHER COMMON NAMES
None known.

PHYSICAL CHARACTERISTICS
Grows to 0.98–1.2 in (25–30 mm). Black, with rakelike head and forelegs.

DISTRIBUTION
Mediterranean region and central Europe.

HABITAT
Steppe, forest-steppe, and semi-desert.

BEHAVIOR
Adults track dung by smell as food for themselves and their offspring. The female stands head down and rolls a dung ball with the second and third pairs of legs; ball is buried as food for larva.

FEEDING ECOLOGY AND DIET
Adults use membranous mandibles to strain fluids, molds, and other suspended particles as food from dung; larvae eat solid dung.

REPRODUCTIVE BIOLOGY
Single humpbacked larva feeds and pupates inside buried dung ball.

CONSERVATION STATUS
Not threatened.

SIGNIFICANCE TO HUMANS
Symbol of the ancient Egyptian sun god Ra. Scarab jewelry is worn today as a good luck charm. Significant recyclers of animal waste. ◆

American burying beetle
Nicrophorus americanus

FAMILY
Silphidae

TAXONOMY
Nicrophorus americanus Olivier, 1790, "Amérique septentrionale."

OTHER COMMON NAMES
None known.

PHYSICAL CHARACTERISTICS
Reaches 0.8–1.4 in (20–35 mm). Black, with orange antennal club. Head and pronotum have central orange spot; and elytra have four wide spots.

DISTRIBUTION
Formerly throughout eastern North America; restricted now to isolated populations in the Midwest, with populations reintroduced to Rhode Island and Massachusetts.

HABITAT
Woodlands, grassland prairies, forest edge, and scrubland.

BEHAVIOR
Provides parental care for its young.

FEEDING ECOLOGY AND DIET
Scavenges and buries carrion.

REPRODUCTIVE BIOLOGY
Mating pair prepares carrion as food for themselves and their larvae.

Ocypus olens
Nicrophorus americanus
Titanus gigantea

CONSERVATION STATUS
Listed as Endangered by IUCN and by the U.S. Fish and Wildlife Service.

SIGNIFICANCE TO HUMANS
Symbolic of habitat destruction and modification throughout eastern North America. ◆

Devil's coach-horse
Ocypus olens

FAMILY
Staphylinidae

TAXONOMY
Staphylinus olens Müller, 1764. Type locality not specified.

OTHER COMMON NAMES
English: Rove beetle.

PHYSICAL CHARACTERISTICS
Grows to 0.9–1.3 in (22–33 mm). Black, with short elytra exposing abdominal segments.

DISTRIBUTION
Lower elevations of Europe, Russia, Turkey, North Africa, and the Canary Islands; established in parts of North America.

HABITAT
Under stones, damp leaves, and moss.

BEHAVIOR
When alarmed, the beetle spreads its powerful jaws and curls its abdomen over its back to emit a foul-smelling brown fluid.

FEEDING ECOLOGY AND DIET
Adults and larvae prey on soil-dwelling invertebrates.

REPRODUCTIVE BIOLOGY
Nothing is known.

CONSERVATION STATUS
Not threatened.

SIGNIFICANCE TO HUMANS
Once a symbol of evil associated with death. ◆

Ma'kech
Zopherus chilensis

FAMILY
Zopheridae

TAXONOMY
Zophorus [sic] *chilensis* Gray, 1832. Type locality not specified.

OTHER COMMON NAMES
English: Jeweled beetle.

PHYSICAL CHARACTERISTICS
Grows to 1.3–1.6 in (34–40 mm). Extremely hard exoskeleton; rough back is white with irregular black blotches.

DISTRIBUTION
Southern Mexico to Venezuela and Colombia.

HABITAT
Found on bark of dead trees.

BEHAVIOR
When disturbed, adults tuck in their legs and play dead.

FEEDING ECOLOGY AND DIET
Larvae presumably feed on fungal hyphae in rotten wood; adults eat cereals in captivity.

REPRODUCTIVE BIOLOGY
Nothing is known.

CONSERVATION STATUS
Not threatened.

SIGNIFICANCE TO HUMANS
Living beetles are adorned with beads, tethered with a gold chain, and worn as living costume jewelry as traditional reminder of an ancient Yucatecan legend. ◆

Resources

Books

Arnett, Ross H., Jr., and Michael C. Thomas, eds. *American Beetles.* Vol. 1, *Archostemata, Myxophaga, Adephaga, Polyphaga: Staphyliniformia.* Boca Raton, FL: CRC Press, 2001.

Arnett, Ross H., Jr., Michael C. Thomas, Paul E. Skelley, and J. Howard Frank, eds. *American Beetles.* Vol. 2, *Polyphaga: Scarabaeoidea through Curculionidae.* Boca Raton, FL: CRC Press. 2002.

Crawford, C. S. *Biology of Desert Invertebrates.* Berlin: Springer-Verlag, 1981.

Crowson, R. A. *The Biology of the Coleoptera.* London: Academic Press, 1981.

Elias, Scott A. *Quaternary Insects and Their Environments.* Washington, DC: Smithsonian Institution Press, 1994.

Evans, Arthur V., and Charles L. Bellamy. *An Inordinate Fondness for Beetles.* Berkeley: University of California Press, 2000.

Evans, Arthur V., and J. N. Hogue. *Introduction to California Beetles.* Berkeley: University of California Press, in press.

Evans, D. L., and J. O. Schmidt, eds. *Insect Defenses: Adaptive Mechanisms and Strategies of Prey and Predators.* Albany: State University of New York Press, 1990.

Klausnitzer, B. *Beetles.* New York: Exeter Books, 1983.

Lawrence, J. F., and E. B. Britton. *Australian Beetles.* Carlton, Australia: Melbourne University Press, 1994.

Lawrence, J. F., and A. F. Newton, Jr. "Families and Subfamilies of Coleoptera (with Selected Genera, Notes, References and Data on Family-Group Names)." In *Biology, Phylogeny, and Classification of Coleoptera.* Papers Celebrating

Resources

the 80th Birthday of Roy A. Crowson, edited by J. Pakaluk and S. A. Slipinski. Warsaw, Poland: Muzeum i Instytut Zoologii PAN, 1995.

Meads, M. *Forgotten Fauna: The Rare, Endangered, and Protected Invertebrates of New Zealand.* Wellington, New Zealand: Department of Scientific and Industrial Research, 1990.

Pakaluk, J., and S. A. Slipinski, eds. *Biology, Phylogeny, and Classification of Coleoptera.* Papers Celebrating the 80th Birthday of Roy A. Crowson. Warsaw, Poland: Muzeum i Instytut Zoologii PAN, 1995.

Stehr, Frederick W., ed. *Immature Insects.* Vol. 2. Dubuque, IA: Kendall/Hunt Publishing Company, 1991.

White, R. E. *A Field Guide to the Beetles of North America.* Boston: Houghton Mifflin, 1983.

Periodicals

Alcock, J. "Postinsemination Associations between Males and Females in Insects: The Mate-Guarding Hypothesis." *Annual Review of Entomology* 39 (1994): 1–21.

Beutel, R. "Über Phylogenese und Evolution der Coleoptera (Insecta), insbesondere der Adephaga." *Abhandlungen des Naturwissenshaftlichen Vereins in Hamburg* 31 (1997): 1–164.

Beutel, R., and F. Haas. "Phylogenetic Relationships of the Suborders of Coleoptera (Insecta)." *Cladistics* 16 (2000): 103–141.

Caterino, M. S., V. L. Shull, P. M. Hammond, and A. P. Vogler. "Basal Relationships of Coleoptera Inferred from 18S rDNA Sequences." *Zoologica Scripta* 31 (2002): 41–49.

Eberhard, William G. "Horned Beetles." *Scientific American* 242, no. 3 (1980): 166–182.

Emlen, D. J. "Integrating Development with Evolution: A Case Study with Beetle Horns." *Bioscience* 50, no. 5 (2000): 403–418.

Farrell, Brian D. "'Inordinate Fondness' Explained: Why Are There So Many Beetles?" *Science* 281 (1998): 555–559.

Hadley, N. F. "Beetles Make Their Own Waxy Sunblock." *Natural History* 102, no. 8 (1993): 44–45.

Lomolino, Mark V., J. C. Creighton, G. D. Schnell, and D. L. Certain. "Ecology and Conservation of the Endangered American Burying Beetle (*Nicrophorus americanus*)." *Conservation Biology* 9, no. 3 (1995): 605–614.

McIver, J. D., and G. Stonedahl. "Myrmecomorphy: Morphological and Behavioral Mimicry of Ants." *Annual Review of Entomology* 38 (1993): 351–379.

Milne, L. J., and M. J. Milne. "The Social Behavior of Burying Beetles." *Scientific American* 235, no. 2 (1976): 84–89.

Murlis, J. "Odor Plumes and How Insects Use Them." *Annual Review of Entomology* 37 (1992): 505–532.

Rettenmeyer, C. W. "Insect Mimicry." *Annual Review of Entomology* 15 (1970): 43–74.

Other

"The Balfour-Browne Club." October 1996 [April 14, 2003]. <http://www.lifesci.utexas.edu/faculty/sjasper/beetles/BBClub.htm>.

"Beetles (Coleoptera) and Coleopterists." [April 14, 2003]. <http://www.zin.ru/Animalia/Coleoptera/eng>.

"Coleoptera Home Page." March 28, 2003 [April 14, 2003]. <http://www.coleoptera.org>.

"*The Coleopterist.*" [April 14, 2003]. <http://www.coleopterist.org.uk>.

"Coleopterists Society." [April 14, 2003]. <http://www.coleopsoc.org>.

"The Japan Coleopterological Society." [April 14, 2003]. <http://www.mus-nh.city.osaka.jp/shiyake/j-coleopt-soc.html>.

"Wiener Coleopterologen Verein (Vienna Coleopterists Society)." September 7, 2001 [April 14, 2003]. <http://www.nhm-wien.ac.at/NHM/2Zoo/coleoptera/wcv_e.html>.

Arthur V. Evans, DSc

Strepsiptera
(Strepsipterans)

Class Insecta
Order Strepsiptera
Number of families 9

Photo: A male *Xenos* species visits a daisy in the succulent karoo of Northern Cape Province, South Africa. Like other members of the family Stylopidae, it has four-segmented tarsi, and the last two segments of the antler-like antennae form a parallel pair of broad plates. (Photo by Mike Picker and Charlie Griffiths. Reproduced by permission.)

Evolution and systematics

The Strepsiptera are thought to have originated in the Carboniferous, but the earliest known fossils are from the lower Cretaceous amber of Lebanon. The phylogenetic position of the strepsipterans is unclear, with four possible phylogenetic placements. The first is as the sister group of the beetle family Rhipiphoridae (Coleoptera), which has similar parasitic development, flabellate (branched) antennae, and forewing reduction. The second is as the sister group of beetles (Coleoptera), because they use only the hind wings for flight. The third possible placement is as the sister group of flies (Diptera), which have hind wing halteres (front wing and hind wing halteres could have switched places by a homeotic mutation) and similarities in their DNA. Fourth, they may be placed outside the endopterygote insects, such as beetles and flies, because the pupal stage is preceded by two pharate larval instars (enclosed within the skin of a previous instar), with external wing buds in females, and larval eyes are carried over to the adult stage. The more than 500 described species of strepsipterans are classified in two suborders: Mengenillidia, with two families (one, Mengeidae, known only from fossil males in Eocene Baltic amber), and Stylopidia, with seven families.

Physical characteristics

Strepsipterans have two different morphologies ("hypermetamorphosis"). The first larval instar, called the triun-gulinid, is free-living and actively host-seeking, and has simple eyes, legs, and two long caudal setae. The second through the last instar larvae (they can have 4 to 7 instars depending on the species) are maggot-like and endoparasitic. Adult males usually are 0.04–0.12 in (1–3 mm) long, with some reaching 0.3 in (7 mm). Females typically are 0.19 in (5 mm) long, but they can range from 0.08 to 1.18 in (2–30 mm). There is extreme sexual dimorphism: adult males look like insects, whereas females look like larvae (larviform) and lack legs, antennae, and external genitalia. Adult males are free living, whereas most females are internal parasites on other insects and remain in the host for their entire lives. Females of the family Mengenillidae are free living.

Males have berry-like eyes, with each lens separated by cuticle or setae; branched antennae; nonfunctional mouthparts; reduced forewings in the form of halteres used for balancing during flight; fan-shaped hind wings with longitudinal veins but no cross veins; and no cerci. Strepsipterans also are called "twisted wing parasites," referring to the peculiar twisted shape of the male hind wings in flight. In the suborder Stylopidia the female head and thorax are united to form a cephalothorax, which protrudes from the body of the host. The female lives within the last larval skin, in which she also pupates, and her whole abdominal cavity is filled with eggs or developing embryos.

Heavily stylopised (parasitized by Strepsiptera insects) *Ammophila* wasp (family Sphecidae), showing three pupae sticking out of the abdomen. Up to five pupae may be found on some *Ammophila* individuals, which become weakened and can no longer fly. (Photo by Mike Picker and Charlie Griffiths. Reproduced by permission.)

Distribution

Although the order is cosmopolitan, most species occur in the tropics.

Habitat

Strepsipterans are found wherever their hosts live. Known hosts include Lepismatidae (Thysanura), Blattidae (Blattodea), Mantidae (Mantodea), Gryllidae, Gryllotalpidae, Tettigoniidae (Orthoptera, Ensifera), Tridactylidae (Orthoptera, Caelifera), Psyllidae (Hemiptera, Sternorrhyncha), Cercopidae, Cicadellidae, Membracidae, Delphacidae, Dictyopharidae, Eurybrachidae, Ricaniidae, Flatidae, Fulgoridae, Issidae, Tettigometridae (Hemiptera, Auchenorrhyncha), Coreidae, Cydnidae, Lygaeidae, Pentatomidae, Scutelleridae (Hemiptera, Heteroptera), Tephritidae, Platysomatidae (Diptera), Formicidae, Vespidae, Sphecidae, Colletidae, Halictidae, and Andrenidae (Hymenoptera).

Behavior

The life cycles of strepsipterans are complex. Eggs are retained within the body of the female until they hatch. The triungulins leave the body of the mother's host and wait for another host. They jump or crawl onto the new host and are transported back to its nest, where they penetrate the host's eggs or developing larvae. The second instar larvae live and feed inside the host's body cavity. Larvae molt four to seven times and reach the last instar at about the same time that the host is ready to pupate. The larvae pupate with their heads and thoraxes protruding between the host's fourth and fifth abdominal segments.

Males use an eversible saclike structure on the head (similar to the ptilinum of some flies) to open the puparium, and leave the host while it is flying; they have only a few hours to find females and mate before dying. Females of the suborder Stylopidia remain in the puparium, protruding from the host for their entire lives, their life spans directly linked to those of their hosts. In the Mengenillidae both sexes abandon the host to pupate externally, and the adults are free living.

Feeding ecology and diet

Parasitic larvae of the second through the last instar and Stylopidia females feed by filtering the host blood; free-living adults do not feed.

Reproductive biology

Female strepsipterans attract males with pheromones. Males fertilize females by injecting sperm into the female's body cavity. Females are sexually mature before molting to the adult stage and give birth to live larvae instead of laying eggs. In some species females are parthenogenetic, reproducing without being fertilized by a male.

Conservation status

No strepsipteran is cited by the IUCN.

Significance to humans

Strepsipterans are not very common, and few people other than entomologists are likely to see them. Although strepsipterans do not kill their hosts, they greatly reduce their nutriment intake; "stylopized" (parasitized by strepsipterans) insects generally are sterile, having been effectively castrated by their parasites. Thus, strepsipterans probably are indirectly beneficial to humans, because they may control population levels of economically important insects. Such insects include *Antestia* stinkbugs (Hemiptera, Pentatomidae) that attack coffee plantations; fruit flies (Diptera, Tephritidae); long-horned grasshoppers (Orthoptera, Tettigoniidae), which defoliate oil palms and coconuts; virus-transmitting leafhoppers (Hemiptera, Cicadellidae), for example, *Perkinsiella vitiensis* on sugarcane and *Nephotettix* species on rice; and plant hoppers (Hemiptera, Delphacidae), such as *Nilaparvata lugens* and *Sogatella furcifera* on rice.

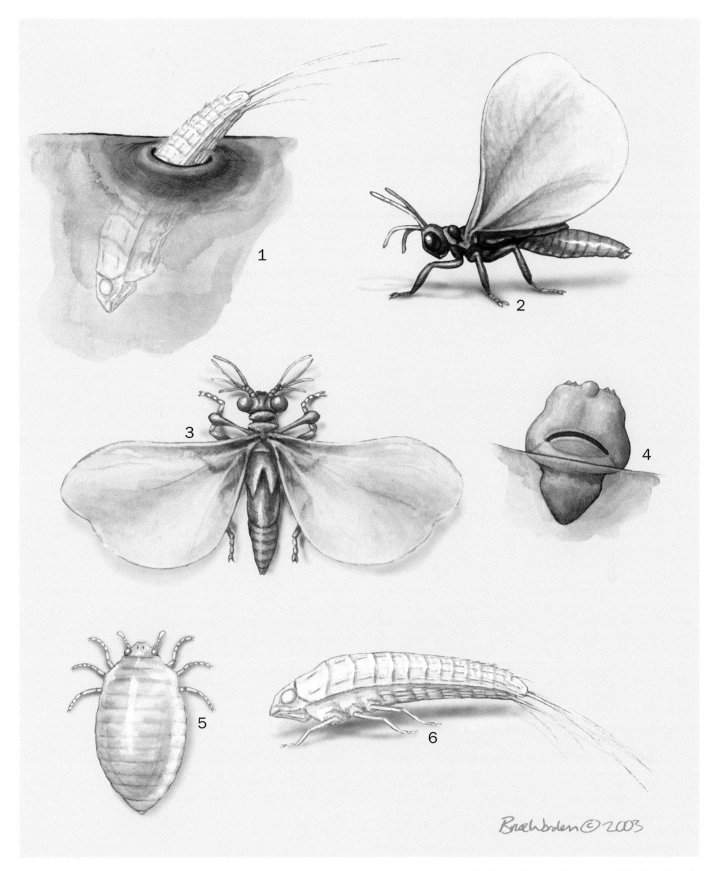

1. *Stichotrema dallatorreanum* adult; 2. *S. dallatorreanum* larva; 3. *Halictophagus naulti adult*; 4. *H. naulti* larva; 5. *Eoxenos laboulbenei* adult; 6. *E. laboulbenei* larva. (Illustration by Bruce Worden)

Species accounts

No common name
Halictophagus naulti

FAMILY
Halictophagidae

TAXONOMY
Halictophagus naulti Kathirithamby and Moya-Raygoza, 2000, Tlaltizapán, Morelos, Mexico.

OTHER COMMON NAMES
None known.

PHYSICAL CHARACTERISTICS
Males have three tarsomeres and seven antennomeres; females are larviform.

DISTRIBUTION
So far known only from its type locality, Tlaltizapán in Morelos, Mexico.

HABITAT
Parasitizes the corn leafhopper *Dalbulus maidis* (Hemiptera, Cicadellidae).

BEHAVIOR
Nothing is known.

FEEDING ECOLOGY AND DIET
Endoparasitic larval instars and females feed on the body fluids of the host; adult males do not feed.

REPRODUCTIVE BIOLOGY
Nothing is known.

CONSERVATION STATUS
Not listed by the IUCN.

SIGNIFICANCE TO HUMANS
The corn leafhopper *Dalbulus maidis* is the most important leafhopper pest of maize in Latin America; *Halictophagus naulti* thus is an important biological control agent. ◆

No common name
Eoxenos laboulbenei

FAMILY
Mengenillidae

TAXONOMY
Eoxenos laboulbenei Peyerimhoff, 1919, Cannes, southern France.

OTHER COMMON NAMES
None known.

PHYSICAL CHARACTERISTICS
Males have well-developed hind wings and branched antennae with six antennomeres. Females lack wings, but they have eyes, unbranched antennae with four or five antennomeres, and legs. Both sexes leave the host to pupate in the ground.

Halictophagus naulti
Eoxenos laboulbenei
Stichotrema dallatorreanum

DISTRIBUTION
Spain, Portugal, southern Italy and southern France, North Africa, and the Canaries.

HABITAT
Parasitizes several species of silverfish (Thysanura, Lepismatidae).

BEHAVIOR
Triungulinids (first instar larvae) are very active and jump onto all instars of their host, to penetrate their body cavities. The last larval stage leaves the host and pupates; pupation lasts from one to three weeks. Males usually live only one to two hours.

FEEDING ECOLOGY AND DIET
Endoparasitic larval instars feed on body fluids of the host; adults do not feed.

REPRODUCTIVE BIOLOGY
During the summer, larval development takes place quickly. Some females never hatch from the puparium and reproduce parthenogenetically; others hatch and are fertilized by males.

CONSERVATION STATUS
Not threatened.

SIGNIFICANCE TO HUMANS
None known. ◆

No common name
Stichotrema dallatorreanum

FAMILY
Myrmecolacidae

TAXONOMY
Stichotrema dallatorreanum Hofenender, 1919, Wogeo, Schouten Islands, Papua, New Guinea

OTHER COMMON NAMES
None known.

PHYSICAL CHARACTERISTICS
The female is larviform and very large, at about 0.98 in (25 mm), with hooked protuberances.

DISTRIBUTION
Papua New Guinea

HABITAT
Parasites of the long-horned grasshoppers *Sexava nubila*, *Segestes decoratus*, and *Segestidea novaeguineae* (Orthoptera: Tettigoniidae), which are found on oil palms.

BEHAVIOR
Male myrmecolacids typically parasitize ants (a pupa of a possible male *S. dallatorreanum* was found on a *Camponotus* ant), whereas females utilize grasshoppers, crickets, katydids, or mantids. Triungulinids (first instar larvae) emerge alive from the adult female, which remains endoparasitic in her host. Because the host, with the female, moves around, more triungulinids are dispersed (the host remains alive and acts as the disperser) onto the surface of oil palm leaves. The triungulinids then seek the footpads of passing long-horned grasshoppers, to which they attach to penetrate their new hosts.

FEEDING ECOLOGY AND DIET
Endoparasitic larval instars and females feed on body fluids of the host.

REPRODUCTIVE BIOLOGY
On entry into the host, the triungulinid molts into a second instar larva that moves up the host's leg to the abdominal region, where it molts four times but without shedding the old cuticle. Adult females are capable of producing about a million triungulinids. This species is parthenogenetic.

CONSERVATION STATUS
Not threatened.

SIGNIFICANCE TO HUMANS
Long-horned grasshoppers defoliate large areas of oil palm plantations in Papua New Guinea; *Stichotrema dallatorreanum* is released there to help control this pest. ◆

Resources

Books
Carvalho, E. Luna de., and M. Kogan. "Order Strepsiptera." In *Immature Insects*, Vol. 2, edited by F. W. Stehr. Dubuque, IA: Kendall/Hunt Publishing, 1991.

Kathirithamby, J. "Strepsiptera." In *The Insects of Australia: A Textbook for Students and Research Workers*, Vol. 2, edited by I. D. Naumann, P. B. Carne, J. F. Lawrence, et al. 2nd edition. Carlton, Australia: Melbourne University Press, 1991.

————. "Strepsiptera of Panama and Mesoamerica." In *Insects of Panama and Mesoamerica*, edited by D. Quintero and A. Aiello. New York: Oxford University Press, 1992.

Periodicals
Bohart, R. M. "A Revision of the Strepsiptera with Special Reference to the Species of North America." *California University Publications in Entomology* 7, no. 6 (1941): 91–160.

Kathirithamby, J. "Review of the Order Strepsiptera." *Systematic Entomology* 14 (1989): 41–92.

Kinzelbach, R. K. "The Systematic Position of the Strepsiptera (Insecta)." *American Entomologist* 36 (1990): 292–303.

Whiting, M. F. "Phylogenetic Position of the Strepsiptera: Review of Molecular and Morphological Evidence." *International Journal of Insect Morphology and Embryology* 27 (1998): 53–60.

Other
Pohl, Hans. "Strepsiptera: Twisted-Wing Parasites." 5 Mar. 2002. [7 May 2003] <http://www.strepsiptera.uni-rostock.de/>.

"Strepsiptera: Biology, Genomics, Natural History, and Phylogeny." [7 May 2003] <http://www.strepsiptera.com/>.

Tree of Life Web Project. "Strepsiptera: Twisted-Wing Parasites." [7 May 2003] <http://tolweb.org/tree?group=Strepsiptera&contgroup=Endopterygota>.

Natalia von Ellenrieder, PhD

Mecoptera
(Scorpionflies and hangingflies)

Class Insecta
Order Mecoptera
Number of families 9

Photo: A scorpionfly (*Panorpa communis*) in flight. (Photo by Kim Taylor. Bruce Coleman, Inc. Reproduced by permission.)

Evolution and systematics

Classified within the order Mecoptera are perhaps the most primitive insects with complete metamorphosis. The fossil record of Mecoptera is rich, dating back to the lower Permian, when they were one of the most abundant insect groups. The modern orders Trichoptera, Lepidoptera, Siphonaptera, Strepsiptera, and Diptera are believed to have descended from a mecopteran ancestor. The order today is made up of remnants of this former diversity, containing about 600 species in nine families: Apteropanorpidae (1 genus, 1 species); Bittacidae (16 genera, 172 species); Boreidae (3 genera, 27 species); Choristidae (3 genera; 8 species); Eomeropidae (1 genus, 1 species); Meropeidae (2 genera, 2 species); Nannochoristidae (2 genera, 7 species); Panorpidae (3 genera, 360 species); and Panorpodidae (2 genera, 9 species). Living examples of Mecoptera vary widely in form and biology. Because of this extreme diversity, the status of the order as a single evolutionary unit is under debate, and it eventually may be divided into several new orders.

Physical characteristics

No single feature unifies the order. Fully winged species share wing structure, having four long, membranous wings,

A scorpionfly (*Panorpa communis*) profile. (Photo by L. West. Bruce Coleman, Inc. Reproduced by permission.)

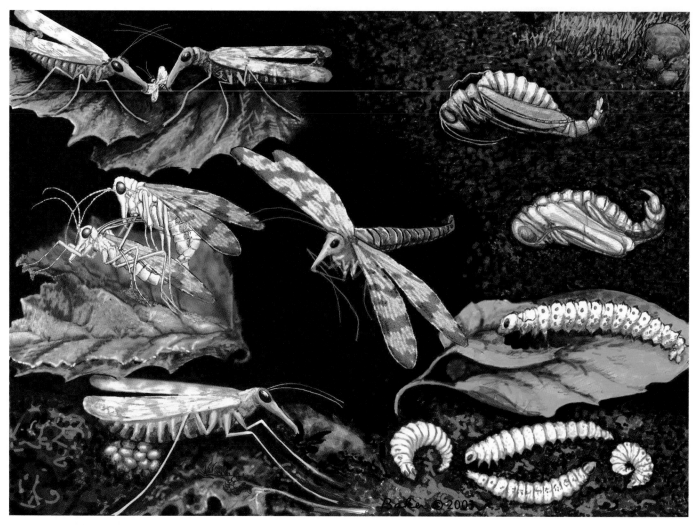

Panorpa nuptialis illustrating the various life stages of mecopterans. (Illustration by Wendy Baker)

with both the front and hind pairs similar in shape and venation (number and shape of veins in an insect's wings). The order name Mecoptera means "long wings." Mecoptera generally have hyaline (clear) wing membranes with dark veins. When a color pattern exists, it is typically a transparent amber coloration or dark brown banding and striping on the wing membranes. Another common feature is the elongation of the mouthparts and lower portions of the head into a rostrum, a useful character for placing short-winged or wingless species within the order. The mandibles are located at the tip of the rostrum.

Many adult body types exist. Males of Panorpidae and Panorpodidae commonly are called scorpionflies. They have enlarged, bulbous genitalia, carried curled above the body, resembling the tail of a scorpion. The hangingflies, family Bittacidae, look like crane flies (order Diptera), in that they have narrow bodies and long, thin legs. Meropeids and eomeropids are cockroach-like, with flattened bodies and tough, densely veined wings. There are three body types of mecopteran larvae: eruciform (caterpillar-like, with false legs on the abdomen), campodeiform (long

and cylindrical, lacking false legs), and scarabaeiform (grublike).

Distribution

Mecoptera are distributed worldwide, including the northern polar regions. Some families are very restricted in distribution. The highest species diversity occurs in the Indomalayan biogeographical region, which encompasses Southeast Asia and Indonesia.

Habitat

Mecoptera are found chiefly in cool, moist habitats. Forests with plentiful shade support the greatest diversity of species. Most larvae develop beneath soil or litter. Nannochoristid larvae, however, are completely aquatic. Adults of Boreidae, called snow scorpionflies, are found on rocks, snow, and ice in the vicinity of moss clumps, within which their larvae develop.

Mating in *Boreus brumalis* is unique, with an unusual posture and the use of the male's specially modified spiny wings to hold the female during the long period of time spent in copulation. (Illustration by Wendy Baker)

A scorpionfly (*Panorpa communis*) on a leaf in northern France. (Photo by J-C Carton. Bruce Coleman, Inc. Reproduced by permission.)

Behavior

Adult Mecoptera tend to be rather secretive, inactive insects, most frequently found resting on the surface of leaves in dense shade. Flight in most species is feeble and brief. Predaceous species typically feed during the day, while opportunistic species have flexible foraging schedules. Reproductive activity in most species takes place only at night.

Feeding ecology and diet

Mecopterans are carnivorous, herbivorous, or omnivorous. Hangingflies are adapted to a predaceous lifestyle. Their hind legs are raptorial, with a single large tarsal claw, used to capture small insect prey. Prey is pierced with the rostrum, and fluids are withdrawn. Snow scorpionflies apparently are an entirely herbivorous group, feeding on mosses as adults and larvae. Most panorpid scorpionflies are omnivores, feeding opportunistically on dead or dying insects but also on plant secretions such as pollen, fruit juice, and nectar. Mecopteran larvae are mostly omnivorous, an exception being the aquatic larvae of Nannochoristidae, which are predaceous on the larvae of midges (order Diptera).

Reproductive biology

Males court nearby females with displays of wing and body movements, and many offer females a nuptial meal.

Common nuptial meals are prey items and salivary secretions. Competition among males often is fierce, and males that are competitively unsuccessful may attempt to force copulation. Females have been shown to prefer males that offer nuptial meals, and feeding on the meal stimulates egg laying and increases fecundity. Mating may last for several hours. Development progresses through four larval instars, a prepupal stage and a pupal stage. Larval development can be as rapid as one month. Adult life span is of similar duration.

Conservation status

No species of Mecoptera is listed by the IUCN. A decline of mecopteran populations in North America, Mexico, and Java has been noted and is attributable to human activity.

Significance to humans

Mecoptera are not known to affect humans in any way. The common name scorpionfly implies that they are in some way dangerous, but no species stings or bites.

1. Black-tipped hangingfly (*Hylobittacus apicalis*); 2. *Panorpa nuptialis*; 3. Snow scorpionfly (*Boreus brumalis*). (Illustration by Wendy Baker)

Species accounts

Black-tipped hangingfly
Hylobittacus apicalis

FAMILY
Bittacidae

TAXONOMY
Bittacus apicalis Hagen, 1861, southern Illinois, United States.

OTHER COMMON NAMES
None known.

PHYSICAL CHARACTERISTICS
Body and appendages glossy yellow to brown. Wings clear with black tips, held outstretched when at rest.

DISTRIBUTION
Eastern United States.

HABITAT
Understory vegetation of moist, shaded woodlands.

BEHAVIOR
This species hangs from low-growing vegetation by the front legs, which are useless for walking. Movement on vegetation is accomplished through a monkey-like swinging motion. Flight is weak and undulating.

FEEDING ECOLOGY AND DIET
Adults are generalized predators, hunting small insects during the daytime. Prey is captured with the long raptorial hind legs, either from a hanging position or while flying. Draining prey of fluids may take up to one hour.

REPRODUCTIVE BIOLOGY
The male captures or steals a prey item and emits pheromones while in flight. He offers the prey to attracted females, who reject the male or terminate copulation early if the prey is small or poor in quality. Females feed on the prey while mating, which takes 20–30 minutes. After mating, the male usually retains the prey and may use it in subsequent courtship attempts (females are not able to entirely drain the prey within the time spent in copulation). Females scatter eggs over the ground from a hanging position.

CONSERVATION STATUS
Not threatened.

SIGNIFICANCE TO HUMANS
None known. ◆

Snow scorpionfly
Boreus brumalis

FAMILY
Boreidae

TAXONOMY
Boreus brumalis Fitch, 1847, eastern New York, United States.

OTHER COMMON NAMES
None known.

Hylobittacus apicalis

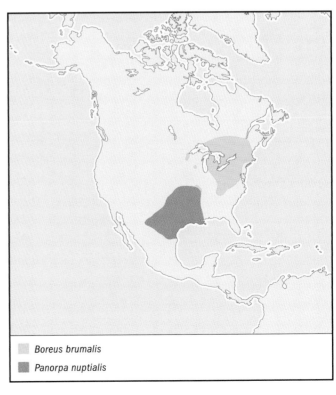

Boreus brumalis

Panorpa nuptialis

PHYSICAL CHARACTERISTICS
Small, dark brown, and compactly built. Males have narrow wings, about half as long as the abdomen, hooked downward and with many spines. Female wings are reduced to tiny pads.

DISTRIBUTION
Northeastern North America.

HABITAT
Larvae live inside clumps of moss growing among rocks or on loamy soil. Adults are found on rocks, soil, snow, and ice near the larval habitat.

BEHAVIOR
Adults are active in fall and winter. Activity at low temperatures is made possible by a substance in the blood that acts like antifreeze. Larvae are scarabaeiform and can be found any time of year within mosses.

FEEDING ECOLOGY AND DIET
Larvae and adults feed on mosses.

REPRODUCTIVE BIOLOGY
The male leaps, grasping an appendage of the female with the claspers at the tip of his abdomen. The male then maneuvers the female onto his back and holds her in place with his hook-like wings. Copulation lasts from one to 12 hours. The female deposits eggs into moss singly or in small clutches. The life cycle probably takes two years.

CONSERVATION STATUS
Not threatened.

SIGNIFICANCE TO HUMANS
None known. ◆

No common name
Panorpa nuptialis

FAMILY
Panorpidae

TAXONOMY
Panorpa nuptialis Gerstaecker, 1863, Texas, United States.

OTHER COMMON NAMES
None known.

PHYSICAL CHARACTERISTICS
Adult body is reddish brown. Wings are amber in color, with striking bands of dark brown. Larva are caterpillar-like, pale, and encircled with rings of dark spots and setae.

DISTRIBUTION
South-central United States and northern Mexico.

HABITAT
Dense vegetation in open fields and pastureland. Larvae are found beneath the soil in these habitats.

BEHAVIOR
Adults are active only during the day and rest vertically on vegetation at night.

FEEDING ECOLOGY AND DIET
Adults and larvae feed primarily on feeble or dead soft-bodied insects.

REPRODUCTIVE BIOLOGY
Males infrequently offer a salivary secretion to the female. Females lay eggs in existing cracks in the soil, probing with the ovipositor until a suitable site is found. Larvae develop in about a month. Winter is passed in the pupal stage. Adults emerge in late fall and live nearly a month. This species may have two generations per year.

CONSERVATION STATUS
Not threatened.

SIGNIFICANCE TO HUMANS
None known. ◆

Resources

Books
Byers, G. W. "Mecoptera." In *The Insects of Australia: A Textbook for Students and Research Workers*, edited by CSIRO. 2nd edition. 2 vols. Carlton, Australia: Melbourne University Press, 1991.

———. "Order Mecoptera." In *An Introduction to the Study of Insects*, edited by D. J. Borror, C. A. Triplehorn, and N. F. Johnson. 6th edition. Philadelphia: Saunders College Publishing, 1989.

Periodicals
Byers, G. W., and R. Thornhill. "Biology of the Mecoptera." *Annual Review of Entomology* 28 (1983): 203–228.

Other
Myer, John R. "Mecoptera." June 13, 2001 [March 12, 2003]. <http://www.cals.ncsu.edu/course/ent425/compendium/mecopt1.html>.

"The Scorpion Flies (Mecoptera)." February 25, 2003 [March 12, 2003]. <http://www.earthlife.net/insects/mecop.html>.

"World Checklist of Extant Mecoptera Species." October 31, 1997 [March 12, 2003]. <http://www.calacademy.org/research/entomology/mecoptera/>.

Jeffrey A. Cole, BS

Siphonaptera
(Fleas)

Class Insecta
Order Siphonaptera
Number of families 16

Photo: Bird flea larva. (Photo by Kim Taylor. Bruce Coleman, Inc. Reproduced by permission.)

Evolution and systematics

Fleas may have evolved as early as 140 million years ago (mya), along with their mammalian hosts. Only five flea species are known from fossil records: three from Baltic amber (35–40 mya: *Palaeopsylla baltica*, *Palaeopsylla dissimilis*, and *Palaeopsylla klebsiana*) and two from Dominican amber (15–20 mya: *Pulex larimerius* and an undescribed species of *Rhopalopsyllus*). The specialized combs, setae, and appendages of these relics are very similar to those of their modern relatives. Molecular and morphological data suggest the small Mecopteran family of snow scorpionflies, or (Boreidae), snow fleas is a sister group of Siphonaptera. Snow fleas are not actually fleas and are not parasitic, as are all members of the Siphonaptera.

Based on empirical evidence, some workers have divided the superfamily Pulicoidea into two subfamilies (Pulicinae and Tunginae), whereas others have considered them as two separate families (Pulicidae and Tungidae). DNA analyses conducted from 2000 to 2003 of taxa assigned to these subfamilies indicate that they are two distinct families. The family placement of many of the 244 genera remains to be defined by molecular studies. Until molecular studies redefine the genetic phylogeny, 16 families belonging to five superfamilies are recognized: Ceratophylloidea (Ancistropsyllidae, Ceratophyllidae, Ischnopsyllidae, Leptopsyllidae, and Xiphiopsyllidae), Hystrichopsylloidea (Chimaeropsyllidae, Coptopsyllidae, Ctenophthalmidae, Hystrichopsyllidae, Pygiopsyllidae, and Stephanocircidae), Malacopsylloidea (Malacopsyllidae

and Rhopalopsyllidae), Pulicoidea (Pulicidae and Tungidae), and Vermipsylloidea (Vermipsyllidae). Of about 2,575 species (including subspecies), some 5% occur on birds, while the remaining 95% parasitize mammals. DNA analyses indicate that fleas originated on mammals, with some crossing over later to avian hosts. Fleas typically do not parasitize amphibians and reptiles.

Physical characteristics

Fleas are wingless, laterally compressed, holometabolous insects, and the adults are adapted to a parasitic mode of life. Their eggs are small, elongated spheres, varying in color from pearly white to dark brown to black; they are about 0.02–0.06 in (0.5–1.4 mm) in diameter. Larvae are wormlike and range from 0.02 to 0.4 in (0.5–10 mm) in length, with well-sclerotized head capsules, three thoracic segments without appendages, and ten abdominal segments. There usually are three larval stadia (intervals between molts), although some species of *Tunga* have only two. A silk pupal case contains an exarate (having the appendages not "glued" to the body) pupa that is 0.008–0.4 (0.2–10 mm) long. Fine debris, which blends with the surroundings, often adheres to the pupal case.

In general, adult fleas range from 0.04 to 0.3 in (1.0–8 mm), excluding engorged Tungidae and Vermipsyllidae. Female adult fleas typically are larger than males, and those of

Female (left) and male dog fleas (*Ctenocephalides canis*). (Photo by Kim Taylor. Bruce Coleman, Inc. Reproduced by permission.)

a few species may achieve lengths of 0.6 in (16 mm) when they are engorged or gravid. Their mouthparts are modified for piercing avian or mammalian hosts and sucking their blood. Adults have evolved highly modified combs and setae on their body and legs. These features provide protection for intersegmental membranes and spiracle openings and are used for grasping hairs and feathers and for preventing dislodgment during the host's preening activities. Legs are specialized to promote mobility through the fur and feathers of the host and for jumping to facilitate host acquisition. Fleas that have adapted to a parasitic life on birds (bird fleas) tend to have more setae that are longer and more slender than those of fleas that parasitize mammals (mammal fleas).

The internal anatomy of fleas may be as important in flea phylogeny as the external morphological features. Females possess either one or two spermathecae. Two are considered a primitive condition, because a blind duct exists in most species with a single spermatheca. Most families have six rectal pads, a proventriculus with spines, and a ventral nerve cord. Tungidae are exceptional in possessing only two rectal pads and no proventricular spines; moreover, the nerve cord is displaced dorsally. The variety of morphological specialization (such as head, thoracic, and abdominal combs; the degree of scleroti-

zation of the exoskeleton; the length and number of setae; the development of mouthparts, tarsal claws, and associated bristles on appendages; spiracle characteristics; development of complex genitalia, particularly in males; diversification of internal anatomy; and display of neosomatic growth) is reflective of the corresponding diversity of the respective host species.

Male and female fleas are sexually dimorphic. In addition to size differences, the eighth sternum, ninth tergum, and aedeagus of males are highly specialized, whereas only the seventh sternum of females is particularly modified. The dorsal area of the male head frequently is grooved longitudinally, and the anterior portion of the head is divided from the posterior portion by a distinct suture or groove. The female head may or may not be divided. Antennae generally are longer in males than in females.

Distribution

The Ceratophyllidae, Hystrichopsyllidae, Leptopsyllidae, Vermipsyllidae, Coptopsyllidae, and Ancistropsyllidae occur predominantly in the boreal continents of North America, Europe, or Asia. Those families restricted to the southern continents of Africa, Antarctica, Australia, or South America include Malacopsyllidae, Rhopalopsyllidae, Stephanocircidae, Pygiopsyllidae, Xiphiopsyllidae, and Chimaeropsyllidae. The remaining three families, Ctenophthalmidae, Ischnopsyllidae and Pulicidae, occur in both the Northern and Southern Hemispheres.

Habitat

Fleas parasitize hosts in virtually every conceivable terrestrial habitat, adapting to the microclimate of the nests, burrows, and body conditions. Such adaptations enable fleas to live in the most extreme environmental conditions. For example, *Glaciopsyllus antarcticus* occurs only in the frigid, subzero conditions of the Antarctic. They proliferate in the microclimate of the nest and in the down of their avian host, the southern fulmar (*Fulmarus glacialoides*). Many species (*Xenopsylla* and *Nosopsyllus*) thrive in the dry conditions of deserts, living in the burrows of their rodent hosts, where the temperature and humidity are optimal for their development. Adult fleas are found on mammalian hosts more frequently and in greater numbers than those species parasitizing birds. Fleas have adapted only to birds that use their nests over and over (swallows, seabirds, and some ground-dwelling birds and cavity dwellers). A few species (*Pulex* and *Ctenocephalides*), especially those inhabiting coastal, semitropical, and tropical regions, are free living, jumping on and off their hosts and proliferating in open environmental conditions, such as floors of homes, pathways, barnyards, animal pens, and pet beds.

Behavior

Perpetuation of each species is dependent on the success of finding a host. Some fleas remain in a quiescent pupal state for an extended time, to survive cold periods or to wait until a host approaches. Vibrations produced by an approaching host may stimulate adults to emerge immediately from their pupal cases. Although the visual acuity of fleas is poor, shadows of approaching hosts illicit a jump response. They also are attracted to the warmth and carbon dioxide emitted by potential hosts. A few species, particularly bat fleas, are negatively geotrophic (moving against gravity), crawling up cave walls to locate bats roosting on the cave ceiling.

Feeding ecology and diet

With the exception of *Uropsylla tasmanica*, the larvae of most fleas are free living, scavenging on dried blood, animal dandruff, and animal excreta in the host's nest or the environment. Larval cat fleas (*Ctenocephalides felis felis*) feed on partially digested blood excreted from the anus of adult fleas. The larvae of *Hoplopsyllus*, *Tunga*, and *Dasypsyllus* are documented facultative ectoparasites, feeding either on host tissues or on organic debris in the nest substrate. *Uropsylla tasmanica* larvae burrow into the skin of their host and are the only known true obligate larval flea parasites. A few species are predaceous on other nest-dwelling organisms.

Depending on the species, adult fleas ingest blood by either tapping into a capillary, or by cutting the tissue and caus-

A gray squirrel flea (*Orchopeas howardi*) among squirrel fur. (Photo by Kim Taylor. Bruce Coleman, Inc. Reproduced by permission.)

ing a pool of blood from which to feed. A few species have been observed to imbibe water. The vast majority of fleas are intermittent feeders, among them, *C. f. felis*, and some attach to their hosts permanently, for example, *Echidnophaga gallinacea* and species of *Tunga*. The nutritional requirements of adults are understood only partly. Male and female fleas require a blood meal as a prerequisite for spermatogenesis or oogenesis. Females feed more rapidly than males and require larger volumes of blood to facilitate egg production. The chemistry of host blood is known to affect the host specificity of some species and is suspected to influence many others.

Reproductive biology

Males assume a position directly beneath the female, each facing the same direction. The occipital groove in the dorsal portion of the male head frequently is developed to accommodate the keel-shaped surface of forward sternites of the female abdomen. The male clasps the sides of the female sternites with suckerlike structures on the inner surface of the antennae. He also may clasp the hind legs of the female between a notch in the hind coxa and the retracted femur. The highly modified ninth tergite (basimere and telomere) attaches to the terminal segments of the female in a "clasping" manner. The posterior margin of the seventh sternum of many females is modified with various lobes and sinuses that facilitate attachment during copulation. The vaginal canal also is modified to accommodate partial insertion of the highly modified apical portion of the aedeagus. Sperm transfer is accomplished during insertion of long penis rods through the vaginal canal, into the bursa copulatrix, through the duct of the spermatheca, and into the spermatheca. This process varies from species to species.

Copulation is vastly different in *Tunga*. The female attaches to the skin of the host and soon becomes enveloped by the host tissues, exposing only the caudal disc (the last four abdominal spiracles, the anus, and the vaginal opening). A darkly sclerotic ring made up of host tissues surrounds the caudal disc. Males locate the female and copulate in situ with a highly modified aedeagal apparatus. After copulation, the female develops massive numbers of eggs that are expelled into the environment wherever the host travels. The breeding cycles of *Cediopsylla* and *Spilopsyllus* are bound to the estrus cycle of their hosts, hares and rabbits. Other genera probably are influenced by the breeding cycle of the host, but studies are lacking.

Neosomy is the expansion of pregenital abdominal segments by secretion of new cuticle without molting. This process takes place in females of *Hectopsylla, Neotunga, Tunga* (Tungidae), *Chaetopsylla, Dorcadia, Vermipsylla* (Vermipsyllidae), and *Malacopsylla* (Malacopsyllidae), as a way to accommodate growth up to 1,000x normal size. Neosomatic growth is especially pronounced in *Tunga, Neotunga,* and *Dorcadia.* Males do not undergo neosomatic growth, because their principal function is to mate. Females expand primarily to accomodate egg production.

Some species lay eggs on their host, others do so indiscriminately in the environment, and still others place them in the lair or nest of the host. The duration of each stage of the life cycle varies for each species. An example is the common cat flea *C. f. felis.* Eggs are laid on the host within 48 hours of a blood meal. The eggs drop to the ground (most often in the lair of a cat or dog), hatch, and pass through three larval stadia. The mature larva spins a silken cocoon and molts within, ultimately emerging as an adult. The developmental cycle is completed in two to three weeks under optimal conditions of temperature and humidity but may take as long as three to four months. Across the order, sex ratios are approximately 1:1, and longevity of adults may range from only a few weeks to more than three years.

Conservation status

There are no flea taxa specifically listed as threatened by the IUCN; however, species that are very host specific (that is, they depend on a single host for their existence) are in danger of perishing if their host is endangered. Attempts to identify such combinations have never been undertaken.

Significance to humans

The bite of the dog, cat, and human flea (*Pulex* complex) may cause annoyance, irritation, extreme itching, hypersensitivity, and secondary infections. Many species of flea transmit diseases to humans and their pets directly through their bite, through rubbing or scratching infected feces into an open wound, or by ingesting infected fleas. These include plague (*Yersinia pestis*), murine typhus (*Rickettsia typhi*), and cat scratch fever (*Bartonella henslae*). Fleas also have the potential to transmit Q-fever (*Coxiella burnetti*), tularemia (*Francisella tularensis*), listeriosis (*Listeria monocytogenes*), salmonellosis (*Salmonella* species), and Carrion's disease (*Bartonella bacilliformis*). Fleas also are efficient vectors of myxomatosis, a viral disease of rabbits. Some serve as the intermediate host of the double-pored tapeworm (*Dipylidium caninum*) of humans and several tapeworms of sylvatic (wild) and commensal rodents (*Hymenolepis diminuta* and *H. nana*). The filarial worm, *Dipetalonema reconditum*, and the protozoan blood parasite, *Trypanosoma lewisi*, may be transmitted to dogs and rats, respectively.

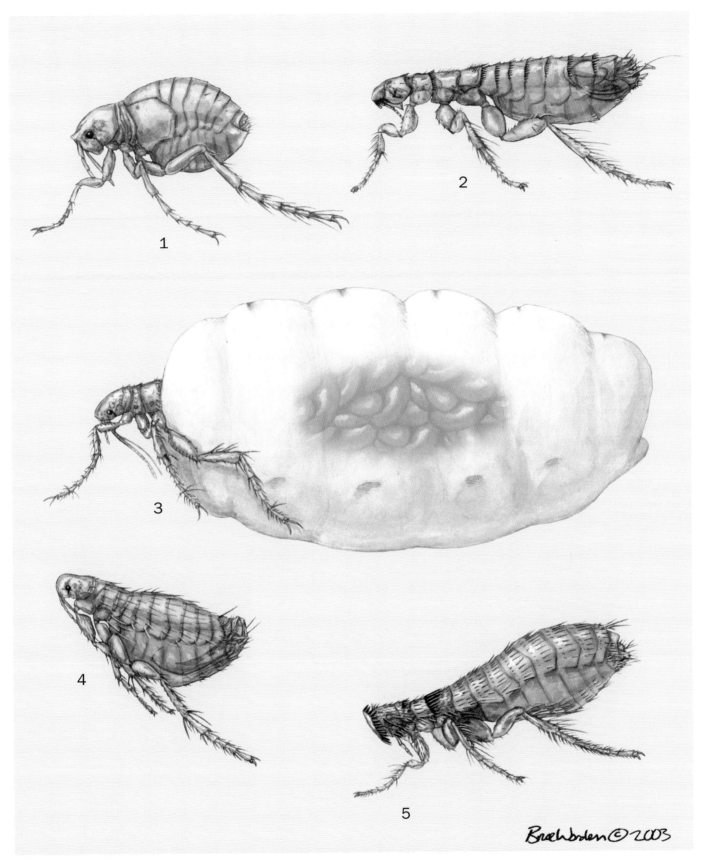

1. Chigoe (*Tunga penetrans*); 2. Bat flea (*Ischnopsyllus octactenus*); 3. Sheep and goat flea (*Dorcadia ioffi*); 4. Oriental rat flea (*Xenopsylla cheopis cheopis*); 5. Helmet flea (*Stephanocircus dasyuri*). (Illustration by Bruce Worden)

Species accounts

Bat flea

Ischnopsyllus octactenus

FAMILY
Ischnopsyllidae

TAXONOMY
Ischnopsyllus octactenus Kolenati, 1856. Type locality not specified.

OTHER COMMON NAMES
None known.

PHYSICAL CHARACTERISTICS
Yellowish-brown in color. Males reach 0.09 in (2.4 mm), and females grow to 0.1 in (2.5 mm). Head, thorax, abdomen, and legs are exceptionally long and slender. The front of the head has two posteriorly directed spatulate ctenidia. Pronotum has a comb of 28 pointed ctenidia. Metanotum and abdominal terga I–VI have ctenidial combs.

DISTRIBUTION
Europe, southern British Isles and Scandinavia, Canary Islands, North Africa and the northern Middle East to Pakistan.

HABITAT
The bat known as Kuhl's pipistrelle (*Pipistrellus kuhli*) rests under the leaves and in the attics of houses, often providing close access for adult fleas.

BEHAVIOR
Bat fleas parasitize bats that roost in areas that bring them into close association with adult fleas. Immature stages develop on the substrate below roosting bats, requiring adult fleas to climb to access the bats or to crawl up on them when baby bats fall from the ceiling and are retrieved by their mothers.

FEEDING ECOLOGY AND DIET
Host preferences include bats of the family Vespertilionidae, particularly *Pipistrellus kuhli*.

REPRODUCTIVE BIOLOGY
Nothing is known.

CONSERVATION STATUS
Not threatened.

SIGNIFICANCE TO HUMANS
None known. ◆

Oriental rat flea

Xenopsylla cheopis cheopis

FAMILY
Pulicidae

TAXONOMY
Xenopsylla cheopis cheopis (Rothschild, 1903), Shendi, Sudan.

OTHER COMMON NAMES
English: Asiatic rat flea, common rat flea, tropical rat flea, Egyptian rat flea.

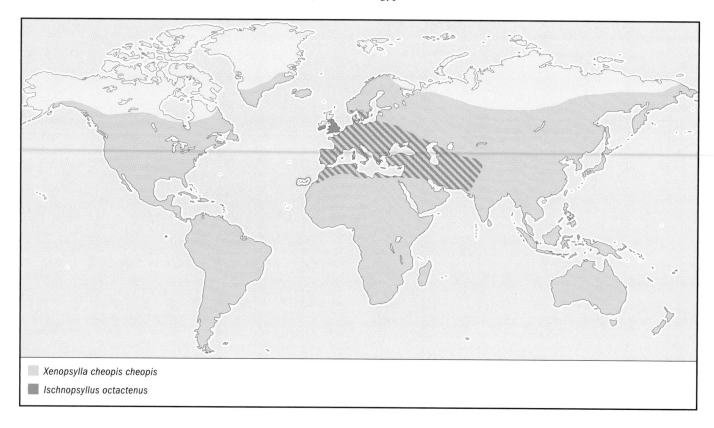

Xenopsylla cheopis cheopis
Ischnopsyllus octactenus

PHYSICAL CHARACTERISTICS
Golden brown in color. Males grow to 0.06 in (1.6 mm) and females to 0.09 in (2.4 mm). Combs are absent. Single row of setae on each abdominal tergite. Oblique row of spiniform setae on the inner aspect of hind coxa.

DISTRIBUTION
Cosmopolitan. Present wherever Norway rats (*Rattus norvegicus*) and roof rats (*Rattus rattus*) are found.

HABITAT
Seaports and unsanitary situations associated with humans that provide food and harborage for rats.

BEHAVIOR
Larvae and fed adults avoid light, whereas unfed adults are attracted to light. Adults can jump 100 times their length.

FEEDING ECOLOGY AND DIET
Prefers commensal rodents of the genus *Rattus* but bites humans readily and is the most efficient vector of plague. The presence of blood and plague bacilli in the proventriculus creates a clot matrix, blocking the gut at temperatures below 80.6°F (27°C). While the proventriculus is blocked, the flea repeatedly attempts to feed, injecting the deadly plague organisms into the host (human or rat). At ambient temperatures of 80.6°F (27°C) or greater, the blockage clears by a combination of enzymes present in the plague bacillus and in the flea's gut. For this reason urban plague epidemics caused by *X. c. cheopis* do not occur during warm seasons.

REPRODUCTIVE BIOLOGY
Males cannot mate until they have fed. Feeding triggers the dissolution of the testicular plug, allowing spermatozoa to pass during copulation. Copulation takes about 10 minutes. Eggs are 0.01–0.02 in (0.3–0.5 mm) in diameter, oval, and white, with a sticky surface. They are laid off the host. Eggs hatch in 2–21 days; three larval stages require about 9–15 days to pupation. Adults may emerge from pupae after a week or more. The life cycle can be completed at a temperature range of 64.4–95°F (18–35°C) but optimally at 80.6°F (27°C), with relative humidity above 60% (optimally 80%).

CONSERVATION STATUS
Not threatened.

SIGNIFICANCE TO HUMANS
X. c. cheopis was responsible for the plague pandemic of Europe in the fourteenth century, which is estimated to have killed 25 million people. ◆

Helmet flea
Stephanocircus dasyuri

FAMILY
Stephanocircidae

TAXONOMY
Stephanocircus dasyuri Skuse, 1893, New South Wales, Australia.

OTHER COMMON NAMES
None known.

PHYSICAL CHARACTERISTICS
Dark reddish-brown color. Males reach 0.114 in (2.9 mm), and females reach 0.118 in (3.0 mm). The head is divided into a forward plate of 14 pointed ctenidia, referred to as a "helmet" or a "crown of thorns," and a rear plate on each side of the head bearing the antenna and genal combs of seven pointed ctenidia. The pronotum has a comb of 22 pointed ctenidia. There are two rows of setae on each abdominal tergite. The dorsal margin of each hind tibia is adorned with a row of setae resembling the teeth of a comb. Three antesensilial bristles occur on each side of the seventh tergite.

DISTRIBUTION
Coastal areas of southern and eastern continental Australia and Tasmania.

HABITAT
Foothill habitats of New South Wales, Queensland, and Tasmania.

BEHAVIOR
Nothing is known.

FEEDING ECOLOGY AND DIET
Wide range in host preferences but found most frequently on quolls (Dasyuridae), potoroos (Macropodidae), and long- and short-nosed bandicoots (Peramelidae).

REPRODUCTIVE BIOLOGY
Nothing is known.

CONSERVATION STATUS
Not threatened.

SIGNIFICANCE TO HUMANS
None known. ◆

Chigoe
Tunga penetrans

FAMILY
Tungidae

TAXONOMY
Tunga penetrans Linnaeus, 1758, America.

OTHER COMMON NAMES
English: Jigger, chigger (not to be confused with the six-legged larval "chigger" mite belonging to the family Trombiculidae), sand flea; Spanish: Chique.

PHYSICAL CHARACTERISTICS
Yellow color, similar to straw. Males and females are about 0.04 in (1.0 mm) long, but gravid females may attain 0.16 in (4.0 mm). Front of head acutely pointed upward. No combs or spinelike setae. Single row of setae on each tergite. Posterior four pairs of spiracles greatly enlarged. Distinct tooth on apex of hind coxa.

DISTRIBUTION
Southern United States, Central and South America, West Indies, and tropical Africa.

HABITAT
Unsanitary situations.

BEHAVIOR
Adults will pass through clothing to feed.

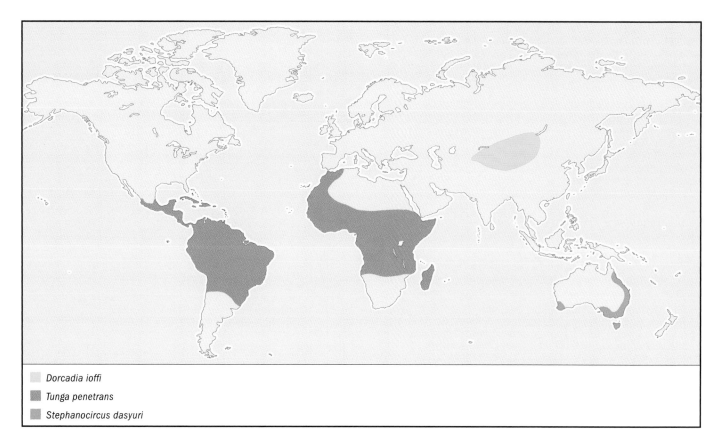

Dorcadia ioffi

Tunga penetrans

Stephanocircus dasyuri

FEEDING ECOLOGY AND DIET
Male and female fleas bite humans intermittently. They prefer the feet, but other areas of the body are not exempt. Only impregnated females permanently attach to the host. They usually select tender areas between the toes, under the nail beds, and along the soles of the feet.

REPRODUCTIVE BIOLOGY
After insemination, the female seeks a host and permanently attaches, is enveloped by the swelling host tissues, and becomes replete with eggs. Eggs are released into the environment and hatch; larvae require about 10–14 days to pupation. Under optimal conditions, adults emerge after about 10–14 days.

CONSERVATION STATUS
Not threatened.

SIGNIFICANCE TO HUMANS
Bites cause extreme irritation. Embedded females may form pustules and cause secondary infections resulting in the sloughing away of toes. Gangrene may ensue and require surgical amputation. Removal of embedded females facilitates healing. ◆

Sheep and goat flea
Dorcadia ioffi

FAMILY
Vermipsyllidae

TAXONOMY
Dorcadia ioffi Smit, 1953, Issyk-Kul, Tien Shan, Turkestan.

OTHER COMMON NAMES
None known.

PHYSICAL CHARACTERISTICS
Distinctly intermixed pale and dark brown markings. Males grow to 0.13 in (3.3 mm), and females reach 0.18 in (4.5 mm); gravid females may attain 0.6 in (16 mm) by neosomatic growth. Combs are absent. Exceedingly long and multisegmented labial palpus.

DISTRIBUTION
Mongolia, Russia, and the Chinese provinces of Qinghai, Xinjiang, Gansu, and Xizang.

HABITAT
Pastures and agricultural areas suitable for domestic sheep, cattle, goats, and wild ungulates.

BEHAVIOR
The legs of engorged females are of little use because of the enormous size of the abdomen. Such females often are seen moving through the host's hair by wormlike contractions.

FEEDING ECOLOGY AND DIET
Although males feed less than females, they both imbibe blood. Males attach to the skin along the sides of the neck, and gravid females have been noted to attach to the inside of the nostrils of wild and domestic even-toed ungulates.

REPRODUCTIVE BIOLOGY
Winter months occur from November to March in the range of this flea. From December through February, animals may have 100–200 adult fleas. The ratio of male to female occur-

ring on hosts is 1:10. Eggs are expelled into the environment in about March. Eggs are oval and darkly pigmented. As spring temperatures increase, eggs hatch. Larval development progresses until pupation in August, and adults emerge in November. The life cycle requires about nine months.

CONSERVATION STATUS
Not threatened.

SIGNIFICANCE TO HUMANS
Parasites of domestic animals; not known to bite humans. ◆

Resources

Books

Hopkins, George Henry Evans, and Miriam Rothschild. *An Illustrated Catalogue of the Rothschild Collection of Fleas (Siphonaptera) in the British Museum (Natural History).* Vols. 1–5. London: British Museum (Natural History), 1953–1971.

Mardon, D. K. *An Illustrated Catalogue of the Rothschild Collection of Fleas (Siphonaptera) in the British Museum (Natural History).* Vol. 6. London: British Museum (Natural History), 1981.

Rothschild, Miriam, and Robert Traub. *A Revised Glossary of Terms Used in the Taxonomy and Morphology of Fleas.* London: British Museum (Natural History), 1971.

Smit, F. G. A. M. *An Illustrated Catalogue of the Rothschild Collection of Fleas (Siphonaptera) in the British Museum (Natural History).* Vol. 7. London: British Museum (Natural History), 1987.

———. "Key to the Genera and Subgenera of Ceratophyllidae." In *The Rothschild Collection of Fleas. The Ceratophyllidae: Key to the Genera and Host Relationships,* edited by Robert Traub, Miriam Rothschild and John F. Haddow. Cambridge, U.K.: Cambridge University Press, 1983.

Traub, Robert, Miriam Rothschild, and John F. Haddow. *The Rothschild Collection of Fleas. The Ceratophyllidae: Key to the Genera and Host Relationships.* Cambridge, U.K.: Cambridge University Press, 1983.

Periodicals

Audy, J. R., F. J. Radovsky, and P. H. Vercammen-Grandjean. "Neosomy: Radical Intrastadial Metamorphosis Associated with Arthropod Symbioses." *Journal of Medical Entomology* 9 (1972): 487–494.

Cavanaugh, D. C.. "Specific Effect of Temperature upon Transmission of the Plague Bacillus by the Oriental Rat Flea, *Xenopsylla cheopis.*" *American Journal of Tropical Medicine and Hygiene* 20 (1971): 264–273.

Humphries, D. A. "The Mating Behavior of the Hen Flea *Ceratophyllus gallinae* (Schrank) (Siphonaptera: Insecta)." *Animal Behavior* 15 (1967): 82–90.

Lewis, Robert E. "Résumé of the Siphonaptera (Insecta) of the World." *Journal of Medical Entomology* 35 (1998): 377–389.

Lewis, Robert E., and David Grimaldi. "A Pulicid Flea in Miocene Amber from the Dominican Republic (Insecta: Siphonaptera: Pulicidae)." *Novitates* 3205 (1997): 1–9.

Rothschild, Miriam. "Fleas." *Scientific American* 213 (1965): 44–53.

———. "Neosomy in Fleas, and the Sessile Life-Style." *Journal of the Zoological Society of London* 226 (1992): 613–629.

———. "Recent Advances in Our Knowledge of the Order Siphonaptera." *Annual Review of Entomology* 20 (1975): 241–259.

Whiting, M. F. "Mecoptera Is Paraphyletic: Multiple Genes and a Phylogeny for Mecoptera and Siphonaptera." *Zoologica Scripta* 31 (2002): 93–104.

Other

"Fleas (Siphonaptera): Introduction." [May 8, 2003]. <http://www.zin.ru/Animalia/Siphonaptera/intro.htm>.

Flea News. [May 8, 2003]. <http://www.ent.iastate.edu/FleaNews/AboutFleaNews.html>.

Fleas of the World. Whiting Lab, Insect Genomics. [Sept. 18, 2003]. http://fleasoftheworld.byu.edu>.

Michael W. Hastriter, MS

Diptera
(Mosquitoes, midges, and flies)

Class Insecta
Order Diptera
Number of families 188

Photo: Non-biting midge larvae among stonewort, showing gills and false legs. (Photo by Kim Taylor. Bruce Coleman, Inc. Reproduced by permission.)

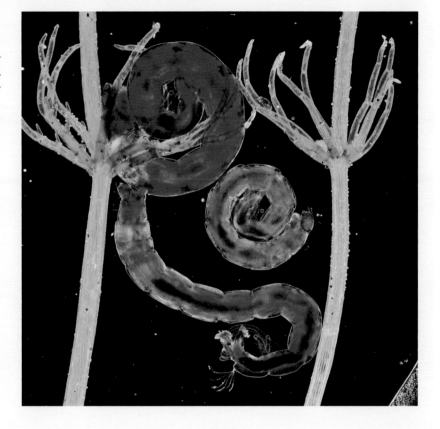

Evolution and systematics

Permotipula patricia from the Upper Triassic of the Mesozoic period from Australia, about 225 million years ago, is the earliest known fossil of a true fly. Since that time, flies have diversified to become one of the four largest orders of insects with about 124,000 species described. It is estimated that an equal number of species, mostly from tropical forests, still await description.

Diptera or true flies are the closest relatives to the Mecoptera or scorpionflies within the Panorpoid-complex (including butterflies and moths, scorpionflies, and fleas), with which they share numerous wing venation characters. Flies are distinguished from other insects by several derived characters, the more important of which are: metathoracic wings transformed into clublike halteres, metathorax reduced in size, labium modified as a labellum, and larval spiracles lacking a closing apparatus.

The order Diptera is divided into two suborders, Nematocera and Brachycera. Nematocera includes generally small, delicate insects with long antennae, regarded as more primitive flies, such as mosquitoes, crane flies, midges, punkies, and no-see-ums. Brachycera includes more specialized, compact, robust flies with short antennae. In older classifications, two divisions were recognized within Brachycera: Orthor-

rhapha and Cyclorrhapha. The Orthorrhapha includes brachyceran flies with free pupae (lacking a puparium) and larvae with incomplete head capsule, such as horse flies and robber flies, and the Cyclorrhapha comprises brachyceran flies with pupae enclosed in the hardened skin of the last larval instar (called puparium) and larvae without head capsule. The Cyclorrhapha are further divided into two groups based on the presence or absence of the ptilinum and associated fissure on the head. The ptilinum is a sac that is everted only once during the emergence of the adult fly to assist in breaking free of the puparium. The Aschiza (flower flies and coffin flies) lack the ptilinum, whereas it is present in the Schizophora. Schizophora in turn comprises a group of flies, the Calypteratae (including house, bottle and flesh flies), with basal lobes in the wings called calypters, and a second group without such lobes, the Acalypteratae (including fruit, vinegar, and oil flies).

These traditional groupings of Diptera have been critically reexamined in recent decades, and many of the categories such as the Nematocera, Orthorrhapha, and Aschiza are not considered natural groups. Instead, these categories consist of a collection of basal lineages from which some natural groups (Brachycera, Cyclorrhapha, and Schizophora, respectively) arose. No consensus has been reached to date on a natural classification for the order.

A male mosquito emerging from pupa skin. (Photo by Dwight R. Kuhn. Bruce Coleman, Inc. Reproduced by permission.)

Physical characteristics

Adults have a mobile head, with large compound eyes that can be contiguous (holoptic condition, found usually in males) or separated (dichoptic condition, most commonly encountered in females) on top. The antennae have six or more segments in nematoceran flies, and five or fewer in brachyceran flies. Mouthparts are adapted for sucking and form a proboscis or rostrum. In predatory species, the mandibles form a pair of piercing stylets, and in cyclorrhaphan flies the labial palps form the labella, membranous sponge-like apical lobes traversed by sclerotized canals called pseudotracheae, through which liquids ascend by capillary action. The major morphological feature that distinguishes flies from other insects is the presence of only one pair of functional wings, hence their scientific name (*di* = two, *pteron* = wing). The mesothorax has become greatly enlarged to contain the powerful flight muscles, and the pro- and metathorax are reduced. The hind wings are modified into halteres, which are small, club-like structures that function as balancing organs during flight. A few other groups of insects have also attained a similar two-winged form, such as males of scale insects (Coccoidea, Hemiptera), and a few flies have lost their wings and halteres as an adaptation to a parasitic life style (e.g., louse-fly families Nyc-

teribiidae and Hypoboscidae) or habitats such as tidal pools (e.g., the midge *Pontomyia* in the Indopacific), snow fields (e.g., the crane fly *Chionea* in Europe), islands (e.g., the crane fly *Limonia hardyana* and the dolichopodid fly *Campsicnemus* from Hawaii), or coffins (e.g., the phorid fly *Conicera*). However, those flies are still recognizable by the structure of their mouthparts, greatly enlarged mesothorax, and legs.

There is a large diversity in leg shape and structure, but all species in this order have five tarsomeres (distal segments), which are used as tactile organs. Legs can be adapted as raptorial organs in predatory forms (e.g., the fore legs of some ceratopogonids, empidids, and ephydrids) or for holding onto the female during copulation, as organs of sexual or combative display (e.g., the ornamented tarsi of some African dolichopodid flies), or as grasping organs in ectoparasites, and can bear combs of setae and brushes of hairs for grooming. The abdomen of flies has a basic number of 11 segments, the last one called the proctiger, represented by only the caudal appendages or cerci and the anus. In nematoceran and orthorrhaphan flies, the abdomen is usually longer and slender, whereas it tends to be robust in the Cyclorrhapha. The male aedeagus or intromitent organ is found on the underside between abdominal segments 9 and 10, and the female genital opening is between segments 8 and 9. Female flies do not have an ovipositor formed by valves;

instead, in the more advanced flies, the terminal segments of the abdomen, from 5 or 6 to the tip, form a "functional ovipositor" in the shape of a tapered telescopic tube. The apical portion of the male abdomen is often flexed—folded ventrally, laterally, or dorsally—and the genitalia is usually twisted 90° to 360° as an adaptation for mating and storing the genitalia when not in use.

Larvae are usually elongated and subcylindrical or fusiform, and often have transverse swollen areas called creeping welts that usually bear a transverse series of microspinules on the ventral surfaces—or on both the ventral and dorsal surfaces—of the first seven abdominal segments. Larvae always lack external wing pads and legs, although some may have fleshy tubercles called prolegs on the prothorax and some abdominal segments. Nematoceran larvae have a sclerotized head capsule and mouthparts that move in a horizontal plane. In Brachyceran larvae, also called maggots, the mandibles move in a vertical plane, the head capsule is reduced (in orthorrhaphan flies) or absent (in cyclorrhaphan flies), and the head possesses an internal cephalo-pharyngeal endoskeleton. Larvae differ in the number of spiracles or openings for their respiratory system. They may be apneustic (without openings); amphipneustic (with an anterior and a posterior pair of spiracles); or metapneustic (with only the posterior pair of spiracles). Aquatic larvae spiracles may be situated at the tip of projections of the body called siphons, which allow them to reach the atmosphere (e.g., in the rat-tailed maggot, the larva of an *Eristalis* syrphid fly, which lives in the bottom of bodies of oxygen-poor water and breaths through a very long siphon, the tail) or pierce underwater air-containing plant tissues (e.g., several mosquito larvae with sawlike siphons).

Diptera pupae have nonfunctional mandibles and one pair of wing pads. The appendages may be free from the body (as in Nematocera) or glued to the body (as in Orthorrhapha brachyceran flies), and the pupa may be free or concealed inside the puparium (as in Cyclorrhapha brachyceran flies).

Distribution

Flies are found on all continents including Antarctica, where two species of midges occur. Most families reach higher species richness in the tropics than in the more temperate areas, although Mycetophilidae, Chironomidae, and Empididae are most diverse at higher latitudes. Some species are very restricted in distribution, such as the wood-boring flies of the family Panthophthalmidae that occur only in Brazilian forests, the wingless fly *Mormotomyia hirsuta*, which is known only from a single fissure in a rock in Kenya where it feeds on bat dung, and the tabanid *Thriambeutes mesembrinoides* from the Usambara Mountains in Tanganyika.

Habitat

Larvae occur in aquatic, semiaquatic, and moist terrestrial environments, as endoparasites of other animals or as miners within plant tissues, but because their cuticle is soft and susceptible to desiccation, only a few live in dry environments.

Non-biting male midge showing plumed antennae. (Photo by Kim Taylor. Bruce Coleman, Inc. Reproduced by permission.)

Larvae of some shore flies (family Ephydridae) live in unusual habitats that would be lethal for other insects, such as hot springs and geysers where the water temperature exceeds 112°F (44.4°C) (*Ephydra brucei*), pools of crude oil (*Helaeomyia petrolei*), and ponds with very high concentrations of salt (the brine fly *Ephydra cinera*). Some bombyliid larvae live inside the nests of various bees, ants, and wasps, on which brood they feed.

The pupa normally occupies the same habitat as the larva. Pupation of most of the fully aquatic species occurs underwater, while semiaquatic species may pupate above water. The pupa may swim to the surface before the adult emerges, or it may remain on the bottom.

Adults are usually terrestrial, active in the daytime, and almost always free-living, the exception being the ectoparasitic adults of the louse-flies (families Hippoboscidae, Streblidae, and Nycteribiidae).

Behavior

Aggregations occur in many flies. Many Bibionidae and Empididae species as well as numerous nematoceran flies form aerial swarms, in which adults—usually males—hover together around some fixed object, such as a tree or bush, a hill top, a rock on a stream, a patch of sunlight in a forest, a building, or a highway (e.g., the lovebug, a bibionid fly that forms mating swarms on roads from Central America to the southern United States). These swarms usually allow males to be more conspicuous to their prospective mates. Other adults form sedentary aggregations for the purpose of mating or for seeking shelter or warmth, such as on roofs, or as the result of a mass emergence. Larvae can also form aggregations, such as leatherjackets and bibionids in the soil and the armyworms

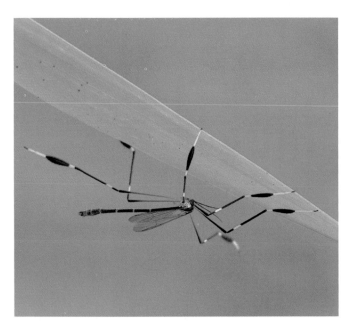

A phantom crane fly (Ptychopteridae) resting on underside of leaf in West Virginia, USA. (Photo by L. West. Bruce Coleman, Inc. Reproduced by permission.)

(*Sciara militaris*), which can be seen marching in long columns across European forests.

Mimicry is known in several flies that imitate ants (e.g., *Sepsisoma*, richardiid flies that mimic *Camponotus* ants; *Microdon mutabilis*, syrphid flies that imitate *Formica lemani* and *Myrmica scabrinodis*); bees and wasps (e.g., several conopid and tephritid fruit flies species that mimic vespid wasps); bumblebees (e.g., *Laphria* asilids that mimic *Bombus* species); and jumping spiders (e.g., fruit flies with banded wings and/or spotted abdomens). The mimicry can be in shape as well as behavior, either as a strategy to deceive the model in order to feed or lay eggs on its brood, or to deceive potential predators.

Feeding ecology and diet

Larvae are phytophagous (feeding on leaves, fruits, or roots of plants), filter organic matter, or are scrapers of algae, predators, parasitoids and saprophagous feeders of decaying organic matter including vegetables, dead animals, or dung. Endoparasitic larvae include those in the family Tachinidae, which parasitize other insects, particularly centipedes and spiders; those in the family Pipunculidae, which parasitize the larvae of cicadas and leafhoppers; and some species of bee flies (family Bombyliidae), which develop in the eggs or larvae of bees and wasps, other flies, beetles, and butterflies.

Adults typically consume liquid food such as nectar and other plant exudates or decomposing organic matter, or they prey on other insects or mollusks; adults of some species feed on little or nothing at all. Females of some groups may take blood meals from vertebrates (e.g., the "biting" mosquitoes, no-see-ums, black flies, and horse flies) as a prerequisite to oviposition.

Reproductive biology

Flies undergo complete metamorphosis, passing usually through four life stages: egg, larva, pupa, and adult. The basic number of larval instars is 4 to 9 for the lower Diptera (Nematocera), and only three in many Cyclorrhapha, as the fourth larval stage develops within the puparium. The length of the egg stage is often brief, encompassing a few days or weeks, and that of the larval stage is variable, ranging from a few days in larvae feeding on short-term resources (such as rotting meat) to more than two years in some larvae that develop in cold and moist habitats.

Mating in flies is often preceded by an initial courtship in which the male or both sexes perform a chain of actions involving body movements or vibrations of the wings. In some dance-flies (family Empididae), males offer a nuptial gift to the female as part of the courtship, such as an insect wrapped up in silk, or just a hollow balloon of silk. Mating usually starts with an initial coupling position of both sexes facing in the same direction, and ends in a final mating position with male and female facing in opposite directions, in a tail-to-tail orientation. Some species are parthenogenetic (e.g., some *Drosophila*, *Lonchoptera*, *Sciara*, and *Diaphorus* flies).

Egg laying is related to the habitat of the larvae. Fruit fly females have a long needle-like ovipositor with which they pierce the skin of the fruit in which the larvae live and feed; parasitic flies show an extraordinary array of strategies to get eggs into their hosts, including laying them on the host's surface, inserting them into the host with a piercing ovipositor, pasting them close or onto the host's nest, hiding them among the host's food, or laying them on another animal that will eventually come in contact with the host, such as a tick or mite. Flies with aquatic larvae oviposit on the water or on rocks or vegetation overlaying the water, and eggs can be laid singly or in the form of rafts.

Conservation status

Several flies are threatened not by direct exploitation but by loss or degradation of their habitats; they are at risk because their ecosystems are at risk. The IUCN Red List includes seven species from this order; three as Extinct, two as Endangered, one as Critically Endangered, and one as Vulnerable. The three Extinct species are the Volutine stoneyian tabanid fly, *Stonemyia volutina* (Tabanidae), from the continental United States; the Ko'okay spurwing long-legged fly, *Campsicnemus mirabilis* (family Dolichopodidae), from the Hawaiian Islands; and Lanai pomace fly, *Drosophila lanaiensis* (Drosophilidae), also from Hawaii. The sugarfoot moth fly, *Nemapalpus nearcticus* (family Psychodidae), from the United States, and the giant torrent midge, *Edwardsina gigantea* (family Blepharoceridae), from Australia, are Endangered. The Tasmanian torrent midge, *Edwardsina tasmaniensis* (family Blepharoceridae), is Critically Endangered due to the establishment of a hydroelectric dam in the Launceston Cataract Gorge at the Dennison River in Australia. Finally, Belkin's dune tabanid fly, *Brennania belkini* (Tabanidae), from Mexico and United States, is considered Vulnerable because of the progressive destruction of its habitat by urban development.

Although not listed by the IUCN, the flower-loving fly, *Rhaphiomidas terminatus* (family Apioceridae), from California, is at risk of extinction. *Rhaphiomidas terminatus* consists of two subspecies, the Delhi sands flower-loving fly, *R. terminatus abdominalis*, and the El Segundo flower-loving fly, *R. terminatus terminatus*. The latter, confined to the El Segundo sand dunes and portions of the sandy alluvial plain of the Los Angeles River—the last remains of which were eliminated by construction for the Los Angeles International Airport in the 1960s—was thought to be extinct but has recently been rediscovered. The Delhi sands flower-loving fly is a rare endemic subspecies restricted to the Delhi series of sand dunes, 98% of which have been converted to residential, agriculture, and commercial uses. The subspecies is listed as Endangered by the United States Fish and Wildlife Service.

Significance to humans

Flies are the most important arthropod vectors of disease in humans and other animals. For example, malaria is believed to have killed more human beings than any other known disease and is still a major cause of illness in many tropical countries. Also, diseases transmitted to humans and livestock by tsetse flies were the main obstacle to European colonization of North Africa. Deer or horse flies can transmit bacterial diseases such as tularemia and anthrax, filariasis (via the Loa Loa worm), and trypanosomiasis (causing surra in caribou, horses, and camels). Mosquitoes are vectors of filariasis (elephantiasis), malaria (caused by four species of the protozoan *Plasmodium*), and viruses including yellow fever and dengue fever; black flies can transmit onchocercosis (via worms); tsetse flies are vectors of trypanosomiasis (causing the fatal disease nagana in cattle and sleeping sickness in humans); and sand flies are vectors of leishmaniasis (via the *Leishmania* protozoan).

In addition to serving as vectors for diseases, flies can cause health problems themselves. Three main dipteran families—Oestridae (bot flies), Sarcophagidae (flesh flies), and Calliphoridae (bottle flies) cause economically important myiasis in livestock and also occasionally in humans. Myiasis is the infestation of live vertebrate animals by dipterous larvae, which at least for a certain period feed on the host's dead or living tissue, liquid body substances, or ingested food.

Still others are nuisance pests or carry filth, such as the eye gnats (genus *Hippelates*, family Chloropidae) and face flies (*Musca autumnalis*, family Muscidae), attracted to the secretions of the eyes of vertebrates; or the house flies (*Musca domestica*, family Muscidae), little-house flies (*Fannia*, family Muscidae), and latrine flies (*Chrysomyia*, family Calliphoridae), which breed in excrement and garbage.

Fruit flies (family Tephritidae) are among the most destructive agricultural pests in the world, eating their way through citrus crops and other fruit and vegetable crops at an alarming rate and forcing food and agriculture agencies to spend millions of dollars on control and management measures. Other agricultural pests include the gall gnats (family Cecidomyiidae), leaf miner flies (family Agromyzidae), and root miner flies (family Anthomyiidae).

Besides their essential roles in our ecosystems, including serving as food items for numerous animals and as pollinators, flies are of some direct benefit to humans. Some are important as biological control agents of weeds and other insects; as indicators of water quality (e.g., midge larvae known as blood worms are indicators of polluted environments); as experimental animals (e.g., much of our knowledge of animal genetics and development has been acquired using the fruit fly, *Drosophila melanogaster* as an experimental subject); and in forensic investigations to establish the time of death, whether the corpse has been moved after death, and the cause of death (e.g., several larvae of families Trichoceridae, Stratiomyidae, Phoridae, Phanidae, Muscidae, Calliphoridae, Sarcophagidae, and others feed on carrion and flesh in different degrees of decomposition, in different situations, and at different times of the year).

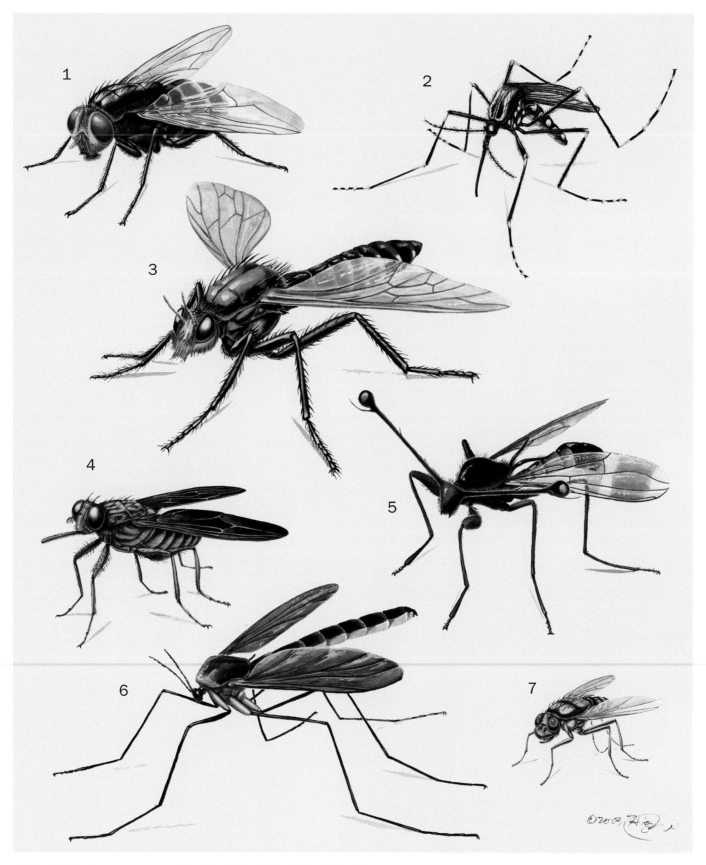

1. New World primary screwworm (*Cochliomyia hominivorax*); 2. Yellow fever mosquito (*Aedes aegypti*); 3. *Dasypogon diadema*; 4. Tsetse fly (*Glossina palpalis*); 5. *Cyrtodiopsis dalmanni*; 6. New Zealand glowworm (*Arachnocampa luminosa*); 7. Petroleum fly (*Helaeomyia petrolei*). (Illustration by Jonathan Higgins)

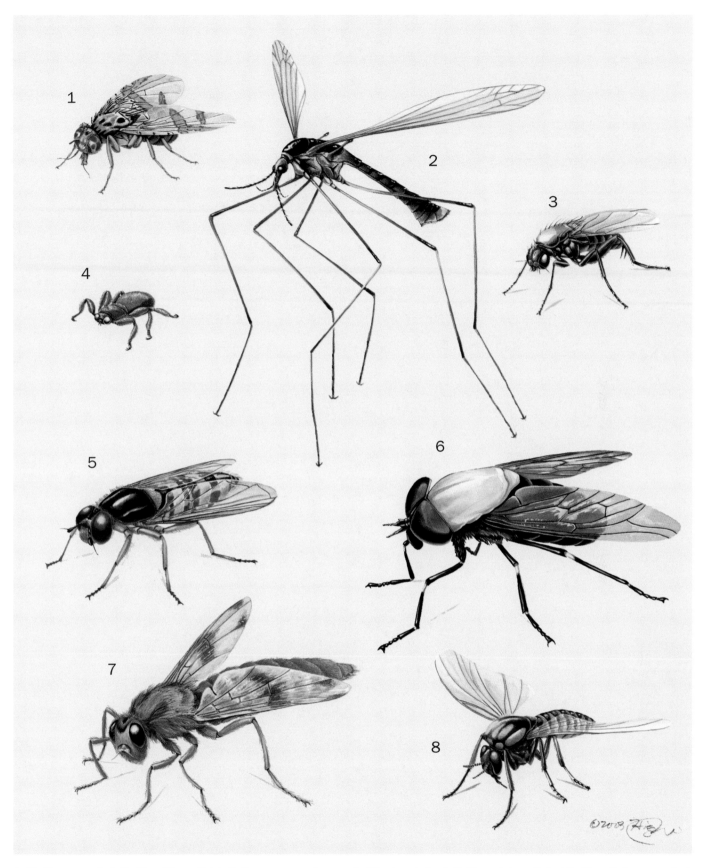

1. Mediterranean fruit fly (*Ceratitis capitata*); 2. European marsh crane fly (*Tipula paludosa*); 3. Fire ant decapitating fly (*Pseudacteon curvatus*); 4. Spider bat fly (*Basilia falcozi*); 5. Chevroned hover fly (*Allograpta obliqua*); 6. Big black horse fly (*Tabanus punctifer*); 7. Horse bot fly (*Gasterophilus intestinalis*); 8. Dawson River black fly (*Austrosimulium pestilens*). (Illustration by Jonathan Higgins)

Species accounts

No common name
Dasypogon diadema

FAMILY
Asilidae

TAXONOMY
Asilus diadema Fabricius, 1781, Italy.

OTHER COMMON NAMES
None known.

PHYSICAL CHARACTERISTICS
Adults are large (about 0.6 in [15 mm]) and sexually dimorphic; males have black abdomen and blackish wings, and females have black abdomen with red markings and brownish wings and are usually larger. Eggs are white, long, oval, and are laid in groups of 1–6 in cocoons made with sand grains glued together and covered inside with a silky lining. Larvae are white to yellowish, elongate, cylindrical, and tapering at each end. Pupae have transverse rows of elongate spines or bristles arising dorsolaterally, and abdominal segment 9 has 1–4 pairs of strongly sclerotized terminal caudal hooks.

DISTRIBUTION
Central region of Mediterranean Europe.

HABITAT
Open, dry, and sandy areas such as steppes, partly eroded hill slopes, sandy hollows, weedy grass plots, dunes, dry meadows, pastures, and olive groves.

BEHAVIOR
Males continuously look for females. Females are more territorial, remaining longer at their foraging positions perched upwards on stems, or sometimes on the ground. They only fly if they see potential prey or if males disturb them. During periods of inactivity, they hang on stems of flowers and grasses or under leaves.

FEEDING ECOLOGY AND DIET
Larvae feed on larvae of scarab beetles (order Coleoptera, family Scarabeidae), and adults mostly on hymenopterans, including honey bees (*Apis mellifera*). With their long, thin legs, the strong spur at the apex of the foretibiae, and long proboscis, they are well adapted for subduing wasps and bees without being stung. After catching a prey in flight, they look for an appropriate perching site before sucking the prey's contents.

REPRODUCTIVE BIOLOGY
Mating is initiated after a short struggle, when the male grasps the female's ovipositor with his genital claspers. Final mating position is end-to-end. The female then flies with the male in copula looking for a suitable place to land. After mating, the female lays the eggs in clutches in the soil, protected inside sand cocoons. The four larval instars and the pupa live in the soil.

CONSERVATION STATUS
Not threatened.

SIGNIFICANCE TO HUMANS
None known. ◆

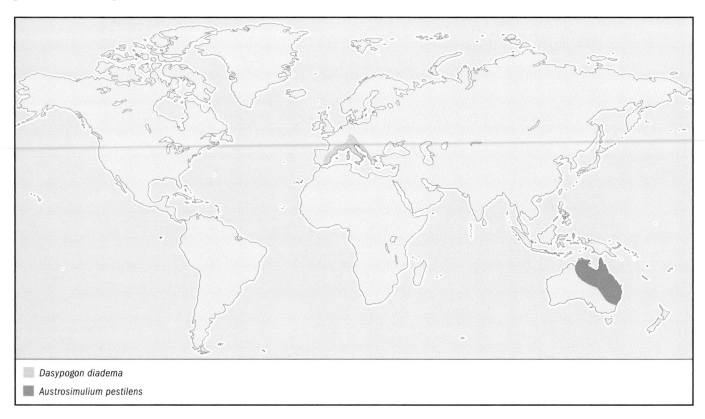

Dasypogon diadema
Austrosimulium pestilens

New World primary screwworm
Cochliomyia hominivorax

FAMILY
Calliphoridae

TAXONOMY
Lucilia hominivorax Coquerel, 1858, French Guiana.

OTHER COMMON NAMES
English: American primary screwworm; Spanish: Gusano barrenador del ganado.

PHYSICAL CHARACTERISTICS
Adults are 0.4–0.6 in (10–15 mm) long, metallic blue to bluish green, with three longitudinal black stripes on the thorax. Face and eyes are orange-brown; palps are short and threadlike, and antennae are plumose to their tips. The larvae are whitish, about 0.4–0.5 mm (10–12 in) long, and have large posterior spiracles, each with three straight slitlike apertures surrounded in part by a prominent dark-pigmented peritreme. They have a ring of spinules on each segment, which gives them the appearance of a screw.

DISTRIBUTION
Native to tropical and subtropical areas of the Americas. The screwworm has been eradicated from the southern United States and Mexico and most of Central America. There is a seasonal spread of the screwworm into the temperate regions of Argentina, Uruguay, and Paraguay during spring and summer, and it is occasionally reported from Chile and southern Argentina from imported animals. It was imported to Libya in North Africa around 1988 and eradicated three years later.

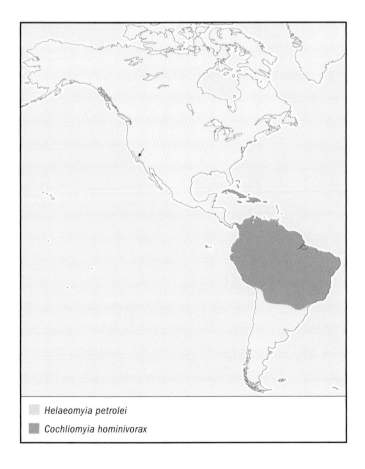

  Helaeomyia petrolei
  Cochliomyia hominivorax

HABITAT
Tropical and subtropical areas; cannot overwinter where soil freezes.

BEHAVIOR
Mature larvae are negatively phototropic (they move away from light), and this behavior facilitates burrowing in the soil to pupate. Adult flies disperse to find mates and oviposition sites. In tropical environments, females tend to disperse only 6.2–12.4 mi (10–20 km) when there is a high density of animals; in arid environments with lower densities of animals, screwworm flies have traveled as far as 186 mi (300 km). In more arid areas, screwworm flies travel along water courses, and in mountainous areas, along the course of valleys, where the climate is warmer and moisture and animal density higher.

FEEDING ECOLOGY AND DIET
Larvae are obligate ectoparasites (external parasites that cannot complete their cycle outside of their host) of mammals and will infest warm-blooded livestock, wildlife, and humans; they are unable to breed in carrion. A cut, abrasion, or other wound in the skin is required for the larvae to invade the host tissue, where they feed on the living tissue. Adults imbibe water and nectar and fluids of exposed wounds.

REPRODUCTIVE BIOLOGY
Individual females lay batches of 200–300 eggs in compact masses on the skin at the edges of fresh wounds or areas where there is a bloody or mucous discharge. Even wounds the size of a tick bite are sufficient to attract oviposition. The eggs hatch in 14–18 hours. Within a day of hatching, the maggots start to feed, burrowing into the living tissue, where they bunch together to feed with their posterior spiracles exposed. In 5–9 days, larvae are fully developed and leave the host to pupate in the soil for about seven days. After 3–5 days of emergence, adult flies are ready to mate. Male screwworm flies will mate several times, while females usually mate only once. Under ideal conditions, the life cycle is completed in 24 days.

CONSERVATION STATUS
Not threatened.

SIGNIFICANCE TO HUMANS
The screwworm is an economically important pest of domestic cattle. Wounds infested by screwworm larvae become increasingly attractive to fly species whose larvae breed on dead organic material, and also to gravid females of the primary screwworm, thus making the syndrome self-perpetuating in endemic areas, where the result is usually the death of the host. ◆

Yellow fever mosquito
Aedes aegypti

FAMILY
Culicidae

TAXONOMY
Culex aegypti Linnaeus, 1762, Egypt.

OTHER COMMON NAMES
None known.

PHYSICAL CHARACTERISTICS
Small in comparison to other mosquitoes, between 0.1–0.15 in (3–4 mm) long, black with a white lyre-shaped patch of scales

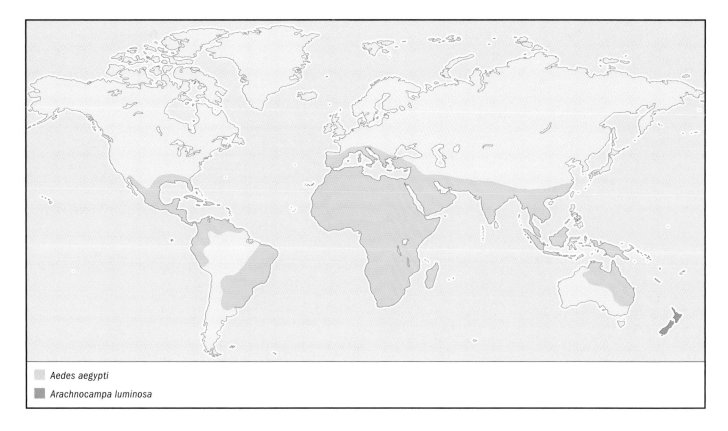

Aedes aegypti

Arachnocampa luminosa

on its thorax and white rings on the legs; wings translucent and bordered with scales. The eggs are white when freshly laid but soon turn black.

DISTRIBUTION
Originally from Africa but has spread by human activity to all tropical and subtropical regions of the world between 45°N and 35°S latitudes.

HABITAT
The yellow fever mosquito is peridomestic, breeding in or near human dwellings in hot, humid areas, and is particularly abundant in towns and cities. Females are early morning or late afternoon feeders but can take a blood meal at night under artificial illumination. Human blood is favored over that of other animals, with preference for the ankle area. Adults usually rest in dark or dimly lit closets, cabinets, or cupboards during the day. Larvae are aquatic.

BEHAVIOR
At rest, the insect turns up its hind legs in a curved fashion and usually cleans them by rubbing one against the other, or crosses them and alternately raises and lowers them.

FEEDING ECOLOGY AND DIET
Males of mosquitoes do not bite humans or animals of any species; they live on fruit juices. By contrast, the females need a blood meal in order to induce the maturation of their eggs. The larvae feed on plankton they filter from the water.

REPRODUCTIVE BIOLOGY
Females lay eggs singly on damp surfaces or clean water held in small containers such as old automobile tires, drums, jars, natural water-retaining cavities in plants, and tree holes. The eggs can resist desiccation for up to one year, and they will hatch when flooded by water. After a few weeks from hatching,

or even 4–10 days if it is warm enough, the larvae reach the pupa stage, which is usually of short duration. Pupae rise to the surface of the water, where the top of the pupal case opens and the new adult emerges. The species is summer-active in higher latitudes and active all year in tropical areas. Adults are killed by temperatures below freezing and do not survive well at temperatures below 41°F (5°C). On average, females live up to a month, but males die sooner.

CONSERVATION STATUS
Not threatened.

SIGNIFICANCE TO HUMANS
Aedes aegypti is the most important domestic vector of the viruses that cause human dengue and urban yellow fever, and for the chikungunya virus in Asia. ◆

No common name
Cyrtodiopsis dalmanni

FAMILY
Diopsidae

TAXONOMY
Diopsis dalmanni Wiedemann, 1830, Java, Indonesia.

OTHER COMMON NAMES
None known.

PHYSICAL CHARACTERISTICS
The most striking characteristic of flies in this family is the position of the eyes on the tip of lateral stalks, which gives them the common name "stalk-eyed flies." In males of *Cyrtodiopsis*

dalmanni, the length from one compound eye to the other is nearly the length of the body; females have shorter eye-stalks.

DISTRIBUTION
Widespread in Southeast Asia.

HABITAT
Found among forest-floor detritus in damp and well-shaded areas near streams in primary or secondary forests.

BEHAVIOR
The long and rigid eye-stalks are used for improved visual orientation and, in males, for sexual competition. The forelegs, adapted for gripping, are used in combats with rivals. The extreme eye-stalk structure requires special foreleg movements to clean the eyes.

FEEDING ECOLOGY AND DIET
Larvae feed on plants. Feeding habits of the adults are still unknown.

REPRODUCTIVE BIOLOGY
Adults stop grazing in the evening, and males stake out rootlets dangling from stream banks. On a disputed rootlet, males face off eye to eye, and the one with the shorter eye span backs down. Several females are found together with a long stalk-eyed male and show continuous sexual receptivity and high rates of multiple mating. Males with exaggerated eye spans can mate with as many as 24 partners in half an hour. After emergence from the puparium, stalk-eyed flies go through a pumping process for about 15 minutes to unfold the eye stalks to their full length simultaneously with the wings.

CONSERVATION STATUS
Not threatened.

SIGNIFICANCE TO HUMANS
None known. ◆

Petroleum fly
Helaeomyia petrolei

FAMILY
Ephydridae

TAXONOMY
Psilopa petrolei Coquillet, 1899, California.

OTHER COMMON NAMES
English: Oil fly.

PHYSICAL CHARACTERISTICS
Adults are small (0.08 in [2 mm]), and their bodies are black and pruinose except cheeks and sides of face, which are grayish. Eyes are hairy and wings are hyaline and tinged with gray on their costal (anterior) half. Larvae are elongate, reaching a length of 0.3–0.4 in (7–10 mm). They breathe through spiracles on their posterior end, which are surrounded by four supporting fans of setae; the fans rest upon the surface of the oil and keep the spiracles above the surface.

DISTRIBUTION
California oilfields.

HABITAT
This is the only insect known from which larvae develop in seepages of crude oil.

BEHAVIOR
Larvae swim slowly, usually near or on the surface of the oil, although they can submerge for a considerable length of time. Adults remain near petroleum pools, hiding in the cracks in the soil, flying about and over the pools, and landing on the margin or on some projecting stone or stick within the pool. They can walk on the surface of the oil as long as no body part other than the tarsi comes in contact with the oil.

FEEDING ECOLOGY AND DIET
The larvae feed on dead insects that have become trapped in the oil pools.

REPRODUCTIVE BIOLOGY
Mating behavior and oviposition are still undescribed. When ready to pupate, the larva leaves the oil and pupates on grass stems on the margins of the pool.

CONSERVATION STATUS
Not threatened.

SIGNIFICANCE TO HUMANS
Crude oil is usually regarded as a very effective insecticide. Thus, petroleum flies are of interest to biotechnologists because they can provide information regarding the ability of organisms to resist the toxic effects of aromatic and petroleum compounds. ◆

Tsetse fly
Glossina palpalis

FAMILY
Glossinidae

TAXONOMY
Nemorhina palpalis Robineau-Desvoidy, 1830, Congo River.

OTHER COMMON NAMES
None known.

PHYSICAL CHARACTERISTICS
Adults are yellowish to brown, with a forward-projecting, piercing proboscis and a hatchet-shaped cell in the center of each wing. The arista arising from the third antennal segment has branched setae. Both sexes have dichoptic eyes. Larvae breathe through a pair of posterior spiracles and in the third stage via a pair of lateral lobes, which contain three air chambers and open through numerous spiracles.

DISTRIBUTION
Western Africa.

HABITAT
Local patches of dense vegetation along banks of rivers and lakes in arid terrain, and also in dense, wet, heavily forested equatorial rainforest.

BEHAVIOR
Larvae show negative phototaxis (avoiding light) and are positively thigmotactic (seeking contact with surfaces).

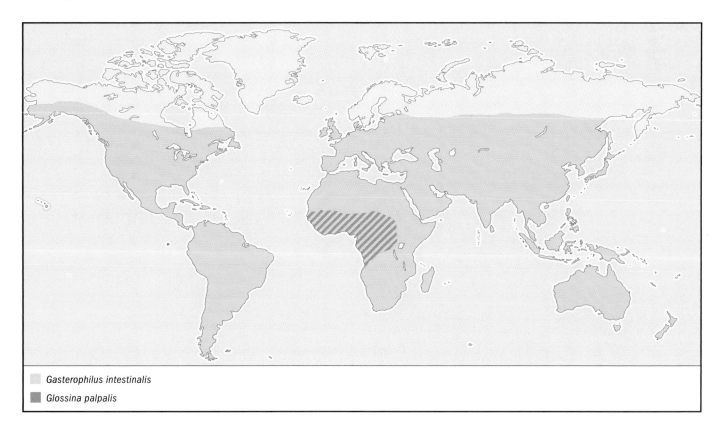

Gasterophilus intestinalis

Glossina palpalis

FEEDING ECOLOGY AND DIET
Adults feed on the blood of any vertebrate they encounter, including reptiles—especially monitor lizards and crocodiles—birds, and mammals. They are not responsive to the conventional vertebrate host odors that other tsetse flies respond to. This species generally feeds while inside dense humid forest habitats, where trailing hosts by sight is easier than by olfaction.

REPRODUCTIVE BIOLOGY
Female tsetse flies reproduce by adenotrophic viviparity, which involves the retention of a single egg that develops to the third larval stage before being deposited. The egg within the uterus hatches in 3–4 days, giving rise to the first-stage larva. The larva feeds from secretions of a pair of uterine glands from the mother. Just before depositing her larvae at the third instar, the adult female actually weighs less than her offspring. The larvae are deposited on the soil and burrow down, where they pupate for 4–5 weeks. The young adult emerges from the puparium using its ptilinum, which it also uses to move up through the soil. Having emerged, both sexes seek a host to gain a blood meal. Males are not fully fertile until several days after emergence, and females are able to mate two to three days after emergence. The first larval offspring is deposited about 9–12 days after the female emerges. Due to the length of development, tsetse flies are relatively long-lived, up to 14 weeks for females, with males having shorter lives of around 6 weeks. Hence the rate of reproduction is extremely slow compared to other dipteran species.

CONSERVATION STATUS
Not threatened.

SIGNIFICANCE TO HUMANS
This species is an important vector of West African trypanosomiasis via the protozoan *Trypanosoma brucei gambiense*, which causes nagana in horses and cattle and sleeping sickness in humans. ◆

New Zealand glowworm
Arachnocampa luminosa

FAMILY
Keroplatidae

TAXONOMY
Bolitophila luminosa Skuse, 1891, New Zealand.

OTHER COMMON NAMES
English: New Zealand fungus gnat, glowing spider bug; Maori: Titiwai.

PHYSICAL CHARACTERISTICS
All stages, except for the eggs, glow. The eggs are small and white and turn brown before larvae emerge. Eggs are coated with a sticky substance and adhere to the substrate. Larvae go through five instars and reach a body length of 1.2–1.6 in (30–40 mm). They are brown but with transparent skin, allowing bluish green light to shine through. Adults are slightly larger than a mosquito, about 0.50 in (15 mm) in length.

DISTRIBUTION
New Zealand.

HABITAT
Damp, sheltered areas with hanging surfaces, dark for a good portion of the day and protected from winds, such as caves; usually associated with streams.

BEHAVIOR
During the day larvae hide in crevices, and during the night they enter their silk tubes and hunt. Bioluminescence is used to attract food in larvae and mates in adults; the hungrier a larvae is, the brighter it glows. If no male is waiting on a female's pupa when the female is ready to emerge, the female will glow brighter and will flash its lights on and off. The flight of adults is slow, and their habits are mostly crepuscular or nocturnal. They are often found on low vegetation, under overhanging rocks and trunks, and along banks of streams.

FEEDING ECOLOGY AND DIET
Glowworm larvae are voracious hunters with large mandibles. They spin a delicate web of horizontal silk tubes from special labial glands in their heads. This web is attached to rocks and branches and suspended from the tubes; they hang from silk threads covered with globules of sticky mucous that can reach 20 in (50 cm). Insects that are attracted to the light of the larvae become caught in the sticky mucous of these "fishing lines." Vibrations are sent up the line and sensed by the larva, which then begins to swallow the line. Certain chemicals within the mucous paralyze the prey, which is finally bitten and killed by the larva. New Zealand glowworms can be cannibalistic, with the larvae often eating other larvae or even adults that happen to fly into the fishing lines. They also feed on mosquitoes, moths, stoneflies, sand flies, caddisflies, midges, ants, spiders, millipedes, and even snails. Adults have reduced mouthparts and do not feed.

REPRODUCTIVE BIOLOGY
Females ready to hatch begin to emit light while still inside the pupa. This attracts males, which land on the pupa and fight with each other in an attempt to dislodge each other from the pupa's surface. This ensures that only the strongest males mate with the females as soon as they hatch. Upon emerging from the pupa, males live for three to five days and females for one or two days. Females lay the eggs one at a time in clusters of 30–40, usually on muddy banks, and can lay as many as four clusters. The eggs hatch after three weeks, and the larvae emerge emitting light and immediately begin to build their homes. The larval stage lasts roughly 6–12 months. When it is time to pupate, the larva arranges its sticky fishing lines into a protective, circular barrier and then hangs in the middle of this circle. Pupation takes about 12 days.

CONSERVATION STATUS
Not threatened.

SIGNIFICANCE TO HUMANS
Caves with gatherings of glowworms constitute a tourist attraction. ◆

Spider bat fly
Basilia falcozi

FAMILY
Nycteribiidae

TAXONOMY
Nycteribia falcozi Musgrave, 1925, Australia.

OTHER COMMON NAMES
None known.

PHYSICAL CHARACTERISTICS
Adults lack wings and their body is compressed dorsoventrally, so that head and legs arise from the dorsal surface of the tho-

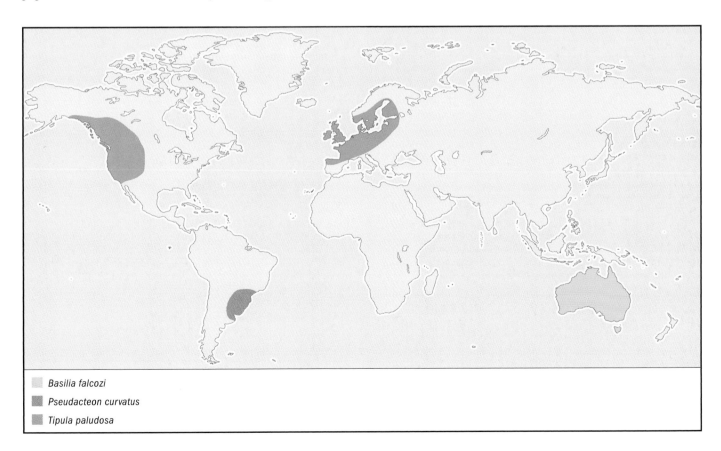

☐ *Basilia falcozi*
■ *Pseudacteon curvatus*
▨ *Tipula paludosa*

rax, giving them a spiderlike appearance. They have long legs with strong claws. The pupae are shining black and flattened.

DISTRIBUTION
Australia.

HABITAT
Ectoparasites on cave-dwelling bats of the family Vespertilionidae.

BEHAVIOR
Adults stay on host except to deposit pupae in roost.

FEEDING ECOLOGY AND DIET
Hematophagous.

REPRODUCTIVE BIOLOGY
Females deposit fully developed pupae singly, gluing them to vertical substrata (e.g., cave walls and trees) in the vicinity of the host bat roosts. This process may be repeated several times during the female's lifetime. Emergence of the adult is triggered by the warmth or contact of a roosting bat.

CONSERVATION STATUS
Not threatened.

SIGNIFICANCE TO HUMANS
None known. ◆

Horse bot fly
Gasterophilus intestinalis

FAMILY
Oestridae

TAXONOMY
Oestrus intestinalis De Geer, 1776, type locality not specified (probably Sweden).

OTHER COMMON NAMES
English: Nit fly; German: Magenfliege, Magendasselfliege.

PHYSICAL CHARACTERISTICS
Adults resemble honeybees in appearance and size (about 0.7 in [18 mm]). They have a hairy head and predominantly yellow-brown hairs dorsally, and whitish ones laterally on the thorax. The abdomen is tricolored, bearing long white hairs on the second segment, predominantly black ones on the third and fourth segments, and dark yellow hairs on the posterior parts. The light yellow eggs are attached to the hair of the host. Larvae are reddish brown and when mature usually around 0.8 in (20 mm) in length. They have two bands of coarse spines with blunt tips per segment (except in the last segment).

DISTRIBUTION
Originally from the Old World, horse bot flies today are virtually cosmopolitan, having spread with humans and their horses around the world.

HABITAT
Associated with their hosts.

BEHAVIOR
Adults are diurnal (active in the daytime) with peak activity occurring in the early afternoon in warm, sunny weather.

FEEDING ECOLOGY AND DIET
Adult mouthparts are greatly reduced; adults do not feed and rely on stored body fats. Larvae infest the digestive tract of horses and other domestic animals such as mules and donkeys, where they feed on tissue in the horse's stomach or gut lining as well as on nutrients from the horse's food. Newly hatched larvae occasionally invade human skin but die in a few days.

REPRODUCTIVE BIOLOGY
Adults are short-lived; females have a life span of one day. Mating occurs in the vicinity of horses where solitary hovering males will establish and defend a territory. They are also known to aggregate at hilltops and other elevated landmarks to mate, where males hover and aggressively pursue passing objects. Males make contact with a female on the wing, couple, and fall to the ground, where copulation is completed. Females lay eggs around the knees of walking, trotting, or standing horses; if this action induces the horses to gallop, the females will pursue them until they stop and then immediately resume oviposition. When the horse licks the area, the heat and wetness from its tongue causes the eggs to hatch. The larvae then enter the mouth and travel to the stomach, where they attach themselves with strong mouth hooks to the walls. Third-stage larvae remain attached for 9 to 11 months before exiting within the horse's feces. Pupation takes place on the ground, lasting for 3–5 weeks.

CONSERVATION STATUS
Not threatened.

SIGNIFICANCE TO HUMANS
Because of their parasitic habit in horses, larvae are more easily found than adults. They cause hide spoilage, ulceration, hemorrhage, severe debilitation, and even death in horses in cases of severe infestations. ◆

Fire ant decapitating fly
Pseudacteon curvatus

FAMILY
Phoridae

TAXONOMY
Pseudacteon curvatus Borgmeier, 1925, Brazil.

OTHER COMMON NAMES
None known.

PHYSICAL CHARACTERISTICS
Small, about the size of their host ant's head, with relatively large eyes, a humped-back thorax, and a hypodermic needle-shaped ovipositor in females. In the field, they appear as minute, fuzzy specks hovering over host ants.

DISTRIBUTION
South America from Brazil to Argentina.

HABITAT
Widely distributed in the natural range of their hosts.

BEHAVIOR
Females are attracted to disturbed mounds, mating flights, or foraging trails of their hosts. Females in attack mode hover 0.1–0.4 in (3–10 mm) above the ants and orient to their movements. Males are not attracted to the ants.

FEEDING ECOLOGY AND DIET
Adults feed on nectar, and the larva on the inner tissues of its host ant's head.

REPRODUCTIVE BIOLOGY
Males and females have multiple matings. The females lay eggs in fire ants of the *Solenopsis saevissima* species-complex. The female stays in front of a chosen worker ant and then quickly moves to the side of the ant and injects a single egg into its thorax. The larva migrates to the head capsule of the worker and passes through three instars, during which time the worker ant appears to behave normally. Just before pupation, the tissue inside the ant's head capsule is consumed, killing the ant in the process. Larval and pupal development each takes 2–3 weeks. The ant's head usually falls off, and the pupa completes its development inside the head capsule, at which point the adult fly emerges from its oral cavity. Sex in this species is apparently determined environmentally, as males are produced from smaller ant workers, whereas females are produced from larger ant workers. Adults are active all year, except during the winter months in the more temperate regions of their range.

CONSERVATION STATUS
Not threatened.

SIGNIFICANCE TO HUMANS
Possible biological control agent for red imported fire ants in the United States. Although the percentage of ants that the flies are able to parasitize and kill is low, they have a large effect on their host's competitive interactions with other ants. The presence of ant decapitating flies disrupts colony-level foraging in the ants, which retreat to their nests, leading to as much as a 50% decrease in resource acquisition. ◆

Dawson River black fly
Austrosimulium pestilens

FAMILY
Simuliidae

TAXONOMY
Austrosimulium pestilens Mackerras and Mackerras, 1948, Queensland, Australia.

OTHER COMMON NAMES
None known.

PHYSICAL CHARACTERISTICS
Adults are small, black, and humped-back, about 0.04 in (1 mm) in length. Eyes are holoptic (in contact with each other) in males and dichoptic (separated) in females. The larvae are 0.008–0.2 in (0.2–4.5 mm) in length, pale with irregular mottling. They have a pair of large cephalic fans on the head, a proleg (false leg) in the thorax, and another one at the tip of the abdomen, armed with a circlet of spines that anchors the larvae to submerged substrates. Pupae have a pair of breathing organs called spiracular gills that look like a collection of threads attached to the back of the head, which enable them to breathe while permanently submerged.

DISTRIBUTION
Queensland and northern New South Wales in Australia.

HABITAT
Larvae require oxygenated, running waters, and they live in both stony and weedy streams. Adults appear along the larval river or streams courses and the surrounding countryside.

BEHAVIOR
After mating, females look for a blood meal in open spaces, seldom entering buildings. Following the blood meal, females rest in trees and shrubs prior to oviposition.

FEEDING ECOLOGY AND DIET
Larvae filter drifting food with their fan-like mouthparts. Adult females require a blood meal before the first batch of eggs is laid. They are pool feeders, imbibing blood from a droplet created from cutting the skin.

REPRODUCTIVE BIOLOGY
Eggs are laid into fresh running water and sink to the bottom, where they can remain quiescent for at least five years on a damp stream bed, until they hatch with the next flood. Larvae construct well-formed silken cocoons when ready to pupate; the pupal stage lasts about two days. Adults usually emerge during daylight hours and float to the surface in a bubble of air. Males tend to emerge before females and form highly visible mating swarms associated with trees close to the water. Females enter these swarms, are captured, and mate, after which they seek a blood meal. Egg-laying swarms are diffuse and occur throughout the daylight hours. Females fly low over the stream and dip their abdomens into the water surface. Adult flies may live up to three weeks.

CONSERVATION STATUS
Not threatened.

SIGNIFICANCE TO HUMANS
Considered a nuisance pest in Australia; females bite cattle, dingoes, goats, horses, humans, and marsupials such as kangaroos and wallabies. ◆

Chevroned hover fly
Allograpta obliqua

FAMILY
Syrphidae

TAXONOMY
Scaeva obliqua Say, 1823, United Sates.

OTHER COMMON NAMES
None known.

PHYSICAL CHARACTERISTICS
Eggs are white, elongate, and oval. The mature larvae are 0.35 in (8–9 mm) in length, elongate and flattened on the dorsum, green with two narrow whitish longitudinal stripes and with transversely wrinkled tegument-bearing papillae. They have mouth hooks on the anterior end of the tapering head, and two posterior respiratory tubes fused together. The pupae are green, darkening at maturity. Adults are 0.25 in (6–7 mm) in length. They have a yellow face, yellow thoracic stripes, and four longitudinal, oblique, yellow stripes or spots on the fourth and fifth abdominal segments. Eyes are holoptic (in contact with one another) in males and dichoptic (separated) in females.

DISTRIBUTION
Canada south to Argentina.

Allograpta obliqua

Tabanus punctifer

Big black horse fly

Tabanus punctifer

FAMILY
Tabanidae

TAXONOMY
Tabanus punctifer Osten Sacken, 1876, Utah.

OTHER COMMON NAMES
English: Klegs, green heads.

PHYSICAL CHARACTERISTICS
Adults are stout and broad-headed, 0.35–1.1 in (9–28 mm) in length, with bulging and brightly colored eyes, gray thorax, black abdomen, and blackish wings. Larvae are cylindrical and have a longitudinally striated tegument.

DISTRIBUTION
Canada from British Columbia south to California, encompassing the western United States from Kansas south to Texas.

HABITAT
Adults are especially common around ponds, streams, and marshes, where the larvae live in shallow water or moist soil.

BEHAVIOR
Adults land stealthily on exposed skin in order to feed.

FEEDING ECOLOGY AND DIET
Adults feed mainly on nectar and flower pollen, and females need a blood meal before they can lay eggs. They have mouthparts adapted for tearing and lapping, and they suck blood mainly from livestock but also from humans. Larvae are predacious, feeding on insect larvae, snails, and earthworms.

REPRODUCTIVE BIOLOGY
Eggs are laid in 3–4 layer masses of 100–1,000 eggs each, covered with a jellylike material, on leaves, rocks, or debris overhanging water or on moist areas. Upon hatching, the larvae fall into the water or onto moist soil, where they undergo 3–4 instars. When ready to pupate, they move to the margin of the pool or drier areas of their habitat.

CONSERVATION STATUS
Not threatened.

SIGNIFICANCE TO HUMANS
Considered a nuisance pest for horses and mules because of the painful bites of females, which occasionally also bite humans. ◆

HABITAT
Found in areas with flowering plants sheltered from wind.

BEHAVIOR
Adults are excellent fliers and can hover and fly backward.

FEEDING ECOLOGY AND DIET
Adults often visit flowers for nectar or may be seen around aphid colonies feeding on honeydew secreted by the aphids. Larvae prey on aphids, but if there are no aphids, they can subsist on plant materials such as pollen.

REPRODUCTIVE BIOLOGY
Adults occur throughout the year in temperate areas, and during spring and summer in the cooler areas of their range. The eggs are laid singly on the surface of a leaf or twig that bears aphids. They hatch in 2–8 days. The larval stage takes 5–20 days, and larvae fasten themselves to a leaf or twig when ready to pupate. The pupal stage takes 8–33 days.

CONSERVATION STATUS
Not threatened.

SIGNIFICANCE TO HUMANS
This flower fly is considered to be beneficial to humans, because adults are agents in the cross pollination of some plants, and the larvae are predators of aphids that attack citrus, subtropical fruit trees, grains, corn, alfalfa, cotton, grapes, lettuce and other vegetables, and ornamentals. When larval populations are high, they may consume 70–100% of the aphid populations. ◆

Mediterranean fruit fly

Ceratitis capitata

FAMILY
Tephritidae

TAXONOMY
Trypeta capitata Wiedemann, 1824, East Indies (probably in error).

OTHER COMMON NAMES
None known.

PHYSICAL CHARACTERISTICS
Adults range from 0.14–0.2 in (3.5–5 mm) in length. The wings are semiopaque, broad, and patterned with yellow, and the eyes

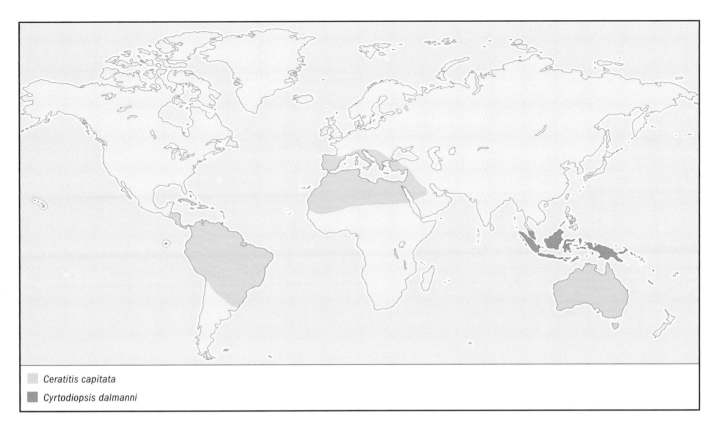

Ceratitis capitata

Cyrtodiopsis dalmanni

are iridescent and multicolored. Males are distinguished by a pair of spatulate projections on the frons, and females by a prominent ovipositor. Larvae are white and cylindrical, with a narrowed anterior end and flattened caudal end, and reach 0.3–0.35 in (7–9 mm) in length. Pupae are cylindrical, approximately 0.12 in (3 mm) in length, and dark reddish brown.

DISTRIBUTION
Native to Africa. Within the last 100 years, has spread throughout most of the world, from Portugal, Spain, Italy, Greece, Jordan, Turkey, parts of Saudi Arabia and most countries along the North African coast as far away as the Americas, the Hawaiian islands, and Australia, mainly due to transportation of infested fruit. It was eradicated from Mexico using the Sterile Insect Technique, which has also been employed effectively in Guatemala, Chile, California, and Florida.

HABITAT
Found wherever appropriate host trees (including citrus, peach, and guava, among many others) grow.

BEHAVIOR
Adults can fly only short distances, but winds may carry them a mile or more away.

FEEDING ECOLOGY AND DIET
Larvae feed within the flesh of fruits where eggs are laid, and adults on exposed sweet substances such as fruit, honeydew, or plant sap.

REPRODUCTIVE BIOLOGY
Eggs are laid beneath the skin of a suitable fruit in batches of 1–10. Females will lay approximately 300 or more eggs during their lifetime; these hatch into first-instar larvae after 2–3 days. The larvae live and feed within the host fruit, undergoing two molts in 6–10 days. When ready to pupate, the third-instar lar-

vae emerge from the fruit and drop to the ground, where they pupate in soil 1–2 in (2.5–5 cm) below the surface. The adults emerge from the pupa approximately 10 days later. They exhibit a "lek" mating system, in which males settle nonrandomly on particular host or nonhost trees and defend individual leaves as mating territories. While on territory, males emit a pheromone attractive to females and, following female arrival, perform a brief courtship display involving wing and head movements before mounting the female.

CONSERVATION STATUS
Not threatened.

SIGNIFICANCE TO HUMANS
The mediterranean fruit fly is a major agricultural pest in temperate and subtropical regions worldwide, attacking over 200 varieties of cultivated fruit crops. ◆

European marsh crane fly
Tipula paludosa

FAMILY
Tipulidae

TAXONOMY
Tipula paludosa Meigen, 1830, Europe.

OTHER COMMON NAMES
English: Leatherjacket (larvae), marsh crane fly; German: Lästlinge Wiesenschnake; Italian: Tipula dei prati.

PHYSICAL CHARACTERISTICS
Adults resemble giant mosquitoes and have a grayish brown body, about 1 in (2.5 cm) in length, two narrow wings, and

very long (0.7–0.9 in [17–25 mm]) brown legs. Larvae are gray, 1.1 in (30 mm) in length, and are known as leatherjackets because of their tough leatherlike skin. Pupae are brown, spiny, and about 1.3 in (33 mm) in length.

DISTRIBUTION
Native to northern Europe; introduced into western Canada and United States.

HABITAT
Found in wet areas of lawns, pastures, fields of forage crops, and grassy banks of drainage ditches, in regions of mild winters, cool summers, and rainfall averaging about 23.5 in (600 mm).

BEHAVIOR
Adults are weak fliers. They are attracted to lights and may enter houses and buildings.

FEEDING ECOLOGY AND DIET
Young larvae feed on humus and rotting vegetable matter, germinating cereal seeds, roots of Gramineae (grass), and the collar area of some young aerial plants.

REPRODUCTIVE BIOLOGY
The marsh crane fly completes one generation per year. Adults are abundant in late August and early September, and each female can lay up to 280 black, shiny eggs, mainly at night, from mid-July to late September. Eggs are laid on the soil surface or at depths of less than 0.4 in (1 cm), and hatch 11–15 days following oviposition. Larvae are found in the upper 1.1 in (30 mm) of soil, throughout the fall and during warm periods in winter. They grow rapidly in spring and reach their full length of about 1.6 in (40 mm) by April or May, pupating approximately in mid-July. Pupae remain underground for about two weeks before working their way to the surface, where the empty pupal case is often left protruding from the soil by the emerging adult. Adults emerge after sunset and mate immediately. Males live about 7 days; females 4–5.

CONSERVATION STATUS
Not threatened.

SIGNIFICANCE TO HUMANS
Of economic importance due to larvae, which can strip the root hairs and girdle the stems of bareroot stock in commercial tree nurseries. ◆

Resources

Books

Colless, D. H., and J. F. McAlpine. "Diptera." In *The Insects of Australia: A Textbook for Students and Research Workers*, vol. 2 (CSIRO), 2nd ed. Carlton, Australia: Melbourne University Press, 1991.

Evenhuis, N. L. *Catalogue of the Fossil Flies of the World (Insecta: Diptera)*. Leiden, Netherlands: Backhuys Publishers, 1994.

———. *Litteratura Taxonomica Dipterorum (1758–1930)*. 2 vols. Leiden, Netherlands: Backhuys Publishers, 1997.

Foote, B. A., F. C. Thompson, G. A. Dahlem, D. S. Dennis, T. A. Stasny, and H. J. Teskey. "Order Diptera." In *Immature Insects*, vol. 2, edited by F. W. Stehr. Dubuque, IA: Kendall/Hunt Publishing, 1991.

Haupt, J., and H. Haupt. *Fliegen und Mücken: Beobachtung, Lebensweise*. Augsburg, Germany: Naturbuch-Verlag, 1998.

McAlpine, J. F., et al., eds. *Manual of Nearctic Diptera*. 3 vols. Ottawa, Canada: Research Branch, Agriculture Canada, 1981–1989.

Oldroyd, H. *The Natural History of Flies*. New York: W.W. Norton and Company, 1965.

Organizations

Arbeitskreis Diptera. Dr. Frank Menzel (Leiter), ZALF e.V., Deutsches Entomologisches Institut, Postfach 100238, Eberswalde, D-16202 Germany. Phone: 03334-589820. Fax: 03334-212379. E-mail: menzel@zalf.de Web site: <http://www.ak-diptera.de>

Dipterists' Forum. Liz Howe, Ger-y-Parc, Marianglas, Benllech, Gwynedd, LL74 8 NS United Kingdom. Web site: <http://www.dipteristsforum.org.uk/>

North American Dipterists' Society. E-mail: aborkent@jetstream.net

The Malloch Society. Graham Rotheray, Research Coordinator, Royal Museums of Scotland, Chamber Street, Edinburgh, EH1 1JF United Kingdom. E-mail: ger@nms.ac.uk

Other

"BIOSIS." Resource guide: Diptera [May 28, 2003]. <http://www.biosis.org/zrdocs/zoolinfo/grp_dipt.htm>.

"The Diptera Site." Systematic Entomology Laboratory, ARS, USDA: Department of Systematic Biology, National Museum of Natural History [May 28, 2003]. <http://www.sel.barc.usda.gov/diptera/diptera.htm>.

"Diptera. Tree of Life Web Project." 2001 [May 28, 2003]. <http://tolweb.org/tree?group=Diptera&contgroup=Endopterygota>.

Web version of the "Catalog of the Diptera of the Australasian and Oceanian Regions." September 23, 1999 [May 28, 2003]. <http://hbs.bishopmuseum.org/aocat/aocathome.html>.

Natalia von Ellenrieder, PhD

Trichoptera
(Caddisflies)

Class Insecta

Order Trichoptera

Number of families 45

Photo: A colorful caddisfly (*Chimara albomaculata* family Philoptomatidae) rests on a leaf in the rainforest of eastern Puerto Rico. (Photo by Rosser W. Garrison. Reproduced by permission.)

Evolution and systematics

The caddisflies that make up the order Trichoptera are small to medium-sized insects that resemble moths (order Lepidoptera) in appearance. Like the Lepidoptera, their sister group, members of the Trichoptera are holometabolous, meaning that they undergo a complete metamorphosis. Members of the two orders have similar structures, and the larvae of both orders spin silk from labial glands. However, there are characteristics that also separate these groups. For example, scales are present on the wings, body, and legs of adult Lepidopterans, but they are not present on adult Trichopterans.

The Trichoptera includes 7,000 species in two suborders: the Annulipalpia, or net spinners and retreat makers, which contains two superfamilies and nine families; and the Integripalpia, or tube case makers, which contains four superfamilies and 32 families. In addition, there are three additional superfamilies that have not been placed into suborders: the Rhyacophiloidea, or primitive caddisflies, containing two families; the Glossosomatoidea, or turtle case makers, containing one family; and the Hydroptiloidea, or purse case makers, also containing one family. This taxonomic division is justified by the deep differences in postembryonic development and larval morphology, among other aspects.

The oldest Trichoptera fossils are from Upper and Middle Triassic; they belong to the Necrotaulidae, a family that became extinct during Cretaceous. The current superfamilies appeared during the Jurassic.

Physical characteristics

Caddisfly larvae resemble caterpillars, with a heavily sclerotized (hardened) head and strong mandibles. The thorax has three pairs of legs; dorsal plates appear on the first or on all three thoracic segments. The abdomen is membranous, sometimes with thread-like gills; the last segment has a pair of anal prolegs. Some larvae (e.g., Annulipalpia and Hydrobiosidae) are free-living and have a lengthened body and long anal prolegs. Other larvae (e.g., Integripalpia) are slow, with plump abdomens, and construct portable cases inside which they live and to which they take hold with their anal prolegs. The larvae of the family Glossosomatidae are of an intermediate type; they make portable cases, and the distal half of the anal prolegs is free from the abdomen. The prolegs are extended through the opening of the case to stabilize the larva on the substrate. The larvae of the family Hydroptilidae are miniscule and similar to free-living larvae in the first larval instar. In the last larval instar, they resemble the case makers as their size increases disproportionately, especially the abdomen, and they spin portable purse-shaped cases.

Caddisfly adults are not as diverse as larvae. They have a small, moth-like appearance, most with very long and slender antennae and reduced mouthparts. Setal warts are present on the head and thorax of adults. The wings are large, covered with hair, and rarely have scales. The wings are held tent-like at rest, and usually the antennae are placed together and directed forward; in most species, the antennae are longer than the body. Adult flight in most species is clumsy and slow. Adults range from 0.048 in (1.2 mm) to 1.76 in (40 mm) in length. Most have a drab coloration, although some brilliantly colored forms are known.

Distribution

Caddisflies are cosmopolitan, although some families are restricted to the Holarctic and Gondwanan regions. No individual species are cosmopolitan in distribution. Most species are restricted in distribution, with high numbers of endemics for several countries.

Caddisfly larvae eating a fish. (Photo by A. Blank. Bruce Coleman, Inc. Reproduced by permission.)

Habitat

Most larvae and pupae live in freshwater, although there are some terrestrial (e.g., *Enoicyla pusilla* [Leptoceridae]) and marine (e.g., *Philansius plebeius* [Chathauiidae]) forms. Adults are generally nocturnal, hiding in cool, moist environments along river banks during the day and becoming active at night. Members of some species are active during the day.

Most freshwater individuals live in cold, unpolluted, well-oxygenated rivers and streams, although some families are adapted to warm, slow-moving waters. The larvae are very sensitive to water temperature, dissolved oxygen, speed of current, illumination, size of silt, and chemical substances. Each species occupies a particular microhabitat in the water body, and several species may coexist in one place without competing. As a river's qualities change from the headwaters to the mouth, there is a correlated replacement of species. Larvae use diverse elements of their natural habitat to build their cases. Those that build nets stick them to the rocks of the riverbed.

Behavior

For the most part, caddisflies are solitary creatures, although individuals in some species have been observed in aggregations. Some adult males defend a small territory such as a branch or stone. The net-making genus *Smicridea* and other members of the family Hydropsychidae defend the area where they deposit their nets; when other individuals approach, the net-makers emit alarm sounds.

All larvae spin silk when they build their nets, retreats, cases, or cocoons. At the end of their larval period, they build a cocoon in which they pupate. When they have completed

their development, the adults cut the cocoon open and swim to the surface. The variety of building mechanisms of the larvae has allowed this order to invade different habitats with great success. The free-living larvae of Annulipalpia build retreats and nets to filter organisms or particles on which they feed; they pupate in a double case built with an external wall is made of loosely stacked stones and a closed silken inner case. The free-living larvae of Rhyacophiloidea build neither nets nor retreats; their pupal cases are similar to those of Annulipalpia. Larvae of Glossosomatidae build a turtle-shaped case, which they replace each time they molt. Before pupation, they remodel the last case, sticking it to the substrate by the border, and build a second closed silken case inside. Larvae of Hydroptilidae build a case in the last larval instar that is also used for pupation.

The case-making larvae of Integripalpia build tubular cases of silk to which they stick parts of plants or minerals arranged in different ways; they remain inside these cases, with only the anterior part of the body protruding; they drag the case when they crawl and use the case to pupate. The adults are short-lived; their primary activity consists of looking for a mate and, in the case of females, laying eggs. To compensate for larval drift that occurs downstream, many species fly upstream for oviposition. Adults exhibit only short dispersal flights.

Communication between the sexes involves chemicals of sexual attraction (pheromones) and visual and auditory signals. Some of these signals have different meanings in different species; for instance, in some species drumming sounds stimulate mating, but in others it precedes an attack. Likewise, some species flap or spread their wings before mating, but for others such a display is an attack sign toward other males. Some pheromones produce swarming in males, which then exhibit dancing behavior in order to attract females.

Feeding ecology and diet

Adults consume only liquids, such as nectar or sap. There is a strong omnivorous tendency among larvae, which feed on detritus, suspended particles, algae, vascular plants, and whole animals or parts of animals.

The free-living larvae are filter-feeders (feeding on particles) or predators, while the case-maker larvae are shredders, grazing on plants or scraping algae from surfaces. Some *Oxyethira* (Hydroptilidae) quickly pierce and suck fluids from filamentous algae, cell by cell. Several types may occur in the same family, each one represented by different genera. Both larvae and adults are an important food source for fishes and other aquatic animals.

Reproductive biology

Mating takes place almost immediately after individuals emerge from their cocoons; they mate in flight, on riparian vegetation, or on the ground. Males are attracted from long distances by female pheromones. When they are close to each other, one or both of them move their wings, producing vibrations or scraping the substrate; others produce shakes in the body or stimulate mating by touching each other with their antennae. In some species the female raises her wings when she is ready to copulate. Mating may last a few minutes to several hours. Both males and females may copulate several times during their short life. Insemination is internal by means of the male's genital organ transferring sperm in free form or via a spermatophore.

Eggs are cemented to stones or plants in the water. Sometimes the female dives for nine to 30 minutes to lay them. In other species, the female carries the eggs at the tip of her abdomen, and she spreads the eggs by dipping her abdomen in the water as she flies upstream. Limnephilidae deposit their eggs on vegetation that hangs over the water. There is no nest-building or parental care.

Adults are active mainly in spring and summer, but some species (such as *Verger bruchinus* from Argentina) are active only during winter.

Conservation status

The 2002 IUCN Red List includes four caddisfly species, all of which are categorized as Extinct: *Hydropsyche tobiasi*, *Rhyacophila amabilis*, *Triaenodes phalacris*, and *T. tridonata*. Caddisflies do not tolerate strong variations in their aquatic habitats; organic or chemical wastes can lead to a decrease in their density or their extinction. At the same time, most species have a restricted distribution, so drastic alterations in a particular region may extinguish the species that inhabit that area. The sensitivity of caddisflies to polluting substances can serve as an indicator of water quality in a given habitat. No direct efforts at conservation are being made, but caddisflies may benefit indirectly when aquatic ecosystems are protected.

Significance to humans

Some legends in central Japan were based on the doll-like cases of the genus *Goera*; these legends revolved around a young girl offered as a sacrifice. Beginning in the 1980s, the visual artist Hubert Duprat utilized caddisflies to create unique sculptural forms. He first removed larvae from their natural habitat, and then he provided the larvae with precious materials, prompting the caddisflies to construct "jeweled" cocoons. Some South American natives use larval cases as earrings and as beads for necklaces.

Caddisfly larvae live in a tubelike case of leaves, twigs, and sand. Most of the life cycle is spent in the tube. (Photo by ©Patrick Grace/ Photo Researchers, Inc. Reproduced by permission.)

Salmon and other fishes are attracted by caddisfly larvae, pupae, and adults, leading to the creation of a series of sophisticated patterns for fly-fishing that imitate the different stages of species from different regions.

Some species gnaw on wood structures in the water, while other species cause damage to rice fields and to aquatic ornamental and commercial plants. Adults are often attracted to lights, where thousands of them may appear simultaneously and cause damage to air conditioners and other devices. Caddisflies also may reduce visibility when they lay eggs on roads; apparently they confuse the shine of roads at night with water. The cementing substance of the eggs and eggs broken by tires can be transformed into a gelatin that is hazardous for drivers.

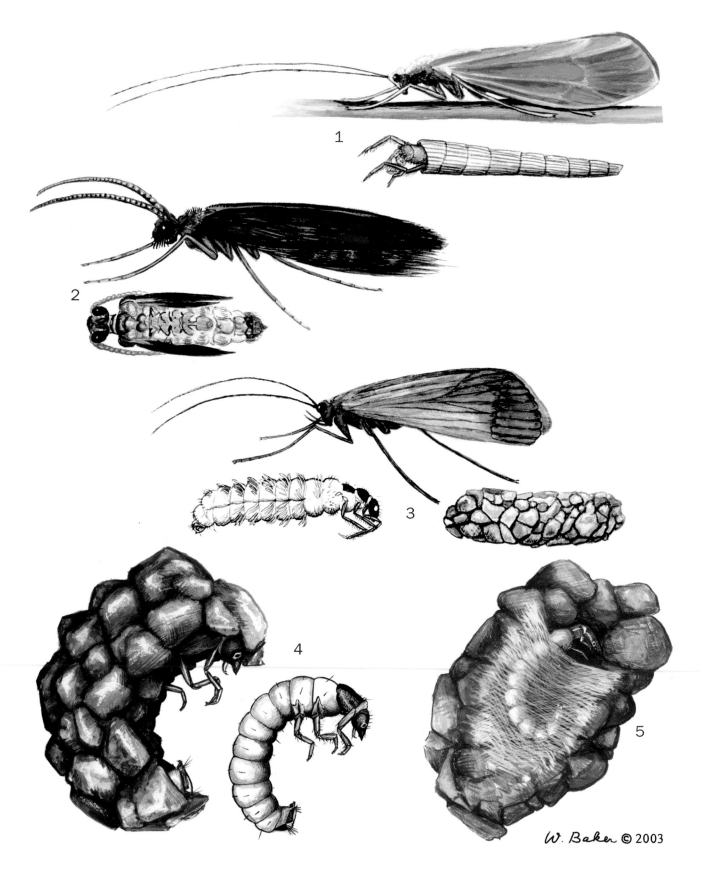

1. *Triaenodes bicolor* adult and larva in larval case; 2. *Abtrichia antennata* adult and dorsal views of larva; 3. October caddisfly (*Dicosmoecus gilvipes*) adult and larva; 4. *Glossosoma nigrior* larva; 5. *Stenopsyche siamensis* larva. (Illustration by Wendy Baker)

Species accounts

No common name
Glossosoma nigrior

FAMILY
Glossosomatidae

TAXONOMY
Glossosoma nigrior Banks, 1911, North Carolina, United States.

OTHER COMMON NAMES
None known.

PHYSICAL CHARACTERISTICS
Larvae are small, up to 0.36 in (9 mm) in length. Adults have a black body and black and white hair; their wings and legs are clear brown.

DISTRIBUTION
Eastern United States and Canada.

HABITAT
On stones in rapids of cool streams.

BEHAVIOR
Larvae make turtle-shaped cases with small pebbles.

Dicosmoecus gilvipes
Glossosoma nigrior

FEEDING ECOLOGY AND DIET
Larvae are scrapers and feed on diatoms and detritus.

REPRODUCTIVE BIOLOGY
Eggs are attached to stones.

CONSERVATION STATUS
Not threatened.

SIGNIFICANCE TO HUMANS
None known. ◆

No common name
Abtrichia antennata

FAMILY
Hydroptilidae

TAXONOMY
Abtrichia antennata Mosely, 1939, Santa Catharina, Brazil.

OTHER COMMON NAMES
None known.

PHYSICAL CHARACTERISTICS
With a forewing length of 0.16 in (4 mm), this is one of the largest species of the family (all Hydroptilidae are small in size, less than 0.24 in [6 mm]). Males have big projections on their heads, and their antennae have enlarged basal articles.

DISTRIBUTION
Southern Brazil, western Uruguay, and eastern Argentina.

HABITAT
Streams and rivers with stone beds.

BEHAVIOR
Mature larvae build oval convex cases and attach them to rocks; the cases have one opening on each end.

FEEDING ECOLOGY AND DIET
Larvae feed on algae growing on rocks.

REPRODUCTIVE BIOLOGY
Nothing is known.

CONSERVATION STATUS
Not threatened.

SIGNIFICANCE TO HUMANS
None known. ◆

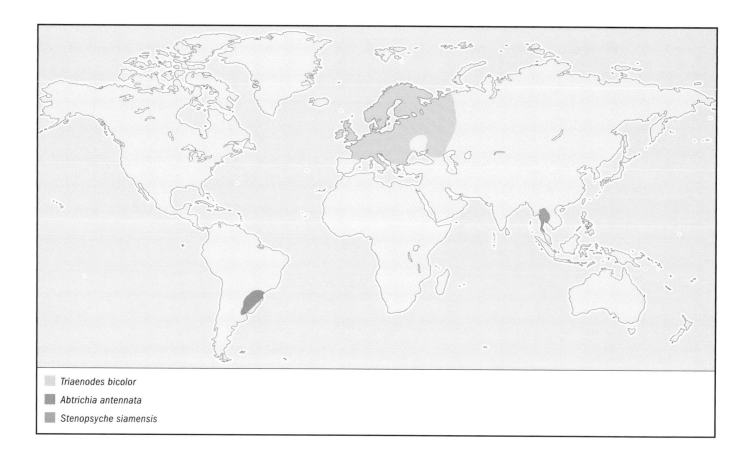

Triaenodes bicolor
Abtrichia antennata
Stenopsyche siamensis

No common name
Triaenodes bicolor

FAMILY
Leptoceridae

TAXONOMY
Leptocerus bicolor Curtis, 1834, Britain.

OTHER COMMON NAMES
Polish: Szuwarek.

PHYSICAL CHARACTERISTICS
Larvae are 0.34–0.52 in (8.5–13.0 mm) in length and ocher yellow in color. Adults have a delicate, brown body with very long antennae.

DISTRIBUTION
Europe east to western Russia.

HABITAT
Vegetation, usually in shallow sections close to the river bank, in stagnant and slowly running waters at a depth of 0.008–0.059 in (0.2–1.5 m).

BEHAVIOR
Larvae construct cases of small, spirally arranged, oblong plant particles, generally 0.6 in (15 mm) long, tapering posteriorly. Larvae attach their remodeled cases to the substrate and use them for pupation.

FEEDING ECOLOGY AND DIET
Green plants.

REPRODUCTIVE BIOLOGY
Females deposit their eggs in a spiral pattern on aquatic plants.

CONSERVATION STATUS
Not threatened.

SIGNIFICANCE TO HUMANS
Considered a pest of cultivated rice fields. ◆

October caddisfly
Dicosmoecus gilvipes

FAMILY
Limnephilidae

TAXONOMY
Stenophylax gilvipes Hagen, 1875, Colorado, United States.

OTHER COMMON NAMES
English: Northern casemaker caddisfly.

PHYSICAL CHARACTERISTICS
Adults are large, dark brown in color, and the veins of the forewing are darker than the membrane. The forewing is 0.84–1.16 in (21–28 mm) in length.

DISTRIBUTION
Western montane North America.

HABITAT
Larvae live on rocks in running waters and shores of lakes from 1,300–6,000 ft (395–1,830 m) in elevation.

BEHAVIOR
Larvae build stout cases of gravel; they use the same cases to pupate.

FEEDING ECOLOGY AND DIET
Larvae graze on diatoms and organic particles.

REPRODUCTIVE BIOLOGY
Individuals have a one-year (and occasionally a two-year) cycle. Adults emerge in late summer.

CONSERVATION STATUS
Not threatened.

SIGNIFICANCE TO HUMANS
Important as fish food, and used by fishermen as bait. ◆

No common name
Stenopsyche siamensis

FAMILY
Stenopsychidae

TAXONOMY
Stenopsyche siamensis Martynov, 1931, Thailand.

OTHER COMMON NAMES
None known.

PHYSICAL CHARACTERISTICS
Larvae are large, more than 1.4 in (35 mm) in length; adults are brown and gray in color.

DISTRIBUTION
Southeast Asia, including Myanmar and Thailand.

HABITAT
Larvae inhabit stream rapids.

BEHAVIOR
Larvae construct retreats and food-catching nets in wide crevices beneath boulders.

FEEDING ECOLOGY AND DIET
Organic detritus and particles.

REPRODUCTIVE BIOLOGY
Eggs are attached to stones.

CONSERVATION STATUS
Not threatened.

SIGNIFICANCE TO HUMANS
Useful as a biological indicator organism. Larvae show abnormalities when water has been contaminated with certain pesticides. ◆

Resources

Books

Ivanov, D. V. "Vibrations, Pheromones, and Communication Patterns in Trichoptera." In *Proceedings of the Eighth International Symposium on Trichoptera*, edited by R. W. Holzenthal and O. S. Flint. Columbus: Ohio Biological Survey, 1997.

————. "Contribution to the Trichoptera Phylogeny: New Family Tree with Consideration of Trichoptera-Lepidoptera relations." In *Proceedings of the Tenth International Symposium on Trichoptera*. Reprinted in *Nova Suppl. Ent. Keltern* 15 (2002): 277–292.

Vieira-Lanero, R. "Las larvas de los Tricópteros de Galicia (Insecta: Trichoptera)." Ph.D. diss. Universidad de Santiago de Compostela, Spain, 2000.

Wiggins, G. *Larvae of the North American Caddisfly Genera (Trichoptera)*. 2nd ed. Toronto and Buffalo, NY: University of Toronto Press, 1996.

Periodicals

Kjer, K. M., R. J. Blahnik, and R. W. Holzenthal. "Phylogeny of Caddisflies (Insecta, Trichoptera)." *Zoologica Scripta* 31(2002): 83–91.

Wiggins, G., and J. Richardson. "Revision and Synopsis of the Caddisfly Genus *Dicosmoecus* (Trichoptera: Limnephilidae: Dicomoecinae)." *Aquatic Insects* 4, no. 4 (1982): 181–217.

Other

"Trichoptera World Checklist." Trichoptera Checklist Coordinating Committee. [7 May 2003]. <http://entweb.clemson.edu/database/trichopt/>.

Elisa Angrisano, PhD

Lepidoptera
(Butterflies, skippers, and moths)

Class Insecta
Order Lepidoptera
Number of families 122

Photo: Swallowtail (*Lamproptera meges*) from Thailand, drinking and expelling water. (Photo by Rosser W. Garrison. Reproduced by permission.)

Evolution and systematics

For such a large group (arguably the second-largest group of insects, with approximately 150,000 described species), the fossil record for lepidopterans is meager. An estimated 600–700 fossil specimens are known, of which the earliest is a small moth, *Archeolepis mane*, from Dorset, England, dating from the early Jurassic period. Other fossils include leaf miners and preserved specimens in Cretaceous amber for primitive moths. More advanced families, including butterflies and noctuid moths, are known mostly from wings or preserved impressions of the insect; one of the most famous is a probable nymphalid butterfly, *Prodryas persephone*, from the rich shale Florisant fossil beds of the Oligocene period in Colorado. Fossils of immature stages are rare, but among them are a probable sphingid-like larva from the Pliocene in Germany and a possible noctuid egg from the late Cretaceous in eastern North America, for example. A factor contributing to the relative scarcity of fossil lepidopterans is their more fragile nature compared with other insect orders.

Lepidoptera is one of the two major orders (the other being the Diptera), along with scorpionflies, caddisflies, and fleas, forming the "panorpoid" complex. Trichoptera (caddisflies) is the sister group of Lepidoptera. Both possess either hairy or scaly wings, caterpillar-like larvae, reduction of mandibles in the adult stage, and similar wing venation. The Lepidoptera are distinguished by no fewer than 27 uniquely derived characters, including the possession of fleshy prolegs with hooklike crochets and a silk-producing spinneret in larvae, loss of the median ocellus, coiled sucking mouthparts, wings covered with a dense layer of overlapping shingle-like deciduous scales, and lack of cerci in adults.

The vast majority of lepidopterans are moths. Most moths have drab, somber colors; are nocturnal; do not have clubbed antennae; and often rest with their wings held rooflike over the back. Skippers (Hesperiidae) are small, mothlike, day-flying butterflies that have a clubbed antenna ending in a curved tip called the apiculus, and they usually hold their wings over their backs. Butterflies, with about 19,000 described species, represent only about 13% of all Lepidoptera, though they probably are the most popular group. They differ from skippers in not having an apiculus at the end of the antenna.

Although the Lepidoptera make up an easily definable group, the higher classification below the ordinal level has undergone significant changes in light of newer phylogenetic classification methods. For many years, such terms as Rhopalocera (butterflies excluding skippers), Heterocera (all moths), and Microlepidoptera (micro-moths) were used to classify these insects. Later, all lepidopterans were classified according to the complexity of wing venation and wing-coupling systems. Most recent classifications use four suborders based on mouthpart morphological features and female reproductive systems. Within the Glossata, Monotrysian lepidopterans have genitalia with a single opening for both copulation and oviposition, while Ditrysian lepidopterans, which include about 98% of all the Lepidoptera, have genitalia with separate openings for copulation and oviposition. The most current classification recognizes about 30 superfamilies with 122 families; all except three occur within the Glossata.

Physical characteristics

Adult lepidopterans vary widely in size and structure. The smallest species are leaf-miner moths in the families Nepticulidae (forewing 0.06 in, or 1.5 mm) and Heliozelidae (forewing 0.07 in, or 1.7 mm); the largest known species is *Thysania agrippina* (Noctuidae) from the American tropics,

Lepidoptera mouthparts (top left) and feeding techniques: Adults feeding on A. tree sap; B. nectar; and C. herbivore dung. D. Caterpillars (larvae) feed on leaves. (Illustrated by Patricia Ferrer)

with a wingspan of up to 11.2 in (280 mm). The smallest known butterfly species probably is *Micropsyche ariana* from Afghanistan or the Western pygmy blue of the United States. Both have a forewing length of 0.20–0.28 in (5–7 mm). The largest is Queen Alexandra's birdwing of New Guinea, with females attaining a forewing length of up to 5.16 in (129 mm).

Adults possess a coiled tongue or proboscis (absent in primitive micro-lepidopterans, which retain mandibles) derived from the galeae of the maxillary palps; moniliform, pectinate, or clubbed antennae; large compound eyes; two lateral ocelli; a pair of erect scaled labial palps. They lack cerci. Wings usually are large compared with the small, elongate bodies and frequently are densely covered with overlapping scales that assume a wide variety of patterns and combination of colors. Wing venation varies. In small micro-moths, it can be reduced to a few veins. In many of the larger macro-moths and butterflies, however, venation consists of an oblong discal cell formed by reduction of the main basal veins (medius and part

of the radius) and a series of mostly unbranched veins radiating from the discal cell. Wing venation is similar in both wings, but the hind wing usually is smaller. The leading margin of the forewing often is crowded with veins providing strength to the leading edge.

Wing coupling is made possible by an extension of a jugal lobe (in primitive ghost moths) on the rear of the forewing, which overlaps and couples with the hind wing base during wing flexing. The most common type of wing coupling involves a small, strong cluster of hairs on the hind-wing base, called the frenulum, which is retained by a retinaculum at the base of the forewing. This mechanism allows for the hind wing to be extended in unison with the forewing in the course of flexing. A loss of the frenulum/retinaculum mechanism occurs in butterflies, skippers (except for one species), and certain genera of moths, resulting in an amplexiform wing-coupling device, where the strong anterior basal lobe of the hind wing overlies the base of the forewing.

Larvae generally are fleshy, soft, elongate animals with a chitinized semicircular head capsule. Antennae are small and inconspicuous. Mouthparts comprise a set of opposable mandibles. A small, erect, silk-producing organ, the spinneret, is present on the labium. The five-segmented legs on the thorax are small and end in a simple tooth, but they are lost in some families. The ventral part of the abdomen possesses differentiating number of fleshy prolegs on segments three through six and the tenth segment. The fleshy tips often bear various series of retractable hooks, called crochets, which allow larvae to grab on to a substrate. The surface of the body possesses setae, the placement of which is important in classifying various families. The body surface can be smooth or possess clusters of hairs, fleshy tubercles, or urticating hairs. Some small micro-moths are specialized leaf miners that have secondarily lost their legs and have a forward-projecting head with rasping mouthparts. Pupae generally are elongate with appendages fastened to the body. Many are encased in a silk cocoon, others are attached to a silk substrate by a small cluster of hooks called cremaster, and still others are accompanied by a single strong silk girdle. Some moth pupae are naked and occur in the ground.

Distribution

Lepidopterans occur on all land masses except Antarctica. The most northerly species may be the lymantriid moth, *Gynaephora groenlandica*, which has been taken on Ward Hunt Island (83°5′ north latitude) in the Canadian Arctic. Lepidopterans are most diverse in the humid tropical zones.

Habitat

Eggs, larvae, and pupae occur in nearly all terrestrial habitats, where they often are found on or near their food plants. Larvae of a few species of moths are associated with aquatic plants in freshwater streams and ponds, and a few others (some species of Lycaenidae) live in ant nests. Adults visit nectar sources, and some species are attracted to carrion, oozing tree sap, or excrement. Adults often rest on foliage, tree trunks or any other substrate. Some species aggregate on shrubs, trees, or cave entrances.

Behavior

Butterflies and moths require a certain body temperature (usually between 77 and 79°F, or 25–26°C) to be able to fly, and they regulate internal temperature according to the environmental temperature. Butterflies of temperate areas increase heat absorption by spreading the wings, angling the exposed surface, and making direct contact with the substrate (dorsal basking) or by folding the wings above the body so that they are perpendicular to the sun's rays (lateral basking). In the tropics, butterflies fly early in the morning and at dusk, seeking refuge in the shade when the temperature climbs too high. Most moths regulate their thoracic temperature through actively preheating the thorax by vibrating their wings and simultaneously contracting antagonistic flight muscles. To diminish heat dissipation, sphingids and other large moths have

Automaris sp. using startling coloration to deter a predator. (Illustration by Patricia Ferrer)

insulating scales and hairs on the thorax and air sacs and diaphragms that separate the thorax from the abdomen so that the abdomen remains cooler.

Most adult butterflies and moths are solitary, but there are cases of gregarious behavior. Certain migratory species crowd together in wintering quarters, others form nocturnal roosts, and still others cluster on damp ground. Lepidopterans display a wide range of types of defensive behavior. Several larvae build protective cases, where they spend all or part of their time. For example, the bagworms of the family Psychidae form a case of silk covered with twigs, leaf fragments, and sand. Other larvae, feeding in exposed situations, show cryptic concealment, blending in by means of their green or brownish color with the leaves or bark where they feed or imitating part of a plant. When discovered, various species display startle or escape responses, such as exposing brightly colored and sharp, acrid osmeteria; regurgitating a brightly colored liquid; or faking sudden death by dropping to the ground.

Several aposematic larvae (larvae with warning coloration) in the families Lymantriidae, Limacodidae, Anthelidae, and others accompany their bright colors with urticating setae, or

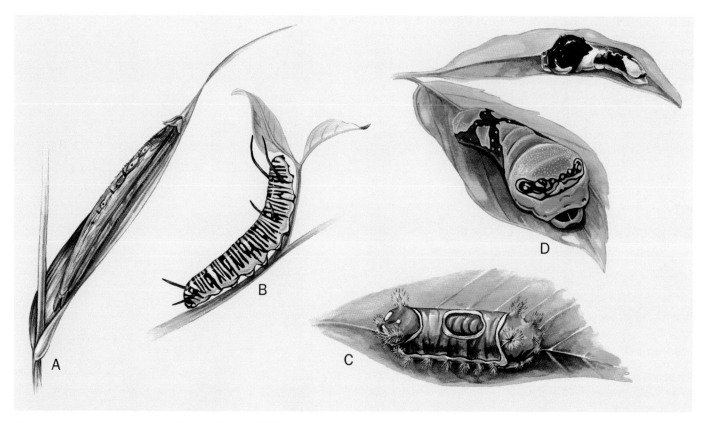

Defensive techniques of some lepidopteran larvae. A. The twig-like larva of the owl butterfly (*Caligo* sp.); B. Aposematic coloration of the African monarch (*Danaus chrysippus*); C. Poisonous spikes of the saddleback caterpillar (*Sibine stimulea*); D. The snake-like larva of the victorine swallowtail (*Papilio victorinus*) and its earlier form, which looks like bird droppings (top). (Illustration by Patricia Ferrer)

spines, which are physically or chemically irritating or cause allergic reactions, making the larvae inedible to predators. Certain larvae even mimic snakes. The same defensive behaviors occur in adults; for example, several species pretend to be dead when they are handled, some arctiids discharge a foul-smelling yellow fluid from specialized thoracic glands, and pyralid and noctuid moths emit ultrasonic sounds to warn bats about their bad taste. Several diurnal lepidopterans exhibit aposematic coloration, announcing their distastefulness or mimicking truly inedible species. This is the case with several species that mimic wasps and bees with their orange-and-black or yellow-and-black abdomens and translucent wings and a few species complexes in which there is Batesian mimetism (e.g., monarch butterflies) and Müllerian mimetism (clusters of similarly marked distasteful species, for example monarchs and viceroys).

Most butterflies and moths disperse, but only about 200 species regularly migrate long distances, returning to the areas where they breed. Migratory butterflies are found among the pierids, nymphalids, lycaenids, and hesperids and moths in the sphingids and noctuids (night flight) and uranids (day flight). In northern parts of North America and Europe, many species display seasonal southward movements: certain butterflies fly south in late summer or autumn and then north from Central or South America or South Africa in spring. A particular individual of a migrating species is not able to undertake a complete round trip; the return trip is accomplished

by the offspring. The best-known example of a migrating butterfly is the monarch butterfly, *Danaus plexippus*, distributed from southern Canada to Paraguay in the New World. In North America three to five generations feed on milkweeds in the summer. In the fall they migrate south to overwintering places in California, Mexico, and Florida, where adults congregate in great numbers on certain kinds of trees. In the spring they begin their journey back north, laying eggs along the way before dying. The subsequent generation completes the return flight the following fall.

Feeding ecology and diet

Lepidopterans are predominantly phytophagous (plant-feeders), and 99% of them exploit higher plants (angiosperms); larvae feed on plant tissues, while adults feed mostly on nectar. The vast majority of these insects specialize in one (monophagy) or a few (oligophagy) food plants; only a small percentage feed on a wide range of food plants (polyphagy). Mouthparts of most adult butterflies and moths are designed for sucking, and almost all of them feed on nectar. Some lepidopterans complement their nectar-based diet with pollen, and others are saprophagous, sucking fluids on rotting fruits, carrion, dung, or droppings. Among the noctuids, for example, there are a few that feed on fruits that have been opened by other animals and others that have a strong proboscis and are able to pierce fruit skin. In the tropics a few species are able to suck the lachrymal fluids of cattle and other mammals, including humans, while others prefer urine, excreta, and cutaneous secretions. *Calpe eustrigiata*, an oriental noctuid moth, has developed hematophagous habits, piercing the skin of mammals to feed on their blood. Lepidopterans also imbibe water and salts from the substrate.

Herbivorous larvae usually specialize in a certain tissue of the plant: leaves, flowers, seeds, buds, or wood. There are several endophytic species, the larvae of which spend their life digging galleries inside plant tissues. Most lepidopteran larvae are exophytic, however, feeding on the outside of plants. Some larvae have unusual diets. Micropterigid larvae feed on mosses, whereas adults, which have chewing mouthparts, feed on pollen. Certain larvae feed on lichens (some arctiids, noctuids, and lycaenids), fungi (some tineids), algae and diatoms on the surface of submerged rocks (aquatic pyralids), or ferns. There are some carnivorous caterpillars that feed on other insects. For example, certain Hawaiian geometrids prey on Diptera, and Asian noctuids eat scale insects. A few lycaenids feed on aphids, others on homopterans, and still others on ant eggs, larvae, and pupae. Some noctuid larvae live in the pitcher of the carnivorous plant *Nephentes*, where they eat the plant's prey, and certain tineids and gelechiids steal prey from spider webs. Several microlepidopterans eat plant and animal derivatives, such as cloth, wool, fur, feathers, grains, dried fungi, paper, rotting wood, and droppings of birds and mammals.

Swallowtail tiger (*Pterourus glaucus*) caterpillar defense display. (Photo by Kerry Givena. Bruce Coleman, Inc. Reproduced by permission.)

Reproductive biology

Most species reproduce bisexually, but some European species of psychids are parthenogenetic. Two mate-location strategies are known among butterflies: territoriality and patrolling. Territorial males perch on a post, from which they defend an area and attack and pursue intruders. Patrolling males fly through proper habitats in search of receptive females. Butterflies rely on visual stimuli to locate their mates and use pheromones produced by the male only secondarily, whereas moths generally locate their mates by pheromones liberated by the female to attract the male. Among butterflies, when the male spots a female of a specific shape or color, he pursues her. The female then drops to the ground, and the male may perform a courtship ritual, moving his antennae and wings around her. He also may liberate an aphrodisiac pheromone from the glands on his wings, legs, thorax, or abdomen (modified scales or brushlike tufts of hairs called androconia), which acts over a short distance and a short period of time. If the female is receptive, mating takes place. In moths the female produces pheromones from the abdominal glands (usually located between segments eight and nine) that act over long distances to attract males. Courtship typically is simple, and mating takes place shortly after the male reaches the female. Male receptors for the pheromone are in his bipectinate antennae and allow him to detect a single molecule of the pheromone.

Females insert the eggs into plant tissues with an ovipositor or wedge-shaped papillae anales, glue them to a substrate, or simply drop them during flight. In some *Monopis* species (tineids) the eggs are retained in the enlarged vagina until they are ready to hatch, and the larvae emerge immediately after eggs are laid. Eggs can be laid singly or in batches; the number of eggs and their size depend on the size of the adult and the degree of alimentary specialization of the larva. Species with specialized monophagus diets produce a larger number of eggs of smaller size, which typically are dropped in flight; species with generalist polyphagus diets, on the other hand, tend to choose oviposition sites carefully. Eggs may pass through a diapause period in some species, during which they remain in a latent state; in temperate regions diapause may allow eggs to survive the winter or, in tropical regions, drought periods.

The duration of the larval stage varies with the species, its feeding ecology, the temperature, and the availability of food; it can range from 15 to 30 days to two years. When mature, the larva searches for a suitable place to pupate: in soil or litter, beneath a stone or bark (i.e., sphingids, notodontids, noctuids, and saturnids), or by rolling up a leaf and sewing the margins together with silk (i.e., tortricids and pyralids). Some larvae spin silk to attach themselves head downward to a support or to construct a protective web (arctiids), silken tube shelters (some pyralids and tineids), or cocoons (silkworm and gypsy moths). Pupae or chrysalids suspend themselves through a thread of silk from the cremaster. Emergence from cocoons takes place through caplike flaps with a secretion of fluid that dissolves the cocoon wall, or the insects may cut or force their way through the wall with sharp structures on the head.

In temperate areas most species spend short and low-temperature winter months as eggs, as larvae in diapause, or as pupae and summer months as adults. In the tropics, where sea-

A silk moth (*Automeris cecrops*) caterpillar in Arizona, USA. (Photo by Bob Jensen. Bruce Coleman, Inc. Reproduced by permission.)

sonal temperature differences are small, some species are present as adults all year. Seasonal variations for species occur in deciduous forests and savannas but are related to rainy and dry seasons. The average life span of an adult is only a few days or weeks, although some danaines and heliconines may live for as long as six months. The timing of flight is correlated with the life cycle of the food plant and whether the species overwinter. Species that overwinter as adults, such as the mourning cloak (*Nymphalis antiopa*) and several *Polygonia* anglewings in North America, fly in early February and March. Species that overwinter as pupae fly next, followed by species that spend the winter as larvae and, finally, those that do so as eggs.

Conservation status

More than any other order of insects, lepidopterans have attracted the attention of conservationists. In all, 284 lepidopterans of 747 total insects are listed on the Red List of the World Conservation Union (IUCN); 25 are on the U.S. Fish and Wildlife Service list of endangered species; and several others, including the giant birdwing butterflies of the Indo-Australian region, are listed by the Convention on International Trade in Endangered Species (CITES). Butterflies, in particular, have been accorded conservation status because they are easily seen, colorful, day-flying insects and

favorites with collectors. Collecting and commercial trafficking of butterflies often have been considered to pose a principal threat to their existence, but, as with many other insects, it is habitat loss that represents the primary threat. Although some subspecies of butterflies (e.g., *Speyeria adiaste atossa* from southern California and *Cercyonis sthenele sthenele* and *Glaucopsyche lygdamus xerces* from the San Francisco Bay Area) have become extinct in recent times, very little is known about the status of moths, because so few people study them. Many endemic species of micro-moths (e.g., *Hyposmocoma* species) from the Hawaiian Islands, for example, are known only from the original specimens collected about a century ago, and it has been postulated that many may be extinct owing to habitat loss and introduction of lepidopteran parasitoids.

Positive steps have been taken to preserve some species considered aesthetically beautiful or rare. One of the best success stories is that of the magnificently colorful birdwing butterflies (*Ornithoptera* and *Troides*) of Papua New Guinea. They represent a prime example of sustainable wildlife management by encouraging butterfly farming, breeding highly desirable species. Survival of the largest and most restricted butterfly in the world, *Ornithoptera alexandrae*, may depend on such a program. Natives have ranched this species and others. The practice advocates maintenance of habitat by encouraging

A monarch chrysalis in its late stage. (Photo by Ed Derringer. Bruce Coleman, Inc. Reproduced by permission.)

A Delaware skipper (*Atrytone logan*) resting. (Photo by Larry West. Bruce Coleman, Inc. Reproduced by permission.)

growth of the butterfly's food plant. Some of the birdwings are offered for sale, and others are released into the environment. Butterfly-breeding farms also have cropped up in many global regions, and they annually ship thousands of living pupae to butterfly zoos throughout the world.

Significance to humans

Butterflies have always been a favorite insect motif in art; they are represented in Egyptian temples, Chinese amulets, Aztec ceramics, and an endless number of paintings, sculptures, gem carvings, textiles, glass, drawings, and poetry, symbolizing joy or sorrow and eternal life or the transience of life. In some cultures, butterflies and moths have a symbolic connection to the soul: the word for butterflies and moths in Russian means "little soul," and in Greek it means simply

"soul." Some European traditions maintain that witches and fairies turn themselves into moths or butterflies to go inside houses. In the pre-Columbian cultures of Central America they were respected in religious and mythical traditions, representing souls of the dead, new plant growth, the heat of fire, sunlight, and various transformations of nature.

Many butterflies and moths fly from one flower to another and are major factors in pollination, and hence reproduction, of angiosperms. Some species are directly beneficial in that they have predatory larvae that feed on aphids and scales (e.g., several pyralids, lycaenids, noctuids, and blastobasids) or are parasites of plant-sucking leafhoppers (epipyropids). To this day, the silkworm is used in the production of textiles; the eggs of a gelechiid moth are employed commercially to raise the parasitoid wasp *Trichogramma*, which is released to control noctuid moths; and the larvae of a pyralid from South America are used in South Africa and Australia to keep in check invasive *Opuntia* cacti. In tropical areas of South America, New Guinea, Australia, Madagascar, and Africa, caterpillars are a complement of the human diet.

Because of their phytophagous nature, many caterpillars are major pests. Examples include leaf-rollers (Tortricidae), leaf-tires and webworms (Pyralidae), leaf miners (Incurvariidae and Gracillariidae), cutworms and armyworms (Noctuidae), underground grass grubs (Hepialidae), borers (Hepialidae, Cossidae, and Sesiidae), forest defoliators (Limacodidae and Geometriidae), stored fibers and foods pests (Tineidae, Gelechiidae, and Pyralidae), and crop pests (Tortricidae, Plutellidae, Gelechiidae, Noctuidae, and Pieridae).

1. European cabbage white (*Pieris rapae*) larva; 2. *Pieris rapae* pupa; 3. *Pieris rapae* adult; 4. Silkworm (*Bombys mori*) larva; 5. *Bombys mori* pupa; 6. *Bombyx mori* adult; 7. Death's head hawk moth (*Acherontia atropos*) larva; 8. *Acherontia atropos* pupa. (Illustration by Patricia Ferrer)

1. Yucca moth (*Tegeticula yuccasella*); 2. Large blue (*Maculinea arion*); 3. Female Queen Alexandra's birdwing (*Ornithoptera alexandrae*); 4. *Micropterix calthella*; 5. Citrus leaf miner (*Phyllocnistis citrella*); 6. Webbing clothes moth (*Tineola bisselliella*) 7. Bagworm (*Thyridopteryx ephemeraeformis*). (Illustration by Michelle Meneghini)

MLM ©2003

1. Blue morpho (*Morpho menelaus*); 2. Corn earworm (*Helicoverpa zea*); 3. Atlas moth (*Attacus atlas*); 4. Gypsy moth (*Lymantria dispar*); 5. *Parargyractis confusalis*; 6. Indian mealmoth (*Plodia interpunctella*). (Illustration by Patricia Ferrer)

Species accounts

Silkworm
Bombyx mori

FAMILY
Bombycidae

TAXONOMY
Phalaena mori Linnaeus, 1758, China.

OTHER COMMON NAMES
English: Silkmoth; French: Ver de la soie, bombyx du mûrier; German: Seidenspinner; Spanish: Gusano de seta; Finnish: Silkkiperhonen.

PHYSICAL CHARACTERISTICS
Caterpillars (1.5 in, or 4 cm) are pale brown, with brown marks on the thorax and a horn on the tail. They pupate in a white to yellow cocoon, the color depending on genetics and diet. The silk forming the cocoon is a single, continuous thread (1,000–3,000 ft, or 300–900 m, long) of a protein secreted from salivary glands. Adults are heavy, rounded, furry, and whitish with pale brown lines. The forewings have a hooked tip, and the wingspan is 1.5–2.5 in (4–6 cm).

DISTRIBUTION
Originally from the north of China, the north of India, Japan, Taiwan, and Korea. Now also bred in Europe and North and South America as a commodity in the textile market.

HABITAT
On mulberry worldwide.

BEHAVIOR
Adults cannot fly. Larvae are so domesticated now that they cannot survive without the assistance of humans.

FEEDING ECOLOGY AND DIET
Caterpillars feed on mulberry leaves; adults have atrophied mouthparts and do not feed.

REPRODUCTIVE BIOLOGY
The female lays 200–500 lemon-yellow eggs that turn black and hatch in spring. In four to six weeks, larvae undergo four molts and then spin a silk cocoon (in a process taking three or more days) to pupate. Adults emerge in three weeks, reproduce, and die within five days. Univoltine (having one generation per year) under natural conditions.

CONSERVATION STATUS
Bred in captivity for thousands of years; no wild colonies remain.

SIGNIFICANCE TO HUMANS
Used to make silk and for education and research. Originally domesticated in China. To harvest silk, cocoons are boiled in water to kill pupae and help unravel thread. Dead pupae sometimes are used as cockroach bait or fish food or to fertilize mulberry trees. ◆

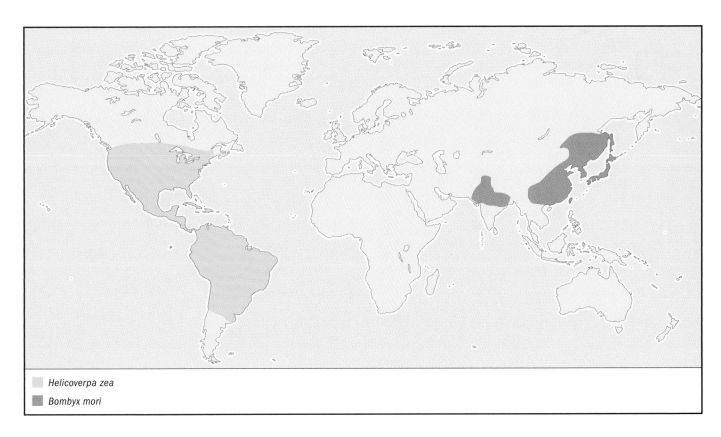

Helicoverpa zea
Bombyx mori

Citrus leaf miner
Phyllocnistis citrella

FAMILY
Gracillaridae

TAXONOMY
Phyllocnistis citrella Stainton, 1856, Calcutta, India.

OTHER COMMON NAMES
German: Citrusmotte; Spanish: Minador de los cítricos, lagarta minadora dos citros; Portuguese: Lagarta minadora dos citros.

PHYSICAL CHARACTERISTICS
Adults are small, with a wingspan of 0.16 in (4 mm) and a length of 0.08 in (2 mm) at rest. The forewing is white with silvery reflections, black and tan markings, and a small apical black spot. Full-grown larvae (0.12 in, or 3 mm) are flattened and translucent green with atrophied prolegs and legs.

DISTRIBUTION
Originally from India, but now associated with citrus groves worldwide.

HABITAT
Can occur wherever citrus is grown. Adults are seldom seen. Larvae make narrow, serpentine mines within the leaves of citrus, visible from the underside.

BEHAVIOR
Adults fly at night and rest on the underside of leaves or tree trunks during the day. Hatched larvae bore into leaves, where they immediately begin to mine soft tissue between the upper and lower surfaces.

FEEDING ECOLOGY AND DIET
Larvae are sap feeders. Hosts include citrus and other Rutaceae and also *Jasiminum*, mistletoe on citrus, and kumquats. Adults feed on nectar.

REPRODUCTIVE BIOLOGY
Mating takes place 14–24 hours after the emergence of the adult. Larvae undergo four molts, and pupation occurs within the mine near the leaf's edge. Total developmental time is about 5–20 days.

CONSERVATION STATUS
Not threatened.

SIGNIFICANCE TO HUMANS
A significant agricultural pest of citrus throughout the world. Current infestations have been subjected to control measures. These measures include biological control by introducing Chalcidoid parasitic wasps from the Old World and pesticide applications as part of an integrated pest management program. ◆

Large blue
Maculinea arion

FAMILY
Lycaenidae

TAXONOMY
Papilio arion Linnaeus, 1758, Nuremberg, Germany.

OTHER COMMON NAMES
French: l'Argus arion; German: Schwarzfleckenbläuling, Schwarzgefleckte Bläuling; Spanish: Hormiguera de lunares;

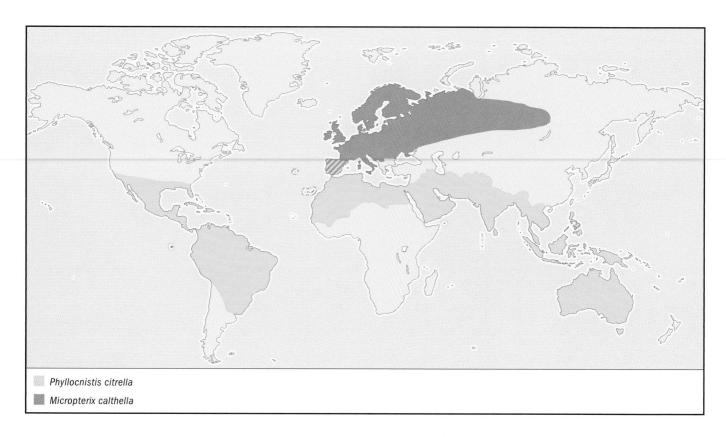

Phyllocnistis citrella

Micropterix calthella

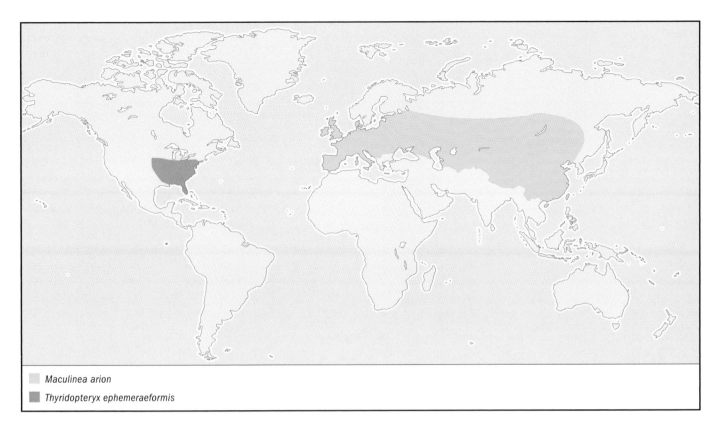

Maculinea arion

Thyridopteryx ephemeraeformis

Dutch: Tijmblauwtje; Finnish: Muurahaissinisiipi; Polish: Modraszek arion; Swedish: Svartfläckig blåvinge.

PHYSICAL CHARACTERISTICS
Adults' wingspan is 0.64–0.8 in (16–20 mm). Upper side of the forewing is bright blue with large black spots; underside is grayish with large black spots and bluish or greenish area near the base. Larvae head is small, the head and legs are hidden, and the body is covered with short setae.

DISTRIBUTION
Western Europe to southern Siberia, Mongolia, and China.

HABITAT
Rough, dry, open grasslands where *Myrmica sabuleti* ants occur; in northern latitudes ants are restricted to warm, south-facing slopes with short grass.

BEHAVIOR
Larvae are predators in underground nests of *Myrmica sabuleti* ants, which they attract through secretions from the honey gland.

FEEDING ECOLOGY AND DIET
Young larvae feed on pollen and seeds of wild thyme and oregano, older larvae eat ant eggs and larvae.

REPRODUCTIVE BIOLOGY
Univoltine, adults fly for three to four weeks between June and August in small, isolated colonies; they are slow to disperse. Females may mate before their wings have dried and lay their eggs singly on flowers of thyme or oregano by the end of their first day as adults. After three weeks of feeding on host plants, caterpillars drop and hide on the ground, where ants mistake them for ant larvae and take them to nests. Pupation occurs in ant nests; emergence of adults takes place the following summer.

CONSERVATION STATUS
Categorized as Lower Risk/Near Threatened by the IUCN. Threatened by changes in modern agriculture and land management. Extinct or declining in northern Europe; now reestablished in England, and still common in Siberia and the Far East.

SIGNIFICANCE TO HUMANS
Study of the relationship between these butterflies and ants applies to practical conservation and habitat management; also allows construction and testing of population models and generates new theories in evolutionary biology. ◆

Gypsy moth
Lymantria dispar

FAMILY
Lymantridae

TAXONOMY
Phalaena (Bombyx) dispar Linnaeus, 1758. Type locality not given but certainly in Europe.

OTHER COMMON NAMES
French: Spongieuse; German: Schwammspinner; Spanish: Bombyx disparate; Swedish: Lövskogsnunna.

PHYSICAL CHARACTERISTICS
Adults are sexually dimorphic. Males are light to dark brown with irregular black markings, a wingspan of 1–1.5 in (25.4–38 mm), and bipectinate antennae. Females are all white with irregular black lines on the wings, a wingspan of 2.24–2.68 in

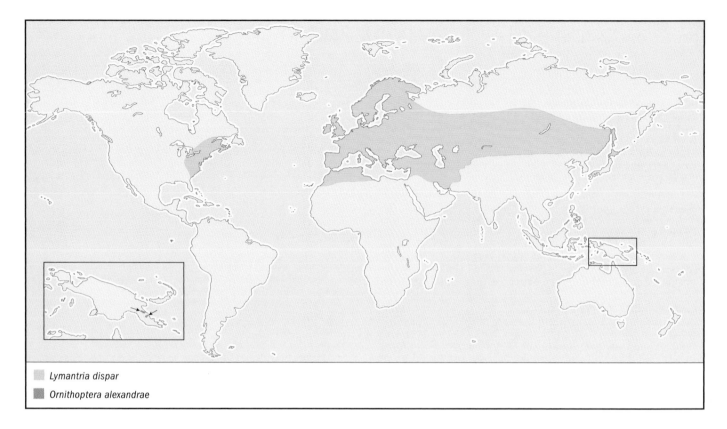

Lymantria dispar
Ornithoptera alexandrae

(56–67 mm), and narrower bipectinate antennae. Full-grown larva (1.48–2.40 in, or 37–60 mm) are gray with a series of hair tufts and five pairs of blue followed by six pairs of red spots. Pupa (0.76–1 in, or 19–25 mm) are brown with small circlets of hairs.

DISTRIBUTION
Palearctic region with the exception of the extreme north and south; introduced into northeastern North America.

HABITAT
Occurs in forests, fields, and even within cities and suburbs.

BEHAVIOR
Males are active fliers in late afternoon or at night; females do not fly (except for Japanese race). Larvae remain on food plants while feeding.

FEEDING ECOLOGY AND DIET
Larvae are voracious feeders of more than 500 different trees and shrubs, including many pines, oaks, poplars, willows, and birches. Adults do not feed.

REPRODUCTIVE BIOLOGY
Males fly to pheromone-emitting females for mating, which occurs in the summer. Females lay clusters of 100–1,000 eggs on any surface, including cars and picnic furniture, allowing for easy dispersal (hence their common name). Eggs overwinter, and larvae hatch in the spring and go through five (males) or six (female) instars, lasting 20–60 days. Adults emerge from the pupal stage (14–17 days) in late July or early August.

CONSERVATION STATUS
Not threatened.

SIGNIFICANCE TO HUMANS
Important forest pest in Europe, Asia, and the northeastern United States. Quarantine laws in the United States and other countries have been enacted to halt further spread of this species. ◆

No common name
Micropterix calthella

FAMILY
Micropterygidae

TAXONOMY
Phalaena calthella Linnaeus, 1761, Sweden.

OTHER COMMON NAMES
German: Dotterblumen-Schabe.

PHYSICAL CHARACTERISTICS
Tiny moths, with a wingspan of 0.32–0.4 in (8–10 mm). Yellow head with a tuft of hairs, functional mandibles, and no proboscis. Metallic bronzy forewings with a purplish patch at the base, similar in venation to the hind wings, with which they are coupled through the jugum. Legs and abdomen are golden brown. Larvae have eight pairs of nonmuscular conical abdominal prolegs ending in a single claw. Pupae have free appendages and articulated mandibles, used to open the cocoon.

DISTRIBUTION
Europe to central Siberia in Asia.

HABITAT
Adults fly in damp and shady habitats, such as trails and the margins of forests. Larvae are secretive, burrowing through thick mats of leaf litter, mosses, or lichens as deep as 4 in (10 cm) in loose soil.

BEHAVIOR
Adults feed on flowers during the daytime and can be attracted to lights at night.

FEEDING ECOLOGY AND DIET
Adults feed on pollen from various plants with exposed flowers, from trees and shrubs to grasses; they have a preference for species of Ranunculaceae, a primitive group of angiosperms. Larvae are external feeders on decayed plant detritus, fresh leaves of angiosperms, liverworts, and possibly fungal hyphae.

REPRODUCTIVE BIOLOGY
Eggs are laid singly or in clusters of two to 45 on the food plant. Larval development takes from 132 to 141 days. It is thought that there are three larval instars, with the last one overwintering. Pupation occurs in early spring inside a silken cocoon on the ground or on detritus; adults emerge in spring and fly from May to June.

CONSERVATION STATUS
Not threatened.

SIGNIFICANCE TO HUMANS
None known. ◆

Corn earworm
Helicoverpa zea

FAMILY
Noctuidae

TAXONOMY
Phalaena zea Boddie, 1850, North America.

OTHER COMMON NAMES
English: Tomato fruitworm, sorghum headworm, vetchworm, cotton bollworm; Spanish: Gusano cogollero, tomatero, bellotero, gusano del fruto.

PHYSICAL CHARACTERISTICS
Adults are medium-sized and sandy-colored, with a wingspan of 1.50 in (37.5 mm). Forewings have faint, irregular markings on the apical half; the hind wings are a dull translucent white with a narrow dark apical band. Full-grown larvae (1.28–1.60 in, or 32–40 mm) are variously colored red, pink, or green with irregular longitudinal dark stripes. Body has spinules.

DISTRIBUTION
Southern Canada to Argentina.

HABITAT
Adults are found on flowers at dusk and can be attracted to lights; larvae are found on ears of corn, tomatoes, and other crops.

BEHAVIOR
Adults fly mainly at dusk. Hatched larvae tunnel into fruit often associated with frass (insect excrement) but can be found on flowers, leaves, and seedlings.

FEEDING ECOLOGY AND DIET
Larvae are serious pests of corn, tomatoes, and cotton. Occasional hosts include bean, broccoli, cabbage, chrysanthemum, eggplant, head cabbage, lettuce, okra, pepper, and more than 100 other plants.

REPRODUCTIVE BIOLOGY
Females can lay eggs on corn silks, on terminal leaflets of tomatoes, and on the crowns of seedlings of lettuce. Larvae pupate about 2–6 in (5–15 cm) beneath the soil and overwinter.

CONSERVATION STATUS
Not threatened.

SIGNIFICANCE TO HUMANS
A major pest of several important food crops, necessitating various forms of integrated pest management, especially in the warmer parts of its range. ◆

Blue morpho
Morpho menelaus

FAMILY
Nymphalidae

TAXONOMY
Papilio menelaus Linnaeus, 1758, Guianas.

OTHER COMMON NAMES
English: Blue morpho butterfly; Spanish: Mariposa azul.

PHYSICAL CHARACTERISTICS
Iridescent blue wings. The female is duller, with a brown edge and white spots surrounding blue. The undersides are brown with bronze eyespots. The wingspan is 6 in (150 mm). Larva are red-brown with bright patches of lime green and reddish-brown and white tufts of hair on the dorsum.

DISTRIBUTION
South America, from the Guianas to Brazil and Bolivia.

HABITAT
Rainforests.

BEHAVIOR
Adults soar in the canopy of the jungle, coming near the ground in clearings; the underside of the wings provides camouflage at rest. The slow beating of the wings gives rise to alternating flashing and disappearance of their iridescent blue coloring, possibly startling potential predators. Males are territorial; their vivid blue is used to intimidate other males. Larvae are active at night and, if threatened, release a strong smell from scent gland located between the forelegs.

FEEDING ECOLOGY AND DIET
Adults imbibe the juices of rotting fruit; larvae feed on the neotropical plant *Erythroxylum pilchrum*.

REPRODUCTIVE BIOLOGY
Nothing is known.

CONSERVATION STATUS
Not threatened.

SIGNIFICANCE TO HUMANS
Owing to the microscopic structure of the wing scales, the wings are iridescent, and their color changes when seen from different angles. This quality makes them useful as models to create counterfeit-proof currency and charge cards. ◆

Acherontia atropos
Morpho menelaus
Parargyractis confusalis

Queen Alexandra's birdwing
Ornithoptera alexandrae

FAMILY
Papilionidae

TAXONOMY
Troides alexandrae Rothschild, 1907, Biagi, southeastern New Guinea.

OTHER COMMON NAMES
None known.

PHYSICAL CHARACTERISTICS
Adults are sexually dimorphic. Males have long (3.24–4.36 in, or 81–109 mm; wingspan of 6.7–7.5 in, or 170–190 mm), iridescent blue and black, ellipsoid forewings and smaller hind wings. The thorax is crimson red on the sides, and the abdomen is brilliant yellow. Females' wings are larger (4.08–5.16 in, or 102–129 mm long; wingspan of 7.1–8.3 in, or 180–210 mm) and brown; the outer half of the hind wings has creamy white, elongate, wedge-shaped spots. The thorax is the same as in the male, and the abdomen is creamy white. Eggs are round, white, and up to 0.20 in (5.1 mm) in diameter. Full-grown larva (up to 4.72 in, or 118 mm) are dark wine red and covered with fleshy tubercles, with a pale yellow saddle mark on fourth abdominal segment. Pupa (up to 3.6 in, or 90 mm) are light brown with irregular patches of yellow.

DISTRIBUTION
Restricted to lowland areas of Northern Province of Papua New Guinea.

HABITAT
Occurs in primary and secondary rainforest; larvae live on the food plant.

BEHAVIOR
Adults are diurnal and fly high during the day, visiting flowers. Larvae remain motionless on the food plant when not eating. If touched, larvae extend the osmeterium. Before pupating, larvae may ring a branch or stem of the food plant, thereby killing it. Larvae often wander for 24 or more hours in search of a proper pupation site, usually under leaves from 4 in (10 cm) off the ground to high in the canopy.

FEEDING ECOLOGY AND DIET
Adults imbibe nectar at flowers of introduced ornamentals, *Caesalpinia* and *Zanthoxylum*, often high up in forest canopy. Larvae feed on *Pararistolochia alexandriana* and *P. meridionaliana*. First-stage larvae eat their eggshells; younger larvae eat young, tender leaves; and older, full-grown larvae consume larger, older leaves.

REPRODUCTIVE BIOLOGY
Low egg fecundity (about 240 eggs per female). Females inspect the food plant carefully before depositing eggs underneath a leaf. Larvae normally undergo five molts, lasting about 70 days. Pupation lasts about 40 days. Marked adults have been seen up to three months after initial marking.

CONSERVATION STATUS
Listed as Endangered by the IUCN and by CITES I and the U.S. Fish and Wildlife Service. These acts effectively ban the commercial exchange of this species. Its major threat is habitat destruction due to the ever encroaching oil palm and timber industries. The best hope for conserving *O. alexandrae* may be the commercial breeding of specimens, since its high demand probably will ensure its survival.

SIGNIFICANCE TO HUMANS
The species is arguably the flagship emblem of butterfly conservation, and it often is illustrated in many books relating to invertebrate species conservation. ◆

European cabbage white
Pieris rapae

FAMILY
Pieridae

TAXONOMY
Papilio rapae Linnaeus, 1758, Sweden.

OTHER COMMON NAMES
English: Small white; French: Petit blanc du chou; German: Kleiner Kohlweissling; Spanish: Blanquita de la col; Polish: Bielinek rzepnik.

PHYSICAL CHARACTERISTICS
Wingspan of 1.75–2.25 in (4.5–5.8 cm). The upper side of wings is white; forewing has black tip. Two submarginal black spots in the female and one in the male. Undersides evenly yellow-green or gray-green. Eggs are pale yellow, bottle-shaped, ridged, and 0.06 in (1.5 mm) high. Larva (up to 0.75 in, or 3 cm) are green with pale yellow line on the back and a line of yellow spots on each side. Pupa sculptured and angular; color influenced by background color.

DISTRIBUTION
Palearctic region with the exception of extreme north and south; introduced into North America and Australia.

HABITAT
Almost any type of open space, including weedy areas, gardens, roadsides, cities, and suburbs.

BEHAVIOR
Female is active during the day, with slow, lumbering, erratic flight; male flies in a straight line. Larva is a slow crawler and sits on the upper surfaces of leaves of food plants in daylight, camouflaged by coloration; when disturbed it regurgitates a poisonous fluid.

FEEDING ECOLOGY AND DIET
Larva feeds on plants containing mustard oil (among them, cabbage, radish, broccoli, mustard, and other Brassicaceae species). Adult feeds on the nectar of several flowers, including mustards, dandelion, red clover, asters, and mints.

REPRODUCTIVE BIOLOGY
Males patrol for females. Females lay single eggs on the undersides of the host plant. Larval stage takes 17 days; pupa overwinters, and adults emerge in summer. Adults breed all year in subtropical areas; there are three to five generations in temperate regions.

CONSERVATION STATUS
Not threatened.

SIGNIFICANCE TO HUMANS
Considered a pest; larvae feed on vegetable gardens and crops. ◆

Yucca moth
Tegeticula yuccasella

FAMILY
Prodoxidae

TAXONOMY
Pronuba yuccasella Riley, 1872, central Missouri.

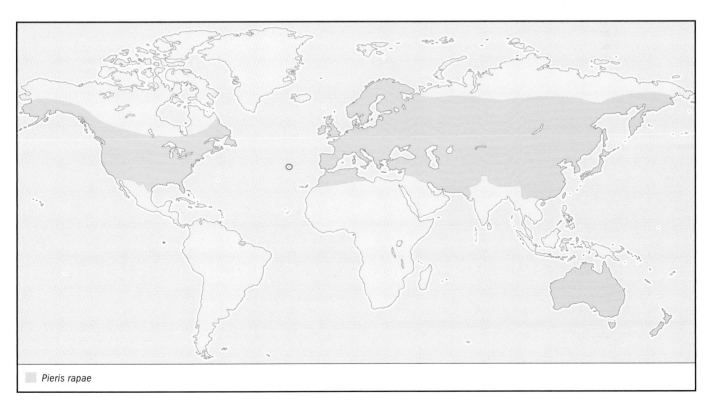

Pieris rapae

Attacus atlas
Tegeticula yuccasella

OTHER COMMON NAMES
English: *Tegeticula* moth.

PHYSICAL CHARACTERISTICS
Adults small and white; wingspan is 0.07–0.11 in (1.8–2.75 mm). Forewings are white, wide, and blunt; hind wings are brownish-gray. Females are larger than males, with a swordlike ovipositor and enlarged maxillary palps. Larvae are pink and more than 0.4 in (10 mm) long. They form a spherical or elongate cocoon, about 0.24–0.32 in (6–8 mm) long.

DISTRIBUTION
Southern Canada south through Mexico.

HABITAT
Associated with yucca plants growing in sand dunes, grasslands, pine forests, and glades.

BEHAVIOR
Adults hide inside yucca flowers during the daytime and fly at night. Between dusk and midnight females gather pollen from flowers using their maxillary palps; they form balls of sticky pollen and push them into the receptive tips of yucca pistils (stigmata). Larval stages grow inside developing yucca fruits.

FEEDING ECOLOGY AND DIET
Larvae feed on developing yucca seeds, consuming a small percentage of the hundreds of seeds within capsules.

REPRODUCTIVE BIOLOGY
Mating occurs inside yucca flowers; the gravid female visits flowers and lays an egg in one of the six locules (cavities within the ovary). The larva remains inside the little feeding cavity of fused seeds within the capsule until the first autumn rains and then drops to the ground, burrows into soil, and overwinters in a silk-and-sand cocoon. Pupation occurs after the onset of spring rains, and the adult emerges in late May or in June. Adults stay close to their home yucca clusters and remain active for less than a week.

CONSERVATION STATUS
Not threatened.

SIGNIFICANCE TO HUMANS
The association between yucca and moth constitutes a classic example of "mutualism" and "co-evolution" and provides a unique opportunity for studying these phenomena. The yucca plant cannot produce seeds without the moth, and without the plant the moths would die off in one generation. ◆

Bagworm
Thyridopteryx ephemeraeformis

FAMILY
Psychidae

TAXONOMY
Sphinx ephemeraeformis Haworth, 1803, Great Britain (apparently in error).

OTHER COMMON NAMES
English: Evergreen bagworm moth.

PHYSICAL CHARACTERISTICS
Adult males have short (0.5 in, or 12.5 mm; wingspan of 1.12 in, or 24 mm), clear wings; hairy black bodies; and feathery antennae. Females resemble maggots, with no functional eyes, legs, wings, mouthparts, or antennae. Larvae live inside a bag

made of silk and bits of needles, bark, or twigs and are up to 1.5–2.5 in (40–65 mm) long.

DISTRIBUTION
Eastern United States.

HABITAT
A wide range of broadleaf and evergreen trees and shrubs (128 species) serve as hosts, including arborvitae and other ornamental conifers, sycamore, and willow.

BEHAVIOR
The female remains inside the bag her entire life; males are nimble fliers and in the fall fly around infested trees in search of a mate. During feeding, caterpillars emerge from the top of the bag and hang on to the host plant with their legs and sometimes with a silken thread; the bottom of the bag remains open to allow fecal material to pass out. Young larvae disperse, walking or using wind currents. During molts and pupation, caterpillars seal the bags.

FEEDING ECOLOGY AND DIET
Larvae feed on conifers, maple, oak, dogwood, and willow, consuming one branch at a time and leaving only the middle rib of leaves.

REPRODUCTIVE BIOLOGY
Univoltine, they overwinter as eggs inside the mother's bag. Eggs hatch in spring. Larvae crawl out in search of food and construct a bag, where they molt four or more times before pupation. Adults emerge in the fall. Males are attracted to the female's bags by pheromones and mate. Female lay 500–1,600 eggs within bag, after which they drop to the ground and die.

CONSERVATION STATUS
Not threatened.

SIGNIFICANCE TO HUMANS
Pest; defoliates evergreen ornamentals. ◆

No common name
Parargyractis confusalis

FAMILY
Pyralidae

TAXONOMY
Cataclysta confusalis Walker, 1865, North America.

OTHER COMMON NAMES
None known.

PHYSICAL CHARACTERISTICS
Adult is small, with a forewing span of 0.20–0.44 in (5–11 mm). Light tan with well-defined darker areas of brown. Hind wing broadly triangular with a marginal row of small, eyelike spots. Larva (0.60 in, or 15 mm) is greenish with brown head capsule, thoracic shields, and filamentous clusters of gills along the side. Pupa (0.32 in, or 8 mm) is brown in feltlike cocoon. All immature stages found on rocks in creeks and streams.

DISTRIBUTION
Northern California east to Montana, Idaho, and Nevada and north to southern British Columbia.

HABITAT
Near and in creeks and streams.

BEHAVIOR
Adults often found at rest on bridge supports crossing streams or on bushes, trees, and other vegetation bordering streams.

FEEDING ECOLOGY AND DIET
Larvae feed on algae and diatoms on substrate (e.g., rock crevices) in streams.

REPRODUCTIVE BIOLOGY
Female swims to the bottom of riffle and lays a few to several hundred eggs. Larvae live under protected silken areas on alga-covered substrate; pupation occurs in these waterproof silken linings. Adults emerge and swim to the surface.

CONSERVATION STATUS
Not threatened.

SIGNIFICANCE TO HUMANS
None known. ◆

Indian mealmoth
Plodia interpunctella

FAMILY
Pyralidae

TAXONOMY
Tinea interpunctella Hübner, 1813, Europe.

OTHER COMMON NAMES
French: Pyrale indienne de la Farine, teigne des fruits secs; German: Kupferrote, Dörrobstmotte; Spanish: Polilla de la fruta seca; Portuguese: Traça dos cereais.

PHYSICAL CHARACTERISTICS
Adults are small, with a wing length of 0.20–0.34 in (5.0–8.5 mm) and a wingspan of 0.62–0.79 in (16–20 mm). Forewing has reddish-brown apical half and ocher basal half; hind wing is dull translucent white. Full-grown larvae are 0.36–0.60 in (9–15 mm), with yellow-brown head capsule and prothoracic shield; remainder of the body is dull white to pinkish.

DISTRIBUTION
Cosmopolitan.

HABITAT
Often seen in pantry areas of homes, in supermarkets, and in feed stores, but adults also occur outdoors and are attracted to light at night. Larvae feed on stored foods.

BEHAVIOR
Adults fly at night. Hatched larvae tunnel into food often associated with frass and silk webbing.

FEEDING ECOLOGY AND DIET
Larvae are significant nuisance pests of dried stored foods; characteristic webbing is associated with larval feeding. Adults often are seen flying in these areas.

REPRODUCTIVE BIOLOGY
Females can lay eggs three to four days after emerging as adults. Larval development can take 13–288 days, depending on temperature and humidity. Generations overlap in home and warehouse situations.

CONSERVATION STATUS
Not threatened.

Plodia interpunctella

SIGNIFICANCE TO HUMANS
A minor to significant pantry pest of stored dried food materials. Infestations can be persistent, requiring cleanliness and exclusion practices such as hermetically sealed containers for food. ◆

Atlas moth
Attacus atlas

FAMILY
Saturniidae

TAXONOMY
Phalaena atlas Linnaeus, 1758, "Citro Asiae, Americae (fixed as Bogor, Djawa, Indonesia)."

OTHER COMMON NAMES
English: Giant comet moth, giant silkmoth; French: Papillon géant; Italian: Farfalla cobra; German: Tropischer Spinner.

PHYSICAL CHARACTERISTICS
Wings are reddish-brown with a triangular transparent spot (membrane without scales). The wingspan is 8 in (20 cm), and the tips of the forewings are curved. Stout, hairy body. Males have feathery antennae; females are larger than males. Large, spherical, reddish eggs. Larvae are bluish-green with shades of pink; they have bumps and are covered with a fine white powder. Pupate in cocoons made of broken strands of silk.

DISTRIBUTION
Tropics of Asia, including India and Southeast Asia.

HABITAT
Lowland to upper mountain forests.

BEHAVIOR
Active at night, irregular flight, attracted to light. Females position themselves to enable maximum dispersal of their sexual attractants by wind. Males up to three miles downwind can detect these scents. Larvae wander, looking for food.

FEEDING ECOLOGY AND DIET
Larvae are polyphagous: feed on a wide range of trees (e.g., Jamaican cherry tree, soursop, cinnamon, rambutan, guava, and citrus). Adults have atrophied mouthparts and do not eat.

REPRODUCTIVE BIOLOGY
After mating, the female lays small groups of up to several hundred eggs on the undersides of leaves; adults die a few hours later. Eggs require eight to 14 days to hatch, depending on temperature. Pupal stage lasts about four weeks.

CONSERVATION STATUS
Not threatened.

SIGNIFICANCE TO HUMANS
Larva are cultivated commercially around the world by collectors. In Taiwan cocoons are used as pocket purses and in northern India to make Fagara silk. ◆

Death's head hawk moth
Acherontia atropos

FAMILY
Sphingidae

TAXONOMY
Sphinx atropos Linnaeus, 1758, Europe.

OTHER COMMON NAMES
English: Bee robber; German: Totenkopfschwärmer; French:
Sphinx têtede-mort, fluturele cap-mort; Spanish: Mariposa de
la muerte; Estonian: Tontsuru; Swedish: Dödskallefjäril;

PHYSICAL CHARACTERISTICS
Adult has skull-like pattern on thorax. Large and heavily built,
with 4.4–4.8 in (11–12 cm) wingspan. Dark forewings and yellow hind wings with black submarginal lines. Proboscis is
short, stout, and hairy. Abdomen has yellow riblike markings.
Larva is 4.8–5.2 in (12–13 cm) and colored yellow, green, or
brown with a large posterior horn. Pupa grows to 3–3.2 in
(7.57ndash;8 cm) long and is mahogany brown and glossy.

DISTRIBUTION
Afrotropical, extending north to Mediterranean; migrant in
central and northern Europe.

HABITAT
Prefers dry and sunny locations; frequents open scrubs with
solanaceous (nightshade) plants and cultivated areas where
potato is grown.

BEHAVIOR
Larvae are inactive, moving only to find a fresh leaf; when disturbed, they click their mandibles and may even bite. Adults
are active from dusk to midnight; during the day they rest on
tree trunks, walls, or leaves on the ground. Attracted to light
and occasionally to blossoms. Frequent beehives, where they
rob honey. In defense, they mimic honeybees' cutaneous fatty
acids and raise their wings, run, and hop around. When disturbed, they emit loud, shrill squeaks, forcing air out the proboscis, followed sometimes by secretion of moldy smell from
glandular hairs in the abdomen.

FEEDING ECOLOGY AND DIET
Larvae prefer solanaceaous plants, especially potato. Short proboscis prevents adult from taking nectar from deep-set flowers;

instead, they imbibe honey, juice from rotten fruits, and sap
from trees.

REPRODUCTIVE BIOLOGY
Breed year-round in Africa; adults migrate to Europe from
May to September but do not overwinter there. Eggs are laid
singly underneath old leaves of host plant. Pupation takes place
in a very fragile cocoon 6–16 in (15–40 cm) deep in the soil in
a smooth-sided cavity.

CONSERVATION STATUS
Not as common as it used to be, owing to use of insecticides.

SIGNIFICANCE TO HUMANS
In Greek mythology, Atropos is the eldest of the three Fates,
who severs the thread of life. The moth is considered a sinister
creature because of its skull design and the loud sound emitted
when it is disturbed. It once was thought to be a harbinger of
war, pestilence, and death, and it has entered modern mythology as an emblem of perverted evil in the book and film *Silence
of the Lambs*, in which the trademark of the serial murderer is a
pupa placed in the mouth of his female victims, whom he later
skins. ◆

Webbing clothes moth
Tineola bisselliella

FAMILY
Tineidae

TAXONOMY
Tinea bisselliella Hummel, 1823, most likely Europe.

Tineola bisselliella

OTHER COMMON NAMES
English: Common clothes moth; French: Mite des vêtements; German: Kleidermotte; Spanish: Polilla de la ropa; Dutch: Kleermot; Finnish: Vaatekoi; Norwegian: Klädesmal.

PHYSICAL CHARACTERISTICS
Adults grow to 0.16–0.28 in (4–7 mm) in length, with a wingspan of 0.6 in (1.5 cm). Golden brown with a tuft of reddish-yellow hairs on the head. Narrow wings fringed with long hairs are reddish-yellow with no markings. Larvae are yellowish with dark brown heads and grow to 0.4 in (1 cm) long; they live in a tube of silk, excreta, and other debris found in the area where they feed.

DISTRIBUTION
Originally from the Old World; now cosmopolitan with the exception of the tropics.

HABITAT
Can inhabit any storage area. Breeds at 50–91.4°F (10–33°C). Its optimum relative humidity is 70%.

BEHAVIOR
Adults tend to fly in darkened areas. Males are the fliers, whereas females generally walk or run. Soon after hatching, larvae begin constructing a silken tube. This case acts as shelter during the day, offering the larva good camouflage, from which it emerges at night to feed.

FEEDING ECOLOGY AND DIET
Caterpillar commonly feeds in dark, protected areas on woolens, furs, feathers, animal bristle brushes, dead insects, dried animal carcasses, pollen, and other dried plant and animal products. Adult mouthparts are atrophied, and they do not feed.

REPRODUCTIVE BIOLOGY
The female lays about 50 eggs in darkened areas, such as natural fibers and cloth; afterward it dies. Larvae pass through five instars, although under adverse conditions there may be as many as 40. Pupation takes place within the silken case. Total life span varies from five to nine months; it may exceed two years if the larva goes into a dormant period. In heated buildings females can mate and lay eggs at any time during the year.

CONSERVATION STATUS
Not threatened.

SIGNIFICANCE TO HUMANS
This household and industrial pest of natural fibers can also infest dried vegetable material. ◆

Resources

Books

Coville Jr., Charles. *A Field Guide to the Moths of Eastern North America.* Boston: Houghton Mifflin, 1984.

Kristensen, N. P. *Lepidoptera: Moths and Butterflies.* Volume 1, *Evolution, Systematics, and Biogeography.* Berlin and Hawthorne, NY: Walter de Gruyter, 1999.

Leverton, Roy. *Enjoying Moths.* London: T. and A. D. Poyser, 2001.

Nielsen, E. S., and I. F. B. Common. "Lepidoptera (Moths and Butterflies)." In: *The Insects of Australia: A Textbook for Students and Research Workers.* Vol. 2. 2nd edition. Carlton, Australia: Melbourne University Press, 1991.

Parsons, Michael. *The Butterflies of Papua New Guinea: Their Systematics and Biology.* London: Academic Press, 1998.

Sbordoni, Valerio, and Saverio Forestiero. *Butterflies of the World.* Westport, CT: Firefly Books, 1998.

Stehr, Frederick W., ed. "Order Lepidoptera." In *Immature Insects.* Vol. 1. Dubuque, IA: Kendall/Hunt Publishing, 1987.

Tyler, H., K. S. Brown Jr., and K. Wilson. *Swallowtail Butterflies of the Americas: A Study in Biological Dynamics, Ecological Diversity, Biosystematics and Conservation.* Gainesville, FL: Scientific Publishers, 1994.

Van-Right, R. I., P. R. Ackery. *The Biology of Butterflies.* Symposium of the Royal Entomological Society of London 11. London: Academic Press, 1984.

Periodicals

Heppner, J. B. "Classification of Lepidoptera." Part 1, "Introduction." *Holarctic Lepidoptera* 5, suppl. 1 (1998): 1–148.

Organizations

Association for Tropical Lepidoptera. P. O. Box 141210, Gainesville, FL 32614-1210 United States. Phone: (352) 392-5894. Fax: (352) 373-3249. E-mail: jbhatl@aol.com Web site: <http://www.troplep.org>.

Idalia Society of Mid-American Lepidopterists. 219 West 68th Street, Kansas City, MO 64113 United States. Phone: (816) 523-2948.

Lepidoptera Research Foundation, Inc.. 9620 Heather Road, Beverley Hills, CA 90210 United States.

Lepidopterists' Society. 1900 John Street, Manhattan Beach, CA 90266-2608 United States.

Lepidopterological Society of Japan. c/o Ogata Hospital, 3-2-17 Imabashi 3, Chuo-ku, Osaka, 541 Japan. E-mail: vem15452@niftyserve.or.jp.

Societas Europaea Lepidopterologica. c/o Zoological Institute, University of Bern, Baltzerstrasse 3, Bern, CH-3012 Switzerland.

Natalia von Ellenrieder, PhD
Rosser W. Garrison, PhD

Hymenoptera

(Sawflies, ants, bees, and wasps)

Class Insecta

Order Hymenoptera

Number of families About 84

Photo: A female sawfly (family Tenthredinidae) guards her eggs in the Amazon of Peru. (Photo by George D. Dodge. Bruce Coleman, Inc. Reproduced by permission.)

Evolution and systematics

Hymenoptera is a worldwide order of at least 100,000 described species, more biologically diverse than any other insect order. The order is divided into two suborders—Symphyta (wood wasps and sawflies) and Apocrita (wasps, bees, and ants)—with 22 superfamilies and about 84 families. The fossil record dates from the Triassic (245–210 million years ago). The Hymenoptera may be the sister group of the Antliophora, made up of Diptera (flies), Siphonaptera (fleas), and Mecoptera (scorpionflies and hangingflies); and the Amphiesmenoptera, comprising the Trichoptera (caddisflies) and Lepidoptera (butterflies and moths).

Physical characteristics

Adult hymenopterans range in size from minute to large, at 0.006–4.72 in (0.15–120 mm) and from slender (e.g., many wasps) to robust (e.g., the bumble bees). The head usually is very mobile. The compound eyes often are large and sometimes strongly convergent dorsally. Fine setae occasionally emerge from between facets, and ocelli may be present, reduced, or absent, especially in forms with reduced wings. The antennae are long and multisegmented, and their surfaces are covered with various sense organs. The mouthparts vary from the generalized biting type to the combined sucking and chewing type (e.g., bees). Mandibles typically are present and are used by the adult to cut its way out of the pupal cell, for defense, for killing and handling prey, and in nest construction.

The first abdominal segment of the Apocrita is attached firmly to the metathorax and usually is separated from the remaining abdominal segments (metasoma) by a narrow waist (petiole). In Apocrita thoracic segments plus the first abdominal segment are called the mesosoma, and the incorporated first abdominal segment is the propodeum, followed by the remainder of abdomen. There are generally two pairs of wings. Venation is most complete in Symphyta and mostly reduced in small Apocrita. The hind wings have rows of hooks (hamuli) along the leading edge that couple with the hind margin of the forewing in flight. The legs frequently are cursorial (adapted for running), sometimes with fossorial (adapted for digging) forelegs; the hind legs are modified to carry pollen.

Sensory structures (sensilla) on the ovipositor enable the female to recognize suitable egg-laying sites. In some ants, bees, and wasps the ovipositor has lost its egg-laying function and is used as a defensive, venomous stinger. Although the smooth stingers of ants and wasps allow for repeated use, the barbed stings of the honeybee can be used only once. As the honeybee struggles to leave the stinger and venom behind, it is disemboweled and soon dies.

Bees have several morphological adaptations associated with pollen collection, including plumose (branched) hairs. Moreover, the hind tibia and basitarsus are enlarged, with long hairs on their outer surfaces. These hairs either form a brush (scopa) or are reduced to a fringe surrounding a bare area or concavity (corbicula, or pollen basket). Leaf cutter bees have a well-developed scopa on the ventral surface of the abdomen.

A tree wasp (*Vespula sylvestris*) worker leaves its nest hole in a dead tree stump. (Photo by Kim Taylor. Bruce Coleman, Inc. Reproduced by permission.)

Distribution

Hymenopterans are found worldwide.

Habitat

Hymenoptera occur in soil and litter or on vegetation. Most are active on bright, sunny days, hunting insects, gathering pollen and nectar, or assembling nest-building materials. Some parasitic species are active at night, when their nocturnal hosts are active.

Behavior

Symphyta lay their eggs on or in leaves, stems, wood, and leaf litter, and females sometimes stand guard over their egg masses. The larvae are almost exclusively phytophagous. Pupation takes place within the plant tissue or in the ground. Most have a single generation a year and overwinter as full-grown larvae. The larvae of many species of Apocrita are parasitoids in the immature stages of other insects (or other invertebrates), while the adults are free living.

Idiobiont parasitoids prevent any further development of the host after initial parasitization. Koinobiont parasitoids allow the host to continue its development and often do not kill and consume the host until the host has reached its maximum size. The development of a secondary parasite, or hyperparasite, at the expense of a primary parasite is more frequent in Hymenoptera. The Apocrita species that have the ovipositor modified into a sting are grouped together into the Aculeata. Female Scoliidae locate and attack large subterranean beetle larvae in their burrows or earthen pupal cells. The sting does not kill the host but only immobilizes it. Then the female lays her eggs and departs, leaving her offspring to develop without further assistance. Aculeate hymenopterans may or may not use their stings to immobilize the host; mutillids bite through the host cell and lay an egg on the mature larvae or pupae of the host.

Somewhat more complex behavior is exhibited by many Pompilidae, in which the female captures and paralyzes her prey and then drags it to a cavity or crevice on the ground. She lays an egg on the host and usually seals the cavity before leaving. Cleptoparasites, such as cuckoo bees of the genus *Chrysis* (actually a metallic blue or green, thick-bodied wasp), construct no nests of their own and instead rely primarily on the food stores of hosts (*Chrysis*). In other aculeates (for example, other Pompilidae), the female wasp prepares a nest before locating prey and can relocate her nest when she returns with prey.

Female *Eumenes* (Vespidae) lay eggs in the empty cell before prey are introduced. The female then provides the cell with prey and seals it before the larva begins to develop. This is called mass provisioning, because the initial food that is amassed must be sufficient to feed the larva during the entire course of its development. Progressive provisioners, such as many Sphecidae and Vespidae, provide additional food at intervals. Large, multicelled nests, in which each cell is stocked with many small prey, are characteristic of many behaviorally advanced species. Bees (*Megachile* and *Xylocopa*) are similar to wasps in this respect, except that they provision larval cells with pollen and nectar rather than arthropod prey.

Truly social or eusocial hymenopterans have a division of labor, with a caste system involving sterile individuals that assist the reproducers, cooperation among colony members in tending the young, and overlap of generations capable of contributing to colony functioning. Among hymenopterans exhibiting primitive eusociality are paper wasps. The highly eusocial hymenopterans comprise the ants, some wasps, and many bees. Common construction materials for nest building are mud, leaves, and masticated wood chips that are formed into a paper- or carton-like nest.

Feeding ecology and diet

Adult parasitoids require carbohydrates in the form of honeydew, nectar, or other plant secretions. Many female parasitoids also feed extensively on the body fluids of hosts, to sustain egg production. Most symphytan larvae are phytophagous (plant eaters); larvae of Siricidae are wood borers and utilize cellulases produced by fungi to feed on wood; the enzymes necessary for processing the wood fragments are acquired from the fungus ingested. Apocritan larvae have diverse feeding habits; they may be parasitic or gall forming, or they may be fed with prey or nectar and pollen by their parents or other colony members. Adult hymenopterans mostly feed on nectar, pollen, or honeydew produced by Homoptera; only a few consume other insects. Many wasps feed their young macerated or paralyzed insects and spiders. Most bees feed young pollen and nectar. Many members of the order visit flowers for nectar or pollen. Leaf cutter ants feed on fungi, which they cultivate in the nest. The ants live in an obligate mutualistic association with a fungus. The fungus has lost its capacity for sexual reproduction; dispersal occurs by means of the queen ant carrying fungal hyphae to a new nest site.

Reproductive biology

Males of parasitic species commonly search for females at their emergence sites and occasionally fight for possession of such sites. Some males form female-attracting swarms. Among most hymenopterans the females produce pheromones, chemicals that attract and sexually stimulate males of the same species. Courtship in the Apocrita is common and complex, involving sequences of antennal contact, leg and wing vibrations, and mandibular movements. Sex in most Hymenoptera species is determined by the fertilization of the egg; fertilized eggs develop into females, and unfertilized eggs usually develop into males. Thus, females determine the sex of their offspring and can manipulate the rate of increase of their populations.

The adult female searches for a host by responding to a series of cues in the environment. When she locates a potential host, she examines it often with her antennae or the tip of the ovipositor, to decide whether it is acceptable as a site for egg deposition. Females of some species lay their eggs on a broad range of similar hosts within a particular habitat, whereas others are highly specific to a single host or a few closely related species. Most parasitoid species lay their eggs on or in the body of the host, and many have a long ovipositor to reach hosts in cocoons, burrows, or other protected situations. In some cases only a single egg is laid on a host (solitary parasitism); in others, several to many eggs may be laid on the same host (gregarious parasitism). The larva may be ectoparasitic, that is, developing externally, or endoparasitic, developing within the host. Sometimes early instars are endoparasitic, and later instars are ectoparasitic. Pupation typically occurs within or beside the host remains.

Hymenoptera are holometabolus insects, that is, they have a life cycle progressing from egg to larva to pupa to adult. The eggs are ovoid or sausage-shaped, with a respiratory stalk in some parasitic species and with hooklike or sucker-like attachment devices in some ectoparasites. The outer covering typically is thin and smooth, sometimes with little yolk. Polyembryony, that is, development of several individuals from one egg, occurs in some parasitic species.

In Symphyta the larva is eruciform. The head is well sclerotized, and there are three pairs of thoracic legs and abdominal prolegs (commonly on the second through eighth and the tenth abdominal segments). There are as many as eight larval instars, with females often having one more than males. In Apocrita the larva is vermiform, apodous, grublike, or maggot-like, and the head capsule is weakly sclerotized. Primitively, they probably have five larval instars, but the number is reduced in many endoparasitoids.

A leaf cutter ant (*Atta*) carries a leaf section to its nest in Corcounado National Park, Costa Rica. (Photo by Carol Hughes. Bruce Coleman, Inc. Reproduced by permission.)

Larval heteromorphosis occurs in many parasitic species. In these species, whereas the final instar larva is vermiform, the first or intermediate instars are of diverse forms. Pupae of the exarate type have free appendages that are not glued to the body; they may form in a cocoon in the host or in special cells. The cocoon is spun with silk from the labial glands. Some hymenopterans may spend a very long period in diapause within the cocoon.

Conservation status

The 2002 IUCN Red List includes 152 hymenopteran species. Of these, 3 are listed as Critically Endangered; 142 as Vulnerable; 6 as Lower Risk/Near Threatened; and 1 as Data Deficient.

Hymenopterans are susceptible to the indiscriminate use of insecticides and to habitat destruction. The effects of insecticides include the eradication of nontarget organisms, including such pollinators as bees and wasps. Without the pollinating services of bees and other insects, we would have few vegetables, fruits, and flowers and little or no clover. To achieve preservation of these key aspects of our lives, hymenopteran biotopes must be conserved and utilized in a sustainable fashion.

Significance to humans

Certain hymenopterans were considered deities in such civilizations as Egypt. To the Greeks the bee *Melitta* was

A female leaf cutter bee (*Megachile centuncularis*) carries a leaf section to her nest. (Photo by Kim Taylor. Bruce Coleman, Inc. Reproduced by permission.)

known as the Goddess Honey Mother. The ant-hunting species of *Dinoponera* was a symbol of virility for several Amazon tribes and was used in initiation rituals. The Mixe people of Oaxaca (Mexico) believe that they become more powerful through the ingestion of ants; ants thus have come to stand for virility, fortitude, and courage. The bee *Melipona beechei* in Mexico represented the spiritual world, merriment, and rain. Wasps and bees are represented frequently in literature, music, theater, cinema, and television. The Greek comic playwright Aristophanes, for example, wrote a play called *The Wasps*, satirizing the Athenians' love of litigation and characterizing jurors as wasps in their harshness. The Russian composer Nikolai Rimski-Korsakov, inspired by the sound that bees produce, composed the "Flight of the Bumblebee," a piece of music for strings set within the opera *The*

Tale of Tsar Saltan, the story of a prince who is turned into a bee. In modern times, several protagonists of the Pokemon cartoons represent hymenopterans.

From the standpoint of human beings, Hymenoptera probably is the most beneficial order of insects; it contains many insects that are of value as parasitoids or predators of other arthropods, including insect pests. They have been employed successfully as pest control agents in several countries. Bees generally are regarded as the most important group of insect pollinators. *Apis mellifera* is of great commercial value as a producer of honey that is used extensively as food and in the manufacture of many products. Beeswax is used in making candles, sealing wax, polishes, certain types of ink, models of various kinds, and in other products such as face and hand creams, lipsticks, and lip salves.

Nest-building Hymenoptera can be domestic nuisances. Bees and wasps inject venom when they sting. Many people are highly sensitive to bee or wasp stings and may suffer anaphylactic shock leading to death or disability as the result. The sting of the female hymenopteran can seem unprovoked, but it is, in fact, an aggressive defense of the nest.

Hymenopterans are food for humans in some parts of the world. They are edible at all stages of growth. Boiling tends to break down the poison, which is basically protein and, at boiling temperatures, the stinger softens. Pounding them before boiling makes them more edible. Ants (except the fire ant) and ant larvae are edible and tasty. Australian aborigines living in arid regions derive sugar from species of *Melophorus* and *Camponotus*, popularly known as honey pot ants. Specialized worker ants (repletes) are fed with nectar by other workers and store it in their huge distended crops. Honey pot ants in the western United States and Mexico belong to the genus *Myrmecocystus*.

Few species of Hymenoptera are harmful. Sawfly larvae cause damage to forests, orchards, and ornamental trees. Wood-boring larvae, in association with fungi, can cause extensive damage to plantations of conifers. A few ants are pests, for example, the leaf cutter ants or the seed harvesters; others protect sap-sucking insects, which are pests themselves.

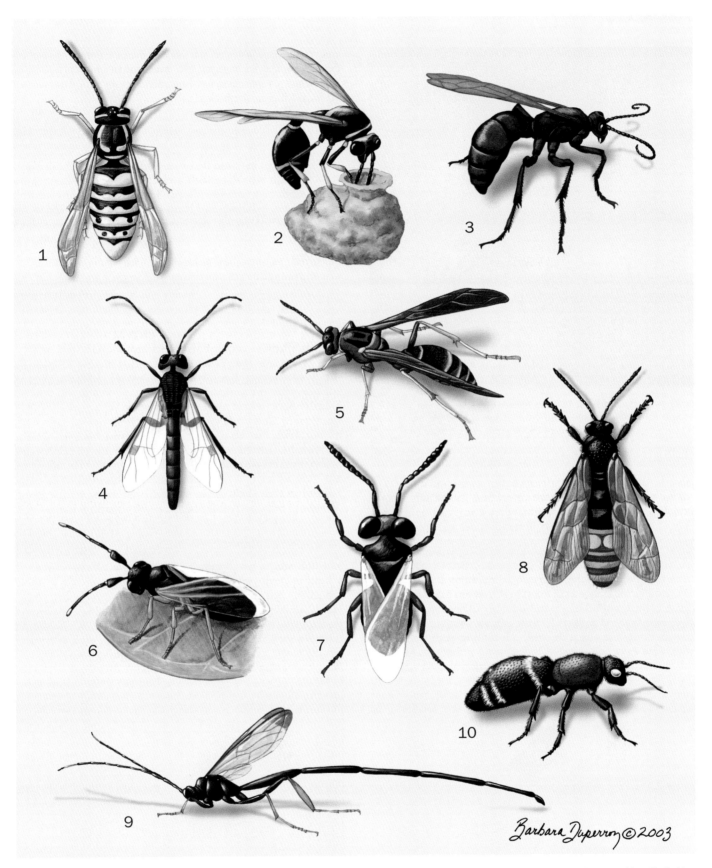

1. Yellow jacket (*Vespula germanica*); 2. Potter wasp (*Eumenes fraternus*); 3. Tarantula hawk (*Pepsis grossa*); 4. *Ibalia leucospoides*; 5. Golden paper wasp (*Polistes fuscatus*); 6. *Apoanagyrus lopezi*; 7. *Trissolcus basalis*; 8. Digger wasp (*Scolia dubia*); 9. *Pelecinus polyturator*; 10. Velvet ant (*Mutilla europaea*). (Illustration by Barbara Duperron)

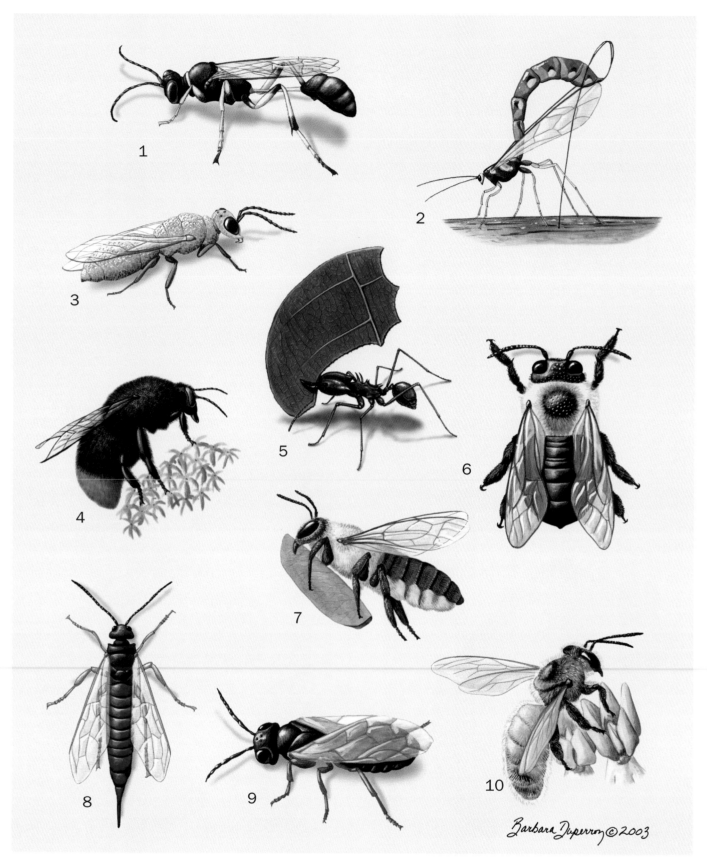

1. Mud dauber wasp (*Sceliphron caementarium*); 2. Cuckoo wasp (*Chrysis coerulans*); 3. *Megarhyssa nortoni*; 4. Large red-tailed bumble bee (*Bombus lapidarius*); 5. Leaf cutter ant (*Atta sexdens*); 6. Large carpenter bee (*Xylocopa virginica*); 7. Alfalfa leaf cutter bee (*Megachile rotundata*); 8. European wood wasp (*Sirex noctilio*); 9. Pear and cherry slug (*Caliroa cerasi*); 10. Honey bee (*Apis mellifera*). (Illustration by Barbara Duperron)

Species accounts

Honeybee

Apis mellifera

FAMILY
Apidae

TAXONOMY
Apis mellifera Linnaeus, 1758.

OTHER COMMON NAMES
English: Common honeybee, European honeybee.

PHYSICAL CHARACTERISTICS
Body length of workers is 0.37–0.62 in (9.5–15.8 mm). The male drone is 0.62 in (15.8 mm), and the queen is 0.75 in (19.5 mm) long. Body is golden brown and black, with pale orange/yellow rings on the abdomen. The head, antennae, and legs are almost black; fine hairs cover the thorax and only lightly cover the abdomen. Wings are translucent. Pollen baskets are made of specialized hairs and are located on the outer surface of the tibiae of the hind legs.

DISTRIBUTION
Worldwide distribution. There are many geographical races distributed throughout Europe, Africa, and parts of western Asia as well as the Americas.

HABITAT
Most colonies live in man-made commercial hives; swarms that escape generally nest in hollow trees. They visit a wide range of native and introduced flowers.

BEHAVIOR
Eusocial. A colony of honeybees consists of a queen, several thousand workers, and, in certain seasons, a few hundred drones. A queen can lay as many as 2,000 eggs in a single day. In her four to five years of life she produces about two million eggs. More than 80,000 bees can live in a single colony. The primary function of the queen is to lays eggs; the workers are sterile females that supply the colony with food, guard the nest, and build the combs. These females live about four or five weeks. The drones are fertile males who fly out at certain times of the year to mate with new queens. The cells in the nest are arrayed in vertical combs, two cell layers thick. The comb consists of adjoining hexagonal cells made of wax secreted by the workers' wax gland. The bees use these cells to rear their brood and to store their food. Honey is gathered in the upper part of the comb, and beneath it, in descending order, are rows of pollen-storage cells, worker brood cells, and drone brood cells. The groundnut-shaped queen cells normally are built at the lower edge of the comb. Honeybee colonies are perennial, with the queen and workers overwintering in the hive. Bee colonies propagate by swarms. Shortly before one or more queens emerge from the queen cells, the old queen leaves the hive with about half of its population. Honeybees have impressive communication abilities, especially with respect to forage sites (the type of flower and the direction and distance from the hive).

FEEDING ECOLOGY AND DIET
Adults and larvae eat a mixture of honey (a sweet, viscid material made of the nectar of flowers in the honey sac) and a type of pollen called beebread. The queen's larvae are fed by the

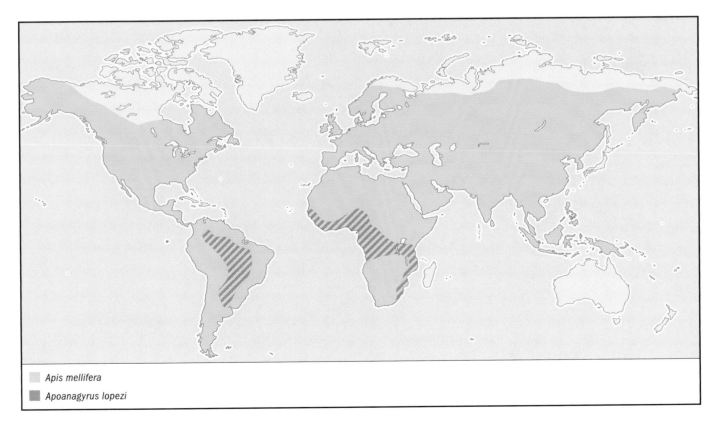

Apis mellifera
Apoanagyrus lopezi

third day with royal jelly (a nutritious secretion of the pharyngeal glands).

REPRODUCTIVE BIOLOGY
Fertilized eggs develop into females, and unfertilized eggs develop into males. Whether a larval honeybee destined to be a female becomes a worker or queen depends on the sort of food it is fed.

CONSERVATION STATUS
Not threatened.

SIGNIFICANCE TO HUMANS
Pollination of plants and production of honey. They are the main bee pollinators worldwide. ◆

Large red-tailed bumblebee
Bombus lapidarius

FAMILY
Apidae

TAXONOMY
Apis lapidaria Linnaeus, 1758.

OTHER COMMON NAMES
English: Red-tailed bumblebee; German: Steinhummel.

PHYSICAL CHARACTERISTICS
Body length is 0.78–0.98 in (20–25 mm). Color is mainly black with a red tail; the male has a yellow collar band.

DISTRIBUTION
Palearctic region as well as Europe, Asia, and North Africa.

HABITAT
Occurs in most grassland areas, collecting nectar from species with short corollas that make good landing platforms, such as daisies, dandelions, and thistles.

BEHAVIOR
Eusocial. The nests are in underground hollows, surrounded by a cover of moss and other materials. The colonies are annual, and only the fertilized queen overwinters. In the spring the queen selects a nest site and begins nest construction; the first brood raised by the queen consists of workers. With the exception of egg laying, the workers take over all duties of the colony, including food gathering and its storage in little sacs resembling honey pots, and care for the larvae. In summer, males and queens are produced, and in the fall all but the queens die.

FEEDING ECOLOGY AND DIET
Adults feed on nectar and pollen; larvae are reared with a mixture of pollen and nectar, which are carried in large pollen sacs on the back legs and in the stomach, respectively.

REPRODUCTIVE BIOLOGY
The queen makes pots of wax and pollen, into which the first eggs are laid. When these eggs hatch, she provides them with honey while making storage cells for honey and more cells for future eggs.

CONSERVATION STATUS
Not threatened.

SIGNIFICANCE TO HUMANS
Valuable as pollinators. ◆

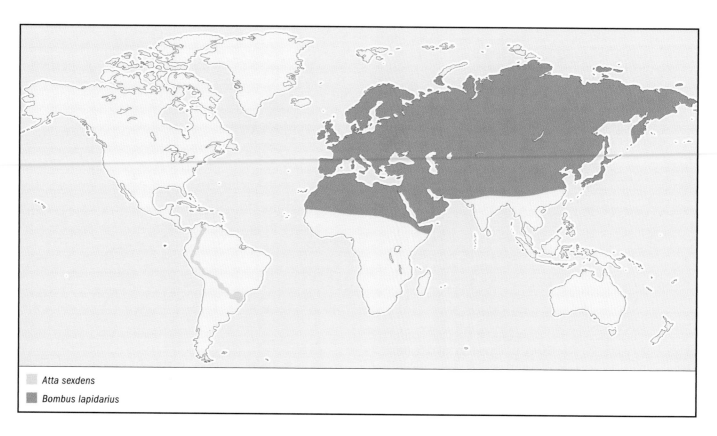

Atta sexdens
Bombus lapidarius

Large carpenter bee
Xylocopa virginica

FAMILY
Apidae

TAXONOMY
Apis virginica Linnaeus, 1771.

OTHER COMMON NAMES
English: Drone carpenter bee.

PHYSICAL CHARACTERISTICS
Body length is about 1.28 in (20 mm). The body is mostly black. The antennal scape is completely dark; the thorax has yellow hair both dorsally and laterally. The abdomen is black, with a slight purplish tint, and the legs have dark brown hair. Males have white areas on the head.

DISTRIBUTION
Eastern United States southward to Texas and northern Florida.

HABITAT
Adults are seen in well-lighted areas.

BEHAVIOR
These bees make their nests in wood. They nest on wood that is not painted and usually construct the nest in the same areas for generations. Carpenter bee tunnels are about 4–6 in (101.6–152.4 mm) long and 0.5 in (12.7 mm) in diameter. The bees frequently reuse old tunnels for pollen storage and overwintering chambers. They prefer rotten or seasoned wood.

FEEDING ECOLOGY AND DIET
Adults feed on nectar; larvae on pollen mixed with nectar.

REPRODUCTIVE BIOLOGY
After excavating the gallery, female bees gather pollen, which is mixed with regurgitated nectar. The pollen mass is put at the end of a gallery, an egg is laid, and the female places a partition or cap over the cell, composed of chewed wood pulp. This process is repeated until a linear complement of six to eight end-to end cells is completed. Bees emerge in the late summer and overwinter as adults, with mating taking place in spring.

CONSERVATION STATUS
Not listed by the IUCN.

SIGNIFICANCE TO HUMANS
The most important damage done by carpenter bees is weakening of structural timbres and gallery excavation on wooden water tanks. They are, however, valuable pollinators of wild and domestic fruit blossoms. Females of carpenter bees may sting (rarely), and male bees may hover or dart at humans who venture into the nesting area. ◆

Cuckoo wasp
Chrysis coerulans

FAMILY
Chrysididae

TAXONOMY
Chrysis coerulans Fabritius, 1805.

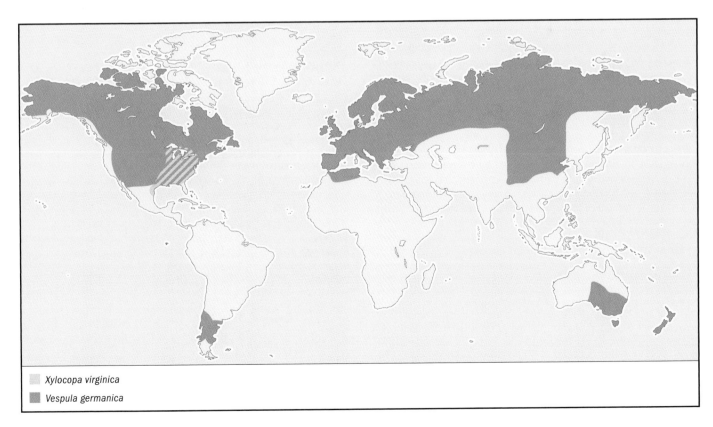

☐ *Xylocopa virginica*
■ *Vespula germanica*

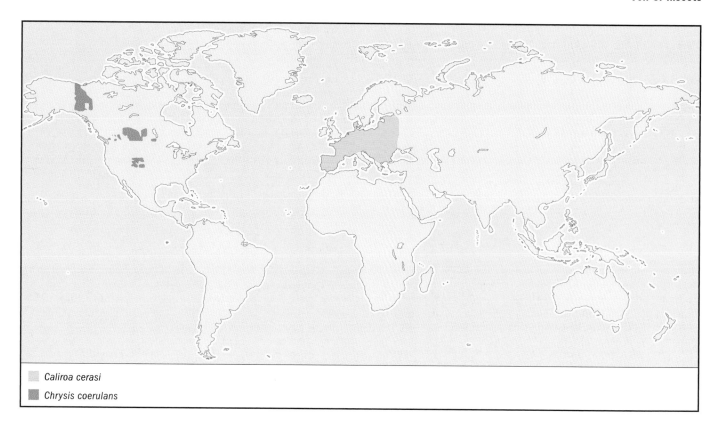

Caliroa cerasi

Chrysis coerulans

OTHER COMMON NAMES
Spanish: Avispas cuclillo.

PHYSICAL CHARACTERISTICS
Body length of 0.47 in (12 mm). Metallic green and blue in color and usually has a coarsely sculptured body. Fairly complete venation of the front wing but no closed cells in hind wing. Abdomen has only three visible segments and is hollowed out ventrally.

DISTRIBUTION
Canada (Yukon and southern Canada) and United States (Colorado).

HABITAT
Flowers, arid and sandy soils, old wood exposed to sun, and pebbles.

BEHAVIOR
Cleptoparasite of *Eumenes fraternus.*

FEEDING ECOLOGY AND DIET
Adults feed chiefly on flowers for nectar and pollen; larvae feed on larvae of the potter wasp. *Eumenes fraternus.*

REPRODUCTIVE BIOLOGY
The female uses her mandibles to penetrate the host nest or to gain access to the host by entering an open cell that is to be provisioned. If the host attacks the female, her concave metasoma allows the female to roll up into an impenetrable ball and play dead. The larva hatches and addresses itself immediately to the consumption of the host egg and the paralyzed caterpillars that were to serve as food for the host larvae.

CONSERVATION STATUS
Not listed by the IUCN.

SIGNIFICANCE TO HUMANS
They are more injurious than beneficial to humans because they attack wasps (*Eumenes*) that are themselves beneficial to humans. However, adult cuckoo wasps do serve as pollinators. ◆

No common name
Apoanagyrus lopezi

FAMILY
Encyrtidae

TAXONOMY
Apoanagyrus lopezi De Santis, 1964, Chacras de Coria, Argentina.

OTHER COMMON NAMES
None known.

PHYSICAL CHARACTERISTICS
Small wasp, 0.027 in (1.4 mm) long. Black or shiny blue and violet coloring, with light areas.

DISTRIBUTION
Argentina, Brazil, Colombia, and Paraguay. It was introduced and established in Africa (Senegal, Gambia, Guinea Bissau, Guinea Conakry, Sierra Leona, Ivory Coast, Ghana, Togo, Benin, Nigeria, Cameroon, Niger, Central African Republic, Gabon, Democratic Republic of the Congo, Angola, Rwanda, Burundi, Tanzania, Malawi, Mozambique, and the Republic of South Africa).

HABITAT
Savanna, tropical rainforest, and highlands.

BEHAVIOR
Parasitoid on the nymphal stages of the cassava mealybug, *Phenococcus manihoti* (Hemiptera: Pseudoccocidae).

FEEDING ECOLOGY AND DIET
Adults feed of nectar and pollen; immature insects are parasitoids.

REPRODUCTIVE BIOLOGY
Females copulate only once, whereas males may do it several times. Parthenogenesis also occurs, in which case all offspring are male. Females deposit about 40 eggs. Females feed on the exudate of wounds inflicted by the ovipositor on nymphal stages of the mealybug that serves as host for the young. The life cycle is complete in 11–25 days: egg, 2 days; larva, 6 days; prepupa, 4 days; pupa, 6 days.

CONSERVATION STATUS
Not threatened.

SIGNIFICANCE TO HUMANS
One of the most successful biological control programs used these wasps to eradicate the cassava mealybug, *Phenacoccus manihoti*. This pest was first reported in the Democratic Republic of the Congo in 1971, and within a few years it had infested nearly all of tropical Africa, devastating the primary sources of nutrition for 200 million people. The introduction of *Apoanagyrus lopezi* immediately produced spectacular control in field trials in Nigeria; by 1990 these parasitoids had been established successfully in 24 African countries and had spread over more than 1 million mi² (2 million km²). The mealybug is now under complete control throughout the whole of its range in Africa. ◆

Leaf cutter ant
Atta sexdens

FAMILY
Formicidae

TAXONOMY
Formica sexdens Linnaeus, 1758, Paramaribo, Surinam.

OTHER COMMON NAMES
Spanish: Hormigas cortadoras, sepes.

PHYSICAL CHARACTERISTICS
The dorsum of the thorax bears three pairs of teeth or spines. The body surface is partly tuberculate and reddish. Major workers or soldiers are 0.43–0.47 in (11–12 mm) in length.

DISTRIBUTION
Costa Rica, Panama, Colombia, Venezuela, Guiana, Ecuador, Peru, Bolivia, Brazil (Rondônia, Acre, Amazonas, Pará, Amapá, Maranhao, Piaui, Ceará, Rio Grande do Norte, Paraiba, Pernambuco, Sergipe, Alagoas, Bahia, Minas Gerais, Goias, and Mato Grosso), Paraguay, and Argentina.

HABITAT
Colonies are found throughout the rainforest floor, tropical deciduous forest, and tropical scrub forest.

BEHAVIOR
Eusocial insects, living in large colonies. Their nests are in the ground, and they bring cut-up leaves from several kinds of plants into them. These materials are chewed and processed to make a nutrient medium on which a certain species of fungus grows. The colony is formed when a queen has been fertilized. There are two major female castes, the reproductive queen (winged) and the workers (wingless). Males are winged. Distinct subcastes are called, according to their size, minor, media, or major workers or soldiers. They exhibit an elaborate range of behaviors, including foraging, cutting living plant tissues, defense of the nest, and care of broods. The nest of leaf cutters expands into labyrinths of chambers located near the surface, which contains the fungus gardens. Large dipper pits hold detritus and waste. A few of these pits contain more soil than organic matter, which is needed as a cover, especially for pathogenic waste. There are ventilation channels in which hot air rises from the refuse chambers and cools, and oxygen-rich air is drawn into the nest.

FEEDING ECOLOGY AND DIET
Feeds on fungus.

REPRODUCTIVE BIOLOGY
The female mates with three to eight males. She stores 206–320 million sperm for ten or more years. A queen can produce up to 150 million daughters, the majority of them workers. The queen establishes the nest and continues to produce eggs for the duration of the nest's existence. At the appropriate time, reproductive males and females make their nuptial flight, mate, and the females attempt to found a new colony.

CONSERVATION STATUS
Not threatened.

SIGNIFICANCE TO HUMANS
Because they attack most kinds of vegetation, including crop plants, they are serious economic pests. Leaf cutter ants dominate the ecosystems in which they occur. These ants are a good source of protein for humans, and they are eaten in parts of Mexico. Indians used the jaws of the soldier ant as sutures to hold together the edges of wounds. ◆

No common name
Ibalia leucospoides

FAMILY
Ibaliidae

TAXONOMY
Ichneumon leucospoides Hochenwarth, 1785.

OTHER COMMON NAMES
None known.

PHYSICAL CHARACTERISTICS
Body length of 0.63–0.67 in (16–17 mm). Body and legs are black. Female abdomen is somewhat elongated; in males, it is pyriform.

DISTRIBUTION
Originally from central Europe, it has been introduced to New Zealand, Australia, Tasmania, and Argentina (Andean Patagonia).

HABITAT
Adults may be seen on the bark of trees.

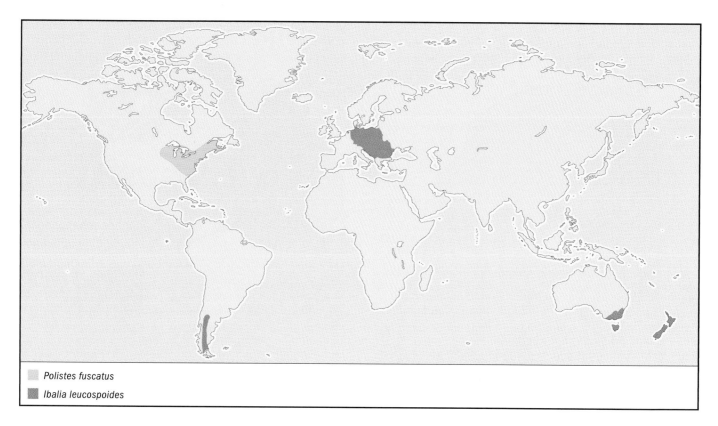

Polistes fuscatus
Ibalia leucospoides

BEHAVIOR
Primary solitary endoparasitoids of hymenopterous, wood-boring larvae (Siricidae).

FEEDING ECOLOGY AND DIET
Adults feed on nectar and honeydew; larvae feed on sirex larvae.

REPRODUCTIVE BIOLOGY
The males emerge first, and mating takes place while the female is in the act of laying eggs. When attacking its host in the host's tunnel, the parasitoid female inserts the ovipositor into the tunnel entrance, and the stalked egg is placed either in the egg of the host or in the newly hatched larva. The mature larva emerges from the body of the host and completes its feeding externally.

CONSERVATION STATUS
Not threatened.

SIGNIFICANCE TO HUMANS
Their biological control action is complementary to that of *Megarhyssa nortoni*, which attacks the last instars of sirex larvae. ◆

No common name
Megarhyssa nortoni

FAMILY
Ichneumonidae

TAXONOMY
Rhyssa nortoni Cresson, 1864, Colorado, United States.

OTHER COMMON NAMES
None known.

PHYSICAL CHARACTERISTICS
Giant wasps, 0.59–1.77 in (15–45 mm) long. The female ovipositor can be twice as long as the body, so the wasp can reach overall lengths in excess of 6.29 in (130 mm). Body is colored black, reddish-brown, and yellow and has a distinctive series of round yellow spots down the side of the abdomen.

DISTRIBUTION
United States. Introduced into New Zealand, Tasmania, and Brazil.

HABITAT
These wasps are found commonly around pine plantations.

BEHAVIOR
Primary solitary endoparasitoids of last instar larvae of hymenopterous wood-boring larvae (Siricidae).

FEEDING ECOLOGY AND DIET
Adults feed on nectar; larvae feed on sirex larvae.

REPRODUCTIVE BIOLOGY
Female sirex wasps lay their eggs in pine trees and introduce a special wood-digesting fungus at the same time. The sirex larvae then bore through the wood, which is digested with the aid of the fungus. It is the smell of this fungus that attracts the female *Megarhyssa* to an infected tree. She uses her long ovipositor to bore through the wood until she encounters a sirex larva and then paralyzes it before laying an egg on it. The parasitoid completes development within the host.

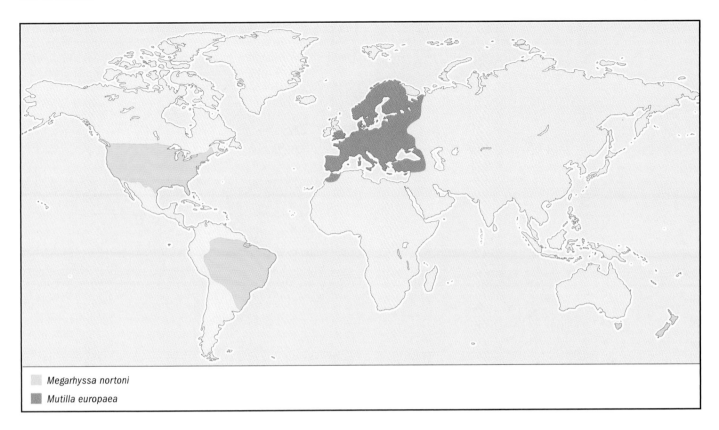

Megarhyssa nortoni

Mutilla europaea

CONSERVATION STATUS
Not threatened.

SIGNIFICANCE TO HUMANS
Megarhyssa nortoni has been utilized for the biological control of siricid forest pests; its action is complementary to the action of *Ibalia leucospoides*, which attacks the first instars of sirex. It is harmless to people. ◆

Alfalfa leaf cutter bee
Megachile rotundata

FAMILY
Megachilidae

TAXONOMY
Apis rotundata Fabritius, 1787.

OTHER COMMON NAMES
English: Leaf cutter bee; Spanish: Abeja polinizadora de la alfalfa.

PHYSICAL CHARACTERISTICS
Moderate size, about 0.31–0.35 in (8–9 mm). A fairly stout-bodied bee. Black with yellow pilosity in the front and at the dorsum of the thorax and whitish pilosity in ventrolateral areas, legs, and mouth. Abdomen has short hairs that are white and yellow along the dorsal ridge. Females have pollen brushes on the ventral side of abdomen.

DISTRIBUTION
Originally from Europe, but it was introduced into North America. It is distributed in the Neartic, particularly in Canada; from Canada it was introduced to Chile.

HABITAT
Grasslands.

BEHAVIOR
Solitary. Cuts leaves and petal pieces to line cells for offspring in nests. When they cut leaves, they leave a characteristic oblong or circular cutout on the edge of the leaf or petal. Several oblong leaf pieces line individual cells in the nest.

FEEDING ECOLOGY AND DIET
Adults feed on nectar; larvae eat pollen and nectar.

REPRODUCTIVE BIOLOGY
The cells are filled with pollen and nectar. A single egg is laid in each cell, which is capped with round leaf pieces. The bee then starts over, collecting oblong leaf pieces.

CONSERVATION STATUS
Not threatened.

SIGNIFICANCE TO HUMANS
Valuable as pollinators of alfalfa crops. ◆

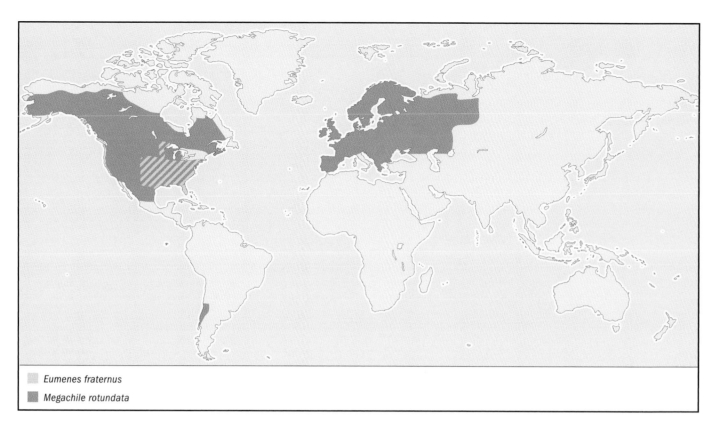

Eumenes fraternus

Megachile rotundata

Velvet ant
Mutilla europaea

FAMILY
Mutillidae

TAXONOMY
Mutilla europaea Linnaeus, 1758.

OTHER COMMON NAMES
English: Cow-killer, mule-killer; Spanish: Avispas ater-
ciopeladas.

PHYSICAL CHARACTERISTICS
Body length of 0.47–0.55 in (12–14 mm). Females are wingless,
and males are winged. Both have downy hairs covering the
body; they are brownish-red in color, except for the abdomen,
which is an iridescent black-blue with white, hairy bands.

DISTRIBUTION
Europe.

HABITAT
Adults commonly are seen on the ground in open areas, especially in sandy places.

BEHAVIOR
Females attack larvae and pupae of bumblebees.

FEEDING ECOLOGY AND DIET
Adults feed on nectar; their grubs eat the food provisions
stored in the invaded nest and also the grubs and pupae of the
host.

REPRODUCTIVE BIOLOGY
Velvet ants have an unusual courtship ritual, with mating accomplished in the air. The winged male carries the flightless
female while copulating. Females lay the eggs with those of the
bumblebee, on which the larvae later feed. The mature larvae
knit a cocoon within the cell cocoon, or puparium, of the host.

CONSERVATION STATUS
Not threatened.

SIGNIFICANCE TO HUMANS
This family must be regarded as predominantly injurious, inas-
much as the majority of insects that they attack are predaceous
or parasitic upon other insects. They also have the habit of attacking such pollinators as honeybees or bumblebees. Females
have a sting that is painful to humans. ◆

No common name
Pelecinus polyturator

FAMILY
Pelecinidae

TAXONOMY
Ichneumon polyturator Drury, 1773, type locality not specified.

OTHER COMMON NAMES
None known.

PHYSICAL CHARACTERISTICS
Body length of 2 in (50.8 mm). Females are shiny black with
an extremely long, narrow abdomen. Males have a much

Pepsis grossa

Pelecinus polyturator

shorter abdomen but are extremely rare and are almost never seen.

DISTRIBUTION
Southern portions of the eastern provinces of Canada, the eastern United States (west to North Dakota, Colorado, and New Mexico) and Mexico south to central Argentina.

HABITAT
Deciduous woodlands.

BEHAVIOR
Larvae are endoparasitoids of June beetles (Scarabaeidae). Females are active in late summer and early fall.

FEEDING ECOLOGY AND DIET
Adults feed of nectar and pollen; immature stages are parasitoids.

REPRODUCTIVE BIOLOGY
Little is known about the reproductive biology. The species has been reared from larvae of Scarabaeidae (Coleoptera), particularly species of *Phyllophaga*, and it appears to be a solitary endoparasitic koinobiont. Males are very rarely collected. Tropical populations are bisexual, and temperate populations consist only of females (geographic parthenogenesis).

CONSERVATION STATUS
Not listed by the IUCN.

SIGNIFICANCE TO HUMANS
The metasoma of female pelecinids is extremely flexible; when handled, they occasionally are capable of inflicting a sting, but it is a mild sensation, like a pinprick. ◆

Tarantula hawk
Pepsis grossa

FAMILY
Pompilidae

TAXONOMY
Sphex grossa Fabritius, 1798, India (in error).

OTHER COMMON NAMES
English: Spider wasp, spider-hunting wasp; Spanish: San Jorge, avispón, matacaballos, halcón de las arañas.

PHYSICAL CHARACTERISTICS
Slender, with long, spiny legs. Body length is 0.94–2 in (24–51 mm). Body and legs are black with a mainly blue-green shine, often with a violet or copper tinge. Antennae are black, usually with orange on the tip of the last segment. Wings typically are black with quite strong blue-violet reflections, sometimes diffuse amber or orange, often with a dark border.

DISTRIBUTION
Southern United States and the West Indies south through Mexico to north-central Peru and the Guianas.

HABITAT
Rainforest to desert.

BEHAVIOR
Adults typically are found on flowers or on the ground in search of prey. The members of this genus make their nests in burrows in the ground and provision them exclusively with mygalomorph spiders (usually Theraphosidae, or tarantulas).

FEEDING ECOLOGY AND DIET
Adults feed on nectar; larvae feed on spiders.

REPRODUCTIVE BIOLOGY
Adults capture and paralyze a spider and then prepare a cell for it in the ground, in rotten wood, or in a suitable crevice in rocks. An egg is laid on the victim; the spider is buried alive with the larva, which hatches within a few days. It pupates after it has consumed the spider.

CONSERVATION STATUS
Not listed by the IUCN.

SIGNIFICANCE TO HUMANS
This species of wasp is especially attracted to the flowers of *Asclepias* species (Asclepiadaceae), and it probably forms the largest group of pollinators. The female wasp's sting can be excruciatingly painful, but they are not known to attack humans without being provoked. It is difficult to assess the economic effect of their predation on spiders, because, in most cases, the economic significance of the spiders themselves is unknown. ◆

No common name
Trissolcus basalis

FAMILY
Scelionidae

TAXONOMY
Telenomus basalis Wollaston, 1858, Madeira Archipelago.

OTHER COMMON NAMES
None known.

PHYSICAL CHARACTERISTICS
Body length of 0.04–0.05 in (1–1.3 mm). Black with downward-pointing elbow-shaped antennae and a flattened abdomen. Wing veins are reduced.

DISTRIBUTION
United States (Florida, Louisiana, Mississippi, and South Carolina), West Indies, Venezuela (Aragua), Brazil (Minas Gerais, Paraná, and Distrito Federal), Australia (Canberra), Egypt, Ivory Coast, Morocco, Republic of South Africa (Pretoria), Zimbabwe (Harare), Portugal, France, and Italy.

HABITAT
Occurs in all crops attacked by bugs, including cotton, grains, soybeans and other legumes, tomatoes and other solanaceous crops, sweet corn, sunflowers, cole crops, cucurbits, and fruits and nuts.

BEHAVIOR
Primarily solitary idiobiont endoparasitoid of pentatomid eggs; completes development from egg to adult within the host egg.

FEEDING ECOLOGY AND DIET
Adults feed on nectar; larvae eat the eggs of bugs.

REPRODUCTIVE BIOLOGY
Adults mate immediately after emerging from host eggs. The female inserts her egg into a host egg. After oviposition, females apply an external marker with the ovipositor to avoid superparasitism. There are three larval instars, and most feeding occurs during the last two. Pupation takes place in the host egg.

CONSERVATION STATUS
Not threatened.

SIGNIFICANCE TO HUMANS
Scelionids have been used quite successfully in classic biological control programs directed principally against pest hemipterans. The species has been introduced into many different countries to control *Nezara viridula* (Pentatomidae). ◆

Digger wasp
Scolia dubia

FAMILY
Scoliidae

TAXONOMY
Scolia dubia Say, 1837.

OTHER COMMON NAMES
English: Blue-winged wasp.

PHYSICAL CHARACTERISTICS
Body length is 0.63 in (16 mm). Hairy and blue-black in color, with two yellow spots (one on each side of the abdomen). Behind the yellow spots, the abdomen is brownish and the hairs on the body more noticeable. The wings are dark blue.

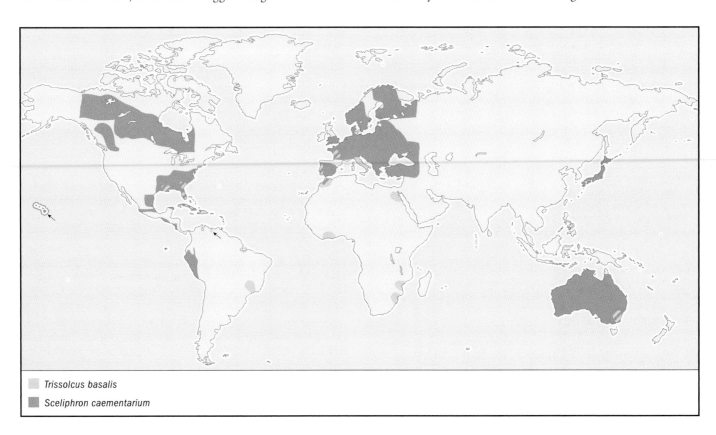

□ *Trissolcus basalis*
■ *Sceliphron caementarium*

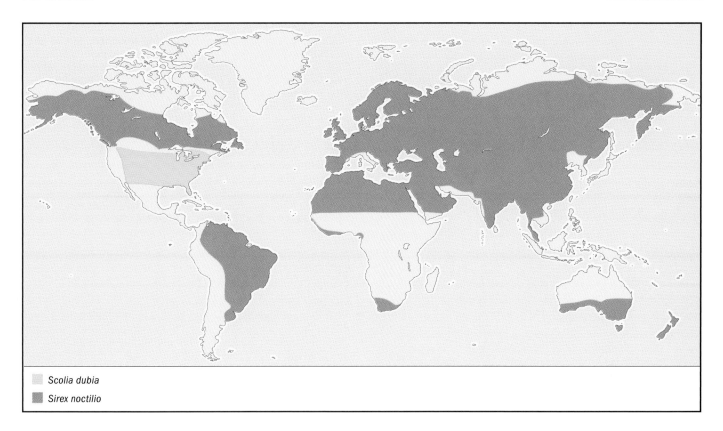

□ Scolia dubia
■ Sirex noctilio

DISTRIBUTION
New England to Florida and west to the Rocky Mountains.

HABITAT
Adults often are found on flowers. These wasps appear in the early morning and fly low over lawns infested with scarabeid larvae. The flight is characteristic: they fly on a horizontal plane only a few inches above the ground and following a circular or figure eight–shaped course. Males become active somewhat earlier than females, and mating takes place largely on the ground or on low vegetation as soon as females appear.

BEHAVIOR
The female dig into soil or fallen leaves with her powerful forelegs to locate a host. On finding a grub, she stings and paralyzes it. She may burrow 0.5 in (1.2 cm) deeper to construct a cell around the host. Then she lays an egg on the outside of the grub. She may sting many grubs without laying eggs on them; such grubs usually do not recover. Unmated individuals of both sexes are gregarious (form aggregations); after mating this gregariousness is abandoned.

FEEDING ECOLOGY AND DIET
Adults feed on nectar derived from the blossoms of various plants; larvae feed on beetle larvae. Green June beetle, May beetle, and Japanese beetle grubs seem to be the primary host.

REPRODUCTIVE BIOLOGY
Females attract males by spreading their wings so that the brilliantly colored body is revealed. Eggs are attached lightly by the posterior end to the body of the host. When the larva hatches, it finds itself provided with food, which has been preserved alive. There are four larval instars; the mature larva spins a cocoon. During the larval stages the entire contents of the beetle larvae are consumed.

CONSERVATION STATUS
Not threatened.

SIGNIFICANCE TO HUMANS
These wasps are natural agents in the control of grubs in the soil. These digger wasps do not sting people unless they are aggravated or captured by hand. ◆

European wood wasp
Sirex noctilio

FAMILY
Siricidae

TAXONOMY
Sirex noctilio Fabricius, 1793.

OTHER COMMON NAMES
English: Wood wasp, sirex; Spanish: Avispa taladradora de la madera.

PHYSICAL CHARACTERISTICS
Male body length 0.7 in (18 mm); female body length 1.18 in (30 mm). Metallic blue-black with a long ovipositor. Male's abdomen predominantly yellowish-red.

DISTRIBUTION
Originally from Eurasia and northern Africa and now established in New Zealand, Tasmania, Australia, South Africa, Canada, and South America.

HABITAT
Larvae live in wood galleries of such conifers as *Abies*, *Larix*, *Picea*, *Pinus*, and *Pseudotsuga*.

BEHAVIOR
The males form swarms over trees, and mating takes place in the swarms. With her well-developed ovipositor, the female inserts eggs into the tissues of the host plant. Wood wasps carry *Amylostereum* fungal spores in invaginated intersegmental sacs connected to the ovipositor. During egg laying, spores and mucus are injected into the sapwood of trees, causing mycelial infection. The infestation produces a locally dry condition around the xylem, which allows for optimal development of the larvae.

FEEDING ECOLOGY AND DIET
Adults do not feed. Larvae rely on fungal hyphae to break down wood into its digestible components.

REPRODUCTIVE BIOLOGY
The larvae are wood boring. Pupation occurs in galleries made by the larva. They usually spend winter in the tunnels of the host plant.

CONSERVATION STATUS
Not threatened.

SIGNIFICANCE TO HUMANS
Wood-boring larvae, in association with fungi, can cause extensive damage to plantations. It is considered a serious pest of forestry plantations of conifers, notably *Pinus* species. In Australia and New Zealand the fungal disease can cause death of fire-damaged trees or those stressed by drought conditions. Biological control programs include the use of natural enemies, such as the parasitic wasps *Rhyssa*, *Megarhyssa*, and *Ibalia* and the nematod *Deladenus siricidicola*. ◆

Mud dauber wasp
Sceliphron caementarium

FAMILY
Sphecidae

TAXONOMY
Sphex caementarium Drury, 1773.

OTHER COMMON NAMES
English: Yellow and black mud dauber.

PHYSICAL CHARACTERISTICS
Large wasps, with a body length of 0.7–1.1 in (20–30 mm). They are predominantly matte black with a few yellow areas on the thorax and first tergum of the abdomen. The legs are partly yellow and the wings deep amber.

DISTRIBUTION
Canada (Alberta, British Columbia, Manitoba, New Brunswick, Nova Scotia, Ontario, Prince Edward Island, and Quebec), the United States (Florida and Michigan), Central America, and the West Indies. Introduced to Australia, the Pacific islands (Hawaii and Japan), French Polynesia, Samoa, Fiji, the Marshall and Mariana Islands, Peru, and Europe.

HABITAT
Lives in wet areas where mud can be found.

BEHAVIOR
Solitary. Adults emerge in the spring after overwintering, staying about nine months as larvae or pre-pupae. Females spend a few days exploring and feeding on nectar. Afterward, the female mates and hunts for a source of mud to begin the construction of her nest. Then she takes up the search for spiders, which she paralyzes with her sting and carries back to the cell. After completion of one cell, she begins another, until several have been constructed and provisioned. The cells have a characteristic ridged appearance at this point. The entire assemblage of cells finally is coated with a thin layer of mud and smoothed over.

FEEDING ECOLOGY AND DIET
Adults feed on nectar; larvae eat spiders.

REPRODUCTIVE BIOLOGY
The female lays an egg on a spider near the base of the cell, and the cell is capped with a mud plug. The larvae develop on spiders, which the adult female finds on foliage and flowers. Pupation takes place within a cocoon inside the cell. After transforming into an adult wasp, it breaks out of the nest.

CONSERVATION STATUS
Not threatened.

SIGNIFICANCE TO HUMANS
Mud daubers generally do not sting unless provoked, because they do not defend their nests. Their nests can be a nuisance in garages, under the eaves of houses, and in other types of buildings. ◆

Pear and cherry slug
Caliroa cerasi

FAMILY
Tenthredinidae

TAXONOMY
Tenthredo cerasi Linnaeus, 1758.

OTHER COMMON NAMES
English: Pear sawfly, cherry sawfly pear slug, pear slugworm, cherry slugworm, pear slug sawfly, slugworm; Spanish: Chape del cerezo, babosita del cerezo, chape del peral, babosita de los perales, babosita de los frutales, babosa del peral; French: cherry slug tenthride du Poirier, celendre noire; Dutch: Zaawesp van kersen en peren.

PHYSICAL CHARACTERISTICS
Adults are small, with a body length of 0.31 in (8 mm). Wide body, with the abdomen strongly joined to the thorax. Black, smooth, and shiny. Wide and hyaline wings. Females have a sawlike ovipositor.

DISTRIBUTION
Presumed to be of European origin. It has been introduced into the New World (Argentina, Colombia, Uruguay, and Chile), Africa, New Zealand, and Tasmania.

HABITAT
Usually found on foliage or flowers.

BEHAVIOR
The larvae are phytophagous, feeding externally and skeletonizing leaves. There is usually a single generation per year; the insect overwinters in a pupal cell or cocoon in the ground or in another protected situation.

FEEDING ECOLOGY AND DIET
Adults feed on nectar; larvae are external feeders on foliage.

REPRODUCTIVE BIOLOGY
The female lays her eggs on leaves in the spring. She slits the leaf tissue with her ovipositor to deposit small, oval, flattened eggs, which hatch in about two weeks. The eggs are laid between the parenchyma and cuticle. The olive green to black larvae are slug-like, with a well-developed head capsule, thoracic legs, and abdominal prolegs. They are covered with a greenish slime. In autumn larvae make small earthen cells in the soil in which to spend the winter, before pupating in early spring.

CONSERVATION STATUS
Not threatened.

SIGNIFICANCE TO HUMANS
The pear and cherry slug is a cosmopolitan pest in cherry-, plum-, and pear-growing regions of the world. ◆

Potter wasp
Eumenes fraternus

FAMILY
Vespidae

TAXONOMY
Eumenes fraternus Say, 1824.

OTHER COMMON NAMES
English: Mason wasp; Spanish: Avispas alfareras.

PHYSICAL CHARACTERISTICS
Body length of 0.51–0.66 in (13–17 mm). Black with yellow markings on the thorax and abdomen. Wings are smoky with violet iridescence. The first abdominal segment is narrow and stalklike and is followed by a broad, bell-shaped second segment. Middle tibia has one apical spur.

DISTRIBUTION
Minnesota, Nebraska, Kansas, Oklahoma, Texas, and eastward to the Atlantic coast of the United States.

HABITAT
Adults are seen on the ground in open areas or at flowers.

BEHAVIOR
Solitary. These wasps make their nests of mud in the form of miniature ceramic-like pots or jugs. The pots are globular with a narrow neck, which has an expanded rim.

FEEDING ECOLOGY AND DIET
Adults feed on nectar; larvae eat caterpillars.

REPRODUCTIVE BIOLOGY
The female lays a single egg suspended from a wall by a slender filament in the empty nest. She then provisions the nest with 1 to 12 caterpillars (which have been paralyzed by her sting) and seals the nest. The hatching wasp maggot consumes the stored prey in the nest. After transforming into an adult, it breaks out of the nest.

CONSERVATION STATUS
Not listed by the IUCN.

SIGNIFICANCE TO HUMANS
These wasps are not aggressive and rarely sting people, because they do not defend their nests. It has been reported that potter wasp nests served as models for the clay vessels made by Indians. ◆

Golden paper wasp
Polistes fuscatus

FAMILY
Vespidae

TAXONOMY
Polistes fuscatus Fabritius, 1793.

OTHER COMMON NAMES
English: Northern paper wasp, umbrella wasp.

PHYSICAL CHARACTERISTICS
Body length of 0.75–1 in (19.5–25.4 mm). Body is dark reddish-brown and yellow, elongated, and slender, with a spindle-shaped abdomen. Wings are amber to reddish-brown. Males have a pale face, and females have a brown face.

DISTRIBUTION
British Columbia to the Canadian Maritime Provinces and south to West Virginia.

HABITAT
Meadows, fields, and gardens on flowers and near buildings.

BEHAVIOR
Primitively eusocial hymenopterans. Their nests consist of a single more or less circular horizontal comb of paper cells, suspended from a support by a slender stalk. Mixing masticated wood pulp with adhesive saliva, the female builds an open comb consisting of a single layer of hexagonal cells. These paper cells act as a nesting chamber. The comb is attached to the substrate by a strong stalk.

FEEDING ECOLOGY AND DIET
Adults feed on nectar and juices from crushed and rotting fruits; larvae feed on insects that have been pre-chewed by adults.

REPRODUCTIVE BIOLOGY
Female lays an egg in each cell. The cells are open on the lower side while the larvae are growing and are sealed when the larvae pupate. After hanging into an adult wasp, they break out of the nest.

CONSERVATION STATUS
Not listed by the IUCN.

SIGNIFICANCE TO HUMANS
Paper wasps are considered beneficial insects owing to their predation on garden pests. Paper wasps will defend their nests if disturbed, and they sting repeatedly. ◆

Yellow jacket
Vespula germanica

FAMILY
Vespidae

TAXONOMY
Vespa germanica Fabricius, 1793.

OTHER COMMON NAMES
English: German wasp; Spanish: Chaqueta amarilla, avispa chaqueta amarilla, avispa carnívora, avispa alemana.

PHYSICAL CHARACTERISTICS

Body length of 0.41–0.80 in (10.5–20.3 mm). Black antennae and yellow and black body, with black arrowhead-shaped markings pointing backward along the top of the abdomen and black spots on either side.

DISTRIBUTION

Worldwide, including Europe, northern Africa, and temperate Asia. It was introduced into Canada, the United States, New Zealand, Tasmania, Australia, Argentina, and Chile.

HABITAT

These wasps thrive in urban areas; their liking for fruit, meat, and sweets brings them into contact with people.

BEHAVIOR

Eusocial. These wasps live in hidden nests, which are either underground or in the wall cavities or ceilings of buildings. A mature nest is the size of a football and is constructed of a papery material. The combs are covered completely with a balloon-like envelope. Adult forms include fertile females (queens), workers (females, usually sterile), and fertile males. In late summer, colonies produce new queens and males. After mating, these new queens go into hibernation. Males and workers do not survive the winter. The following spring the queen emerges from hibernation and searches for a suitable nesting site. She then collects wood or other vegetable fiber from shallow cells. Females cooperate in nesting, exhibiting reproductive division of labor.

FEEDING ECOLOGY AND DIET

Adults forage for meats, sweets, and fruits. The larvae are fed by the workers with masticated portions of animal matter and, at times, with fruit juices, nectar, and honeydew.

REPRODUCTIVE BIOLOGY

The queen lays an egg in each cell and then protects the resulting larvae, feeding them daily. After 12–18 days the larvae spin cocoon caps over their cells and transform into pupae. When adults emerge about 12 days later, they serve as the first brood of workers, and the queen resumes egg laying.

CONSERVATION STATUS

Not listed by the IUCN.

SIGNIFICANCE TO HUMANS

Because yellow jackets forage for meats, sweets, ripe fruit, and garbage, they pose a threat to humans. They are a particular problem in picnic areas and orchards and around garbage containers. The sting is painful but not usually serious, although some people suffer a bad allergic reaction. Stings inside the throat are potentially fatal and require immediate medical attention. ◆

Resources

Books

Artigas, Jorge N. *Entomología Económica: Insectos de Interés Agrícola, Forestal, médico y Veterinario.* 2 vols. Concepción, Chile: Ediciones Universidad de Concepción, 1994.

Borror, Donald J., Charles A. Triplehorn, and Norman F. Johnson. *An Introduction to the Study of Insects.* Philadelphia: W. B. Saunders, 1989.

Clausen, C. P. *Entomophagous Insects.* New York: Hafner, 1972.

Correa-Ferrêira, B. S. *Utilizaçao do Parasitóide de Ovos* Trissolcus basalis *(Wollaston) no Controle de Percevejos da Soja.* Centro Nacional de Pesquisa de Soja, Circular Técnica no. 11. Londrina, Brazil: EMBRAPA Soja, 1993.

Costa-Neto, Eraldo M. *Manual de Etnoentomología.* Manuales & Tesis Sociedad Entomológica Aragonesa (SEA) 4. Zaragoza, Spain: Sociedad Entomológica Aragonesa, 2002.

CSIRO, eds. *The Insects of Australia: A Textbook for Students and Research Workers.* Carlton, Australia: Melbourne University Press, 1991.

Danks, H. V., and J. A. Downes, eds. *Insects of Yukon: Biological Survey of Canada (Terrestrial Arthropods).* Ottawa, Canada: Canadian Museum of Nature, 1997.

De Santis, L., and P. Fidalgo. *Catálogo de los Himenópteros Calcidoideos de América al Sur de los Estados Unidos.* Serie de la Academia Nacional de Agronomía y Veterinario no. 13. Buenos Aires, Argentina: Editorial Hemisferio Sur, 1994.

Gullan, P. J., and P. S. Cranston. *The Insects: An Outline of Entomology.* Oxford and Malden, MA: Blackwell Science, 2000.

Hanson, P. E., and I. D. Gauld. *The Hymenoptera of Costa Rica.* Oxford: Oxford University Press, 1995.

Hölldobler, Bert, and Edward O. Wilson. *The Ants.* Cambridge, MA: Harvard University Press, 1990.

Samways, M. J. *Insect Conservation Biology.* London and New York: Chapman and Hall, 1994.

Johnson, N. F. *Catalog of World Species of Proctotrupoidea, Exclusive of Platygstridae (Hymenoptera).* Memoirs of the American Entomological Institute no. 51. Gainesville, FL: American Entomological Institute, 1992.

Llorente Bousquets, Jorge, and Juan J. Morrone, eds. *Biodiversidad, Taxonomía y Biogeografía de Artrópodos de Mexico: Hacia una Síntesis de su Conocimiento. Mejicana.* Vol. 3. Mexico: Universidad Autónoma de México, 2002.

Loiácono, M. S., N. B. Diaz, and L. De Santis. *Estado Actual del Conocimiento de Microhimenopteros Chalcidoidea, Cynipoidea, y "Proctotrupoidea" en Argentina.* Monografías Tercer Milenio, vol. 2. Zaragoza, Spain: Sociedad Entomológica Aragonesa (SEA) & CYTED, 2002.

Vardy, C. R. *The New World Tarantula-Hawk Wasp Genus* Pepsis *Fabricius (Hymenoptera: Pompilidae).* 2 parts. Backhuys. Netherlands: Zoologische Verhandelingen, 2000–2002.

Periodicals

De Santis, L. "Catálogo de los Himenópteros calcidoideos de América al Sur de los Estados Unidos." Second supplement. *Acta Entomológica Chilena* 15 (1989): 9–90.

De Santis, L., and N. P. Ras. "Control Biológico de la Cochinilla *Phenacoccus manihoti* en África (Insecta)." *Academia Nacional de Agronomía y Veterinaria* 42, no. 7 (1988): 5–11.

Johnson, N. F. "Systematics of New World *Trissolcus* (Hymenoptera: Scelionidae): Species Related to *T. basalis*)."

Resources

Canadian Entomologist 117, no. 4 (1985): 431–445.

Johnson, N. F., and L. Musetti. "Revision of the Prototrupoid Genus *Pelecinus* Latreille." *Journal of Natural History* 33 (1999): 1513–1543.

Kempf, W. W. "Catálogo Abreviado das Formigas da Regiao Neotropical." *Studia Entomologica* 15 (August 1972): 3–334.

Loiácono, M. S., and C. B. Margaría. "Ceraphronoidea, Playgastroidea, and Proctotrupoidea from Brazil (Hymenoptera)." *Neotropical Entomology* 31, no. 4 (2002): 551–560.

Other

"Chrysis.net" [May 6, 2003]. <http://chrysis.net>.

"Avispa Taladradora de la Madera. *Sirex noticlio* Fab." [May 6, 2003]. <http://gaf.utalca.cl/sirex.htm>.

"Apoidea" [May 6, 2003]. <http://www.bionet.nsc.ru/szmn/Hymenop/Apoidea.htm>.

Biblioteca de Atualizaçao. Ciêcias Entendendo a Natureza. "Vespa contra Vespa um Exemplo de Controle Biológico" [May 6, 2003]. <http://www.editorasaraiva.com.br/eddid/ciencias/biblioteca/artigos/megarhyssa.html>.

"Chapter 1: Honeybees of the Genus *Apis*" [May 6, 2003]. <http://www.fao.org/docrep/X0083e/X0083E02.htm>.

"European Wood Wasp, *Sirex noctilio* F.—Siricidae" [May 6, 2003]. <http://www.faculty.ucr.edu/legneref/biotact/ch-49.htm>.

The Natural History Museum. "*Bombus* " [May 6, 2003]. <http://www.nhm.ac.uk/entomology/bombus/ml.html>.

"Wild West Yorkshire Nature Diary: Ducks on the Water" [May 6, 2003]. <http://www.wildyorkshire.co.uk/naturediary/docs/apr00/apr08.html>.

Marta Loiácono, DSc
Cecilia Margaría, Lic

For further reading

Abe, T., D. E. Bignell, and M. Higashi, eds. *Termites: Evolution, Sociality, Symbiosis, Ecology.* Dordrecht, The Netherlands: Kluwer Academic Publishers, 2000.

Adams, J., ed. *Insect Potpourri: Adventures in Entomology.* Gainesville, FL: Sandhill Crane Press, Inc., 1992.

Agosta, William. *Thieves, Deceivers, and Killers: Tales of Chemistry in Nature.* Princeton, NJ: Princeton University Press, 2001.

Agosti, D., J. D. Majer, L. E. Alonso, and T. R. Schultz, eds. *Ants: Standard Methods for Measuring and Monitoring Biodiversity.* Washington, DC: Smithsonian Institution Press, 2000.

Akre, R. D., G. S. Paulson, and E. P. Catts. *Insects Did It First.* Fairfield, WA: Ye Galleon Press, 1992.

Andersen, Nils M. *The Semiaquatic Bugs (Hemiptera, Gerromorpha): Phylogeny, Adaptations, Biogeography, and Classification.* Entomonograph 13. Klampenborg, Denmark: Scandinavian Science Press, 1982.

Asahina, S. *Blattaria of Japan.* Tokyo: Nakayama-Shoten, 1991. (Japanese, with parts in English).

Aspöck, Horst, Ulrike Aspöck, and H. Hölzel. *Die Neuropteren Europas.* 2 vols. Krefeld, Germany: Goecke and Evers, 1980.

Aspöck, Horst, Ulrike Aspöck, and Hubert Rausch. *Die Raphidiopteren der Erde.* 2 vols. Krefeld, Germany: Goecke und Evers, 1991.

Bell, W. J., and K. G. Adiyodi, eds. *The American Cockroach.* London: Chapman and Hall, 1982.

Berenbaum, M. R. *Bugs in the System: Insects and Their Impact on Human Affairs.* Reading, MA: Addison-Wesley Publishing Company, 1995.

Berner, Lewis, and Manuel L. Pescador. *The Mayflies of Florida.* Rev. ed. Tallahassee: Florida A&M University Press, 1988.

Bey-Bienko, G. Y. "Fauna of the U.S.S.R." In *Insects.* Moscow: Institute of Zoology, Academy of Sciences of the USSR, 1950.

Bolton, B. *A New General Catalogue of the Ants of the World.* Cambridge, MA: Harvard University Press, 1995.

Borror, Donald J., Charles A. Triplehorn, and Norman F. Johnson. *An Introduction to the Study of Insects* 6th edition. Philadelphia: W. B. Saunders, 1989.

Boudreaux, H. B. *Arthropod Phylogeny with Special Reference to Insects.* New York: John Wiley and Sons, 1979.

Bourke, A. F. G., and N. R. Franks. *Social Evolution in Ants.* Princeton, NJ: Princeton University Press, 1995.

Bragg, Philip E. *Phasmids of Borneo.* Kota Kinabalu, Borneo: Natural History Publications, 2001.

Brock, Paul D. *The Amazing World of Stick and Leaf Insects.* Orpington, U.K.: Amateur Entomologists' Society, 1999.

———. *Stick and Leaf Insects of Peninsular Malaysia and Singapore.* Kuala Lumpur: Malaysian Nature Society, 1999.

———. *A Complete Guide to Breeding Stick and Leaf Insects.* Havant, U.K.: T. F. H. Kingdom Books, 2000.

Brown, J. H., and M. V. Lomolino. *Biogeography.* 2nd ed. Sunderland, MA: Sinauer Associates, Inc., 1998.

Buchman, S. L., and G. P. Nabhan. *The Forgotten Pollinators.* Washington, DC: Island Press, 1996.

Camazine, S., J.-L. Deneubourg, N. R. Franks, J. Sneyd, G. Theraulaz, and E. Bonabeau. *Self-Organization in Biological Systems.* Princeton, NJ: Princeton University Press, 2001.

Carpenter, F. M. *Treatise on Invertebrate Paleontology.* Part R, *Arthropoda.* Vol. 3, *Superclass Hexapoda.* Boulder, CO: Geological Society of America, 1992.

Chapman, R. F. *The Insects: Structure and Function.* 4th edition. Cambridge, U.K.: Cambridge University Press, 1998.

Choe, J. C., and B. J. Crespi, eds. *The Evolution of Social Behavior in Insects and Arachnids.* Cambridge, U.K., Cambridge University Press, 1997.

Chopard, L. "Ordre de Dermaptères." In *Traite de Zoologie.* Vol. 9, edited by P. P. Grassé. Paris: Masson and Cie, 1949.

Christiansen, Kenneth A., and Peter F. Bellinger. *The Collembola of North America North of the Rio Grande: A Taxonomic Analysis.* Grinnell, IA: Grinnell College, 1981.

Clausen, C. P. *Entomophagous Insects.* New York: Hafner, 1972.

Coleman, David C., and D. A. Crossley. *Fundamentals of Soil Ecology*. San Diego, CA: Academic Press, 1996.

Collins, N. M., and J. A. Thomas, eds. *The Conservation of Insects and Their Habitats*. London: Academic Press, 1991.

Contreras-Ramos, A. *Systematics of the Dobsonfly Genus Corydalus Latreille (Megaloptera: Corydalidae)*. Thomas Say Monographs. Lanham, MD: Entomological Society of America, 1998.

Corbet, P. S. *Dragonflies: Behavior and Ecology of Odonata*. Ithaca, NY: Cornell University Press, 1999.

Costa-Neto, Eraldo M. *Manual de Entomología*. Manuales & Tesis Sociedad Entomológica Aragonesa (SEA) 4. Zaragoza, Spain: Sociedad Entomológica Aragonesa, 2002.

Coville, Charles Jr. *A Field Guide to the Moths of Eastern North America*. Boston: Houghton Mifflin, 1984.

Cox, C. B., and P. D. Moore. *Biogeography: An Ecological and Evolutionary Approach*. Oxford, U.K.: Blackwell Science, 1998.

Craw, R. C., J. R. Grehan, and M. J. Heads. *Panbiogeography: Tracking the History of Life*. Oxford Biogeography Series 11. New York: Oxford University Press, 1998.

Crawford, C. S. *Biology of Desert Invertebrates*. Berlin: Springer-Verlag, 1981.

Croizat, L. *Panbiogeography*. Vols. 1, 2a, and 2b. Caracas, Venezuela: [n.p.], 1958.

———. *Space, Time, Form: The Biological Synthesis*. Caracas, Venezuela: [n.p.], 1964.

Crowson, R. A. *The Biology of the Coleoptera*. London: Academic Press, 1981.

Crozier, R. H., and P. Pamilo. *Evolution of Social Insect Colonies*. Oxford, U.K.: Oxford University Press, 1996.

CSIRO, eds. *The Insects of Australia: A Textbook for Students and Research Workers*. 2nd edition. Carlton, Australia: Melbourne University Press, 1991.

Danks, H. V., and J. A. Downes, eds. *Insects of Yukon: Biological Survey of Canada (Terrestrial Arthropods)*. Ottawa, Canada: 1997.

Dent, David. *Insect Pest Management*. 2nd edition. New York: CABI Publishing, 2000.

Domínguez, Eduardo, ed. *Trends in Research in Ephemeroptera and Plecoptera*. New York and London: Kluwer Academic/Plenum Publishers, 2001.

Dunkle, S. W. *Dragonflies Through Binoculars: A Field Guide to Dragonflies of North America*. New York: Oxford University Press, 2000.

Ehrmann, Reinhard. *Mantodea: Gottesanbeterinnen der Welt*. Münster, Germany: NTV, 2002.

Elias, Scott A. *Quaternary Insects and Their Environments*. Washington, DC: Smithsonian Institution Press, 1994.

Evans, Arthur V., and Charles L. Bellamy. *An Inordinate Fondness for Beetles*. Berkeley: University of California Press, 2000.

Evans, Arthur V., and J. N. Hogue. *Introduction to California Beetles*. Berkeley: University of California Press, in press.

Evans, D. L., and J. O. Schmidt, eds. *Insect Defenses: Adaptive Mechanisms and Strategies of Prey and Predators*. Albany: State University of New York Press, 1990.

Evenhuis, N. L. *Catalogue of the Fossil Flies of the World (Insecta: Diptera)*. Leiden, The Netherlands: Backhuys Publishers, 1994.

———. *Litteratura Taxonomica Dipterorum (1758–1930)*. 2 vols. Leiden, Netherlands: Backhuys Publishers,1997.

Field, L., ed. *The Biology of Wetas, King Crickets and Their Allies*. Oxford, U.K.: CAB International, 2001.

Fry, R., and D. Lonsdale. *Habitat Conservation for Insects: A Neglected Green Issue*. Middlesex, U.K.: The Amateur Entomologists' Society, 1991.

Gangwere, S. K., et al., eds. *The Bionomics of Grasshoppers, Katydids and Their Kin*. Oxford, U.K.: CAB International, 1997.

Gillott, Cedric. *Entomology*. 2nd edition. New York: Plenum Press, 1995.

Goddard, Jerome. *Physician's Guide to Arthropods of Medical Importance*. 2nd edition. Boca Raton, FL: CRC Press, 1996.

Gordon, D. G. *The Compleat Cockroach: A Comprehensive Guide to the Most Despised (and Least Understood) Creature on Earth*. Berkeley, CA: Ten Speed Press, 1996.

———. *The Eat-A-Bug Cookbook: 33 Ways to Cook Grasshoppers, Ants, Water Bugs, Spiders, Centipedes, and Their Kin*. Berkeley, CA: Ten Speed Press, 1998.

Gordth, Gordon, and David Headrick, compilers. *A Dictionary of Entomology*. New York: CABI Publishing, 2001.

Grösser, Detlef. *Wandelnde Blätter*. (Book on Leaf Insects). Frankfurt, Germany: Edition Chimaira, 2001.

Grasse, P. P. *Termitologia*, 3 vols. Paris: Masson and Cie, 1982–1986.

Grimaldi, D. A. *Amber: Window to the Past*. New York: American Museum of Natural History, 1996.

Grissell, E. *Insects and Gardens*. Portland, OR: Timber Press, 2001.

Gullan, P. J., and P. S. Cranston. *The Insects: An Outline of Entomology*. Malden, MA: Blackwell Science, 2000.

Gupta, A. P. *Arthropod Phylogeny*. New York: Van Nostrand Reinhold, 1979.

Gwynne, D. T. *Katydids and Bush-Crickets: Reproductive Behavior and Evolution of the Tettigoniidae*. Ithaca, NY: Cornell University Press, 2001.

Hölldobler, Bert, and E. O. Wilson. *The Ants*. Cambridge, MA: Harvard University Press, 1990.

Hamel, D. R. *Atlas of Insects on Stamps of the World*. Falls Church, VA: Tico Press, 1991.

Hamilton, W. D. *Narrow Roads of Gene Land*. Vol. 1, *The Evolution of Social Behaviour*. Oxford, U.K.: W. H. Freeman/Spektrum, 1996.

Handlirsch, A. *Die fossilen Insekten und die Phylogenie der rezenten Formen*. Leipzig, Germany: Engelmann, 1906–1908.

Hanson, P. E., and I. D. Gauld. *The Hymenoptera of Costa Rica*. Oxford, U.K.: Oxford University Press, 1995.

Helfer, Jacques R. *How to Know the Grasshoppers, Crickets, Cockroaches and Their Allies*. New York: Dover Publications, 1987.

Hennig, W. *Phylogenetic Systematics*. Urbana: University of Illinois Press, 1966.

———. *Insect Phylogeny*. New York: John Wiley and Sons, 1981.

Holldobler, B., and E. O. Wilson. *Journey to the Ants: A Story of Scientific Exploration*. Cambridge, MA: Belknap Press of Harvard University Press, 1994.

Hopkins, George Henry Evans, and T. Clay. *A Checklist of the Genera and Species of Mallophaga*. London: British Museum of Natural History, 1952.

Humphries, C. J., and L. R. Parenti. *Cladistic Biogeography: Interpreting Patterns of Plant and Animal Distributions*. Oxford Biogeography Series 12. Oxford, U.K.: Oxford University Press, 1999.

Imadaté, G. *Fauna Japonica Protura (Insecta)*. Tokyo: Keigaku, 1974.

Janetschek, Heinz. *Handbuch der Zoologie eine Naturgeschichte der Staemme des Tierreiches*. 2nd edition. Berlin: Walter de Gruyter, 1970.

Keller, L., ed. *Queen Number and Sociality in Insects*. Oxford, U.K.: Oxford University Press, 1993.

Kim, K. C., H. D. Pratt, and C. J. Stojanovich. *The Sucking Lice of North America*. University Park: Pennsylvania State University Press, 1986.

Kirby, P. *Habitat Management for Invertebrates: A Practical Handbook*. Sandy, Bedfordshire, U.K.: Royal Society for the Protection of Birds, 1992.

Klausnitzer, B. *Beetles*. New York: Exeter Books, 1983.

———. *Insects: Their Biology and Cultural History*. New York: Universe Books, 1987.

Kofoid, C. A., et al., eds. *Termites and Termite Control*. Berkeley: University of California Press, 1934.

Krishna, K., and F. M. Weesner, eds. *Biology of Termites*. 2 vols. New York: Academic Press, 1969–1970.

Kristensen, N. P., ed. *Lepidoptera: Moths and Butterflies*. Berlin and Hawthorne, NY: Walter de Gruyter, 1999.

Krzeminska, E., and W. Krzeminski. *Les fantomes de l'ambre: Insectes fossiles dans l'ambre de la Baltique*. Neuchâtel, Switzerland: Musée d'Histoire Naturelle de Neuchâtel, 1992.

Landolt, Peter, and Michel Sartori, eds. *Ephemeroptera and Plecoptera: Biology, Ecology, Systematics*. Fribourg, Switzerland: Mauron, Tinguely, and Lachat SA, 1997.

Lawrence, J. F., and E. B. Britton. *Australian Beetles*. Carlton, Australia: Melbourne University Press, 1994.

Ledger, J. A. *The Arthropod Parasites of Vertebrates in Africa South of the Sahara*. Johannesburg: South African Institute for Medical Research, 1980.

Leverton, Roy. *Enjoying Moths*. London: T. and A. D. Poyser, 2001.

Lewis, T., ed. *Thrips as Crop Pests*. Wallingford, U.K.: CAB International, 1997.

Llorente Bousquets, Jorge, and Juan J. Morrone, eds. *Biodiversidad, taxonomía y biogeografía de artrópodos de Mexico: hacia una síntesis de su conocimiento*. Mexico: Universidad Autónoma de México, 2002.

Loiacono, M. S., N. B. Diaz, and L. De Santis. *Estado actual del conocimiento de microhimenopteros Chalcidoidea, Cynipoidea, y "Proctotrupoidea" en Argentina*. Monografías Tercer Milenio, vol. 2. Zaragoza, Spain: Sociedad Entomológica Aragonesa (SEA) y Cyted, 2002.

Lubbock, J. B. *Monograph of the Collembola and Thysanura*. London: Ray Society, 1873.

Maynard Smith, J., and E. Szathmáry. *The Major Transitions in Evolution*. Oxford, U.K.: W. H. Freeman, 1995.

Maynard, E. A. *A Monograph of the Collembola or Springtail Insects of New York State*. Ithaca, NY: Comstock Publishing Co., Inc., 1951.

McAlpine, G. F., ed. et al., eds. *Manual of Nearctic Diptera*. 3 vols. Ottawa, Canada: Research Branch, Agriculture Canada, 1981–1989.

McEwen, P. K., T. R. New, and A. E. Whittington, eds. *Lacewings in the Crop Environment*. Cambridge, U.K., and New York: Cambridge University Press, 2001.

McGavin, George C. *Bugs of the World*. London: Blandford Press, 1993.

Meads, M. *Forgotten Fauna: The Rare, Endangered, and Protected Invertebrates of New Zealand*. Wellington, New Zealand: Department of Scientific and Industrial Research, 1990.

Menzel, P., and F. D'Aluisio. *Man Eating Bugs: The Art and Science of Eating Insects*. Berkeley, CA: Ten Speed Press, 1998.

Michener, C. D. *The Bees of the World.* Baltimore: Johns Hopkins University Press, 2000.

Morrone, Juan J. *Biogeografía de America Latina y el Caribe.* Vol. 3. Zaragoza, Spain: M&T Manuales; and Tesis SEA, Sociedad Entomológica Aragonesa, 2001.

Needham, James C., Minter J. Westfall, and Michael L. May. *Dragonflies of North America.* Rev. edition. Gainesville, FL: Scientific Publishers, 2000.

Nelson, G., and N. I. Platnick. *Systematics and Biogeography: Cladistics and Vicariance.* New York: Columbia University Press, 1981.

New, T. R. *An Introduction to Invertebrate Conservation Biology.* Oxford, U.K.: Oxford University Press, 1995.

———. *Butterfly Conservation.* 2nd edition. Melbourne, Australia: Oxford University Press, 1997.

Nosek, Josef. *The European Protura: Their Taxonomy, Ecology and Distribution, with Keys for Determination.* Geneva, Switzerland: Museum d'Histoire Naturelle, 1973.

Oldroyd, H. *The Natural History of Flies.* New York: W. W. Norton and Company, 1965.

Otte, D. *The North American Grasshoppers.* 2 vols. Cambridge, MA: Harvard University Press, 1981–1984.

———. *The Crickets of Hawaii: Origin, Systematics & Evolution.* Philadelphia: The Orthopterists' Society, 1994.

Otte, D., and Paul Brock. *Phasmida Species File. A Catalog of the Stick and Leaf Insects of the World.* Philadelphia: Orthopterists' Society, 2003.

Pakaluk, J., and S. A. Slipinski, eds. *Biology, Phylogeny, and Classification of Coleoptera.* Papers Celebrating the 80th Birthday of Roy A. Crowson. Warsaw, Poland: Muzeum i Instytut Zoologii PAN, 1995.

Parsons, Michael. *The Butterflies of Papua New Guinea: Their Systematics and Biology.* London: Academic Press, 1998.

Poinar, G. O. *Life in Amber.* Stanford, CA: Stanford University Press, 1992.

———. *The Amber Forest.* Princeton, NJ: Princeton University Press, 1999.

Preston-Mafham, K. *Grasshoppers and Mantids of the World.* London: Blandford Press, 1998.

Prete, F. R., H. Wells, P. H. Wells, and L. E. Hurd, eds. *The Praying Mantids.* Baltimore: Johns Hopkins University Press, 1999.

Price, Peter W. *Insect Ecology.* 3rd edition. New York: John Wiley and Sons, 1997.

Price, Roger D., Ronald A. Hellenthal, Ricardo L. Palma, Kevin P. Johnson, and Dale H. Clayton. *The Chewing Lice: World Checklist and Biological Overview.* Illinois Natural History Survey Special Publication no. 24. Champaign-Urbana: Illinois Natural History Survey, 2003.

Quintero, D., and A. Aiello, eds. *Insects of Panama and Mesoamerica.* New York: Oxford University Press, 1992.

Rasnitsyn, A. P., and D. L. J. Quicke. *History of Insects.* Dordrecht, The Netherlands: Kluwer, 2002.

Rentz, D. C. F. *Grasshopper Country: The Abundant Orthopteroid Insect Fauna of Australia.* Sydney: University of New South Wales Press, 1986.

Resh, V. H., and R. T. Cardé. *Encyclopedia of Insects.* San Diego, CA: Academic Press, 2003.

Resh, V. H., and D. M. Rosenberg. *The Ecology of Aquatic Insects.* New York: Praeger Publishers, 1984.

Rohdendorf, B. B., ed. *Fundamentals of Paleontology.* Vol. 9: *Arthropoda, Tracheata, Chelicerata.* Washington, DC: Smithsonian Institution and National Science Foundation, 1991.

Romoser, William S., and John G. Stoffolano, Jr. *The Science of Entomology.* 4th edition. Boston: WCB/McGraw-Hill, 1998.

Ross, A. *Amber: The Natural Time Capsule.* London: British Museum of Natural History, 1998.

Ross, K. G., and R. W. Matthews, eds. *The Social Biology of Wasps.* Ithaca, NY: Comstock Publishing Associates, 1991.

Rothschild, Miriam, and Robert Traub. *A Revised Glossary of Terms Used in the Taxonomy and Morphology of Fleas.* London: British Museum of Natural History, 1971.

Sakai, S. *Dermapterorum Catalogus Praeliminaris.* 4 parts. Tokyo: Department of Biology and Chemistry, Daito Bunka University, 1970–1973.

Salmon, J. T. *An Index to the Collembola.* Vol. 1. Bulletin no. 7. Wellington, New Zealand: Royal Society of New Zealand, 1964.

Samways, M. J. *Insect Conservation Biology.* London and New York: Chapman and Hall, 1994.

Sands, W. A. *The Identification of Worker Castes of Termite Genera from Soils of Africa and the Middle East.* London: CAB International, 1988.

Sbordoni, Valerio, and Saverio Forestiero. *Butterflies of the World.* Westport, CT: Firefly Books, 1998.

Schaefer, Carl W., ed. *Studies on Hemipteran Phylogeny.* Thomas Say Publications in Entomology. Lanham, MD: Entomological Society of America, 1996.

Schaefer, Carl W., and A. R. Panizzi, eds. *Heteroptera of Economic Importance.* Boca Raton, FL: CRC Press, 2000.

Schmitt, M. *Wie sich das Leben entwickelte: Die faszinierende Geschichte der Evolution.* Munich, Germany: Mosaik, 1994.

Schuh, R. T., and J. A. Slater. *True Bugs of the World (Hemiptera: Heteroptera): Classification and Natural History.* Ithaca, NY: Cornell University Press, 1995.

Seeley, T. D. *The Wisdom of the Hive.* Cambridge, MA: Harvard University Press, 1995.

Silsby, J. *Dragonflies of the World.* Collingwood, Australia: CSIRO Publishing, 2001.

Smit, F. G. A. M. *An Illustrated Catalogue of the Rothschild Collection of Fleas (Siphonaptera) in the British Museum (Natural History).* Vol. 7. London: British Museum of Natural History, 1987.

Speight, Martin R., M. D. Hunter, and A. D. Watt. *Ecology of Insects: Concepts and Applications.* Oxford, U.K.: Blackwell Science, 1999.

Stark, B. P., S. W. Szczytko, and C. R. Nelson. *American Stoneflies: A Photographic Guide to the Plecoptera.* Columbus, OH: Caddis Press, 1998.

Stehr, Frederick W., ed. *Immature Insects.* Vol. 2. Dubuque, IA: Kendall/Hunt Publishing Company, 1991.

Stewart, K. W., and B. P. Stark. *Nymphs of North American Stonefly Genera (Plecoptera).* 2nd ed. Columbus, OH: Caddis Press, 2002.

Storozhenko, S. Y. *Systematics, Phylogeny and Evolution of the Grylloblattidan Insects (Insecta: Grylloblattida).* Valdivostok: Dalnauka, 1998. (In Russian.)

Taylor, R. L. *Butterflies in My Stomach; or, Insects in Human Nutrition.* Santa Barbara, CA: Woodridge Press, 1975.

Thornhill, Randy, and John Alcock. *The Evolution of Insect Mating Systems.* Cambridge, MA: Harvard University Press, 1983.

Traub, Robert, Miriam Rothschild, and John F. Haddow. *The Rothschild Collection of Fleas: The Ceratophyllidae: Key to the Genera and Host Relationships.* Cambridge, U.K.: Cambridge University Press, 1983.

Tuxen, S. L. *The Protura: A Revision of the Species of the World with Keys for Determination.* Paris: Hermann, 1964.

———. *Fauna of New Zealand.* New Zealand: Science Information Publishing Centre, 1986.

Tyler, H., K. S. Brown Jr., and K. Wilson. *Swallowtail Butterflies of the Americas: A Study in Biological Dynamics, Ecological Diversity, Biosystematics and Conservation.* Gainesville, FL: Scientific Publishers, 1994.

Uvarov, B. P. *Grasshoppers and Locusts: A Handbook of General Acridology.* Vol. 1. Cambridge, U.K.: Cambridge University Press, 1966.

Van-Right, R. I., P. R. Ackery. *The Biology of Butterflies.* Symposium of the Royal Entomological Society of London 11. London: Academic Press, 1984.

Vickery, V. R., and D. K. M. Kevan. *A Monograph of the Orthopteroid Insects of Canada and Adjacent Regions.* Vol. 1. Ste. Anne de Bellevue, Quebec, Canada: Lyman Entomological Museum and Research Laboratory, 1983.

Wallace, A. R. *The Geographical Distribution of Animals.* London: MacMillan and Co., 1876.

Weitschat, W., and W. Wichard. *Atlas of Plants and Animals in Baltic Amber.* Munich, Germany: Pfeil, 2002.

Westfall, Minter J. Jr., and Michael L. May. *Damselflies of North America.* Gainesville, FL: Scientific Publishers, 1996.

White, R. E. *A Field Guide to the Beetles of North America.* Boston: Houghton Mifflin, 1983.

Wiggins, G. *Larvae of the North American Caddisfly Genera (Trichoptera).* 2nd edition. Buffalo, NY and Toronto: University of Toronto Press, 1996.

Williams, D. F., ed. *Exotic Ants: Biology, Impact, and Control of Introduced Species.* Boulder, CO: Westview Press, 1994.

Wilson, E. O. *The Insect Societies.* Cambridge, MA: Belknap Press of Harvard University Press, 1971.

———. *Sociobiology: The Abridged Edition.* Cambridge, MA: Belknap Press of Harvard University Press, 1980.

———. *Success and Dominance in Ecosystems: The Case of the Social Insects.* Oldendorf/Luhe, Germany: Ecology Institute, 1990.

Wooten, Anthony. *Insects of the World.* New York: Blanford Press, 1984.

The Xerces Society. *Butterfly Gardening.* San Francisco: Sierra Club Books, 1998.

Organizations

American Mosquito Control Association
P. O. Box 234
Eatontown, NJ 07727-0234
Phone: (732) 544-4645
E-mail: amca@mosquito.org
<http://www.mosquito.org>

American Zoo and Aquarium Association
8403 Colesville Road, Suite 710
Silver Spring, MD 20910
<http://www.aza.org>

Animal Behavior Society, Indiana University
2611 East 10th Street, no. 170
Bloomington, IN 47408-2603

Arbeitskreis Diptera
Dr. Frank Menzel (Leiter), ZALF e.V., Deutsches
Entomologisches Institut, Postfach 100238
Eberswalde D-16202
Germany
Phone: 03334-589820
E-mail: menzel@zalf.de
<http://www.ak-diptera.de>

Asociación Europea de Coleopterología
<http://www.ub.es/aec/>

Association for Tropical Lepidoptera
Phone: (352) 392-5894
<http://www.troplep.org>

Bee Improvement and Bee Breeders Association
50 Station Road
Cogenhoe, Northampton NN7 1LU
United Kingdom
Phone: 01604-890117
E-mail: membership@bibba.com
<http://www.angus.co.uk/bibba/>

British Dragonfly Society
<http://www.dragonflysoc.org.uk>

The British Entomological and Natural History Society
c/o The Pelham-Clinton Building
Dinton Pastures Country Park
Davis Street
Hurst, Reading
Berkshire RG10 0TH
United Kingdom
<http://www.benhs.org.uk/benhs.html>

The Cambridge Entomological Club
Museum of Comparative Zoology
26 Oxford Street
Cambridge, MA 02138
<http://entclub.org>

The Coleopterists Society
<http://www.coleopsoc.org/>

The Dipterist's Club of Japan
<http://www.osk.3web.ne.jp/~yonetsua/DCJE.html>

The Dipterists' Forum
Ger-y-Parc, Marianglas
Benllech
Gwynedd LL74 8NS
United Kingdom

Dragonfly Society of the Americas
2091 Partridge Lane
Binghamton, NY 13903

Entomological Society of America
9301 Annapolis Road
Lanham, MD 20706-3115
Phone: (301) 731-4535
<http://www.entsoc.org>

Entomological Society of Canada
393 Winston Avenue
Ottawa, ON K2A 1Y8
Canada
Phone: (613) 725-2619
E-mail: entsoc.can@sympatico.ca
<http://www.esc-sec.org/>

The Entomological Society of China
19 Zhong Guan Cun Lu
Haidian, Beijing
China
Phone: 86-10-62552266
E-mail: zss@panda.ioz.ac.cn
<http://panda.ioz.ac.cn/ioz/esc.html>

Gesellschaft Deutschsprachiger Odonatologen
<http://www.libellula.org>

Idalia Society of Mid-American Lepidopterists
219 West 68th Street
Kansas City, MO 64113
Phone: (816) 523-2948

International Association of Neuropterology
<http://www.neuroptera.com>

International Isoptera Society
<http://www.cals.cornell.edu/dept/bionb/isoptera/homepage
.html>

The International Society of Hymenopterists
<http://iris.biosci.ohio-state.edu/ish/ishhome.html>

International Union for the Study of Social Insects
<http://www.iussi.org/website>

Lepidoptera Research Foundation, Inc.
9620 Heather Road
Beverly Hills, CA 90210

The Lepidopterists' Society
<http://alpha.furman.edu/~snyder/snyder/lep/>

Lepidopterogical Society of Japan, c/o Ogata Hospital
3-2-17 Imabashi 3, Chuo-ku
Osaka 541
Japan

The Malloch Society
Graham Rotheray, Research Coordinator
Royal Museums of Scotland, Chamber Street
Edinburgh EH1 1JF
United Kingdom
E-mail: ger@nms.ac.uk

North American Butterfly Association
4 Delaware Road
Morristown, NJ 07960
<http://www.naba.org/>

North American Dipterists' Society
E-mail: aborkent@jetstream.net

North American Pollinator Protection Campaign
c/o The Coevolution Institute
423 Washington St., 4th floor
San Francisco, CA 94111-2339
<http://www.nappc.org>

The Orthopterists' Society
<http://www.orthoptera.org>

The Phasmid Study Group
40 Thorndike Road
Slough, Berkshire SL2 1SR
United Kingdom
Phone: 01753-579447
<http:www.stickinsect.org.uk>

Royal Entomological Society
41 Queen's Gate
London SW7 5HR
United Kingdom
Phone: 020 7584-8361
E-mail: reg@royensoc.co.uk
<http://www.royensoc.co.uk/>

Sociedad Colombiana de Entomología
SOCOLEN Carrera 50 # 27-70
Bloque C Nivel 7
Bogotá D.C.
República de Colombia
<http://www.socolen.org.co/>

Sociedad Mexicana de Entomología
E-mail: sme@colpos.mx
<http://www.geocities.com/RainForest/Vines/7352/>

Sociedad Venezolana de Entomología
E-mail: sve@sve.org.ve
<http://www.sve.org.ve/>

Societas Europaea Lepidopterologica
c/o Zoological Institute, University of Bern
Baltzerstrasse 3
Bern CH-3012
Switzerland

The Xerces Society
4828 SE Hawthorne Blvd.
Portland, OR 97215
<http://www.xerces.org>

Dr. Fritz Dieterlen
Zoological Research Institute and
A. Koenig Museum
Bonn, Germany

Dr. Rolf Dircksen
Professor, Pedagogical Institute
Bielefeld, Germany

Josef Donner
Instructor of Biology
Katzelsdorf, Austria

Dr. Jean Dorst
Professor, National Museum of
Natural History
Paris, France

Dr. Gerti Dücker
Professor and Chief Curator,
Zoological Institute, University of
Münster
Münster, Germany

Dr. Michael Dzwillo
Zoological Institute and Museum,
University of Hamburg
Hamburg, Germany

Dr. Irenäus Eibl-Eibesfeldt
Professor and Director, Institute of
Human Ethology, Max Planck
Institute for Behavioral Physiology
Percha/Starnberg, Germany

Dr. Martin Eisentraut
Professor and Director, Zoological
Research Institute and A. Koenig
Museum
Bonn, Germany

Dr. Eberhard Ernst
Swiss Tropical Institute
Basel, Switzerland

R. D. Etchecopar
Director, National Museum of
Natural History
Paris, France

Dr. R. A. Falla
Director, Dominion Museum
Wellington, New Zealand

Dr. Hubert Fechter
Curator, Lower Animals, Zoological
Collection of the State of Bavaria
Munich, Germany

Dr. Walter Fiedler
Docent, University of Vienna, and
Director, Schönbrunn Zoo
Vienna, Austria

Wolfgang Fischer
Inspector of Animals, Animal Park
Berlin, Germany

Dr. C. A. Fleming
Geological Survey Department of
Scientific and Industrial Research
Lower Hutt, New Zealand

Dr. Hans Frädrich
Zoological Garden
Berlin, Germany

Dr. Hans-Albrecht Freye
Professor and Director, Biological
Institute of the Medical School
Halle a.d.S., Germany

Günther E. Freytag
Former Director, Reptile and
Amphibian Collection, Museum of
Cultural History in Magdeburg
Berlin, Germany

Dr. Herbert Friedmann
Director, Los Angeles County
Museum of Natural History
Los Angeles, California, U.S.A.

Dr. H. Friedrich
Professor, Overseas Museum
Bremen, Germany

Dr. Jan Frijlink
Zoological Laboratory, University of
Amsterdam
Amsterdam, The Netherlands

Dr. H. C. Karl Von Frisch
Professor Emeritus and former
Director, Zoological Institute,
University of Munich
Munich, Germany

Dr. H. J. Frith
C.S.I.R.O. Research Institute
Canberra, Australia

Dr. Ion E. Fuhn
Academy of the Roumanian Socialist
Republic, Trajan Savulescu Institute of
Biology
Bucharest, Romania

Dr. Carl Gans
Professor, Department of Biology,
State University of New York at
Buffalo
Buffalo, New York, U.S.A.

Dr. Rudolf Geigy
Professor and Director, Swiss Tropical
Institute
Basel, Switzerland

Dr. Jacques Gery
St. Genies, France

Dr. Wolfgang Gewalt
Director, Animal Park
Duisburg, Germany

Dr. H. C. Viktor Goerttler
Professor Emeritus, University of
Jena
Jena, Germany

Dr. Friedrich Goethe
Director, Institute of Ornithology,
Heligoland Ornithological Station
Wilhelmshaven, Germany

Dr. Ulrich F. Gruber
Herpetological Section, Zoological
Research Institute and A. Koenig
Museum
Bonn, Germany

Dr. H. R. Haefelfinger
Museum of Natural History
Basel, Switzerland

Dr. Theodor Haltenorth
Director, Mammalology, Zoological
Collection of the State of Bavaria
Munich, Germany

Barbara Harrisson
Sarawak Museum, Kuching, Borneo
Ithaca, New York, U.S.A.

Dr. Francois Haverschmidt
President, High Court (retired)
Paramaribo, Suriname

Dr. Heinz Heck
Director, Catskill Game Farm
Catskill, New York, U.S.A.

Dr. Lutz Heck
Professor (retired), and Director,
Zoological Garden, Berlin
Wiesbaden, Germany

Dr. H. C. Heini Hediger
Director, Zoological Garden
Zurich, Switzerland

Dr. Dietrich Heinemann
Director, Zoological Garden, Münster
Dörnigheim, Germany

Dr. Helmut Hemmer
Institute for Physiological Zoology,
University of Mainz
Mainz, Germany

Dr. W. G. Heptner
Professor, Zoological Museum,
University of Moscow
Moscow, Russia

Dr. Konrad Herter
Professor Emeritus and Director
(retired), Zoological Institute, Free
University of Berlin
Berlin, Germany

Dr. Hans Rudolf Heusser
Zoological Museum, University of
Zurich
Zurich, Switzerland

Dr. Emil Otto Höhn
Associate Professor of Physiology,
University of Alberta
Edmonton, Canada

Dr. W. Hohorst
Professor and Director, Parasitological
Institute, Farbwerke Hoechst A.G.
Frankfurt-Höchst, Germany

Dr. Folkhart Hückinghaus
Director, Senckenbergische Anatomy,
University of Frankfurt a.M.
Frankfurt a.M., Germany

Francois Hüe
National Museum of Natural History
Paris, France

Dr. K. Immelmann
Professor, Zoological Institute,
Technical University of Braunschweig
Braunschweig, Germany

Dr. Junichiro Itani
Kyoto University
Kyoto, Japan

Dr. Richard F. Johnston
Professor of Zoology, University of
Kansas
Lawrence, Kansas, U.S.A.

Otto Jost
Oberstudienrat, Freiherr-vom-Stein
Gymnasium
Fulda, Germany

Dr. Paul Kähsbauer
Curator, Fishes, Museum of Natural
History
Vienna, Austria

Dr. Ludwig Karbe
Zoological State Institute and
Museum
Hamburg, Germany

Dr. N. N. Kartaschew
Docent, Department of Biology,
Lomonossow State University
Moscow, Russia

Dr. Werner Kästle
Oberstudienrat, Gisela Gymnasium
Munich, Germany

Dr. Reinhard Kaufmann
Field Station of the Tropical Institute,
Justus Liebig University, Giessen,
Germany
Santa Marta, Colombia

Dr. Masao Kawai
Primate Research Institute, Kyoto
University
Kyoto, Japan

Dr. Ernst F. Kilian
Professor, Giessen University and
Catedratico Universidad Austral,
Valdivia-Chile
Giessen, Germany

Dr. Ragnar Kinzelbach
Institute for General Zoology,
University of Mainz
Mainz, Germany

Dr. Heinrich Kirchner
Landwirtschaftsrat (retired)
Bad Oldesloe, Germany

Dr. Rosl Kirchshofer
Zoological Garden, University of
Frankfurt a.M.
Frankfurt a.M., Germany

Dr. Wolfgang Klausewitz
Curator, Senckenberg Nature
Museum and Research Institute
Frankfurt a.M., Germany

Dr. Konrad Klemmer
Curator, Senckenberg Nature
Museum and Research Institute
Frankfurt a.M., Germany

Dr. Erich Klinghammer
Laboratory of Ethology, Purdue
University
Lafayette, Indiana, U.S.A.

Dr. Heinz-Georg Klös
Professor and Director, Zoological
Garden
Berlin, Germany

Ursula Klös
Zoological Garden
Berlin, Germany

Dr. Otto Koehler
Professor Emeritus, Zoological
Institute, University of Freiburg
Freiburg i. BR., Germany

Dr. Kurt Kolar
Institute of Ethology, Austrian
Academy of Sciences
Vienna, Austria

Dr. Claus König
State Ornithological Station of Baden-
Württemberg
Ludwigsburg, Germany

Dr. Adriaan Kortlandt
Zoological Laboratory, University of
Amsterdam
Amsterdam, The Netherlands

Dr. Helmut Kraft
Professor and Scientific Councillor,
Medical Animal Clinic, University of
Munich
Munich, Germany

Dr. Helmut Kramer
Zoological Research Institute and A.
Koenig Museum
Bonn, Germany

Dr. Franz Krapp
Zoological Institute, University of
Freiburg
Freiburg, Switzerland

Dr. Otto Kraus
Professor, University of Hamburg,
and Director, Zoological Institute and
Museum
Hamburg, Germany

Dr. Hans Krieg
Professor and First Director (retired),
Scientific Collections of the State of
Bavaria
Munich, Germany

Dr. Heinrich Kühl
Federal Research Institute for
Fisheries, Cuxhaven Laboratory
Cuxhaven, Germany

Dr. Oskar Kuhn
Professor, formerly University
Halle/Saale
Munich, Germany

Dr. Hans Kumerloeve
First Director (retired), State
Scientific Museum, Vienna
Munich, Germany

Dr. Nagamichi Kuroda
Yamashina Ornithological Institute,
Shibuya-Ku
Tokyo, Japan

Dr. Fred Kurt
Zoological Museum of Zurich
University, Smithsonian Elephant
Survey
Colombo, Ceylon

Dr. Werner Ladiges
Professor and Chief Curator,
Zoological Institute and Museum,
University of Hamburg
Hamburg, Germany

Leslie Laidlaw
Department of Animal Sciences,
Purdue University
Lafayette, Indiana, U.S.A.

Dr. Ernst M. Lang
Director, Zoological Garden
Basel, Switzerland

Dr. Alfredo Langguth
Department of Zoology, Faculty of
Humanities and Sciences, University
of the Republic
Montevideo, Uruguay

Leo Lehtonen
Science Writer
Helsinki, Finland

Bernd Leisler
Second Zoological Institute,
University of Vienna
Vienna, Austria

Dr. Kurt Lillelund
Professor and Director, Institute for
Hydrobiology and Fishery Sciences,
University of Hamburg
Hamburg, Germany

R. Liversidge
Alexander MacGregor Memorial
Museum
Kimberley, South Africa

Dr. Konrad Lorenz
Professor and Director, Max Planck
Institute for Behavioral Physiology
Seewiesen/Obb., Germany

Dr. Martin Lühmann
Federal Research Institute for the
Breeding of Small Animals
Celle, Germany

Dr. Johannes Lüttschwager
Oberstudienrat (retired)
Heidelberg, Germany

Dr. Wolfgang Makatsch
Bautzen, Germany

Dr. Hubert Markl
Professor and Director, Zoological
Institute, Technical University of
Darmstadt
Darmstadt, Germany

Basil J. Marlow, BSc (Hons)
Curator, Australian Museum
Sydney, Australia

Dr. Theodor Mebs
Instructor of Biology
Weissenhaus/Ostsee, Germany

Dr. Gerlof Fokko Mees
Curator of Birds, Rijks Museum of
Natural History
Leiden, The Netherlands

Hermann Meinken
Director, Fish Identification Institute,
V.D.A.
Bremen, Germany

Dr. Wilhelm Meise
Chief Curator, Zoological Institute
and Museum, University of Hamburg
Hamburg, Germany

Dr. Joachim Messtorff
Field Station of the Federal Fisheries
Research Institute
Bremerhaven, Germany

Dr. Marian Mlynarski
Professor, Polish Academy of
Sciences, Institute for Systematic and
Experimental Zoology
Cracow, Poland

Dr. Walburga Moeller
Nature Museum
Hamburg, Germany

Dr. H. C. Erna Mohr
Curator (retired), Zoological State
Institute and Museum
Hamburg, Germany

Dr. Karl-Heinz Moll
Waren/Müritz, Germany

Dr. Detlev Müller-Using
Professor, Institute for Game
Management, University of Göttingen
Hannoversch-Münden, Germany

Werner Münster
Instructor of Biology
Ebersbach, Germany

Dr. Joachim Münzing
Altona Museum
Hamburg, Germany

Dr. Wilbert Neugebauer
Wilhelma Zoo
Stuttgart-Bad Cannstatt, Germany

Dr. Ian Newton
Senior Scientific Officer, The Nature
Conservancy
Edinburgh, Scotland

Dr. Jürgen Nicolai
Max Planck Institute for Behavioral
Physiology
Seewiesen/Obb., Germany

Dr. Günther Niethammer
Professor, Zoological Research
Institute and A. Koenig Museum
Bonn, Germany

Dr. Bernhard Nievergelt
Zoological Museum, University of
Zurich
Zurich, Switzerland

Dr. C. C. Olrog
Institut Miguel Lillo San Miguel de
Tucumán
Tucumán, Argentina

Alwin Pedersen
Mammal Research and Arctic Explorer
Holte, Denmark

Dr. Dieter Stefan Peters
Nature Museum and Senckenberg
Research Institute
Frankfurt a.M., Germany

Dr. Nicolaus Peters
Scientific Councillor and Docent,
Institute of Hydrobiology and
Fisheries, University of Hamburg
Hamburg, Germany

Dr. Hans-Günter Petzold
Assistant Director, Zoological Garden
Berlin, Germany

Dr. Rudolf Piechocki
Docent, Zoological Institute,
University of Halle
Halle a.d.S., Germany

Dr. Ivo Poglayen-Neuwall
Director, Zoological Garden
Louisville, Kentucky, U.S.A.

Dr. Egon Popp
Zoological Collection of the State of
Bavaria
Munich, Germany

Dr. H. C. Adolf Portmann
Professor Emeritus, Zoological
Institute, University of Basel
Basel, Switzerland

Hans Psenner
Professor and Director, Alpine Zoo
Innsbruck, Austria

Dr. Heinz-Siburd Raethel
Oberveterinärrat
Berlin, Germany

Dr. Urs H. Rahm
Professor, Museum of Natural History
Basel, Switzerland

Dr. Werner Rathmayer
Biology Institute, University of
Konstanz
Konstanz, Germany

Walter Reinhard
Biologist
Baden-Baden, Germany

Dr. H. H. Reinsch
Federal Fisheries Research Institute
Bremerhaven, Germany

Dr. Bernhard Rensch
Professor Emeritus, Zoological
Institute, University of Münster
Münster, Germany

Dr. Vernon Reynolds
Docent, Department of Sociology,
University of Bristol
Bristol, England

Dr. Rupert Riedl
Professor, Department of Zoology,
University of North Carolina
Chapel Hill, North Carolina, U.S.A.

Dr. Peter Rietschel
Professor (retired), Zoological
Institute, University of Frankfurt a.M.
Frankfurt a.M., Germany

Dr. Siegfried Rietschel
Docent, University of Frankfurt;
Curator, Nature Museum and
Research Institute Senckenberg
Frankfurt a.M., Germany

Herbert Ringleben
Institute of Ornithology, Heligoland
Ornithological Station
Wilhelmshaven, Germany

Dr. K. Rohde
Institute for General Zoology, Ruhr
University
Bochum, Germany

Dr. Peter Röben
Academic Councillor, Zoological
Institute, Heidelberg University
Heidelberg, Germany

Dr. Anton E. M. De Roo
Royal Museum of Central Africa
Tervuren, South Africa

Dr. Hubert Saint Girons
Research Director, Center for
National Scientific Research
Brunoy (Essonne), France

Dr. Luitfried Von Salvini-Plawen
First Zoological Institute, University
of Vienna
Vienna, Austria

Dr. Kurt Sanft
Oberstudienrat, Diesterweg-
Gymnasium
Berlin, Germany

Dr. E. G. Franz Sauer
Professor, Zoological Research
Institute and A. Koenig Museum,
University of Bonn
Bonn, Germany

Dr. Eleonore M. Sauer
Zoological Research Institute and A.
Koenig Museum, University of Bonn
Bonn, Germany

Dr. Ernst Schäfer
Curator, State Museum of Lower
Saxony
Hannover, Germany

Dr. Friedrich Schaller
Professor and Chairman, First
Zoological Institute, University of
Vienna
Vienna, Austria

Dr. George B. Schaller
Serengeti Research Institute, Michael
Grzimek Laboratory
Seronera, Tanzania

Dr. Georg Scheer
Chief Curator and Director,
Zoological Institute, State Museum of
Hesse
Darmstadt, Germany

Dr. Christoph Scherpner
Zoological Garden
Frankfurt a.M., Germany

Dr. Herbert Schifter
Bird Collection, Museum of Natural
History
Vienna, Austria

Dr. Marco Schnitter
Zoological Museum, Zurich
University
Zurich, Switzerland

Dr. Kurt Schubert
Federal Fisheries Research Institute
Hamburg, Germany

Eugen Schuhmacher
Director, Animals Films, I.U.C.N.
Munich, Germany

Dr. Thomas Schultze-Westrum
Zoological Institute, University of
Munich
Munich, Germany

Dr. Ernst Schüt
Professor and Director (retired), State
Museum of Natural History
Stuttgart, Germany

Dr. Lester L. Short , Jr.
Associate Curator, American Museum
of Natural History
New York, New York, U.S.A.

Dr. Helmut Sick
National Museum
Rio de Janeiro, Brazil

Dr. Alexander F. Skutch
Professor of Ornithology, University
of Costa Rica
San Isidro del General, Costa Rica

Dr. Everhard J. Slijper
Professor, Zoological Laboratory,
University of Amsterdam
Amsterdam, The Netherlands

Bertram E. Smythies
Curator (retired), Division of Forestry
Management, Sarawak-Malaysia
Estepona, Spain

Dr. Kenneth E. Stager
Chief Curator, Los Angeles County
Museum of Natural History
Los Angeles, California, U.S.A.

Dr. H. C. Georg H.W. Stein
Professor, Curator of Mammals,
Institute of Zoology and Zoological
Museum, Humboldt University
Berlin, Germany

Dr. Joachim Steinbacher
Curator, Nature Museum and
Senckenberg Research Institute
Frankfurt a.M., Germany

Dr. Bernard Stonehouse
Canterbury University
Christchurch, New Zealand

Dr. Richard Zur Strassen
Curator, Nature Museum and
Senckenberg Research Institute
Frandfurt a.M., Germany

Dr. Adelheid Studer-Thiersch
Zoological Garden
Basel, Switzerland

Dr. Ernst Sutter
Museum of Natural History
Basel, Switzerland

Dr. Fritz Terofal
Director, Fish Collection, Zoological
Collection of the State of Bavaria
Munich, Germany

Dr. G. F. Van Tets
Wildlife Research
Canberra, Australia

Ellen Thaler-Kottek
Institute of Zoology, University of
Innsbruck
Innsbruck, Austria

Dr. Erich Thenius
Professor and Director, Institute of
Paleontolgy, University of Vienna
Vienna, Austria

Dr. Niko Tinbergen
Professor of Animal Behavior,
Department of Zoology, Oxford
University
Oxford, England

Alexander Tsurikov
Lecturer, University of Munich
Munich, Germany

Dr. Wolfgang Villwock
Zoological Institute and Museum,
University of Hamburg
Hamburg, Germany

Zdenek Vogel
Director, Suchdol Herpetological
Station
Prague, Czechoslovakia

Dieter Vogt
Schorndorf, Germany

Dr. Jiri Volf
Zoological Garden
Prague, Czechoslovakia

Otto Wadewitz
Leipzig, Germany

Dr. Helmut O. Wagner
Director (retired), Overseas Museum,
Bremen
Mexico City, Mexico

Dr. Fritz Walther
Professor, Texas A & M University
College Station, Texas, U.S.A.

John Warham
Zoology Department, Canterbury
University
Christchurch, New Zealand

Dr. Sherwood L. Washburn
University of California at Berkeley
Berkeley, California, U.S.A.

Eberhard Wawra
First Zoological Institute, University
of Vienna
Vienna, Austria

Dr. Ingrid Weigel
Zoological Collection of the State of
Bavaria
Munich, Germany

Dr. B. Weischer
Institute of Nematode Research,
Federal Biological Institute
Münster/Westfalen, Germany

Herbert Wendt
Author, Natural History
Baden-Baden, Germany

Dr. Heinz Wermuth
Chief Curator, State Nature Museum,
Stuttgart
Ludwigsburg, Germany

Dr. Wolfgang Von Westernhagen
Preetz/Holstein, Germany

Dr. Alexander Wetmore
United States National Museum,
Smithsonian Institution
Washington, D.C., U.S.A.

Dr. Dietrich E. Wilcke
Röttgen, Germany

Dr. Helmut Wilkens
Professor and Director, Institute of
Anatomy, School of Veterinary
Medicine
Hannover, Germany

Dr. Michael L. Wolfe
Utah, U.S.A.

Hans Edmund Wolters
Zoological Research Institute and A.
Koenig Museum
Bonn, Germany

Dr. Arnfrid Wünschmann
Research Associate, Zoological Garden
Berlin, Germany

Dr. Walter Wüst
Instructor, Wilhelms Gymnasium
Munich, Germany

Dr. Heinz Wundt
Zoological Collection of the State of
Bavaria
Munich, Germany

Dr. Claus-Dieter Zander
Zoological Institute and Museum,
University of Hamburg
Hamburg, Germany

Dr. Fritz Zumpt
Director, Entomology and
Parasitology, South African Institute
for Medical Research
Johannesburg, South Africa

Dr. Richard L. Zusi
Curator of Birds, United States
National Museum, Smithsonian
Institution
Washington, D.C., U.S.A.

Glossary

Abdomen—The posterior of the three main body divisions.

Acaricide—A substance that kills mites and ticks.

Aculeate—An ant, bee, or wasp that possesses a stinger.

Adeagus—Part of the male genitalia used in the transfer of spermatozoa, which is inserted into the female during copulation; its shape is often used in distinguishing related species.

Adecticous—A pupa that does not have moveable mandibles.

Aestivation—A period of dormancy that is entered into when conditions are not favorable, particularly during very warm or very dry seasons.

Alate—An insect with wings.

Allopatric—Biologically relating to or taking place in separate areas.

Alloparental—The raising and caring of offspring by individuals other than the biological parents.

Ametabolous—Development in which little or no external metamorphic changes are noticeable in the larval to adult transition.

Anal—Relating to or being close to the anus.

Anaplasmosis—An infectious disease spread by organisms of the genus *Anaplasma*.

Androconia—Modified scales present on males containing glandular structures that produce an odor to attract the opposite sex.

Annulae—Ring-like segments, markings, or divisions.

Antennae—Pair of segmented appendages located on the head that perform sensory functions. Often referred to as "feelers."

Antibiosis—A provocative association between organisms that is detrimental, inhibitive, and preventative to one or more of them.

Apiary—A colony of bee hives, often kept for the purpose of collecting honey.

Apterous—An insect without wings.

Arboreal—An insect that lives in, on, or among trees.

Arista—A large bristle on the dorsal apical antenna of flies.

Arolium—A small pad located between the claws, or at the base of the claw.

Arrhenotoky—Parthogenetic production of male offspring.

Bivouac—A mass encampment made up of ant workers within which the queen and brood live while their colony is in a stable environment.

Bivoltine—The production of two broods or generations in one season or year.

Bot—Larvae (maggots) of the fly family Oestridae, which are obligate endoparasites of mammals.

Brood—A group made up of members of a species that have hatched or become adult at approximately the same time and live together in a limited area.

Budding—The development of new colonies by one or more reproductive females and a group of workers.

Castes—Hierarchical groups of order present among populations of social insects that define the division of labor.

Caudal—Referring or pertaining to the posterior end of the body.

Cephalic—Referring or pertaining to the anterior end of the body.

Cephalothorax—The body region that consists of the head and thoracic segments.

Cerci—Paired appendages present at the posterior end of the abdomen.

Chaetae—Articulate or non-articulate hairs or bristles.

Chorion—The shell or covering of an insect egg.

Chrysalis—The pupa; an enclosed casement where insects pass the pupal stage and develop without ingesting food.

Clavate—Club-shaped.

Cocoon—A protective casing in which the pupa forms.

Colony—A locally isolated population.

Commensalism—Symbiotic relationship between two or more species in which no group is injured and at least one group benefits.

Communal—Cooperation between females of one species in production and building, but not in caring for the brood.

Conspecific—Belonging to the same species.

Cosmopolitan—Occurring throughout most of the world.

Crepitate—To make a series or sharp, crackling noises.

Cursorial—Adapted for habitual running.

Cuticle—The noncellular outer layers of the body.

Dealate—A sexually mature adult that drops or forcefully removes its wings.

Decticous—A pupa with mandibles that are moveable.

Dentate—Having teeth, or structures that function as teeth.

Diapause—A period of time in which development is suspended or arrested and the body is dormant.

Dichoptic—An organism with eyes that are apart or separate on the top of the head.

Drone—A male bee.

Dulosis—Enslavement of an individual of one species by another group in order to raise the pupae or larvae of the conquered individual.

Ecdysis—Molting or shedding of the exoskeleton.

Eclosion—Hatching or emerging from the egg or pupa.

Ectoparasite—A parasite that lives on the exterior body of its host.

Empodium—A bristle- or pad-like growth between the claws of the foot.

Endemic—Belonging to or being from a particular geographical region.

Endopterygote—The internal development of wings.

Endocuticle—The innermost layer of the cuticle.

Endoparasite—A parasite that lives inside the body of its host.

Epicuticle—The surface layers of the cuticle.

Epigaeic—A group that lives or forages primarily above ground.

Eusocial—A group that produces a division of labor and cooperates in rearing its young.

Exocuticle—Hard and darkened layer of the cuticle lying between the endocuticle and epicuticle.

Exoskeleton—The external plates of the body wall.

Exopterygote—An insect whose wings gradually develop on the outside of the body and has no pupal stage.

Fossorial—Adapted to digging.

Frass—Insect excrement.

Glabrous—An organism or structure that is smooth and without hairs.

Gregarious—To live in a community or group.

Grub—A scarabaeiform larva.

Gynandromorph—An individual that exhibits both male and female characteristics.

Hematophagous—An organism that feeds or subsides on blood.

Hemimetabolous—An insect that undergoes simple metamorphosis with egg, larval, and adult stages.

Hermaphrodite—An insect that has both male and female sexual organs.

Holometabolous—An insect that undergoes a complete metamorphosis with egg, larval, pupal, and adult stages.

Holoptic—An insect whose eyes are touching or almost touching.

Host—The organism in or on which a parasite lives.

Hyaline—Translucent, clear, and colorless.

Hyperparasite—A parasitic insect whose host is another parasite.

Imago—The adult stage in which an insect reproduces.

Inquiline—An animal that lives in the nest or abode of another animal species.

Instar—A stadium between two successive molts.

Larva—The immature and wingless form of an insect that hatches from the egg and increases in size as it progresses through several molts until it transforms into a pupa, chrysalis, or adult.

Maggot—A fly larva without legs or a well-developed head capsule.

Malpighian tubes—Excretory tubes of insects that arise between the median and posterior portions of the digestive tract.

Mandible—The first pair of jaws in insects.

Maxillae—The second pair of jaws located behind the mandibles in insects.

Meconium—Fluid ejected by an individual after emerging from its pupa or chrysalis.

Metamorphosis—A change in physical form or substance.

Moult—Shedding of the exoskeleton.

Mycetoxylophagous—An organism that bores into rotten wood.

Myiasis—Condition arising from infestation and invasion by fly larvae.

Myrmecophilous—"Ant-loving"; applied to insects that live in ant nests.

Neotenics—An adult that retains characteristics of immature stages.

Nit—The egg of a louse.

Nocturnal—An organism that is active mostly at night.

Nymph—Larva of hemimetabolous insects.

Obligate ectoparasites—External parasites that cannot complete their cycle when removed from their host.

Ocellus—Simple form of eye present in an insect, consisting of a single beadlike lens.

Ootheca—The cover or case that surrounds a mass of eggs.

Oviparous—An organism that lays eggs.

Ovipositor—The apparatus through which the female lays eggs.

Ovoviviparous—An organism that produces young that hatch out of their egg while still within their mother.

Parasite—An organism that lives in or on the body of another living organism, feeding off of its host.

Parthenogenesis—Development of an egg without fertilization.

Phoresy—Nonparasitic relationship between two organisms in which one uses the other as a means of transportation.

Phytophagous—An organism that solely feeds upon plants.

Polyembryony—The production of several embryos from a single egg.

Polyphagous—An organism that consumes a variety of foods.

Predaceous—An organism that preys on other organisms.

Predator—An animal that attacks and feeds on animals that are usually smaller and weaker than itself.

Prognathous—An insect whose jaws are directed forward and head is in the plane of the main body axis.

Proleg—A fleshy, stumpy appendage that is not a leg present on the thorax or abdomen of some insect larvae.

Pseudovipositor—Terminal abdominal segment of females from where eggs are layed. Also known as oviscapt.

Ptilinum—An inflatable sac located on the head of some flies that assists them in emerging from their puparium.

Pupa—Stage that comes between the larval and adult periods, in which an organism does not feed and undergoes metamorphosis.

Puparium—Case in which a pupa is enclosed.

Pupiparous—Insects that give birth to fully-grown larvae and pupate almost immediately.

Queen cell—The cell in which a queen honey bee develops from egg to adult.

Raptorial—A structure (jaw or leg) adapted to seize and grasp prey.

Release calls—Alarm calls produced when an animal is seized by a predator.

Reniform—Kidney shaped.

Reticulate—A surface that appears netted or as a network of veins.

Saprophytic—An organism that lives on dead or decaying organic matter.

Scopa—A brush; a dense tuft of hair in which bees collect pollen; fringe of long, dense, and sometimes modified scales along the caudal margin of abdominal segment viii present in male Lepidoptera; an inflated, often pilose, apicoventral flange running most of the length of a gonostylus in male Symphyla (Hymenoptera).

Segment—Subdivision of a body or appendage between areas of flexibility associated with muscle attachments.

Sensu stricto—In the "strict sense."

Seta—A bristle.

Social—An organism that lives in organized communities or groups.

Soldier—Member of a worker subcaste that functions to protect the colony.

Solitary—An organism occurring singularly or in pairs, but never in colonies.

Stadium—An interval between molts in a developing insect. See **Instar**.

Stridulation—A shrill, creaking noise made by rubbing body structures together.

Synanthropic—Associated with human habitation.

Synonym—Two or more names that have been given to the same species.

Tarsomeres—A subdivision or segment of the tarsus.

Tarsus—Leg segment attached to the apex of the tibia and bearing the pretarsus, consisting of 1–5 tarsomeres or segments.

Tegmen—The leathery forewing of an orthopteran.

Telson—Terminal region of the abdomen that bears the anus; found in the embryos of many insects but rarely in adults.

Thelytoky—Parthenogenesis in which only females are produced from unfertilized eggs.

Troglophilous—An organism that lives in caves.

Univoltine—A group that produces only one generation per year.

Vermiform larva—A legless, worm-like larva without a well-developed head.

Vibrissae—A pair of large bristles that is present just above the mouth in some organisms.

Viviparous—An organism that produces live young.

Workers—Sterile males and females that perform a colony's work.

Xylophagous—An organism that bores into sound wood.

Insects family list

Entognatha [Class]
  Collembola [Order]
    Arthropleona [Suborder]
      Actaletidae [Family]
      Brachystomellidae
      Coenaletidae
      Cyphoderidae
      Entomobryidae
      Hypogasturidae
      Isotomidae
      Microfalculidae
      Neanuridae
      Odontellidae
      Oncopoduridae
      Onychiuridae
      Paronellidae
      Poduridae
      Protentomobryidae
      Tomoceridae

    Neelipleona [Suborder]
      Neelidae [Family]

    Symphypleona [Suborder]
      Dicyrtomidae [Family]
      Mackenziellidae
      Sminthuridae

  Protura [Order]
    Acerentomoidea [Suborder]
      Acerentomidae [Family]
      Protentomidae

    Eosentomoidea [Suborder]
      Eosentomidae [Family]
      Sinentomidae

  Diplura [Order]
    Anajapygidae [Family]
    Campodeidae
    Dinjapygidae
    Evalljapygidae
    Heterojapygidae
    Japygidae
    Parajapygidae

    Procampodeidae
    Projapygidae

Insecta [Class]
  Microcoryphia [Order]
    Meinertellidae [Family]
    Machilidae

  Thysanura [Order]
    Lepidothrichidae [Family]
    Nicoletiidae
    Lepismatidae
    Maindroniidae

  Ephemeroptera [Order]
    Schistonota [Suborder]
      Ameletopsidae [Family]
      Ametropodidae
      Baetidae
      Behningiidae
      Coloburiscidae
      Ephemeridae
      Euthyplociidae
      Heptageniidae
      Isonychiidae
      Leptophleibiidae
      Oligoneuriidae
      Oniscigastridae
      Palingeniidae
      Polymitarcyidae
      Potamanthidae
      Siphlonuridae

    Pannota [Suborder]
      Baetiscidae [Family]
      Caenidae
      Ephemerellidae
      Leptohyphidae
      Neoephemeridae
      Prosopistomatidae
      Tricorythidae

  Odonata [Order]
    Anisoptera [Suborder]
      Aeshnidae [Family]

      Austropetaliidae
      Cordulegastridae
      Corduliidae
      Gomphidae
      Libellulidae
      Neopetaliidae
      Petaluridae

    Anisozygoptera [Suborder]
      Epiophlebiidae [Family]

    Zygoptera [Suborder]
      Amphipterygidae [Family]
      Calopterygidae
      Chlorocyphidae
      Coenagrionidae
      Dicteriadidae
      Euphaeidae
      Hemiplebiidae
      Isostictidae
      Lestidae
      Lestoideidae
      Megapodagrionidae
      Perilestidae
      Platycnemididae
      Platystictidae
      Polythoridae
      Protoneuridae
      Pseudolestidae
      Pseudostigmatidae
      Synlestidae

  Plecoptera [Order]
    Antarctoperlaria [Suborder]
      Austroperlidae [Family]
      Diamphipnoidae
      Eustheniidae
      Gripopterygidae

    Arctoperlaria [Suborder]
      Capniidae [Family]
      Chloroperlidae
      Leuctridae
      Nemouridae
      Notonemouridae

Peltoperlidae
Perlidae
Perlodidae
Pteronarcyidae
Scopuridae
Styloperlidae
Taeniopterygidae

Blattodea [Order]
Blaberidae [Family]
Blattellidae
Blattidae
Cryptocercidae
Nocticolidae
Polyphagidae

Isoptera [Order]
Hodotermitidae [Family]
Kalotermitidae
Mastotermitidae
Rhinotermitidae
Serritermitidae
Termitidae
Termopsidae

Mantodea [Order]
Acanthopidae [Family]
Amorphoscelidae
Chaeteessidae
Empusidae
Eremiaphilidae
Hymenopodidae
Iridopterygidae
Liturgusidae
Mantidae
Mantoididae
Metallyticidae
Sibyllidae
Tarachodidae
Thespidae
Toxoderidae

Mantophasmatodea [Order]
Mantophasmatidae [Family]

Grylloblattodea [Order]
Grylloblattina [Suborder]
Grylloblattidae [Family]

Dermaptera [Order]
Arixeniina [Suborder]
Arixeniidae [Family]

Forficulina [Suborder]
Anisolabididae [Family]
Apachyidae
Chelisochidae
Diplatyidae
Forficulidae
Labiduridae
Pygidicranidae
Spongiphoridae

Hemimerina [Suborder]
Hemimeridae [Family]

Orthoptera [Order]
Caelifera [Suborder]
Acrididae [Family]
Charilaidae
Cylindrachetidae
Eumastacidae
Lathiceridae
Lentulidae
Ommexechidae
Pamphagidae
Pauliniidae
Pneumoridae
Proscopiidae
Pyrgomorphidae
Rhipipterygidae
Tanaoceridae
Tetrigidae
Tridactylidae
Trigonopterygidae
Xyronotidae

Ensifera [Suborder]
Cooloolidae [Family]
Gryllacrididae
Gryllidae
Gryllotalpidae
Haglidae
Myrmecophilidae
Rhaphidophoridae
Schizodactylidae
Stenopelmatidae
Tettigoniidae

Phasmida [Order]
Anareolatae [Suborder]
Diapheromeridae [Family]
Phasmatidae

Areolatae [Suborder]
Aschiphasmatidae [Family]
Bacillidae
Heteronemiidae
Phylliidae
Pseudophasmatidae

Timematodea [Suborder]
Timematidae [Family]

Embioptera [Order]
Anisembiidae [Family]
Australembiidae
Clothodidae
Embiidae
Embonychidae
Notoligotomidae
Oligotomidae
Teratembiidae

Zoraptera [Order]
Zorotypidae [Family]

Pscoptera [Order]
Pscocomorpha [Suborder]
Amphipsocidae [Family]
Archipsocidae
Asiopsocidae
Caeciliidae
Calopsocidae
Cladiopsocidae
Dolabellapsocidae
Ectopsocidae
Elipsocidae
Epipsocidae
Hemipsocidae
Lachesillidae
Mesopsocidae
Myopsocidae
Peripsocidae
Philotarsidae
Pseudocaeciliidae
Psilopsocidae
Psocidae
Ptiloneuridae
Stenopsocidae
Trichopsocidae

Troctomorpha [Suborder]
Amphientomidae [Family]
Compsocidae
Liposcelidae
Manicapsocidae
Musapsocidae
Pachytroctidae
Sphaeropsocidae
Troctopsocidae

Trogiomorpha [Suborder]
Lepidopsocidae [Family]
Prionoglarididae
Psoquillidae
Psyllipsocidae
Trogiidae

Phthiraptera [Order]
Amblycera [Suborder]
Boopiidae [Family]
Gyropidae
Laemobothriidae
Menoponidae
Ricinidae
Trimenoponidae

Anoplura [Suborder]
Echinophthiriidae [Family]
Haematopinidae
Hoplopleuridae
Linognathidae
Pediculidae
Polyplacidae
Pthiridae

Ischnocera [Suborder]
Heptapsogasteridae [Family]

Philopteridae
Trichodectidae

Rhyncophthirina [Suborder]
Haematomyzidae [Family]

Hemiptera [Order]
Auchenorrhyncha [Suborder]
Acanaloniidae [Family]
Achilidae
Achilixiidae
Aetalionidae
Aphrophoridae
Cercopidae
Cicadellidae
Cicadidae
Cixiidae
Clastopteridae
Delphacidae
Derbidae
Dictyopharidae
Eurybrachyidae
Eurymelidae
Flatidae
Fulgoridae
Gengidae
Hylicidae
Hypochthonellidae
Issidae
Kinnaridae
Lophopidae
Machaerotidae
Meenoplidae
Membracidae
Nogodinidae
Ricaniidae
Tettigarctidae
Tettigometridae
Tropiduchidae

Coleorrhyncha [Suborder]
Peloridiidae [Family]

Heteroptera [Suborder]
Acanthosomatidae [Family]
Aenictopecheidae
Aepophilidae
Aerophilidae
Alydidae
Anthocoridae
Aradidae
Belostomatidae
Berytidae
Canopidae
Ceratocombidae
Cimicidae
Colobathristidae
Coreidae
Corixidae
Cydnidae
Dinidoridae

Dipsocoridae
Enicocephalidae
Gelastocoridae
Gerridae
Hebridae
Helotrephidae
Hermatobatidae
Hydrometridae
Hyocephalidae
Hypsipterygidae
Idiostolidae
Joppeicidae
Leptopodidae
Lestoniidae
Lygaeidae
Macroveliidae
Malcidae
Medocostidae
Megarididae
Mesoveliidae
Microphysidae
Miridae
Nabidae
Naucoridae
Nepidae
Notonectidae
Ochteridae
Omaniidae
Pachynomidae
Paraphrynoveliidae
Peloridiidae
Pentatomidae
Phloeidae
Piesmatidae
Pleidae
Plataspididae
Plokiophilidae
Polyctenidae
Pyrrhocoridae
Reduviidae
Rhopalidae
Saldidae
Schizopteridae
Scutelleridae
Stemmocryptidae
Stenocephalidae
Termitaphididae
Tessaratomidae
Thaumastellidae
Thaumastocoridae
Thyreocoridae
Tingidae
Urostylidae
Veliidae
Velocipedidae
Vianaididae

Sternorrhyncha [Suborder]
Aclerdidae [Family]
Adelgidae
Aleyrodidae

Aphididae
Asterolecaniidae
Beesoniidae
Calophyidae
Carsidaridae
Cerococcidae
Coccidae
Conchaspididae
Dactylopiidae
Diaspididae
Eriococcidae
Halimococcidae
Homotomidae
Kermesidae
Kerriidae
Lecanodiaspididae
Margarodidae
Ortheziidae
Phacopteronidae
Phenacoleachiidae
Phoenicococcidae
Phylloxeridae
Pseudococcidae
Psyllidae
Stictococcidae
Triozidae

Thysanoptera [Order]
Terebrantia [Suborder]
Adiheterothripidae [Family]
Aeolothripidae
Fauriellidae
Heterothripidae
Melanthripidae
Merothripidae
Thripidae
Uzelothripidae

Tubulifera [Suborder]
Phlaeothripidae [Family]

Megaloptera [Order]
Corydalidae [Family]
Sialidae

Raphidioptera [Order]
Inocelliidae [Family]
Raphidiidae

Neuroptera [Order]
Ascalaphidae [Family]
Berothidae
Chrysopidae
Coniopterygidae
Dilaridae
Hemerobiidae
Ithonidae
Mantispidae
Myrmeleontidae
Nemopteridae
Neverorthidae
Nymphidae

Osmylidae
Polystoechotidae
Psychopsidae
Rhachiberothidae
Sisyridae

Coleoptera [Order]
 Adephaga [Suborder]
  Amphizoidae [Family]
  Carabidae
  Dytiscidae
  Gyrinidae
  Haliplidae
  Hygrobiidae
  Noteridae
  Rhysodidae
  Trachypachidae

 Archostemata [Suborder]
  Crowsonelliedae [Family]
  Cupedidae
  Micromalthidae
  Ommatidae

 Myxophaga [Suborder]
  Hydroscaphidae [Family]
  Lepiceridae
  Microsporidae
  Torridincolidae

 Polyphaga [Suborder]
  Acanthocnemidae [Family]
  Aderidae
  Agyrtidae
  Alexiidae
  Anobiidae
  Anthicidae
  Anthribidae
  Archeocrypticidae
  Artematopidae
  Attelabidae
  Beliidae
  Belohinidae
  Biphyllidae
  Boganiidae
  Boridae
  Bostrichidae
  Bothrideridae
  Brachypsectridae
  Brachypteridae
  Brentidae
  Bruchidae
  Buprestidae
  Byrrhidae
  Byturidae
  Callirhipidae
  Cantharidae
  Caridae
  Cavognathidae
  Cerambycidae
  Cerophytidae

Ceratocanthidae
Cerylonidae
Chaetossomatidae
Chalcodryidae
Chelonariidae
Chrysomelidae
Ciidae
Clambidae
Cleridae
Cneoglossidae
Coccinellidae
Colydiidae
Corylophidae
Cryptophagidae
Cucujidae
Curculionidae
Dascillidae
Dasyceridae
Decliniidae
Dermestidae
Derodontidae
Diphyllostomatidae
Discolomatidae
Drilidae
Dryopidae
Elateridae
Elmidae
Endomychidae
Erotylidae
Eucinetidae
Eucnemidae
Eulichadidae
Geotrupidae
Glaphyridae
Glaresidae
Helotidae
Heteroceridae
Histeridae
Hobartiidae
Hybosoridae
Hydraenidae
Hydrophilidae
Ithyceridae
Jacobsoniidae
Laemophloeidae
Lamingtoniidae
Lampyridae
Languriidae
Latridiidae
Leiodidae
Limnichidae
Lucanidae
Lutrochidae
Lycidae
Lymexylidae
Megalopodidae
Melandryidae
Meloidae
Melyridae
Micropeplidae
Monommatidae

Monotomidae
Mordellidae
Mycetophagidae
Mycteridae
Nemonychidae
Nitidulidae
Nosodendridae
Ochodaeidae
Oedemeridae
Omalisidae
Omethidae
Orsodaenidae
Passalidae
Passandridae
Perimylopidae
Phalacridae
Phengodidae
Phloeostichidae
Phloiophilidae
Phycosecidae
Plastoceridae
Pleocomidae
Podabrocephalidae
Prionoceridae
Propalticidae
Prostomidae
Protocucujidae
Psephenidae
Pterogeniidae
Ptiliidae
Ptilodactylidae
Pyrochroidae
Pythidae
Rhinorhipidae
Rhipiceridae
Rhipiphoridae
Salpingidae
Scarabaeidae
Schizopodidae
Scirtidae
Scraptiidae
Scydmaenidae
Silphidae
Silvanidae
Smicripidae
Sphaeritidae
Sphindidae
Staphylinidae
Stenotrachilidae
Synchroidae
Synteliidae
Telegeusidae
Tenebrionidae
Tetratomidae
Throscidae
Trachelostenidae
Trictenotomidae
Trogidae
Trogossitidae
Ulodidae
Zopheridae

Strepsiptera [Order]
  Mengenillidia [Suborder]
    Mengeidae [Family]
    Mengenillidae

  Stylopidia [Suborder]
    Bohartillidae [Family]
    Calliharixenidae
    Corioxenidae
    Elenchidae
    Halictophagidae
    Myrmecolacidae
    Stylopidae

Mecoptera [Order]
    Apteropanorpida [Family]
    Bittacidae
    Boreidae
    Choristidae
    Eomeropidae
    Meropeidae
    Nannochoristidae
    Panorpidae
    Panorpodidae

Siphonaptera [Order]
    Ancistropsyllidae [Family]
    Ceratophyllidae
    Chimaeropsyllidae
    Coptopsyllidae
    Ctenophthalmidae
    Hystrichopsyllidae
    Ischnopsyllidae
    Leptopsyllidae
    Malacopsyllidae
    Pulicidae
    Pygiopsyllidae
    Rhopalopsyllidae
    Stephanocircidae
    Tungidae
    Vermipsyllidae
    Xiphiopsyllidae

Diptera [Order]
  Brachycera [Suborder]
    Acartophthalmidae [Family]
    Acroceridae
    Agromyzidae
    Anthomyiidae
    Anthomyzidae
    Apioceridae
    Asilidae
    Asteiidae
    Athericidae
    Aulacigastridae
    Bombyliidae
    Braulidae
    Calliphoridae
    Camillidae
    Campichoetidae
    Canacidae

Carnidae
Chamaemyiidae
Chloropidae
Chyromyidae
Clusiidae
Coelopidae
Conopidae
Cryptochetidae
Ctenostylidae
Curtonotidae
Cypselosomatidae
Diastatidae
Diopsidae
Dolichopodidae
Drosophilidae
Dryomyzidae
Empididae
Ephydridae
Eurychoromyiidae
Fanniidae
Fergusoninidae
Gasterophilidae
Glossinidae
Helcomyzidae
Heleomyzidae
Hippoboscidae
Huttoninidae
Ironomyiidae
Lauxaniidae
Lonchaeidae
Lonchopteridae
Megamerinidae
Micropezidae
Milichiidae
Mormotomyiidae
Muscidae
Mydidae
Mystacinobiidae
Nemestrinidae
Neriidae
Neurochaetidae
Nothybidae
Nycteribiidae
Odiniidae
Oestridae
Opomyzidae
Otitidae
Pallopteridae
Pantophthalmidae
Pelecorhynchidae
Periscelididae
Phoridae
Piophilidae
Pipunculidae
Platypezidae
Playstomatidae
Pseudopomyzidae
Psilidae
Pyrgotidae
Rhagionidae
Rhinophoridae

Richardiidae
Ropalomeridae
Sarcophagidae
Scathophagidae
Scenopinidae
Sciadoceridae
Sciomyaidae
Sepsidae
Somatiidae
Sphaeroceridae
Stratiomyidae
Streblidae
Syringogastridae
Syrphidae
Tabanidae
Tachinidae
Tachiniscidae
Tanypezidae
Tephritidae
Teratomyzidae
Tethinidae
Therevidae
Vermileonidae
Xenasteiidae
Xylomyidae
Xylophagidae

  Nematocera [Suborder]
    Anisopodidae [Family]
    Axymyiidae
    Bibionidae
    Blephariceridae
    Canthyloscelidae
    Cecidomyiidae
    Ceratopogonidae
    Chaoboridae
    Chironomidae
    Culicidae
    Deuterophlebiidae
    Dixidae
    Mycetophilidae
    Nymphomyiidae
    Pachyneuridae
    Perissommatidae
    Psychodidae
    Ptychopteridae
    Scatopsidae
    Sciaridae
    Simuliidae
    Synneuridae
    Tanyderidae
    Thaumaleidae
    Tipulidae
    Trichoceridae

Trichoptera [Order]
    Anomalopsychidae [Family]
    Antipodoeciidae
    Arctopsychidae
    Atriplectididae
    Barbarochthonidae

Beraeidae
Brachycentridae
Calamoceratidae
Calocidae
Chathamiidae
Conoesucidae
Dipseudopsidae
Ecnomidae
Glossosomatidae
Goeridae
Helicophidae
Heliocopsychidae
Hydrobiosidae
Hydropsychidae
Hydroptilidae
Hydrosalpingidae
Kokiriidae
Lepidostomatidae
Leptoceridae
Limnephilidae
Limnocentropodidae
Molannidae
Odontoceridae
Oeconesidae
Petrothrincidae
Philopotamidae
Philorheithridae
Phryganeidae
Phryganopsychidae
Plectrotarsidae
Polycentropodidae
Psychomyiidae
Rhyacophilidae
Sericostomatidae
Stenopsychidae
Tasimiidae
Uenoidae
Xiphocentronidae

Lepidoptera [Order]
Aglossata [Suborder]
Agathiphagidae [Family]

Glossata [Suborder]
Acanthopteroctetidae [Family]
Adelidae
Aganaidae
Agnoxenidae
Alucitidae
Anomosetidae
Anthelidae
Apatelodidae
Arctiidae
Argyresthiidae
Arrhenophanidae
Axiidae
Batrachedridae
Blastobasidae
Blastodacnidae
Bombycidae
Brachodidae

Brahmaeidae
Bucculartricidae
Callidulidae
Carposinidae
Carthaeidae
Castniidae
Catapterigidae
Cecidosidae
Cercophanidae
Choreutidae
Chrysopolomidae
Colephoridae
Copromorphidae
Cosmopterigidae
Cossidae
Crinopterygidae
Cyclotornidae
Dalceridae
Depressariidae
Dioptidae
Douglasiidae
Drepanidae
Dudgeoneidae
Elachistidae
Endromidae
Epermeniidae
Epicopeiidae
Epipyropidae
Eriocottidae
Eriocraniidae
Ethmiidae
Eupterotidae
Galacticidae
Gelechiidae
Geometridae
Glyphipterigidae
Gracillariidae
Hedylidae
Heliozelidae
Helliodinidae
Hepialidae
Herminiidae
Hesperiidae
Heterogynidae
Hibrildidae
Holcopogonidae
Hyblaeidae
Hypetrophidae
Immidae
Incurvariidae
Lasiocampidae
Lathrotelidae
Lecithoceridae
Lemoniidae
Limacodidae
Lophocoronidae
Lycaenidae
Lymantriidae
Lyonetiidae
Megalopygidae
Metarbelidae

Mimallonidae
Mnesarchaeidae
Momphidae
Neopseustidae
Neotheoridae
Nepticulidae
Noctuidae
Notodontidae
Nymphalidae
Oecophoridae
Opostegidae
Oxytenidae
Palaeosetidae
Palaephatidae
Papilionidae
Pieridae
Plutellidae
Prodoxidae
Prototheoridae
Pseudarbelidae
Psychidae
Pterophoridae
Pterothysanidae
Pyralidae
Ratardidae
Roeslerstammiidae
Saturniidae
Schreckensteiniidae
Scythrididae
Sematuridae
Sesiidae
Sombrachyidae
Sphingidae
Symmocidae
Thaumetopoeidae
Thyretidae
Thyrididae
Tineidae
Tineodidae
Tischeriidae
Tortricidae
Uraniidae
Urodidae
Yponomeutidae
Ypsolophidae
Zygaenidae

Heterobathmiina [Suborder]
Heterobathmiidae [Family]

Zeugloptera [Suborder]
Micropterigidae [Family]

Hymenoptera [Order]
Apocrita [Suborder]
Agaonidae [Family]
Andrenidae
Anthophoridae
Aphelinidae
Apidae
Apozygidae
Aulacidae

Austroniidae
Bethylidae
Braconidae
Bradynobaenidae
Ceraphronidae
Chalcididae
Charipidae
Chrysididae
Colletidae
Ctenoplectridae
Cynipidae
Diapriidae
Dryinidae
Elasmidae
Embolemidae
Encrytidae
Eucharitidae
Eucoilidae
Eulophidae
Eupelmidae
Eurytomidae
Evaniidae
Figitidae
Formicidae
Gasteruptiidae
Halictidae
Heloridae
Ibaliidae
Ichneumonidae

Leucospidae
Liopteridae
Megachilidae
Megalyridae
Megaspilidae
Melittidae
Monomachidae
Mutillidae
Mymaridae
Mymarommatidae
Ormyridae
Oxaeidae
Pelecinidae
Perdeniidae
Perilampidae
Platygasteridae
Plumariidae
Pompilidae
Proctotrupidae
Pteromalidae
Rhopalosomatidae
Roproniidae
Rotoitidae
Sapygidae
Scelionidae
Sclerogibbidae
Scolebythidae
Scoliidae
Sierolomorphidae

Signiphoridae
Sphecidae
Stenotritidae
Stephanidae
Tanaostigmatidae
Tetracampidae
Tiphiidae
Torymidae
Trichogrammatidae
Trigonalyidae
Vanhorniidae
Vespidae

Symphyta [Suborder]
Argidae [Family]
Anaxyelidae
Blasticotomidae
Cephidae
Cimbicidae
Diprionidae
Megalodontidae
Orussidae
Pamphiliidae
Pergidae
Siricidae
Tenthredinidae
Xiphydriidae
Xyelidae

• • • • •

A brief geologic history of animal life

A note about geologic time scales: A cursory look will reveal that the timing of various geological periods differs among textbooks. Is one right and the others wrong? Not necessarily. Scientists use different methods to estimate geological time—methods with a precision sometimes measured in tens of millions of years. There is, however, a general agreement on the magnitude and relative timing associated with modern time scales. The closer in geological time one comes to the present, the more accurate science can be—and sometimes the more disagreement there seems to be. The following account was compiled using the more widely accepted boundaries from a diverse selection of reputable scientific resources.

Geologic time scale

Era	Period	Epoch	Dates	Life forms
Proterozoic			2,500-544 mya*	First single-celled organisms, simple plants, and invertebrates (such as algae, amoebas, and jellyfish)
Paleozoic	Cambrian		544-490 mya	First crustaceans, mollusks, sponges, nautiloids, and annelids (worms)
	Ordovician		490-438 mya	Trilobites dominant. Also first fungi, jawless vertebrates, starfishes, sea scorpions, and urchins
	Silurian		438-408 mya	First terrestrial plants, sharks, and bony fishes
	Devonian		408-360 mya	First insects, arachnids (scorpions), and tetrapods
	Carboniferous	Mississippian	360-325 mya	Amphibians abundant. Also first spiders, land snails
		Pennsylvanian	325-286 mya	First reptiles and synapsids
	Permian		286-248 mya	Reptiles abundant. Extinction of trilobytes. Most modern insect orders
Mesozoic	Triassic		248-205 mya	Diversification of reptiles: turtles, crocodiles, therapsids (mammal-like reptiles), first dinosaurs, first flies
	Jurassic		205-145 mya	Insects abundant, dinosaurs dominant in later stage. First mammals, lizards, frogs, and birds
	Cretaceous		145-65 mya	First snakes and modern fish. Extinction of dinosaurs and ammonites, rise and fall of toothed birds
Cenozoic	Tertiary	Paleocene	65-55.5 mya	Diversification of mammals
		Eocene	55.5-33.7 mya	First horses, whales, monkeys, and leafminer insects
		Oligocene	33.7-23.8 mya	Diversification of birds. First anthropoids (higher primates)
		Miocene	23.8-5.6 mya	First hominids
		Pliocene	5.6-1.8 mya	First australopithecines
	Quaternary	Pleistocene	1.8 mya-8,000 ya	Mammoths, mastodons, and Neanderthals
		Holocene	8,000 ya-present	First modern humans

*Millions of years ago (mya)

Index

Bold page numbers indicate the primary discussion of a topic; page numbers in italics indicate illustrations.

A

Abedus spp., 3:62
Abtrichia antennata, 3:378, 3:379, 3:380
Acacia spp., 3:49
Acalypteratae, 3:357
Acanthopidae, 3:177
Acanthoxia spp., 3:203
Acerentomidae, 3:93, 3:94
Acerentomoidea, 3:93
Acherontia atropos. See Death's head hawk moths
Acheta spp., 3:203
Acheta domesticus, 3:203
Acids, 3:21, 3:24
Acorn-nesting ants, 3:68
Acrida spp., 3:203
Acridoidea, 3:201
Acrocinus longimanus. See Harlequin beetles
Acromis sparsa, 3:62
Actornithophilus spp., 3:250
Aculeata, 3:406
Acyrthosiphon pisum. See Pea aphids
Adelgidae, 3:54
Adephaga, 3:315, 3:316, 3:318
See also Beetles
Adiheterothripidae, 3:281
Aedes spp., 3:76
Aedes aegypti. See Yellow fever mosquitos
Aedes trisseriatus, 3:78
Aeolothripidae, 3:281
Aepophilidae, 3:54
African butterflies, 3:37
African goliath beetles, 3:13
African scarabs, 3:43
Agacris insectivora, 3:204
Agathiphagidae, 3:55
Agricultural pests, 3:75
    citrus leaf miners, 3:394
    Diptera, 3:361
    Lepidoptera, 3:389
    Orthoptera, 3:207–208
    See also Pests
Agriculture, pollinators in, 3:78–79
Agriocnemis femina. See Southeast Asian damselflies
Agulla spp. *See* Snakeflies
Alaskan shore bugs, 3:263
Alaus oculatus. See Eyed click beetles
Alderflies, 3:7, **3:289–295**, 3:290, 3:292, 3:293
Alena (Aztekoraphidia) schremmeri. See Schremmer's snakeflies
Aleurodicus dugesi. See Giant whiteflies

Aleurothrixus antidesmae, 3:259
Alfalfa leaf cutter bees, 3:80, 3:410, 3:417, 3:418
Alfalfa springtails. *See* Lucerne fleas
Alimentary canal, 3:20–21
Alkali bees, 3:79
Allograpta obliqua. See Chevroned hover flies
Alloperla roberti, 3:143
Allopteridae, 3:305
Alydidae, 3:264
Amber, fossil insects in, 3:8, 3:9
Amblycera, 3:249, 3:250, 3:251
Ambrosia beetles, 3:63, 3:322
Ameletopsidae, 3:55
American beetles, 3:322
American burying beetles, 3:322, 3:325, 3:332–333
American cockroaches, 3:76, 3:149, 3:150, 3:151, 3:152, 3:154, 3:157–158
American hover flies, 3:43
American primary screwworms. *See* New World primary screwworms
American walkingsticks, common, 3:226, 3:228, 3:230
Ametabolous, 3:33
Amino acids, 3:24
Amitermes spp., 3:166
Amitermes laurensis, 3:165
Amitermes medius, 3:165
Amitermes meridionalis, 3:165
Amitermes vitosus, 3:165
Amitermitinae, 3:165
Ammonia, 3:21
Ammophila spp., 3:336
Amorphoscelidae, 3:177, 3:178
Amphipsocidae, 3:243
Amphizoidae, 3:54
Ampulicidae, 3:148
Anabrus simplex. See Mormon crickets
Anarcroneuria spp., 3:142
Anareolatae, 3:221
Ancistrona vagelli. See Wandering seabird lice
Ancistropsyllidae, 3:347, 3:348
Angel insects. *See* Zorapterans
Anglewings, 3:388
Anisomorpha buprestoides, 3:222–223
Anisopodidae, 3:55
Anisoptera. *See* Dragonflies
Anisozygoptera, 3:133
Anistominae, 3:54
Annulipalpia, 3:375, 3:376
Anobiid beetles. *See* Deathwatch beetles
Anopheles spp., 3:76
Anoplolepis gracilipes. See Yellow crazy ants

Anoplura. *See* Sucking lice
Antarctoperlaria, 3:141, 3:142
Antennae, 3:18, 3:27
    See also Physical characteristics; specific species
Anthelidae, 3:386
Anthophiloptera dryas. See Balsam beasts
Antioch dunes shieldbacks, 3:207
Antipaluria urichi, 3:235, 3:236–237
Antlions, 3:9, 3:63, 3:306, 3:307, 3:310, 3:311, 3:312–313
Antonina graminis. See Rhodesgrass mealy bugs
Ants, **3:405–418**, 3:409, 3:410
    aphids and, 3:6, 3:49, 3:61, 3:406
    behavior, 3:59, 3:65, 3:71, 3:406
    biomes, 3:57
    butterflies and, 3:72
    conservation status, 3:87–88, 3:407
    distribution, 3:67
    ecosystem services of, 3:72
    education and, 3:83
    evolution, 3:405
    feeding ecology, 3:5–6, 3:63, 3:406–407
    habitats, 3:406
    humans and, 3:407–408
    medicinal uses, 3:81
    as pests, 3:75, 3:76
    physical characteristics, 3:27 3:405
    plants and, 3:49
    reproduction, 3:36, 3:59, 3:69, 3:407
    scale insects and, 3:49
    social structure, 3:49, 3:67, 3:68–69, 3:70, 3:72, 3:406
    species of, 3:418
    taxonomy, 3:405
    See also specific types of ants
Anurogryllus spp., 3:207
Aphidae, 3:54
Aphididae, 3:264
Aphids, **3:259–260, 3:267–268**
    ants and, 3:6, 3:49, 3:61
    biological control of, 3:80
    defense mechanisms, 3:66
    eusocial, 3:68
    feeding ecology, 3:63
    humans and, 3:57
    migrating, 3:66
    pea, 3:265, 3:267–268
    as pests, 3:75
    reproduction, 3:34, 3:37–38, 3:39, 3:44, 3:59
    See also Hemiptera; Sternorrhyncha
Aphrophoridae, 3:55
Apicotermitinae, 3:168

INDEX

INDEX

INDEX

INDEX

INDEX